MW00855743

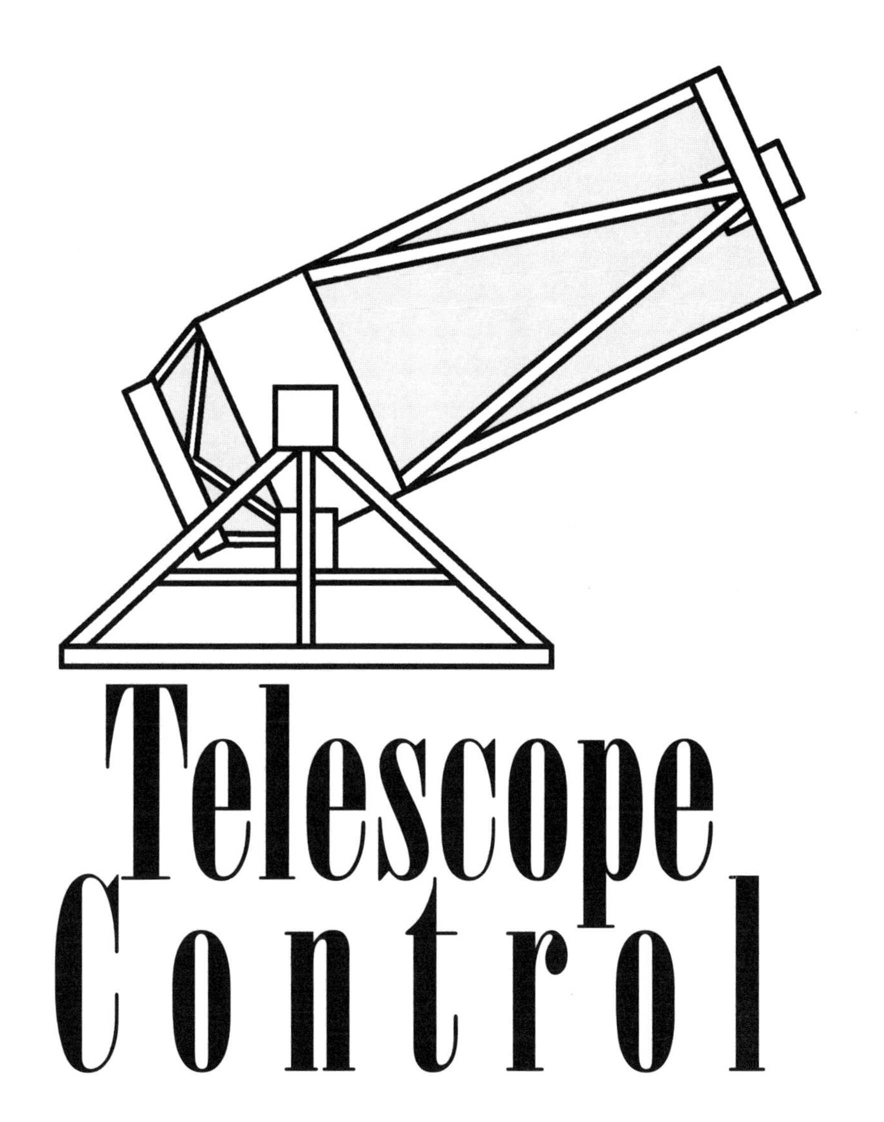
Telescope
Control

Telescope Control

Mark Trueblood & Russell Merle Genet

Published by:

P.O. Box 35025 • Richmond, Virginia 23235 • ☎(804) 320-7016 • www.willbell.com

Published by Willmann-Bell, Inc.
P.O. Box 35025, Richmond, Virginia 23235

Second English Edition

Printed in the United States of America

Library of Congress Cataloging-in-Publication Data.
Trueblood, Mark, 1948-
Telescope control / Mark Trueblood, Russell Merle Genet. –2nd English ed.
p. cm.
Includes bibliographical references and index.
ISBN 0-943396-53-0
1. Telescopes–Data processing. 2. Microcomputers. I. Genet, Russell. II. Title
QB88.T79 1997 97-28071
522'.2'0285–dc21 CIP

97 98 99 00 01 02 03 04 05 9 8 7 6 5 4 3 2

Dedications and Acknowledgements

I dedicate this book to my wife, Pat, who supported this effort with her love, who prepared the index for the second edition, and who helped prepare the manuscript for both the first edition and my Master's thesis; to my parents, who encouraged my interest in astronomy and telescope building at an early age; to my good friend Andrew J. Tomer, who rekindled my interest in astronomy and in the problems of building and computerizing telescopes, and who assisted greatly in the preparation of Chapter 12; and to Irvin M. Winer, who, as a professor of Physics at Wesleyan University, took an unusually personal interest in his students, who was not afraid to pursue offbeat or unpopular areas of science if they held promise of leading to Truth, and who had a refreshingly honest approach to coming to terms with himself, those around him, and the Universe at large. It is after Irv, who died in middle age in early 1982, that I have named my observatory (at Andy Tomer's suggestion).

In recognition of the large amount of material from my Master's thesis (completed in 1983) that serves as the foundation of this book, I acknowledge the help, encouragement, guidance, and good suggestions on the preparation of my thesis offered by my advisor, Prof. Michael F. A'Hearn, and by my examination committee, Prof. Andrew S. Wilson and Prof. David Zipoy, all from the Astronomy Program of the University of Maryland. Warm thanks are also given to Joan B. Dunham, who introduced me to the wonders of Kalman filtering and who critiqued the sections on least squares and Kalman filtering; David W. Dunham, who has cajoled me into going on enough lunar and minor planet observing expeditions that I have become interested in pursuing them further, and who provided information on the process of predicting minor planet occultations; George Kaplan, who suggested how to treat azimuth tilt and field rotation and supplied many of the equations in the epoch J2000.0 form; Louis Boyd, who supplied text and figures describing his cloud sensor; and to my friend and co-author Russell M. Genet, who offered many fine suggestions on control theory and mechanical error sources during the preparation of the thesis.

Despite all this help, mistakes are the full responsibility of the authors, and we encourage readers to let our publisher, Willmann-Bell, know of any errors so they may be corrected in future printings.

Mark Trueblood
Winer Mobile Observatory

I also dedicate this book to my wife, Joyce, who lets me pursue my strange interests unhindered; to the pioneers of microcomputer telescope control, particularly Louis Boyd and Kent Honeycutt; and to the other pioneers in this field whom I would have liked to have known. Those who do the pioneering are usually too busy to write books about things, yet it is they who solve the hard problems when it is not yet known if solutions even exist. It was in 1981 that I decided a book on microcomputer control of telescopes was needed. Perry Remaklus of Willmann-Bell, Inc. kindly agreed, after seeing an outline, to publish such a book. Six months later, with no appreciable work yet done on the book, I received a large package from Mark Trueblood. The note said "...contains a number of chapters from my thesis on microcomputer control of telescopes. Do you suppose it might be a suitable basis for a book on this topic?" I always wondered how Darwin felt when he received a package from Wallace. "Dear Darwin...." I immediately recognized that it was a goodly part of a book in the rough, and proposed to Mark that we co-author this book. I am delighted that he accepted, as the final product is much better than I could have done myself, I learned a lot from Mark, and acquired a good friend in the process.

Another good friend I acquired was Louis Boyd, who not only developed an automatic microcomputer-controlled telescope, but, more importantly, shared his knowledge freely. Louis' many ideas in software and hardware form the basis for the Automatic Photoelectric Telescope (APT) described in this book. It has been a particular pleasure to exchange ideas with him since 1981.

An astronomer who has been influential, with respect to my work in the area of microcomputer control of telescopes, is Kent Honeycutt, of Indiana University. His microcomputer control system at the Goethe-Link Observatory was among the very first in operation.

I totally retired from engineering, astronomy, and things robotic in 1993, adopting the relatively carefree life of a gypsy science writer and lecturer. It was, however, a pleasure to come out of retirement, if only briefly, to revise this book with my friend of many years, Mark Trueblood, and to recount the many further advances Louis Boyd, Kent Honeycutt, and now many others have continued to make to this fascinating symbiosis of the human and the machine.

Russell Merle Genet
Fairborn Institute

Foreword

Astronomy and microcomputers were a natural pair from the start. The data gathering routine at the telescope is characterized by many time-robbing interruptions due to the sun, moon, and the clouds. It is therefore important to ensure efficient use of the clear dark hours that remain to the observer by automating the process as far as is practical. The data gathering is also very number-intensive, calling for just the kind of help the computer can best provide. Fatigue and sleeping are important problems in late night observing for astronomers, but not for computers. One of the major benefits of a computerized telescope and instrument is the error-free data logging it can provide, even at 4:00 AM. As a pure science, astronomy seems always to be underfunded. It is therefore significant that microcomputers are inexpensive, at least compared to earlier generations of computers. Though the advantages of computers to astronomy have been evident for decades, only since the advent of the microcomputer has it been realistic to automate small and moderate size telescopes.

This book is a testimony to the rather amazing progress that has been attained recently in applying microcomputers to small telescopes. Much of this progress has been possible because of the diligence of these authors and others in making available to a wide audience the often hard-won knowledge of how to implement a telescope automation project on a small budget. Practical knowledge turns out to be a very important ingredient in the conduct of astronomy, perhaps otherwise the most impractical of the sciences. The readers who choose to apply the content of this book will soon appreciate the generous use of examples and construction hints.

The development of small telescope automation, and indeed small telescope astronomy, has benefitted from a delightful union of amateur and professional interests. The work is mutually beneficial to the two groups, and in some cases the merging is so complete that one can scarcely make a distinction. Amateurs in all of the experimental physical sciences except astronomy have practically disappeared because of the enormous expense of modern research equipment. Astronomy was headed down this same sad path until the rapid development of microcomputer automation. The potential of microcomputers in astronomy to preserve the ability of individuals to do useful science "just for the fun of it" should not be underestimated. Indeed this preservation may be one of the most important and pleasing results of the publication of books such as this volume.

R. Kent Honeycutt
Indiana Univerity

Table of Contents

Chapter 1

Introduction

1.1 Overview

This is the second edition of *Microcomputer Control of Telescopes.* In 1985, we wrote that book because small computers based on microprocessor integrated circuits had become powerful enough to control a telescope and simplify the task of finding objects in the sky. Furthermore, inexpensive stepper motors and interface circuits were available that made it possible to build a complete telescope control system that offered good performance at an affordable cost. It was time to announce to the amateur telescope making community and to commercial vendors of amateur telescopes that a new era had dawned that would forever change the way telescopes are made and used.

The evolution of technology has made some of the material in the first edition obsolete. Microprocessor chips now form the heart of powerful personal computers, high performance workstations, and even massively parallel supercomputers. The more powerful of today's chips can execute over one billion instructions per second and nearly 100 million floating point operations per second (MFLOPS), rivaling the performance of the Cray 1 supercomputer. It is no longer possible to distinguish between a "microcomputer" and a "regular" computer. Affordable personal computers and single board computers have made it nearly mandatory to include in modern commercial telescope control products a powerful processor chip and catalogs holding thousands of objects. "Microcomputer control" is now the *only* way to control telescopes using a computer, so we changed the book's title to reflect current computer technology.

The market in computers suitable for telescope control has also matured. The emphasis in *Microcomputer Control of Telescopes* was on the Digital Equipment Corporation LSI-11 and the RT-11 real-time operating system, single board computers using the Motorola 6809 and the OS9 operating system, and computers built using the Z-80 chip and the CP/M operating system. Today's personal computer market is dominated by the IBM PC and its clones running Microsoft Windows. However, the market continues to evolve, and we expect to see a different situation in years hence.

Computer control has spread to so many industries that many system components, such as motor controller boards for the PC-clone, Macintosh, and UNIX workstation are now available commercially. In the previous edition of this book, we included several circuit schematic diagrams for custom electronics to perform functions that can now be performed by off-the-shelf circuitry. Many of these boards are affordable by amateurs and can save a great deal of time in developing a system. In this book, you will see fewer electronic schematics and more examples of how to use off-the-shelf boards and subsystems to configure your control system quickly.

Equally dramatic has been the development of graphical user interfaces and associated revolutions in software engineering that make it easier to use a control system. Software development tools are available that make it straightforward to prototype, develop, and modify telescope control systems. These changes have prompted an update to the original book to include new material as well as to remove obsolete material.

But some things never change. The reasons for controlling a telescope and the basic concepts of control system design remain the same. Amateur astronomers continue to demonstrate the knowledge, interest, and ability to carry out research in photoelectric photometry and other fields.

This book still fills a gap in the books that are available on using personal computers in astronomical applications. Most of the other books stress computing that can be done at the leisure of both the hobbyist and the computer, and computing that uses only the basic computer and standard peripheral devices (disks, printers, etc.) as they come from the computer store. Image processing and orbit computing are examples of this type of computing.

This book is primarily concerned with how to connect a non-standard computer peripheral device (a telescope) to a computer and how to program the computer to perform time-critical computations. This is only one example of the more general problem of real-time control, so if viewed in this larger context, this book should find an audience among those interested in *any* real-time control application, such as robotics. One of us (Genet, 1982b) wrote a previous book on real-time control with the TRS-80, but it was primarily devoted to data logging and instrument control in astronomy, not telescope control.

Since the early 1980s, basic research budgets for professional astronomers have not kept pace with the growth in astronomical knowledge.

Answering the growing volume of questions makes increasing research efforts desirable, despite the constant number of research-quality instruments. Although space-borne telescopes have opened up new areas of research, the limited number of such telescopes will keep the main burden of astronomical research on ground-based telescopes.

One way to perform more research on the same number of telescopes is to increase the research throughput per telescope. This can be accomplished by computerizing the telescope and associated data acquisition instruments to minimize non-observing time. This unproductive time includes activities such as slewing to the next object's coordinates, acquiring the object, adjusting the tracking rate, and logging the data. The low cost of PCs and associated system software and interfaces makes it practical to computerize smaller telescopes using modern hardware and control techniques. Much useful research can be done on telescopes of modest size, especially if they are controlled by a computer.

The use of computerized telescope facilities owned by amateur astronomers can help produce greater quantities of useful research. Cooperative efforts between amateurs and professionals in the area of photoelectric photometry have already proven beneficial to both parties. The International Amateur-Professional Photoelectric Photometry (IAPPP) group has demonstrated this since its inception in 1980, and we expect even greater benefits for basic research in the future. CCD cameras priced for the amateur market have made it possible for amateurs to produce good photometric results in unprecedented quantities. In addition, professionals who are interested in computerizing a small telescope or an observing instrument may find this book useful.

Several astronomers, both amateur and professional, have built useful telescope and instrument control systems using personal computers and inexpensive position feedback hardware. Such systems can be incorporated into the manual control systems of existing telescopes without disturbing the functioning of the manual controls. Programs can then be added to provide high accuracy tracking, computer control of data acquisition, real-time data reduction, and even an interactive graphics display of the acquired data. This would allow an observer doing photometry, for example, to correct the data for extinction immediately after the observing session, and to combine his corrected measurements with data gathered on previous evenings to display the developing light curve.

The field of telescope control is emerging from its infancy and is beginning to mature. In *Microcomputer Control of Telescopes*, we predicted how telescope control might progress in the future. In terms of the number of computer-controlled telescopes, we suggested that:

1. The greatest number would be commercial systems made by existing telescope manufacturers, such as Celestron and Meade Instruments. We expected that these would be self-contained systems that operate from a fixed program in EPROM. We hoped that these systems would be able to communicate, via an RS-232 serial line, with external personal computers, so that the latter could provide high-level supervision and simultaneously log data from instruments, such as photometers.

2. The next greatest number would be complete research-grade telescope systems using commercial telescope mounts specifically designed for computer control. We cited the DFM Engineering mount as the first of these.

3. The third greatest number would be built entirely by amateurs, except for the personal computer used to control the system. Enterprising and resourceful amateurs would develop a wide variety of telescope mounts, control systems, and control programs, and their ideas would significantly advance the state-of-the-art.

4. Various specialty systems would evolve over time. Systems described in detail in this book include specialty systems. One is an automatic photoelectric telescope (APT). The APT is not made unusual by its use of either an equatorial mount or open-loop steppers, but by closing the position loop with a photometer. This approach is unconventional, but not unprecedented. The other is a mobile (trailer-mounted) alt-azimuth telescope—an unusual system on both counts (in 1985).

These predictions were surprisingly accurate. The major telescope vendors now offer computer control systems that seem to increase in sophistication every few months. Some do have interfaces to PCs that permit a higher level of integration with custom instruments. Amateurs have developed and continue to develop their own systems, some so good that their inventors have begun to market them. And specialty systems continue to be developed as new telescope design techniques lower the price of telescopes.

This book is intended for a wide audience of varying interests. The most common reader of this book is expected to be an amateur astronomer with a personal computer, a telescope under 10 inches in aperture, and the desire to use his computer to point his telescope with an accuracy that consistently places target objects into the field of view of a low-power eyepiece. For this reader, the system developed by Tomer using stepper motors in an open-loop control system is described in detail. For those with more demanding requirements, or for those who are just curious, more advanced concepts, equipment, and systems are presented. In Parts 4 and 5, we describe in the greatest detail the systems with which we are most familiar. While they are somewhat unusual systems, the principles involved and problems encountered are similar to those that will be faced by developers of any telescope control system. In the final analysis, each system is unique.

Some readers of this book may be new to astronomy. They may have tried to find objects with a small telescope, only to discover how difficult it can be, even for relatively bright objects. These newcomers may be looking to computer control as a substitute for finding objects by hand. We would like to remind these people that technology can sometimes fail, and that it can be very expensive to obtain very high reliability. We are also sympathetic to those who want to get the most out of an observing session—who want to spend the greatest proportion of time possible in observing and to minimize the time spent searching.

For professional observatories, there is no reasonable alternative to computer control, even on smaller telescopes. As a result, commercial systems built for professional observatories are designed to be extremely reliable. They are also quite expensive. Telescopes and control systems for the amateur market are designed to sell for a particular price, and therefore cannot be as reliable. To all those just getting interested in amateur astronomy, we say the best way to start is with a good set of 7x50 binoculars, a magnitude 6 star chart, and a lounge chair. Read the popular amateur astronomy magazines to learn what is interesting during the current month, then go out every clear night and observe. If you pay your dues and learn the sky, you will have that much more fun with a computer controlled telescope, and you will better appreciate what the technology is doing for you if have learned to star-hop your way to target objects. You will also be able to salvage an observing session if the control system malfunctions.

1.2 Organization of Topics

The topic of this book is the use of a personal computer to enhance the normal functions of telescope pointing and tracking. We wish to emphasize the distinction between full *automation* of all telescope control functions, so that human intervention is not necessary, and merely adding a computer to the control system, or *computerization*. Most of the telescope control systems described in this book are not fully automatic. Those that are, are described in a separate part.

This book covers most of the problems that must be solved in designing a telescope control system and proposes solutions to them, then assesses the ability of several currently available system components to handle the demands the performance requirements place on them. Systems developed or being developed by Louis Boyd, commercial firms, and by the authors are described to illustrate the process of building your own system.

To design a telescope control system, you should first decide what the system will do. Next, choose a design approach to making the system work in the desired fashion. The design choices and problems in the selected approach should be foreseen and analyzed. A good approach for very complex systems would then be to construct the system theoretically to see what will be required to make it work. At this stage, specific pieces of hardware can be examined to see which ones, if any, are capable of meeting the system and component performance requirements. Finally, the control system is built and integrated with the telescope. It is our experience that the integration phase can be the longest and the most difficult phase if the earlier design work was not done properly.

Part I of this book is concerned with the benefits one can expect from computerization, and explores the reasons for computerizing a telescope. It also covers the theory of how to develop a telescope control system. This part answers the question "How do I go about building a control system for my telescope?".

Part II is concerned with things you must consider when designing a telescope control system. It covers elementary aspects of modern control theory, error sources, and practical aspects of system design. This part answers the question "What problems do I have to solve in my design?".

Part III discusses the components that are part of telescope control systems. In doing any design, one must know what elements are available for making the end product—that is, one must peek inside the Erector Set box to see what parts are available to build something. This part answers the question "What items can I use in my control system?".

Any manual or textbook is easier to understand if the reader can see detailed examples worked out. Part IV gives several examples of telescope control systems, some simple and inexpensive to implement and some more complex and more expensive. Because fully automatic or robotic telescopes are fundamentally different from telescopes operated in real-time by humans, we discuss such systems in a separate part of the book, Part V.

It is our hope that this combination of some control theory, detailed information about available hardware, examples of working systems, and practical advice (with a little bit of our preaching) will get you started on your own system. The Appendices contain a great deal of valuable information on sources of hardware. We recommend that you become very familiar with what hardware and software are currently available and engineer your system to suit your particular observing program, rather than blindly copy one of the systems presented in this book. That way, you will have a system that works the way you want it to, not the way somebody else thinks it should.

One final word about how this book was written. Each chapter was assigned to one of the authors, while each author reviewed and commented on the entire book. Some differences of opinion were settled through discussion, while others were not. Thus some opinions in this book may be held by only one author. For example, Genet is most comfortable with electronic hardware but dislikes complex software, while Trueblood is at home with software. It may not be accurate, therefore, to say "Trueblood and Genet tell us such and such" when only one of us may believe such and such. To help you sort out who thinks what, Genet took the lead on Chapters 7 and 15–18, and Trueblood on the rest (with considerable help from Andy Tomer on Chapter 12).

You, the reader, should think about your specific telescope control project and decide for yourself whether an opinion expressed in this book applies to your project.

Part I

Why and How to Computerize a Telescope

Chapter 2

Why Control Telescopes With Computers?

2.1 Reasons for Computerized Telescope Control

Given that there are many telescopes, computers, and inventive amateur and professional astronomers, it is only human nature that successful attempts (and a few unsuccessful ones!) will be made to get a computer to control a telescope. The reasons given will be various, but in many—perhaps even most—cases, the real reason will be that they were there and it seemed like a fun thing to do. While some will be content to just use their computers for analysis, projecting star charts, and star war games, many simply will not be able to resist the challenge of computerized telescope control. It is towards these hardy souls, more than any others, that this book is aimed.

There are situations in which computer control of a telescope is essential. A good example is the automatic South Pole telescope (Giovane et al., 1983). No one will be there to operate it. Another example is daytime infrared astronomy, which requires accurate computerized pointing, as the stars are often not visible to the human eye.

A number of amateur astronomers have noted that the necessity of getting up for work each morning seriously cuts into the clear-night observing time. Being familiar with computers, a few amateurs have used them to control their telescopes and related instruments as a way out of this dilemma. The work by Skillman (1981) and by Boyd (Boyd, Genet, and Hall, 1984) is particularly worthy of mention. Skillman's system gathers photometric data from a short-period variable star and nearby comparison star hour after hour, while Boyd solved the problem of automatically locating and measuring many different stars in a single night. It would only be fair to mention that this problem was solved earlier in the mid-1960s using a large remote computer in Tucson to control the 50-inch telescope at Kitt Peak (Goldberg, 1983), and using a minicomputer in a pioneering effort by McNall (1968) and others at the University of Wisconsin. A few telescopes at other observatories have had similar capabilities added to them or designed in from the start over the last few decades.

Computers can be used to control or assist in a wide or narrow range of observing functions. A good example of a truly simple and low-cost application is that of Rafert (1983). At a cost of less than $100 in non-computer parts, two stepper motors were added to the 24-inch telescope at Appalachian State University, and controlled by a Commodore microcomputer. The objective was not to point the telescope anywhere in the sky automatically, but simply to move back and forth between variable and comparison during photometry. Initial telescope pointing and final centering (each time) was still performed by the astronomer. The astronomer was relieved, however, from moving the telescope between the stars and rough-centering it in the finder. Those readers who are experienced photometrists will appreciate how much effort such a simple telescope rough-pointing system can save.

Besides greater data gathering efficiency, computer control of a telescope and its instrumentation can increase the reliability of the data. Data logged manually towards the end of a late night observing session tend to be error-prone. An additional benefit of freeing the astronomer from the data logging and telescope pointing functions is that he is free to think about the astronomy while observing. This allows him to make prompt decisions about the wisdom of beginning or continuing an observation, what the next object to be observed should be, considering the observing conditions, and so on. In some observing programs, such as long period variable star photometry, even these functions can be assumed by the computer, which makes possible the automatic photoelectric telescope described in Part V. This frees the astronomer from all data gathering functions, and allows him to devote his complete attention to designing the observing program, interpreting the reduced data, and publishing the results, as well as sleeping at night.

When a computer is introduced into the control system of a telescope, it should be combined with other elements of the observing program consistently, enhancing the program overall. A'Hearn (1984) points out that some research programs, such as comet photometry, require a great deal of judgement and intervention by a human observer during the process of gathering data. A system he built in the early 1980s using an Apple II computer and photometer control software written in FORTH was quite flexible, and accepted human judgement as an input into the observing process. The roles of the computer and the observer were carefully defined in this system.

It might be good to point out that there are a number of reasons why one might *not* want to control a telescope with a computer. We credit the best argument in this direction to Douglas S. Hall. He pointed out that a number of perfectly good photometrists who regularly gathered high-quality data of considerable scientific value were totally ruined when exposed to computers. Instead of observing, they were spending all their time trying to control their photometers and telescopes with computers. However, with great insight into dynamic processes, he later pointed out that as the number of computer enthusiasts was far greater than the number of photoelectric photometrists, the loss of good photometrists to computers would be more than offset by the conversion of computer enthusiasts into good photometrists!

Another reason for not wanting to computerize a telescope is that there is certain merit to the old dictum KISS (keep it simple, stupid!). There is not much to go wrong with setting circles. Equipment failures can be very time consuming and frustrating—even expensive. Computer-controlled telescopes are usually complicated—definitely not KISS! However, any astronomer who has paid good money to buy this book will not be dissuaded by any of these cautions.

2.2 Cost/Benefit Reasons

Many computerized systems of various types have been developed in response to the expectation that the expense of system development and maintenance will be outweighed by an increase in productivity or more efficient utilization of resources. The system user has a problem that the proposed system should solve. Often the problem is felt as too much demand for a scarce resource, for example, observers requesting more telescope time than there is available. In the early days of computer control, whether computerized systems at large observatories were cost effective was a topic of controversy. Racine (1975) argued that non-observing time (time spent acquiring the object, setting up instruments, etc.) that can be reduced by computerization is a small fraction of total telescope time, so the reduction does not enhance science throughput enough to justify the cost. Boyce (1975) argued that time spent collecting photons can be as low as 7% of total telescope time, and that computer control can improve this low duty cycle. Hill (1975) claimed "for about 10% of the price of the observatory we had increased our output rate by about a factor of four and we expect that to increase." He did not indicate how output rate was measured. Four times as much data gathered does not necessarily mean four times as much useful science was done, and it must be the useful science done that is the measure of system benefit.

The proportion of non-observing time depends on the type of observing being done and how the telescope is being used. For example, UBV photometry may have a large non-observing-to-observing time ratio that may exceed 1. Much of this non-observing time is spent setting photometer controls and pointing the telescope, both of which can be done faster and more accurately by computer. On the other hand, the time spent pointing a telescope for a long plate exposure may be a small fraction of the duration of the exposure. Factors such as time spent removing and attaching equipment (which cannot be controlled by a computer) and whether an astronomer can be doing something useful while the computer controls the telescope, as opposed to controlling the telescope himself, vary from one observatory to the next. Before a small observatory, presumably working with a limited budget, decides to computerize its telescope, it would be wise to make a candid assessment of the observing programs supported by the observatory and how much they would benefit from any proposed computerization project.

One might argue the value of computerizing a small telescope. However, Abt (1980) has shown that for telescopes at Kitt Peak ranging in aperture from 0.4 m to 2.1 m, the most cost effective telescopes were the 0.4-m telescopes, in terms of both the number of published papers and the number of times these papers were cited per dollar of annual expense. Although computerization can bring about greater cost savings on larger telescopes per dollar spent on the computerization project, the dramatic price drops in recent years for fast and reliable computer hardware have made computerization of smaller telescopes economically feasible. One should not, therefore, dismiss the concept of computerizing small telescopes because of the history of high costs incurred by projects undertaken at large observatories in the early 1970s.

The reasons for using a computer in a telescope control system are many and varied. We suggest that you develop a very clear idea of what benefits you expect from a computerized control system and what it will cost before proceeding with the construction of your system. An important cost element that is often overlooked is software, which can be several times the cost of the hardware.

2.3 Modern Observing Methods

Changes in the way we use telescopes and in the detectors we use have increased the need for computer-based telescope control systems. The success of amateur-built robotic telescopes in the 1980s has led professional astronomers to optimize new large telescopes for remote control (e.g., the 3.5-meter telescopes at Apache Point and Kitt Peak). Professional astronomers have also created a commercial market for smaller robotic telescopes. Such ground-based telescopes not only produce more and better data for the dollar, but they help prepare students for observing with space-based telescopes.

The CCD has completely replaced the photographic plate on professional telescopes in all but a few applications, wide-field surveys and astrometry being among the few exceptions (though the accuracy of semiconductor masks and larger chip sizes makes CCDs more attractive for astrometry). It is important to know about the conditions under which an image or spectrum is taken, including where the telescope was pointing and how well it was operating. This need for ancillary data makes it important to control the telescope with a computer, to collect engineering data from the telescope to annotate the digital science data disk file from the instrument.

In the amateur world, CCD cameras are rapidly overtaking film as the preferred method of imaging, due to the much higher quantum efficiency of CCDs that dramatically shortens exposure times. The image produced by the CCD is automatically digital in form, making it easy to enhance and manipulate on a computer. The CCD response to incident light is also very linear, making it possible to use the same instrument for both imaging and accurate photometry. Most CCD cameras plug right into a PC, so the amateur has a PC there anyway controlling the CCD—why not add telescope control as well?

Another development in the amateur marketplace is the embedded computer. Many telescopes and observing aids are sold with single-board or single-chip computers built into them. Amateurs are becoming accustomed to using these systems as part of their observing routine without giving any thought to the computer or the software that simplifies their observing. Many of these products have a port that permits them to be connected to, and commanded by, a computer.

All of these trends in the way we observe make it increasingly imperative to control the telescope and its instruments with computers. Greater power and sophistication of both computer hardware and operating system software, coupled with decreasing prices, have lowered the cost of computer control.

Despite these improvements, it still takes about two staff-years of effort to develop telescope control system software from scratch (see Appendix A). This means that the investment in a telescope control system can still be high, making it necessary to plan the development carefully. In the next chapter, we discuss modern engineering methods for developing computer systems that can help you identify costs and benefits at an early stage.

Chapter 3

Modern Systems Engineering

In this chapter, we present a way to design computer systems that helps you to understand the benefits and costs of the system early in the development cycle, and then to build them with the features you want in the most efficient manner.

3.1 The Project Life Cycle

The history of developing systems employing the kind of high technology used in computerized telescope control systems is checkered. The power and adaptability of digital computers have encouraged the construction of sophisticated, and therefore complicated, systems. In the roughly four decades in which large computerized systems have been developed, there are few customers who have been satisfied with both the performance and the cost of the system that was delivered.

To help you avoid building a system that you or your customer doesn't like, we present an approach for converting the user's goal into a computerized control system. This process can be represented as a hierarchical structure, as follows:

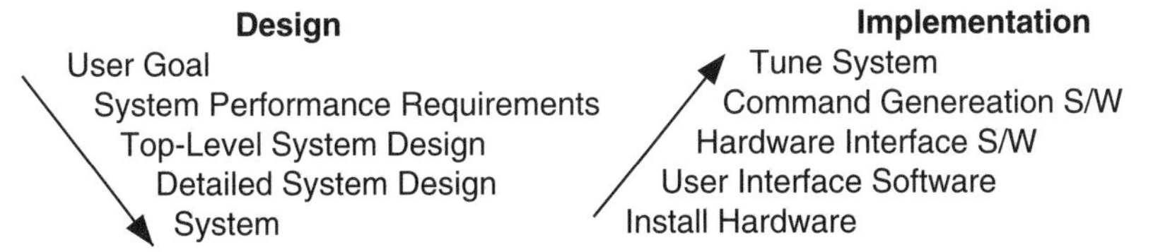

Since the leap from the user goal to the system is too large to manage effectively, we have added intermediate levels, each level representing a definable and controllable step toward the fully working system. The step from one level to the next level requires an increase in detail or a change of focus.

Design your system from the top down. Start by defining your goals, then write down system requirements that state specifically "what" the system will do, but don't worry at this stage "how" it will do it. Functional (what the system will do) and performance (how fast it will do it) requirements are formulated with the user goal and the system operational environment in mind. The top-level system design is the least detailed "how" document, and identifies all hardware and software functions without describing them in complete detail. The overall software organization is described, and major software modules are identified and described. The detailed system design is a set of hardware schematics and software module specifications that give sufficient details to build the system. Examples of system requirements and design are given in Chapter 13.

Build the system from the bottom up. In the early stages of development, ignore minor software features such as apparent place and mechanical corrections. These are simply equations that can easily be added later. The hard work is always the interfaces—to motors, encoders, instruments, and the observer. The best approach to developing the system is "build a little, test a little." The typical computer hobbyist is used to sitting down at the computer and writing a fairly complex program all at once in an evening, or a few evenings, without doing any testing of the program while it is being developed. That approach is adequate for some kinds of computing, but it simply will not work when complex software and hardware interact, and is doomed to failure. It is easier to find a problem in 200 lines of software than in 20,000.

Often it is easiest to start programming the user interface first. But don't try to get the entire user interface going all at once. Instead, try to build a little diagnostic program that lets you enter a motor speed and direction command and that sends the command to the motor. You will probably need this diagnostic program to check out your system when you have a problem, so it is not wasted effort to build it. Don't write the code for all the interfaces then try to debug everything at once. Keep hammering away until you get a single motor shaft to spin. After you have that working, keep adding features one at a time (testing each one before adding the next) until all the hardware talks to the computer.

After you have the basic interfaces working, you can add the rest of the user interface (menus or typed-in commands), then make motors spin at the right speed. When you are at the stage of putting in control equations, first put in the equations to give you basic slew and track speeds, and try them out. Once you have them working, keep adding software models and corrections (one at a time, testing each one) until you attain the level of accuracy you need. Tune the system for the performance you want, then stop. One of the hardest things for a system developer to do is to admit that it is "good enough," stop developing, and put the telescope to good use.

3.2 Defining Your System's Requirements

The system requirements should reflect your goals and your observing program. What kind of observing do you do? What kind of environment will your system operate in, and how does that affect its design? If you are building a portable system, how rapidly must you be able to set up and tear down the system? If you observe solar system objects, do you need a way to read in an ephemeris to point the telescope accurately? What accuracy do you need for pointing and tracking for the kind of observing you do? Your answers to these questions should not be vague. Keep thinking about your observing program and the kind of performance you need until you can answer each question with a number.

System performance requirements should be set keeping in mind not only the observing program and the operational environment, but also the budget. This requires a good knowledge of the available technology and, specifically, a knowledge of how useful the technology is in the operational environment. We help direct you to an understanding of the available technology in Part III.

The functional requirements and available project budget should be the driving forces behind all tradeoffs and design decisions that are made. The first version of the functional requirements is written without benefit of a top-level design. As the project proceeds in its development, the functional requirements should be reviewed periodically and updated, if necessary, keeping in mind that the design is derived from the requirements, not the other way around. Don't add unnecessary bells and whistles to your design, then try to justify them by "fixing up" the requirements. The following telescope and control system characteristics can serve as a starting point for developing your own system requirements:

3.2.1 Portability

Will your telescope be permanently mounted or portable? The answer to this question is usually determined by the amount of light pollution where you live and the type of observing you do. If you observe occultations, which require the observer to travel to a narrow path to see the event, or if you want a portable telescope for other reasons, think about how big the telescope aperture needs to be, how you will transport it, how much time you want to spend setting it up and tearing it down, and how you will power the control system. How much weight can you haul or tow with the vehicle you intend to use? Should you plan on using a notebook computer, or will you provide power for a desktop system? If you use a notebook, how will you interface the motors to the computer?

3.2.2 Setup Time

Is there something about your observing program that requires particularly rapid setup and teardown time? If the telescope is to be portable, how will you tell the control system about the telescope's alignment and orientation to the sky? How accurate must your position sensing be, and how long do you have each night to calibrate the system to achieve that level of accuracy?

3.2.3 Optics

Do the observations you want to make require optics of a certain minimum or maximum size? Do they require a particular optical design? What is the minimum field size? The best plate scale? If you are doing photometry, how many photons per second in what band from what magnitude star do you need to achieve the desired signal to noise ratio? Remember, bigger is *not* always better! Bigger optics rapidly increase the cost of the mount and control system.

3.2.4 Telescope Pointing Accuracy

Finding objects would be much easier if everyone could point their telescopes with 1 arc second ($1''$) accuracy. The problem is, it can double the cost of a project to build the last 5% of performance and reliability into the system. Can you make do with 1 arc minute ($1'$) accuracy that would put the target in the middle of a low-power eyepiece? Do you really need anything better than that? Answering this last question honestly can save a great deal of money and time in building your system.

3.2.5 Telescope Pointing Time

It can be very expensive to build a telescope that slews rapidly. You also need hefty motors to move a massive mount quickly. Large motors can have disadvantages, such as generating a great deal of heat near the optical path, ruining the seeing. A rapid slew rate can be dangerous if someone at a star party steps into the path of a rapidly moving telescope. On the other hand, reasonably rapid slewing is required for portable telescopes to sight enough stars during the setup period to align the telescope for accurate pointing and tracking. Most observers will be happy with a slew rate of 1°–3° per second. Satellite tracking telescopes and some specialty telescopes (e.g., those observing gamma ray bursters and other transient phenomena) need to slew and track at rates up to 90° per second. Don't stand in the way of one of these scopes!

3.2.6 Long Term Tracking Accuracy

How long will you be looking at one object? If you will be doing photoelectric photometry using 10–20 second integrations, you may need a long term tracking accuracy of only 10″ in 20 seconds. Long-exposure images of faint objects require better accuracy, perhaps 1″ in 10 minutes. If you will be guiding long exposures or using an autoguider (e.g., SBIG ST-4) you may be able to tolerate less accuracy in your control system (measured without using the guiding input), then use the autoguider or your inputs on a hand paddle to obtain the desired tracking accuracy.

3.2.7 Short Term Tracking Accuracy

How much "wiggle" can you tolerate in your control system? Will your data be ruined if your telescope wanders randomly within a 2″ circle over a period of a few seconds? High speed photometry data probably would be unaffected, but spectroscopy using a slit 0.″5 wide would suffer.

3.2.8 Data Input and Control Devices

What environment will the keyboard or other data input and control devices (and other equipment) be exposed to? What ranges of temperature and humidity must they be able to withstand? Will they be subjected to windblown dust or sand? Do your human interface devices need to be lit with a red lamp or otherwise be easy to use in the dark without impairing the observer's dark adaptation? Do they need to be conveniently positioned within arm's reach of the eyepiece? Or will they be used in a warm room? Do manual slew commands need to be reviewed by the control system software before they are executed to prevent damage to the telescope and drive? Do you need a manual override switch to permit observations beyond the limit switches?

3.2.9 Computer Environment

What will be the temperature and humidity in the room where the computer is located? Will you need a protective enclosure that can keep the equipment within its operating ranges throughout some specified ranges of temperature and humidity? Should disk drives and other equipment sensitive to foreign particles be protected against wind-driven dust and sand? Do they need protection from shock? How likely is your observatory to be struck by lightning?

3.2.10 Commands

How do you want to enter commands to the computer? Are you a "real programmer" that uses a "dumb terminal" and enters commands using the keyboard? Do you want pull-down menus and a mouse-driven "point and click" interface? Do you want a hand paddle for guiding or for initializing the system?

3.2.11 Extraneous Light Control

Do you have special requirements for baffling and controlling extraneous light? Can you use an open truss, or do you need a closed tube? Some observing programs, such as infrared imaging, require that you do *not* have baffling.

As you design your system your design will evolve, due to your learning more about what it will take to meet your requirements and learning more about what kinds of motors, interfaces, and other items are available and how much they cost. As your design evolves, you should update the requirements and add additional requirements. This is a normal part of the iterative design and build process.

3.3 Designing Your System

After the functional and performance requirements have been formulated, major design decisions need to be made to arrive at a top-level system design. These decisions might include the following:

1. Method of transport
2. Telescope mount type
3. Optical system
4. Drive design approach
5. Control system approach
6. Computational requirements

7. Computer hardware and software.

Each of these topics is addressed below.

3.3.1 Method of Transport

For a portable telescope, you might consider three telescope transport options:

1. Make the telescope tear down into pieces that can be packed into a van or small truck.

2. Mount the telescope permanently inside a van or small truck, and have a roll-off roof section or side panels that fold down to reveal the telescope.

3. Mount the telescope permanently on a trailer and tow the trailer to the observing site.

If you have a stringent requirement for rapid setup, this could eliminate the first option. It takes two people almost half an hour to set up a C-14, a relatively lightweight and compact 14-inch telescope. Your optics requirement might dictate a large telescope that would be difficult to set up and tear down at each site.

3.3.2 Telescope Mount Type

There are many ways to mount a telescope optics assembly, including the equatorial, altitude-azimuth (alt-az), and altitude-altitude (alt-alt) mounts. Each of these has several variations. For example, equatorial designs include the German, fork, English yoke, cross axis, Springfield, horseshoe, "Porter's folly," and Dobsonian on a Poncet platform, just to name a few.

Think about your requirements as you pick a mount design. Will the telescope be in a fixed observatory used primarily for photography? Perhaps some form of equatorial mount would be best. If you give frequent star parties or stay glued to the eyepiece for long periods of time, maybe you should consider a Springfield mount (which leaves the eyepiece stationary no matter where the telescope is pointed). Do you require light weight and low center of mass for portability? Perhaps an alt-az would be best, if you can tolerate or correct for the field rotation. Does your observing program require access to the zenith and to the pole? You should consider an alt-alt mount.

Each mount type impacts the control system differently. A properly aligned equatorial mount requires only one axis to be driven at a constant rate to track an object. The alt-az and alt-alt designs require two axes to be driven at varying speeds over a wide speed range to track an object, and the fields rotate, which can be a real disadvantage in photography unless you have motors (and software to control them) to "de-spin" the rotating fields. It is usually less expensive to design a control system for an equatorial mount, but other mounts may be better suited to your requirements.

3.3.3 Optical System

The optical design is usually determined by the resolution, limiting magnitude, and field size and flatness that are required—and the budget. When retrofitting a control system to an existing telescope, the optical assembly is usually treated as a given.

3.3.4 Drive Design Approach

Unless you have an unusual slew or tracking requirement and a large budget, your drive will have at least one stage of "gear" reduction. Gears always introduce pointing and tracking errors. The advantages and disadvantages of different gearing methods are discussed in later chapters. Different motors have their advantages and disadvantages to consider, as discussed in Chapter 8.

3.3.5 Control System Approach

Your requirements for pointing and tracking accuracy will help you decide if you need feedback in your control system, and if so, what kind. We discuss the role of various types of feedback in Chapter 4, and various devices for providing feedback in Chapter 9.

3.3.6 Computational Requirements

One of the criteria for selecting a control computer is the speed of its processor. To determine the computational requirements, you need to know the rate at which calculations must be performed. In a digital servo, the rate of performing calculations is determined in part by the rate at which errors accumulate during the interval between the calculations and subsequent commands to the motors. Since most of the equations of interest in Chapters 5 and 6 depend on hour angle, elevation, or some other parameter that changes with time, they can be used to assess error growth rates.

To meet various pointing and tracking accuracy requirements, the system as a whole must have a certain maximum error. Each part of the telescope and control system contributes in some way to the overall system error. To make sure you will meet your accuracy requirements, develop an error budget for your system, with each error source having a line in the budget. The simplest way of doing this is to give each error source an equal weight, and then determine the error for each calculation.

There are eight significant calculations discussed in Chapter 5 in the reduction from mean to apparent place, and nine calculations in Chapter 6 for mechanical and alignment corrections for an alt-az mount. If it is assumed these 17 calculations produce random errors that accumulate in quadrature (they do not affect each other), then the root sum squares of the individual errors should be less than or equal to the short term tracking error maximum. This means that

$$\sqrt{17e^2} \leq \text{tracking error requirement}$$

for equally weighted errors. Think about the time to perform each error correction calculation once and the number of times per second it must be performed to stay within the error budget.

3.3.7 Computer Hardware and Software

Selecting computer hardware and software can be the most difficult design decision you will make, and can have the greatest impact on the cost and performance of the system you build. There are many factors to consider, some of which are discussed below.

3.3.7.1 Processor Speed

Use the computational requirements for the control equations as a starting point. As discussed below and in Part III, there are many other things (besides computing apparent place and mechanical corrections) your processor must do to control a telescope. If your estimates show that roughly 100% of a particular processor is required to do a job, since such estimates are often a bit low, and since "collisions" among real-time tasks that need the processor make the actual CPU loading vary considerably, choose a processor that will be loaded on average only about 50% or less.

Another consideration is that the system will evolve over time, with new functions being added (functions are rarely removed from a system). This means you should buy a computer that allows you to upgrade the processor easily and inexpensively. Many computers use a "mother" board for the memory, internal bus, and external interface boards, and a "daughter" board for the CPU. Upgrading the computer consists of swapping the current daughter board with a new one holding a faster CPU. Some even permit multiple daughter boards to be installed.

3.3.7.2 Arithmetic Hardware

In the past, hardware for performing floating point arithmetic was an option. Intel offered separate arithmetic chips for the 8088 through the 80386, but they began placing floating point hardware in the main CPU module starting with the 80486. Hardware arithmetic options are extremely cost-effective, considering the high density of arithmetic instructions in the instruction mix of telescope control software. The penalty for performing floating point arithmetic in software is typically a factor of five to ten in processor utilization. For this reason, most computers come with floating point hardware as a standard feature, though there is still an aftermarket in high performance floating point hardware.

3.3.7.3 Interrupt Hardware

It is easy to design a computer system that generates a few hundred interrupts per second, most of which occur in moving data one byte at a time between memory and interface cards to motors, encoders, modems, and instruments. CCD cameras can generate lots of interrupts if they do not use block transfers or Direct Memory Access (DMA). Many early-generation microprocessors sampled all devices in a round-robin fashion to see which device needed I/O servicing when an interrupt flag was raised. This polling of devices consumes too much time when several hundred interrupts per second are generated in a real-time system.

A better method is to use vectored interrupts. A small section of main memory (typically a few hundred bytes) is dedicated to interrupt vectors. Each "vector" is a memory location storing the address within the operating system program of a particular interrupt service routine. The vectors are loaded with the interrupt service routine addresses when the device drivers are loaded, usually when the operating system is booted. When a device requires interrupt servicing (typically when a data transfer between memory and a peripheral device is completed) it uses an interrupt signal line, often shared with other devices, to signal the processor that it needs servicing.

When the processor is ready to service an interrupt, it manipulates control lines to transfer from the device to the processor a vector address that is stored on the peripheral interface card. The contents of the corresponding address in the dedicated area of main memory are then loaded into the CPU's program counter, which causes the processor to begin executing the interrupt service routine. Such vectored interrupt hardware is built into almost all modern processors, including the Intel 80x86, Motorola 680x0, and the IBM/Apple/Motorola Power PC.

3.3.7.4 Standard Bus Structure

A telescope control system requires computer interface cards to be built or purchased for non-standard devices (e.g., telescope motors). Almost all computers have interface cards available for printers or disk drives, or these features reside on the processor board, but only a few have cards available for stepper motors and shaft encoders. By choosing a computer that uses a bus for which standard off-the-shelf interface cards are available to interface these non-standard devices, you can spend your time building a telescope control system and doing astronomy, instead of spending additional time building and debugging custom interface cards. Furthermore, there is often free or inexpensive software, such as device drivers or dynamic link libraries, that are available for these cards. Such software can save you weeks or months of development effort. Using a standard bus also usually permits selection from a wide range of processors, which permits the kind of processor growth mentioned above. The most popular buses are discussed in Chapter 11.

3.3.7.5 Development Environment

The system development environment consists of the computer and purchased software needed to develop and integrate the control system. The computer hardware configuration for software development should support the following requirements:

1. Rapid entry and correction of source code
2. Rapid production of legible hard copy of source code, link maps, and file listings
3. Rapid compiling, assembling, and linking of programs
4. Sufficient mass storage (hard disk space) to hold the operating system, development tools (e.g., compilers and linkers), the programs you develop, and all the files needed to run the telescope, including catalogs and space for data
5. Ability to copy ("back up") source code in case of hard disk failure
6. Adequate memory for executing editing, copying, compiling, and other utility programs, as well as for running the control system
7. Integration of the operational hardware configuration with the software development system, to allow software testing and integration with the final hardware configuration.

PCs are sold equipped with high resolution screens, keyboards, and good editors for entering code. Very sophisticated text editors and document processors are available for a fraction of the cost of the PC that make entering and editing source code easy.

There are several small, fast, high quality dot matrix printers available for about $100–$400 capable of speeds of 200 characters per second or more with line widths of 80 or 132 columns, and with graphics capability. Laser printers are beginning to be reasonably priced, and bubble-jet printers cost less than laser printers, yet offer print quality nearly as good. You should always keep a paper copy of all the software you write, no matter how good your disk and backup are.

Although early PCs came with one or two floppy disks and no hard drives, today 300 MB hard drives are very inexpensive, and even larger drives well over 1 GB are quite affordable. You can never have too much disk storage. Operating systems, development tools, astronomical catalogs, and CCD cameras all require large amounts of storage.

As hard disk drives become larger in capacity and less expensive, the problem of disk backup gets worse. Although disk drives are now very reliable, occasionally problems do occur, and you do not want to be the victim of a disk crash without having a backup. You might want to consider buying a quarter-inch tape cartridge (QIC) drive capable of storing 500 MB, or even the more expensive Digital Audio Tape (DAT) and Exabyte 8mm tape drives. Magneto-optical read-write optical disks are now available in the 3.5-inch format with capacities of a few hundred megabytes. A reliable backup drive is absolutely essential to making your project a success.

Modern Windows™ operating systems require a large amount of memory, at least 16 MB and preferably 32 MB or even more. Newer processors can address 256 MB of memory or more, but many of the processors in general use have a lower limit. Memory prices have fallen dramatically in the last few years, so that it is more economical to purchase a large amount of memory than it is to lower programmer productivity by causing the programmer to run out of available memory at crucial moments. Not having enough memory can also drastically slow modern operating systems.

You should develop the software on the same machine that will be used to control the telescope, due to the large number of tests with the telescope control interfaces that you will need to make. You will need a machine with adequate cabinet space, bus slots, and power supply reserve for peripherals, so that both software development peripherals and telescope control peripherals can be installed, powered up, and functioning simultaneously. It should take no more than five minutes, and preferably even less time, to make a simple change to the source code and generate and load a new program version.

The most critical component of the development environment is the compiler and associated programmer productivity tools. The main goals in selecting a software development environment are as follows:

1. Minimize the time to learn a new environment by using tools that you know and enjoy using;
2. Generate efficient code that executes quickly in a real-time control environment;
3. Minimize the initial software development costs;
4. Minimize the total system life cycle costs.

Development languages and tools are discussed in Chapter 11.

3.3.7.6 Operating System

The operating system is a program that manages the computer resources, including access to peripheral devices and the processor. The operating system usually includes device drivers that handle the flow of data between main memory and a peripheral device, and some form of task scheduler that determines when programs are allowed to execute.

The development environment is closely tied to the operating system, so often you choose both as a pair rather than picking an operating system then independently picking a development environment. This means the operating system must help you develop the software and run it to control your telescope. Most likely, you will need to run more than one program at the same time, and these programs need to be able to talk to each other. Make sure you choose an operating system that can run more than one program ("process," "task", or "thread") at a time, and that provides a means for these programs to communicate with each other. This topic is covered extensively in Chapter 11.

When the wide array of available computers is evaluated for suitability to the task of real-time telescope control, each should be evaluated not only on its ability to perform all required calculations in the required time, but also on the availability of software and interfaces to the telescope to work in harmony with the computer to develop the telescope control software and support its execution in a real-time control environment.

The key to computer selection is to think out every feature of both hardware and software that is needed to make the system work, and then buy a computer that does not lack any of these features. Many have made the mistake of waiting until the computer was brought home to discover that the hardware or software lacked some feature that was assumed to be included. Most computer vendors will allow a prospective buyer to peruse the manuals before making a large investment in hardware and software.

Every system developer has strengths and weaknesses. Those who know electronics and hardware may be uneasy developing software. Those who are comfortable developing software may be inept with a soldering iron. The best system you can design is the one that emphasizes your strengths and minimizes your weaknesses. Try to get help from experts in the areas where you need it to obtain a measure of balance in your control system.

In this chapter, we have tried to provide an overview of how to design and build your control system. In the next part, we go into greater detail about the things you need to consider in developing your design.

Part II

Telescope Control System Design Considerations

Chapter 4

An Introduction to Control Theory

4.1 Control Options

After you specify your system performance requirements, the next step is to select the overall design approach. In this and the next three chapters, we discuss the problems you must solve in your design.

For telescopes to perform useful work, they must be pointed quite accurately in the sky, and then must track objects over an extended period of time with considerable accuracy. For example, in photoelectric photometry with a typical small telescope, a star might be centered in a $60''$ diaphragm with an accuracy of at least $15''$, and must stay centered with this accuracy for a few minutes of time. This sort of accuracy is routinely obtained with Celestron 8-inch telescopes on modest mounts in backyard observatories.

Pointing and tracking accuracies that keep objects centered within a couple of arc seconds for hours is not at all uncommon when these same mounts are engaged in astrophotography, with the astronomer making constant drive rate corrections manually. When one recalls that there are $1{,}296{,}000''$ in a circle, this sort of accuracy may be surprising. It is achieved in simple systems by having a sharp-eyed observer at the controls. The observer detects deviations of the system from its desired position (errors), and activates the controls to reduce these errors. In this system, the observer's eye is the error sensor, the observer's brain is the computer that determines the appropriate response, and the observer's fingers make the response using the manual telescope control system. If this series of events is thought of as a "loop," then it is the observer that "closes the loop" in this case.

All telescope pointing and tracking systems are closed-loop systems in some sense (when feedback from humans and photometers are counted), as there appears to be no way to achieve the high accuracies demanded by most astronomers other than by detecting and correcting for deviations from desired functioning. There are, however, several ways to close the loop in telescope control systems. In this book we will be concerned with those systems in which, for one reason or another, a computer is involved somewhere in this process. We have chosen to classify the methods by which the loop is closed into two broad types.

The first type is the *classical* closed-loop servo system. This is the one treated in some detail in this chapter and in Trueblood's example in Part IV. The classical system is one in which there are angle position sensors on each axis of the telescope (e.g., optical shaft angle encoders), and a computer that compares the actual position (given by the encoders) with the desired position (the corrected place of the star) and issues appropriate commands to bring these two closer together.

There are two key points here. First, in the classical system, only angular position at the telescope axis (or in many systems, at the final worm, which is one step removed from the axis) is sensed, along with, perhaps, motor velocity. No direct sensing of the star itself is performed. This means that no random or systematic error between the encoder and the image plane, such as tube or mount flexure or atmospheric refraction, is directly sensed. Instead, systematic errors (that are thought to be understood) are modeled in software, and model constants are measured by making a special calibrating observation of stars at sky positions selected for each type of error. It is then expected that after such calibration and modeling, the residual errors that remain after the modeled errors have been removed will be small enough to be ignored.

The second key point in the definition of the classical system is that the computer is almost continuously in the loop, updating the control commands and recalculating telescope flexures, atmospheric refraction, and other systematic errors. As the computer must also do other tasks, this means that a multi-tasking interrupt-driven system is almost a necessity, and the computer must take a break regularly, perhaps several times per second, to make the calculations contained in the control algorithm and to issue the resulting commands to the motors.

We have chosen to call the second type of telescope control system "nonclassical" or "other approaches," where "other" in this case means other than the classical approach. The detailed discussion of these systems is deferred until Part V. However, a brief introduction is appropriate here. Generally, it is the goal of nonclassical systems to get around some of the difficulties of the classical system, although new difficulties are introduced in the process.

The main difficulty with the classical approach is that it senses errors at the telescope axes, not the stars themselves. Since the ultimate goal is to point the telescope at something in the sky, any unmodeled errors are ignored in the classical approach, or one has to try to correct for them some other way—a task not easily accomplished, and one that can load down the computer. The nonclassical approaches generally try to detect the error at the star itself, and as the error is detected and corrected here, no complex calculations are needed, and the computer is not loaded down. However, star sensors are generally more difficult to implement than shaft angle sensors, so one has essentially traded one set of problems for another set.

There are, however, important exceptions to this. First, if the sensor is the astronomer's eye, then implementation may be rather straightforward and inexpensive. Examples of this are the stepper motor telescope control system developed by Rafert mentioned earlier, and Tomer's system described in Part IV.

Second, if for some other reason, an electronic star sensor of one type or another, such as a photoelectric photometer, must be part of the instrumentation at all times, then one can use this sensor to close the loop directly on the star itself. A number of classical systems have star sensors such that once the star is acquired, the star sensor, not the shaft angle encoders, assumes control of the telescope.

A good example of this is the Vienna 1.5-m telescope (Stoll and Jenkner, 1983). In such a system, the job of the classical control system is to point the telescope just accurately enough to permit the star tracker to lock onto the guide star. Lock-on range in some systems can be as large as $1'$–$2'$. Some telescopes with permanent photoelectric sensors have purposely extended their acquisition range to the point where optical shaft angle encoders and correction computations of all kinds have been eliminated entirely. The telescope moves from one area of the sky to another totally "open-loop" (i.e., without any position feedback), and when it arrives in the area of the sky where it "thinks" the star ought to be, it then searches for it with the photometer, acquires it, and then centers it—all under computer control. Such a system is described in Part V.

4.2 The Role of Position Feedback

The purpose of a telescope control system is to point the telescope at the target object and track it with an accuracy dictated by the observing requirements. In the classical control approach, this is done by controlling the angular motion of the telescope about its axes. The system determines where the telescope ought to be at any particular time, then sends commands to the drive motors to move the telescope so that it arrives at the desired position at the proper time. This kind of control system is an example of the general class of servo control systems.

The simplest form of control system is the open-loop system. As shown in Figure 4-1, the open-loop system finds the difference between the desired position and the current position and computes an error signal that forms a command to the telescope drive motor.

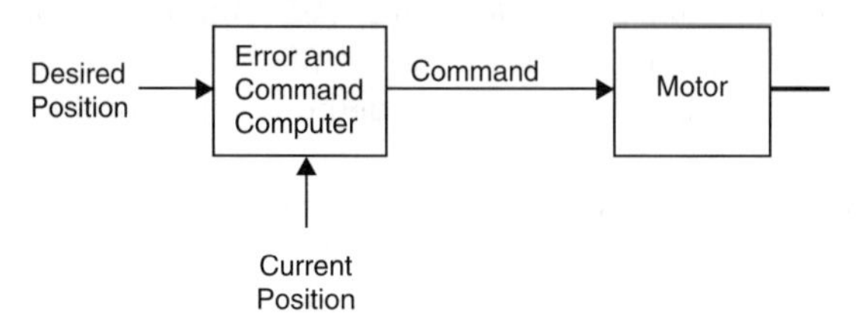

Fig. 4-1 *Typical open-loop servo.*

An example of an open-loop system using analog electronics is the typical variable frequency drive corrector used to control the RA synchronous motor on a small telescope. The observer sets a switch or adjusts a potentiometer ("pot") to obtain the desired drive rate, which is indicated by a mark near the frequency knob.

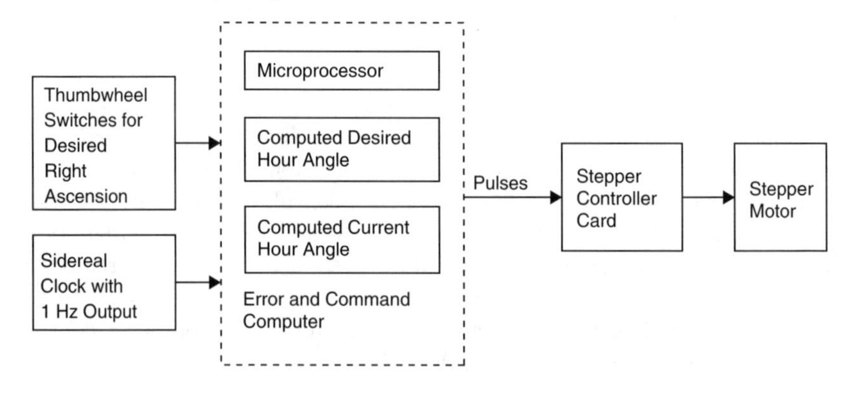

Fig. 4-2 *A digital open-loop Right Ascension drive.*

An example of an open-loop telescope drive with digital electronics is using a computer to control a stepping motor. Figure 4-2 shows a block diagram of a digital open-loop right ascension drive for an equatorially mounted telescope. The computer computes the desired hour angle from the desired right ascension stored in the thumbwheel switches and from the sidereal time in the sidereal clock register. By counting the pulses it sends to the stepper and by using a stored constant representing the gear ratio between the motor and the RA shaft, it can compute the current telescope position, after a previous calibration on one or more stars of known position. The computer then adjusts the rate at which it sends pulses to the stepper controller based on the difference between the computed desired hour angle and the computed current hour angle. After each pulse is sent, the current hour angle register is updated.

This system can be as complex as is needed. The computer can correct the desired hour angle for precession, nutation, aberration, refraction, polar axis misalignment, and other systematic errors down to arc second accuracy, since typical computers available today have adequate CPU power to perform these calculations in real-time. However, achieving arc second accuracy in calculating the desired hour angle does not guarantee arc second accuracy in pointing the telescope.

The problem with all open-loop systems is that once a command is given, there is no way to know if the command was executed accurately. Software models of systematic errors are either not accurate enough, or depend on knowing the positions of all the parts in the tube, mount, and drive train at each instant. These systematic errors include backlash and other gearing errors, flexure, and mechanical misalignments. Such structural element position information is not generally available to the computer. If it were, the algorithms needed to model all systematic errors to very high accuracy would tax the resources of most computers to compute them in real time. Even if these systematic errors were modeled perfectly, random displacements due to wind gusts or other error sources would still introduce significant errors. One can take precautions to size the motors properly and to design a good mount and drive. Such precautions will reduce errors sufficiently for most small telescope applications, but for high-accuracy control of large telescopes, it is conventional to close the loop somehow.

There are other reasons for using a closed-loop servo. If the telescope is slewed by hand (by disconnecting a clutch and moving the tube without using motors), all current computed pointing information in an open-loop system is in error, perhaps by as much as 180°. This is also true when a motor that is not controlled in fixed increments (steps) is used for turning the telescope axes, since the computer has no way of knowing how much the motor has moved the telescope. Thus some situations require a means to sense the telescope position independently of the commands sent to the motors.

Achieving high accuracy pointing in open-loop (stepper-based) systems requires special care. The drive system must be free of backlash and other gear errors. Slewing acceleration and slew speeds must be set so that no steps are lost for any reason. Finally, the telescope structure needs to be quite rigid to avoid excessive flexure if the computation of such corrections is to be avoided (although an open-loop system can compute and correct for flexure). Many of the telescope mounts and drives described in this book were specifically designed to meet these stringent open-loop requirements.

Any system that meets these requirements has an unusually good mount, and can just as easily be used in a closed-loop system. If a well designed telescope and mount are used, and all the corrections discussed in Part II are computed often enough with adequate precision, one could achieve very good accuracy. How good is an open issue, since to our knowledge, no very accurate completely open-loop telescope control system has been attempted. Depending on the amount of work put into tuning such a system, pointing accuracies of 10″ might be possible.

In contrast, several professional observatories have built closed-loop servo systems for telescope control that regularly achieve 5″ overall pointing error or better in all areas of the sky. In the classical closed-loop approach, a device is mounted on the telescope axis to feed the true angular position of that shaft back to the current position register. In this case, the computer uses the actual shaft angle position to compute the error signal.

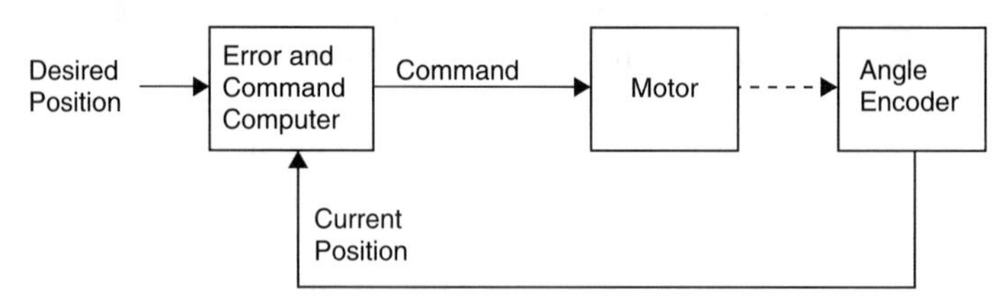

Fig. 4-3 *Typical closed-loop servo.*

As shown in Figure 4-3, the closed-loop system is identical to the open-loop system except that feedback is used to generate the current position information, instead of computing it. In the example given above of a variable frequency drive corrector, if the observer sees the stars slowly drifting through the eyepiece field, he usually corrects the drive rate. In this case, he is "closing the loop" and providing position feedback to the system (in the nonclassical sense). Even in systems in which electronic feedback is used for automatic drive rate corrections, multiple feedback loops can exist, with the operator in the outermost loop guiding the telescope with a hand paddle.

Any closed-loop control system is called a *servomechanism,* or *servo.* This term is used regardless of whether DC servo motors, steppers, or some other type of actuating device is used. A household heating furnace controlled by a thermostat is a common example of a servo.

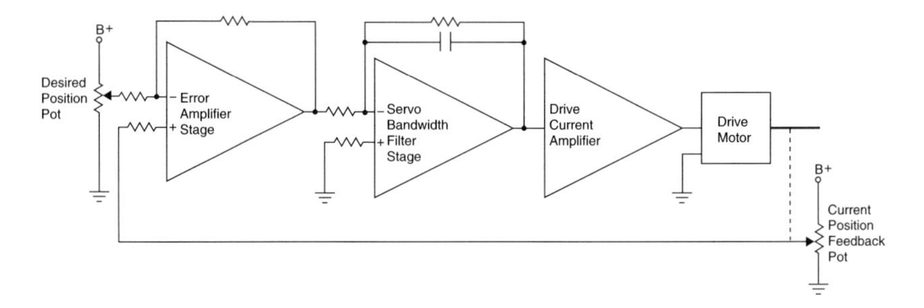

Fig. 4-4 *An analog closed-loop declination drive.*

An example of a closed-loop servo controlling a declination drive on an equatorial mount using analog electronics is shown in Figure 4-4. The observer dials in the desired declination angle by adjusting a linear pot with a pointer knob and a calibrated dial. The feedback is provided by a linear pot turned by the declination axis, and the error computer is an operational amplifier (op-amp) that produces an output proportional to the difference between the *desired* and *current* voltages generated by the two pots. The output voltage is proportional to the error, and can be used to drive a DC motor to point the telescope. Tracking in RA can be accomplished using the same circuit with a modification that adds a constant voltage offset to the error (command) voltage. This approach can give quite acceptable performance for very low cost—assuming high accuracy is not required.

The purpose of the filter in the circuit shown in Figure 4-4 is to limit the bandwidth of the servo and adjust the relative phase between the forward and feedback paths. This is required to help prevent overshoot and oscillation, both of which are the result of high gain in the amplifier and inertia in the motor and telescope. By limiting the maximum rate at which the control voltage can change to that at which the motor can reasonably respond to a step change, the servo performance can be tailored to the mechanical properties of the system. The filter helps to damp the system to permit it to find the desired position in the least time.

An example of a closed-loop servo controlling a declination drive using digital electronics is shown in Figure 4-5. The error and command computer uses the desired position input from the thumbwheel switches as the reference input and the binary shaft encoder output as the current position feedback. The computer converts the shaft encoder output in the form of a binary code to an angle in degrees and compares it to the desired position stored in the thumbwheel switches. The difference between these two is the error, which the computer converts to the number of stepping motor steps required to reduce the error to zero. The computer then computes a step rate profile designed to ramp the stepping motor up to some speed, sustain that speed, then ramp down so that the motor has moved the required number of steps to reduce the error to zero in the shortest possible time. Actually, the error is not reduced to zero, but to half of the angle the controlled shaft moves through in one motor step, or the resolution of the shaft encoder, whichever is greater.

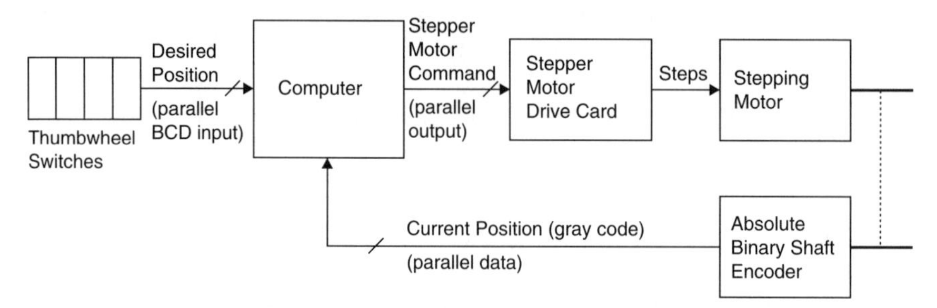

Fig. 4-5 *A digital closed-loop declination drive.*

In computing the number of steps required to minimize the error, the computer can correct the desired position input for precession, nutation, aberration, and refraction to obtain the apparent position in the sky from the mean catalog position entered in the thumbwheel switches, and it can correct the current position feedback number to reflect flexure, nonperpendicularity of the rotation axes, misalignment of the polar axis, nonlinearity of the encoder, misalignment of the optical and mechanical axes, and other sources of systematic error discussed in Chapter 6. The corrected desired and current positions then are used to determine the number of steps required to minimize the error. In this way, systematic errors can be modeled and removed from the error signal. The overall system accuracy is then determined by the accuracy (and expense) of the encoders used for the position feedback, as well as the accuracy of the systematic error models.

Note that there is no servo bandwidth filter explicitly shown in Figure 4-5. The computer fills this function by sampling the desired position and current position inputs periodically, rather than continuously as in the analog circuit. The rate at which these inputs are sampled, and the rate at which commands are sent to the stepper motor drive card determine the servo bandwidth. These rates are under direct control of the software, so the system can be tuned without changing the hardware.

4.3 Equation of Motion

Any servo is essentially a damped harmonic oscillator, an example of which is an automobile body suspended on its chassis with springs and shock absorbers. If your car has old, worn out shock absorbers, and if you push down on a fender suddenly, the car will bounce up and down on its springs several times. This continuous bouncing that dies out slowly is *underdamped* motion. If the shocks are new, when the car is pushed down, it comes all the way back up quickly, then stops. This is *critically* damped motion. If the shocks are too stiff, the body rises slowly from being pushed down, and eventually comes all the way back up. This is *overdamped* motion.

The entire ensemble of the telescope and its control system behaves as a damped harmonic oscillator. In addition, the mechanical telescope structure and the electronics in the control system, taken separately, both act as damped harmonic oscillators. All damped harmonic oscillators satisfy the general equation

$$J\frac{d^2\Theta_o}{dt^2} + F\frac{d\Theta_o}{dt} + K\Theta_o = K\Theta_i \tag{4.1}$$

where $\Theta_i =$ the commanded input angle

$\Theta_o =$ the resulting output angle of the telescope axis

$K =$ the system torque constant

$J =$ the total moment of inertia of the driven component

$F =$ the system damping.

Note that if one merely substitutes electric charge for angle, inductance for inertia, resistance for damping, and capacitance for torque constant, we have the equation for an electronic oscillator.

This equation gives the relationship between an input command angle Θ_i, and how the servo reacts to that command, given by the output angle Θ_o. J includes the motor rotor moment of inertia, J_m, and the load moment, J_l. If a gear train of reduction N:1 is placed between the motor and the load, the load moment is reduced by the factor $1/N^2$. F includes the motor damping, $F_m = T_s/S_o$, where T_s is the motor stall torque and S_o is the motor's unloaded free (maximum) speed. Figure 4-6 shows the speed versus torque curve for a typical servo motor.

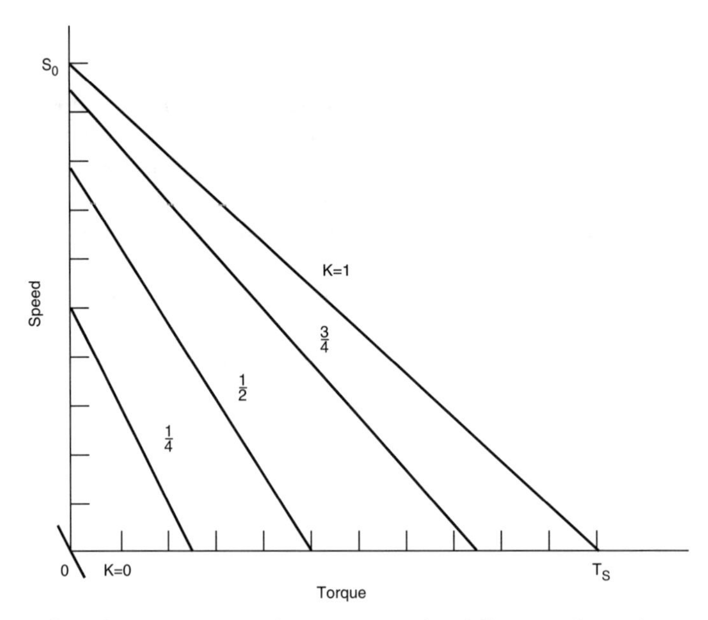

Fig. 4-6 *Speed versus torque characteristics for different values of control voltages. Courtesy Reston Publishing Co.*

The system damping, F, contains sources of damping besides the motor damping, including the effect of viscous damping cups attached to the motor, or velocity damping obtained by attaching a tachometer-generator to the motor and feeding back its signal to the error generator. The latter technique reduces the error signal magnitude by an amount proportional to the motor velocity. K is given by $K_s K_a K_v N$, where K_s is the gain or transfer function of the device generating the commanded input angle signal (such as the volts/radian generated by a pot turned by the observer to command the servo to point the telescope), K_a is the gain of the servo amplifier, K_v is the motor torque constant, and N is the gear ratio between the motor shaft and the load.

To understand the behavior of the servo, consider the homogeneous quadratic equation

$$J\frac{d^2}{dt^2} + F\frac{d}{dt} + K = 0 \tag{4.2}$$

which has roots

$$-\frac{F}{2J} \pm \sqrt{\frac{F^2}{4J^2} - \frac{K}{J}}. \tag{4.3}$$

The following three separate, physically meaningful cases result:

Case 1. Negative, real, and unequal roots.

$$\frac{F^2}{4J^2} > \frac{K}{J}$$

Define the damping ratio $DR = F/2\sqrt{KJ}, > 1$ in this case. This is an overdamped system.

Case 2. Negative, real, and equal roots.

$$\frac{F^2}{4J^2} = \frac{K}{J} \quad \text{or} \quad DR = 1$$

This is a critically damped system.

Case 3. Conjugate and complex roots with negative real parts.

$$\frac{F^2}{4J^2} < \frac{K}{J} \quad \text{or} \quad DR < 1$$

This is an underdamped system.

The damping ratio DR is the ratio F/F_c of actual damping to critical damping. Thus for critical damping, the damping constant F_c should equal $2\sqrt{KJ}$.

When a new position is entered into the servo, in the underdamped case, Θ_o will attain the desired position rapidly, then overshoot. The error signal will reverse sign, and drive the motor in the opposite direction. This oscillation continues for a time determined by the damping ratio. When $F = 0$, the servo oscillates forever at a natural frequency of $\omega_n = \sqrt{K/J}$. For the critically damped and overdamped systems, Θ_o approaches the desired value monotonically and exponentially. Of the three cases being considered, a system that is very slightly underdamped settles to within some defined tolerance of the desired position and stays there sooner than the other two systems. Therefore, the system should be designed to be slightly underdamped to obtain the most rapid response, with an overshoot equal to the error tolerance. If the time to achieve the desired position is not of utmost importance, and oscillations must be tightly controlled, the control system should be overdamped. Most telescope control systems are overdamped, in part to compensate for the typical underdamping of the mechanical structure, and to protect the drive from damage. Figure 4-7 shows the system responses for the three cases.

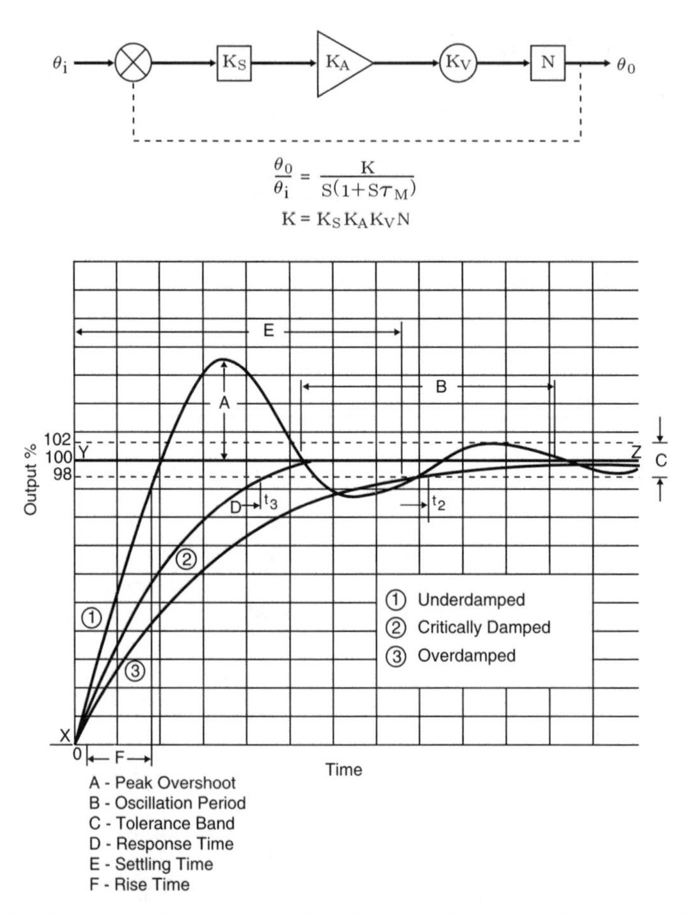

Fig. 4-7 *System response curves for three conditions of damping. Courtesy Reston Publishing Co.*

4.4 The Role of Velocity Feedback

In an underdamped system, the damping ratio is less than 1, which indicates the gain, K, is too large for a given system damping F and for a motor matched to the inertia J. It is desirable to have a large system gain K to keep the system dead band low. Dead band is the range of values over which the input signal Θ_i may vary without causing the output Θ_o to respond. For a DC servo motor, it is the threshold voltage at which the motor just starts to turn as it overcomes internal friction. The dead band is given by $2V_c/(K_sK_a)$, where V_c is the maximum control voltage applied to the motor. By using a large value of K_a, the dead band is reduced. A telescope pointing system must have a very small dead band, since to achieve, say, tens of arc seconds pointing accuracy or better in Θ_o, any change in Θ_i that gives no response in Θ_o produces an error in Θ_o. The large gain K_a needed to reduce the dead band also tends to produce an underdamped system, so the total system damping F must be increased to compensate. Note that the dead band may not be uniquely defined in terms of absolute positions if the system has substantial hysteresis.

To demonstrate the method of increasing F with velocity feedback, one begins with a servo with no damping, which is the extreme case of underdamping. With a step input, the high gain forces the output to rise rapidly, as shown in Figure 4-8(a). At the moment of the step input, the error is greatest, but decreases as the output moves to the desired position, as shown in Figure 4-8(b). When the output reaches the desired position, the motor speed is highest, so that the motor inertia, along with the load inertia, carries the output through the desired point into the characteristic underdamped oscillations.

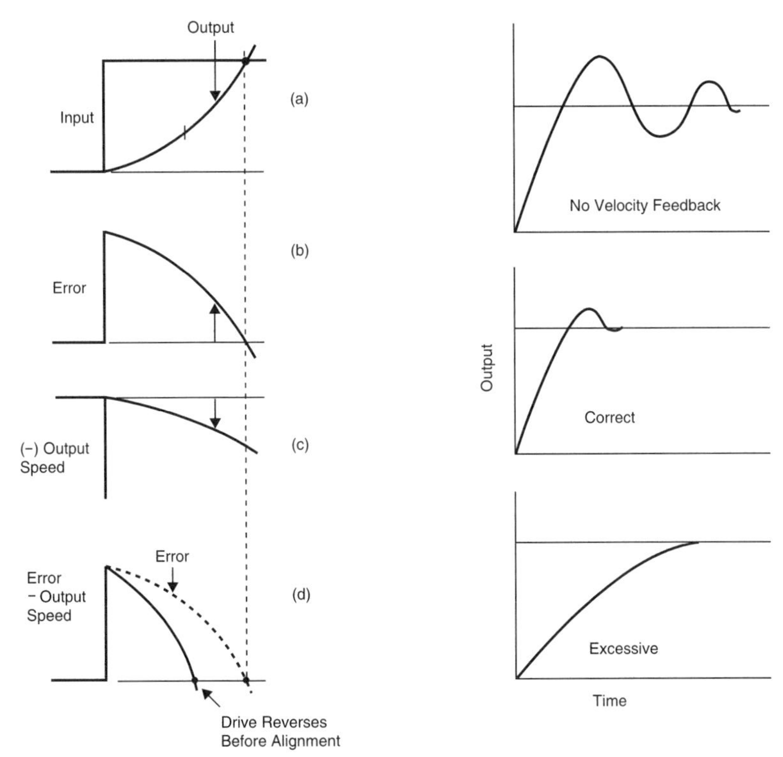

Fig. 4-8 *Effect of velocity feedback. Courtesy Reston Publishing Co.*

The velocity feedback needed to achieve proper damping can be provided for a motor using a tachometer-generator. This is a small DC electric generator that, when coupled to the motor shaft, produces a voltage proportional to the motor speed. This motor speed signal is first inverted to provide negative feedback, then combined with the position error signal and the input control signal to produce a modified error signal, as shown in Figure 4-9.

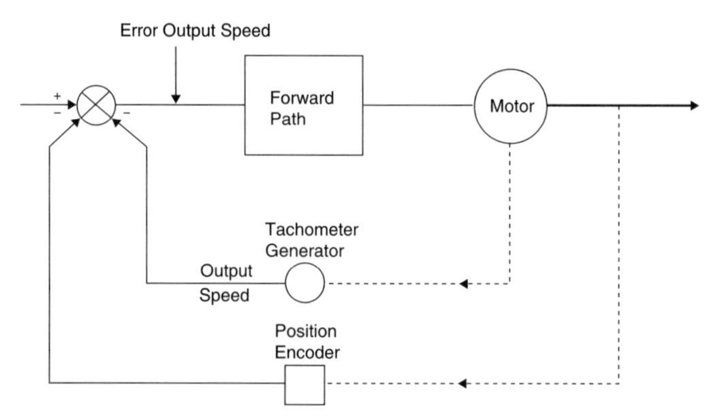

Fig. 4-9 *Servo diagram showing velocity feedback.*

Figure 4-8(c) shows the velocity feedback signal produced by the tachometer-generator, and Figure 4-8(d) shows the modified error signal that is fed to the forward path. If the gains of the position and velocity feedbacks are properly adjusted, the servo will not overshoot the desired position and oscillate, but rather will be properly damped. A tachometer-generator is used in analog systems, but in digital systems using a digital processor as the error computer, the processor can compute shaft speed easily and rapidly using the system clock and only position feedback. To avoid excessive noise from the differencing operation, the position feedback device must have high resolution.

The phase of the feedback signal with respect to the forward path signal is important in determining the stability of the servo. Figure 4-10 shows that oscillations are enhanced, rather than attenuated, if the feedback is out of phase with the forward path. Electronic servo circuits that oscillate can be improved by placing phase-compensating filters in the feedback path. This can also be accomplished in software in a digital servo by adjusting the software execution timing.

Despite the availability of sophisticated structural analysis programs, it is often not worth the trouble to attempt to predict the behavior of a

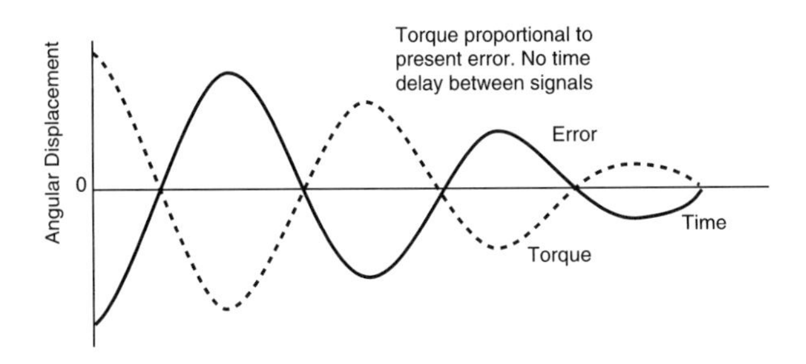

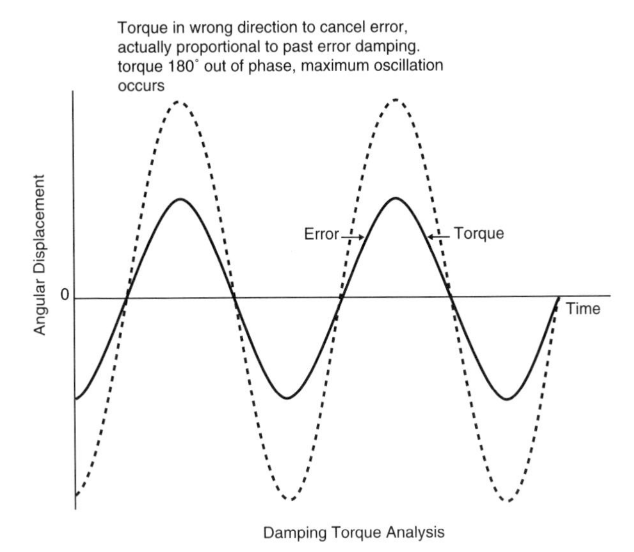

Fig. 4-10 *The effects of feedback phase on servo error. Courtesy Reston Publishing Co.*

telescope control system in advance, except for very large, very expensive telescopes where the price of a mistake is higher than the expense in modeling the system. Instead, a better approach is to use good design in all aspects of the telescope and the control system, and to tune the system after it is built. Being aware of the nature of the problems that might be encountered, the telescope designer can then diagnose and solve these problems as they manifest themselves in the completed telescope. One example of the tuning that is required is adjusting the servo bandwidth. Because of the disparity between typical tracking and slewing rates, two different filters may be required in the control system, each adjusted to give the best performance for its mode of operation. Many servo motor controllers available off-the-shelf have the means to adjust their servo parameters to adapt to the system into which they are installed. Some others offer PC software to help you tune the system.

The forward loop integral should be tuned after the system is built. The system described by the equation of motion given by Equation (4.1) has a servo lag phase error proportional to the size of a new step function input and inversely proportional to the bandwidth of the servo. That is, the servo filter that limits the servo bandwidth also limits how rapidly the servo can respond to a new input command. Since the servo cannot respond immediately to the new command, a phase delay is introduced between the input and the output. One can compensate for this delay by measuring the characteristic time constant for the servo and adding a correction term to the error signal. This can be done with additional circuitry in an analog servo, or with software in a digital servo, and has the effect of adding a forward path integral term to the equation of motion. This permits the use of a lower system gain without increasing the dead band.

These are only two examples of the kind of adjustments that must be made to a control system after it is built to achieve the desired level of performance. The process of tuning is made easier when the telescope, mount, drive, and control system all work together with each other, rather than against each other.

4.5 Matching the Telescope and the Control System

Several projects were undertaken during the last few decades to retrofit a computerized control system to an existing telescope. Many of these projects were quite successful in meeting their design goals and functional requirements, but a few projects failed because the telescope and control system were not well matched. Often, the approach taken in those projects that failed was to rely on the computer software to correct mechanical design deficiencies in the telescope tube or truss, mount, or drive train. Recent projects in which both the telescope and the control system were designed for each other tended to enjoy much greater success. Based on this history, equal attention must be paid to the design of the telescope tube or truss and mount, the drive train, and the computerized control system. Each of these segments of the total system will be treated briefly in the following paragraphs.

First, consider the mechanical parts of the telescope, taken as a whole. If the rotor of the motor that drives the telescope is locked, and then the tube or truss is struck with a quick tap, the telescope will oscillate with a characteristic natural frequency. This mechanical behavior is described by the same Equation (4.1) used to characterize the electronic and electromechanical behavior of a servo circuit. Mechanically, telescopes have very little damping. The oscillations die out exponentially after some number of periods of the natural resonant frequency of the mechanical structure. Since the locked rotor frequencies of telescopes of conventional design tend to be on the order of 1 Hz, it can take a relatively long time for oscillations to die out. This causes a problem for any control system, since momentary gusts of wind, or short, frequent pointing commands cause the telescope to vibrate in a manner that can damage the drive mechanism, smear long-exposure images, and lengthen the time after a pointing command is given before reliable data can be gathered.

This situation can be remedied either by

1. increasing the amount of damping, or

2. raising the natural frequency of the structure.

In the first approach, the number of oscillations the structure makes is reduced, which shortens the time it takes them to damp out. In the second approach, the period of each oscillation is reduced, which also reduces the total damping time by reducing the time it takes for the structure to make a given number of oscillations. Returning to the automobile example again, increasing the damping is analogous to using stiffer shock absorbers. Although this works well for automobile suspensions, it is just not practical to place shock absorbers (or other means of increasing the damping) throughout a telescope's tube, mount, and drive structures. This leaves the second approach—raising the natural frequency of the structure.

The natural frequency of a mechanical structure is determined by the stiffness of its elements and the moment of inertia about the axis being rotated by the drive. A structure can be made stiffer by adding more mass to strengthen key structural elements. This by itself would raise the natural frequency, but it also tends to increase the moment of inertia, which lowers the natural frequency. There is some tradeoff point where the natural frequency is maximized by balancing added stiffness with added inertia. It is this tradeoff point that should be sought in the telescope design, not just minimum moment of inertia, or maximum stiffness. With proper design, a locked rotor natural frequency as high as 10 Hz can be achieved on large telescopes, and even higher frequencies on smaller telescopes.

The moment of inertia about any axis is minimized when all mass elements attached to that axis are as close to the axis as possible. Thus symmetrical mounts, such as fork and yoke types, will tend to have lower (more desirable) moments of inertia, while the moments of inertia of other mount types, such as the German equatorial and cross axis mounts, tend to be significantly higher. Asymmetrical mounts also require large counterweights that add to the moment of inertia but do not increase the mount stiffness.

Next, consider the telescope drive system. Most telescopes of conventional design use a worm gear drive. The problem with this kind of drive is that it forces the control servo to provide considerable damping in the overall system, and tends to limit the slewing speed. To see why this is so, consider what happens when the telescope is at maximum slew speed, and the slew motor is suddenly stopped. The energy of the mass of the telescope and its mount are dissipated by driving the worm gear into the worm, most likely damaging both gears. The use of non-back-drivable gears forces the control system to provide a very gradual deceleration from slew rate to track rate to prevent this kind of damage.

The fact that longer ramp times are required for decelerations when worm gears are used often means the peak slew rate of the drive is never reached before it is time to ramp down to the tracking rate. For some observing programs, this may not be important, but for others, it can be critical. For example, the research program of an automatic photoelectric telescope (described further in Part V) depends on minimizing the time to acquire stars. This is achieved using frequent short motions to execute spiral search patterns.

A better approach to drive design is to eliminate gears altogether. Since the drive contributes to the locked rotor natural frequency of the telescope, it should be very stiff. Although band and chain drives have several advantages over a worm gear drive, and have been used successfully on small computerized telescopes, they are not adequately stiff for larger telescopes. The chain drive also suffers from considerable periodic error. The band drive does not have this drawback, but on larger telescopes, the band would have to be undesirably wide before it would have the required stiffness (Melsheimer, 1984). Some telescopes use helical gears instead of worm gears. Helical gears are back-drivable, and good helical gears have very small errors, since they tend to be averaged out by having several teeth in contact at once. However, the best approach is to use a disk and roller in friction contact. The friction drive is both back-drivable and completely free of tooth-to-tooth gear errors. And it can be made stiff enough for telescopes up to about 4 meters in aperture. For these reasons, it is the drive train approach taken on the portable telescope described in Part IV.

The natural frequency of interest when moving the telescope is not really that with a locked rotor, but that of the system with the motor powered up and moving the telescope. Suppose the tube and mount were both perfectly stiff, and gusts of wind hit the telescope tube at regular intervals. If the motor and (back-drivable) drive are not sufficiently stiff and cannot

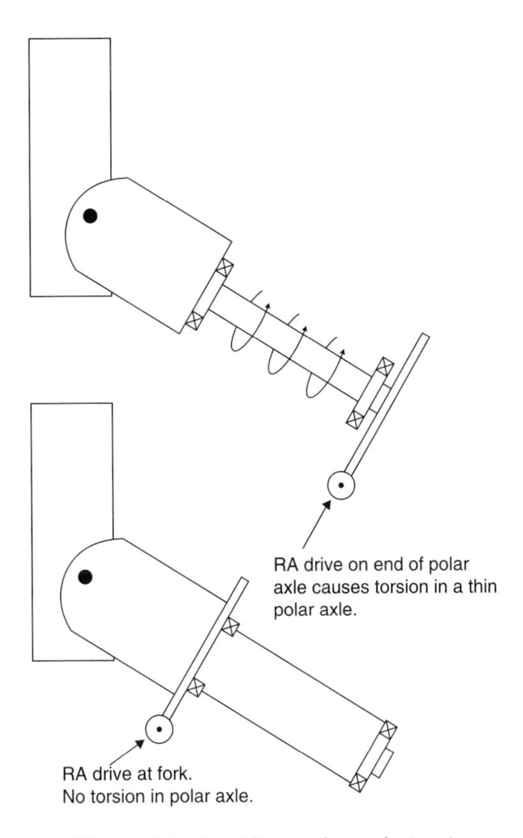

Fig. 4-11 *Avoiding polar axle torsion.*

develop sufficient torque to overcome such load variations, and if the wind gusts come at the right frequency, the motor would lose ground when a gust comes along, then make it up again when the gust subsides, then lose ground again when the next gust arrives. The result is telescope oscillation, rather than proper tracking. This means that drive trains should be designed with high inherent stiffness, and the motor should have adequate torque to overcome varying torque loads.

The traditional drive design, especially on smaller telescopes, has been to drive the tube assembly using a long, thin (2 inches in diameter or smaller) axle driven at the end opposite to that attached to the telescope tube. This approach is to be avoided because it is not very stiff. As shown in Figure 4-11, motor torque applied at one end of the axle tends to wind the axle up like a spring, with the telescope tube oscillating on the opposite end in response to varying wind and friction loads. A much better approach is to attach a drive disk directly to the bottom of a fork mount in RA, and to the tube assembly in declination. The result will be a telescope mount with a high natural frequency that is a pleasure to use and easy to control by computer.

The final element to be considered is the control system. Depending on the telescope design and the requirements imposed by the observing program, position feedback, velocity feedback, both, or no feedback can be used in the servo. Despite the tendency to think of tracking as a time ordered sequence of pointing commands, in reality, pointing (slewing) and tracking are quite different operations. One may want to use some form of feedback for one operation, and go open-loop in the other. Appendix A summarizes the servo approaches used in successful projects in the early 1970s.

In his systems built in the early 1980s that were available commercially, Melsheimer (1983, 1984) used a hybrid system of a stepper for tracking and a DC servo motor for slewing. When the control system was commanded to slew the telescope, the computer issued a speed command to the DC servo motor through its control amplifier. A tachometer on the motor provided velocity feedback to the servo amplifier, which used this feedback to adjust the voltage on the motor to ensure it was turning at the commanded speed. Separate angle position encoders on the telescope axes were used to provide accurate position feedback. Therefore, both position and velocity servo loops were used for pointing. An open-loop system employing stepper motors controlled the tracking.

In his APT, Boyd performs open-loop slews, then closes the pointing loop by using the photometer or a separate CCD camera to lock onto the target star. He then uses simple open-loop tracking, since the amount of time he spends on any one object is relatively short. Boyd regularly achieves better than $12'$ pointing accuracy using stepper motors and a chain drive. Both the short-term and long-term tracking accuracy are adequate for UBV photometry using a 10-second integration period, and a $1'$ diaphragm. Only precession corrections are computed in this system.

If modeling drive train errors is to be avoided in systems requiring high accuracy, or if DC servo motors are used instead of stepper motors, some form of position feedback is usually employed. (In a well designed drive, drive train errors capable of being modeled are usually small enough to be ignored.) For reasons that are explained in Chapter 9, incremental optical shaft angle encoders are often the most cost-effective means of providing both position and velocity feedback. These devices send out a given number of "ticks" or pulses per revolution of the encoder shaft. If an incremental encoder shaft is attached to the axis of a telescope, and if that axis is driven at a constant rate, then the encoder will put out a stream of pulses at a constant rate. Position feedback can be obtained by feeding these pulses (and rotation direction information also provided by the encoder) to a counter that counts up when the telescope axis is turning in one direction, and counts down when the axis is turning in the opposite direction.

The pointing accuracy that is required determines the resolution of the encoder. For example, if 5″ pointing accuracy is required in slewing, then the encoder resolution and its gearing to the telescope axis can be chosen so that the encoder sends out at least one pulse for every 5″ rotation of the telescope axis. Close attention should be paid to the absolute accuracy specification of the encoder, and the method of gear reduction used (if any). Note that to achieve better than 5″ accuracy, expensive encoders with 20 or more bits of resolution and accuracy are usually attached directly to the telescope axis being controlled. A mechanically geared-up encoder of lower resolution (and cost) could, conceivably, achieve these accuracies, provided great care is taken in the gearing.

Most small- to medium-sized telescopes with encoders use them for closed-loop pointing, but still track objects open-loop. The reason for this is that closing the tracking rate (velocity) loop requires encoders of much higher resolution (and expense) than is required for the position servo. Also, it has been found that the accuracy of open-loop tracking in well-designed telescopes is more than adequate for most purposes. Even though the tracking is open-loop, tracking rate corrections (such as for refraction) can be included in the control software.

The encoder resolution requirements for a velocity servo can be made apparent with an example. Suppose that the research program to be pursued requires the telescope to drift from the object due to tracking errors no more than 1″ in ten minutes. With 600 seconds in ten minutes, and a nominal tracking rate of about 15″ per second in RA, an accuracy of 1″ in 9000″, or roughly 0.01%, is required. Now consider what would happen if an encoder with 5″ resolution is used to provide position feedback by sending pulses to a counter read by the control computer.

The first problem is that a deviation in position of 1″ cannot be detected by an encoder having only 5″ resolution. The second problem is that sufficient tracking rate information cannot accumulate in the required amount of time. At the nominal 15″ per second tracking rate, the encoder sends out pulses at the rate of three per second, so the computer need not read the pulse counter any more frequently. The computer could execute a loop every one-third second in which the actual telescope position (determined from the counter) is compared to the desired telescope position (computed using the star's RA and sidereal time), and a new motor command is generated to compensate for the detected error. The problem with this approach is that the changes in motor speed commanded by the computer would often be greater than 0.01%, so the speed of the RA axis would change every one-third second by more than this amount to compensate for errors that accumulated over the last one-third second.

The effect of these relatively large and frequent speed variations on a perfectly stiff telescope and mount would be to jiggle the image at the program loop rate (3 times per second). Since the telescope and mount are not perfectly stiff, additional oscillations could result. If the program loop execution rate is the same as the natural frequency of the telescope, these oscillations would be reinforced, and the image would suffer perceptibly. Therefore, the frequency of speed changes in a closed-loop system should be much higher than (or much lower than) the natural frequency of the telescope. Furthermore, the resolution of the velocity feedback encoder should be high enough to keep the amplitude of speed changes within the allowed tracking rate error. Short term tracking rate requirements for many telescopes imply an encoder resolution of about $0\farcs3$ to $0\farcs5$ with very high tick-to-tick accuracy. Such high resolution encoders are very expensive.

It is clear, then, that the various elements of a telescope, both mechanical and electronic, should be directed toward the particular observing program, and matched to make an integrated system. The characteristics of two of these elements, motors and position encoders, are discussed in greater detail in Part III. Those readers who wish to pursue control theory beyond the level presented here are directed to Miller (1977) for a good introduction to the subject that is more intuitive than mathematical, and to Goldberg (1964) for a more complete and theoretical discussion.

Chapter 5

Systematic Errors I — Astronomical Corrections

When an astronomer plans an evening's observations, he or she usually obtains an object's right ascension and declination from a catalog that uses the coordinate system of some standard epoch. This is the *mean* position of the object. This standard epoch is generally different from the one at the time the observation is to be made. In addition, the object's catalog position is not corrected for effects that depend on the time and location of the observation. Thus the catalog position is usually not where the observer would point a telescope to find the desired object.

Taff (1991, p. 73) defines the *true* position of an object as the point on the celestial sphere where it would be seen from the center of mass of the solar system when the mean place is corrected for precession, nutation, and proper motion.

He defines the *apparent* position of an object as the point on the celestial sphere where an object would be seen from the center of the Earth when the true place is corrected for annual aberration, annual parallax, and orbital motion (if the object is a star in a multiple system). The *topocentric* position of an object is where it would be seen from a point on the surface of the Earth when the apparent place is corrected for diurnal aberration, diurnal parallax, and atmospheric refraction. This is the place to which you must point a perfect telescope to see the object. The topocentric place must be computed once or several times a night, depending on the location of the object in the sky, the desired pointing accuracy, and the length of the observation.

Although these definitions are useful in distinguishing the effects of different corrections, in fact they are somewhat artificial. As discussed later in this chapter, to perform a full correction to topocentric place, it is often

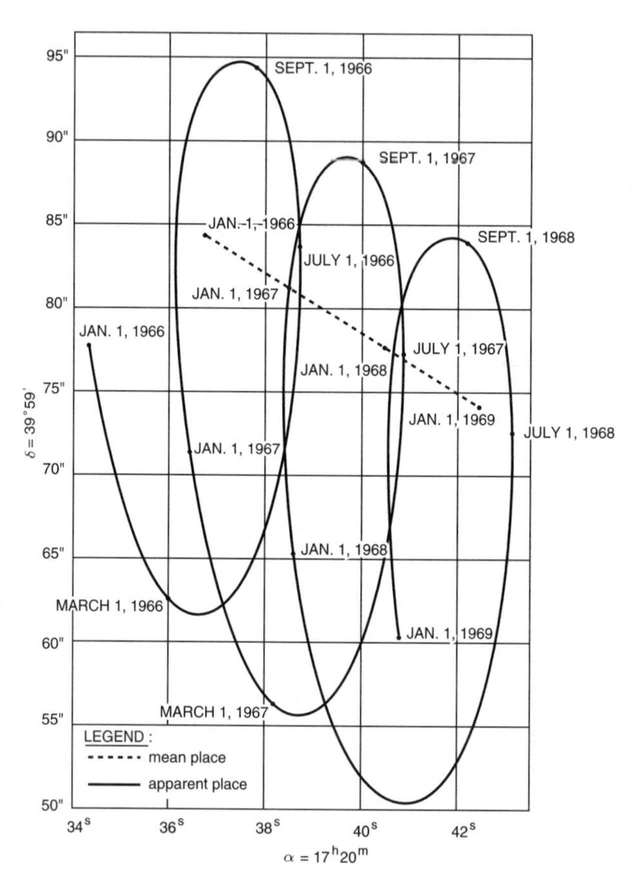

Fig. 5-1 *The apparent place variation of the star CG 23487. Courtesy Ungar and Mueller, 1969.*

more accurate to mix the order in performing certain corrections. To obtain the highest accuracy, one does *not* in fact first compute the true position, then the apparent position, then the topocentric position.

Throughout this book, we use "corrected place" for both the place to point a telescope after making several corrections, and for the coordinates that result from making a single correction, with the understanding that you will determine which corrections you will need to make (which may not be all those for topocentric place in its strict definition) to point your telescope to where the star appears to be when you observe it. To see how large corrections for apparent place can be, Figure 5-1 shows the difference between mean and apparent place for star GC 23487 over the period 1966 to 1969.

In this chapter, we describe the calculations used to convert mean to topocentric place. We also mention briefly the physics behind each correction, but this chapter is no substitute for the several good books on spherical astronomy. Throughout this chapter, we reference these books, which are listed in the Bibliography. We strongly suggest that you read one or more of them to get a better perspective on this subject than we can provide in one chapter.

Since the control computer in a servo must use an object's topocentric coordinates to control the telescope, but the astronomer usually enters the mean position taken from the catalog, the control software must convert mean coordinates to topocentric coordinates before sending commands to the telescope. The corrections to be performed vary according to the ephemeris or catalog used as the source of an object's position.

The corrections that must be applied to the catalog mean position to reduce it to topocentric position are listed in the following table, in order of decreasing size.

Table 5-1

Correction	Maximum Magnitude
1. Precession	40′ (roughly 50 years × 50″ per year)
2. Refraction	30′ (2′ at a zenith distance of 60°)
3. Annual Aberration	20″
4. Nutation	17″
5. Solar Parallax	9″/distance in A.U.
6. Stellar Parallax	1″/distance in parsec
7. Orbital Motion	varies (order of 1″)
8. Proper Motion	varies (a few arc seconds if you use an old catalog)
9. Diurnal Aberration	0.″3
10. Polar Motion	0.″1 (random)

Not all of the corrections listed above need be computed. Only those individual corrections leaving an error just smaller than your required pointing accuracy are needed. For a control system of moderate accuracy, expected only to bring an object within the finder on a typical small telescope, no corrections of any sort are needed if the field of view is 2° or more. To center an object in the finder and to bring it well within the field of view of a small telescope, only precession and refraction corrections are necessary. In most cases, observing is done well above the horizon, and as refraction errors are slight (less than 2′ at a zenith distance of 60°, 3′ at 70°, and 6′ at 80°), refraction corrections may not be required.

In a simple control system that corrects only for precession, one could calculate the coordinates at the current epoch for all of one's favorite objects just once, when the observing list is put together. Such coordinates would be good for several years before they would require updating. This might be simpler to do than make calculations each time the object is observed.

If refraction calculations are made and one is not making long exposures near the horizon, then the calculation can be made once just before moving to the object and need not be repeated.

In any event, it makes sense to correct for only those errors that are significant with respect to the requirements and capabilities of your system. To minimize processor loading, the corrections that are made should be computed only as often as is really required, unless doing it more often just happens to be simpler. Tomer's system (described in Part IV) corrects only for precession and refraction. It is an excellent example of matching the requirements, mechanical assembly, computer, and software to create a balanced system at reasonable cost.

For each correction in this chapter, the dynamic or physical principle is discussed, the equations that embody the physical principle are presented, the range of values of the correction is computed to assess whether the correction need be done at all to achieve a desired level of accuracy, and algorithms for computing the correction are given which we have chosen for their speed and accuracy when used on a personal computer. In Chapter 14, Trueblood demonstrates how to use these algorithms to assess the CPU loading of the control computer.

5.1 Precession

5.1.1 The Physical Basis of Precession

The rotating Earth in the gravitational field of the Sun and Moon is subject to torques that cause it to precess with a period of roughly 26,000 years. This long-period motion of the mean pole of the Earth about the pole of the ecliptic is called *luni-solar precession.* This is analogous to the precession of a top in a gravitational field, as shown in Figure 5-2. Luni-solar precession produces the motion of the equinox that appears in Figure 5-3. In addition to the motion of the mean pole, there is a short period motion of the true pole about the mean pole called *nutation*, which is discussed below; this is illustrated in Figure 5-4. The planets also perturb the orbit of the Earth, causing the plane of the ecliptic to tilt such that the equinox precesses roughly 12″ per century and the obliquity of the ecliptic decreases at roughly 47″ per century (Gurnette and Woolley, 1961). This latter effect is called *planetary precession,* and its effect on the obliquity is shown in Figure 5-5. The general precession that results from both of these motions is

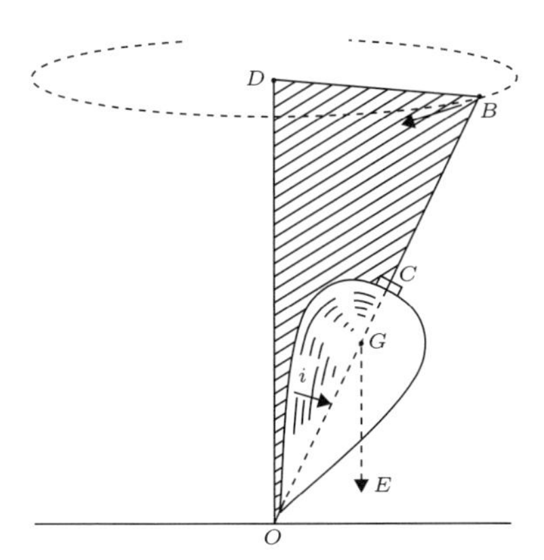

Fig. 5-2 *Precession of a rotating body in a gravitational field. Courtesy Cambridge University Press.*

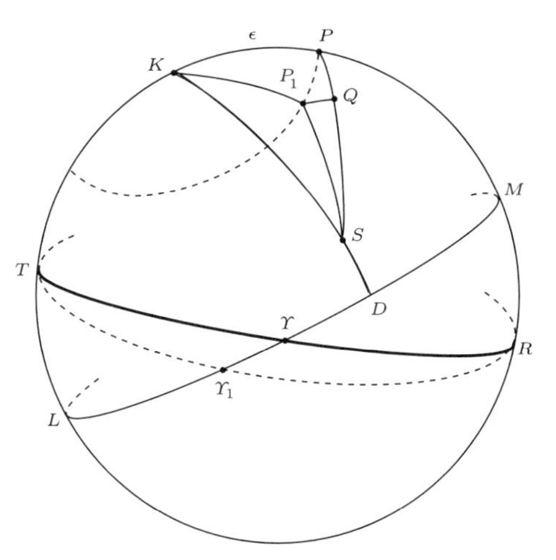

Fig. 5-3 *Equinox motion due to luni-solar precession. Courtesy Cambridge University Press.*

depicted in Figure 5-6. The motion of the Earth's pole due to both precession and nutation is analogous to that shown in Figure 5-4.

5.1.2 Magnitude of the Precession Corrections

Smart (1979, p. 238) combines both luni-solar and planetary precession to obtain an expression for the general precession that is actually observed to occur per year in the unprecessed right ascension (α) and declination (δ) to yield the precessed coordinates (α',δ') of an object. He gives the

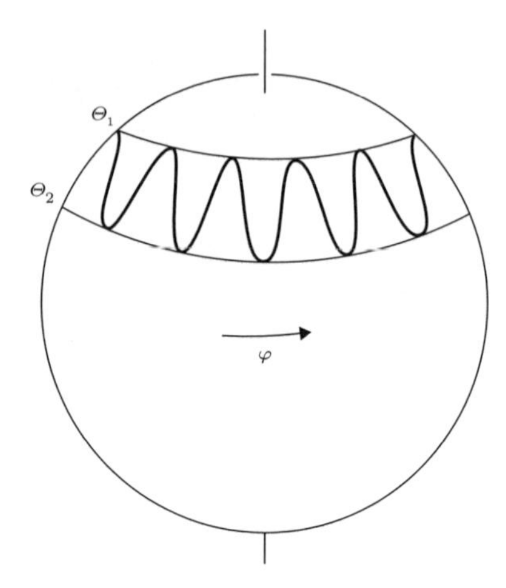

Fig. 5-4 *Nutation of the pole of a precessing rotating body. Courtesy Academic Press.*

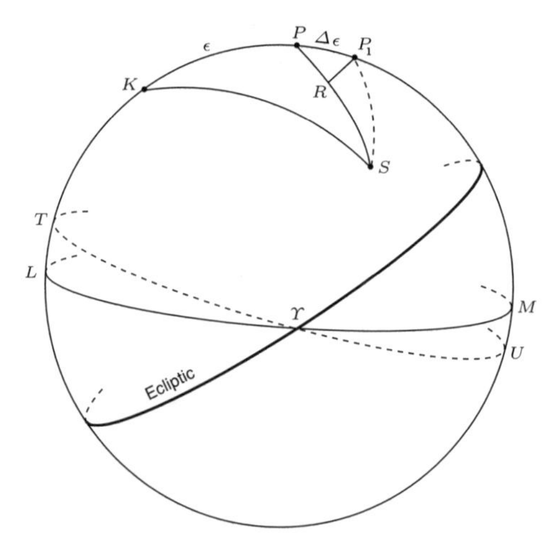

Fig. 5-5 *Obliquity Change Due to Planetary Precession. Courtesy Cambridge University Press.*

following expressions for the annual changes in these quantities:

$$\alpha' - \alpha = (\psi \cos \epsilon - \lambda') + \psi \sin \epsilon \sin \alpha \tan \delta \qquad (5.1a)$$

$$\delta' - \delta = \psi \sin \epsilon \cos \alpha \qquad (5.1b)$$

where $\alpha' - \alpha =$ the annual change in right ascension

$\delta' - \delta =$ the annual change in declination

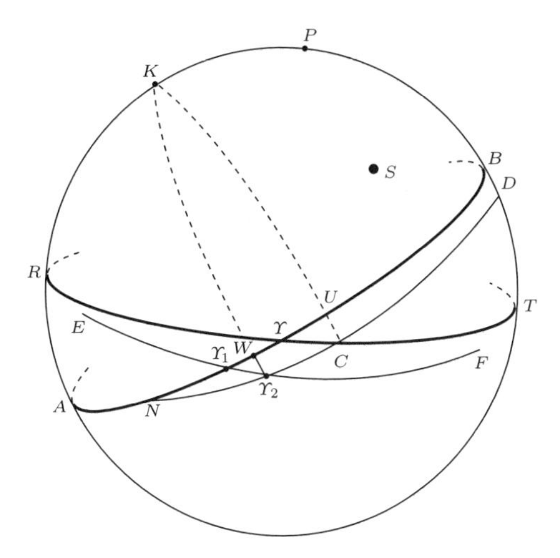

Fig. 5-6 *General Precession. Courtesy Cambridge University Press.*

$\psi =$ the annual luni-solar precession:

$$= 50''\!.2564 + 0''\!.000222t \tag{5.1c}$$

$\epsilon =$ the angle of the obliquity:

$$= 23°27'8''\!.26 - 0''\!.4685t \tag{5.1d}$$

$\lambda' =$ the annual planetary precession:

$$= 0''\!.12 \text{ per year} \tag{5.1e}$$

$t =$ the number of years since 1900. (5.1f)

Equations (5.1) are given to indicate the magnitude of the precession corrections, which is determined solely by the elapsed time between the two epochs involved. Thus there is virtually no limit to the magnitude of the correction, as expressed by these equations, which account only for the secular effect of precession. In reality, the Earth's pole describes a circle every 26,000 years as it precesses, which defines a cone described by the Earth's rotational axis with its apex centered at the Earth's center. The apex angle of the cone is the maximum angle of precession. Since this angle is several degrees, if the time between epochs is sufficiently large, precession can dominate all other sources of systematic error.

In actual practice, standard star catalogs tend to be updated every 50 years or so. Using the value of ψ as an order of magnitude estimate for the variation of an object's right ascension and declination over 50 years, $\psi(= 50''/\text{year}) \times 50$ years yields about 42 arc minutes. This rough approximation overstates the typical effect of precession, since most accurately computed precession corrections for a 50-year period are on the order of a few arc minutes. Most astronomers want to set their telescopes with an accuracy considerably better than a few arc minutes, so a correction for precession is almost always necessary.

5.1.3 Computing General Precession

There are two methods of computing precession corrections in general use:

1. closed algebraic evaluation of equations for the changes in right ascension and declination, and

2. rotation of the equatorial coordinate system using matrices to represent the rotations.

Many computerized systems developed to date employ the mathematically elegant matrix rotation method. However, as is shown below, the evaluation of closed algebraic expressions imposes a lighter burden on the processor when computing either precession alone, or both precession and nutation. These two methods are explained and compared for computational efficiency in the following paragraphs.

Expressions using the first method are presented in Gurnette and Woolley (1961, p. 38):

$$\alpha' - \alpha = M + N \sin \frac{\alpha' + \alpha}{2} \tan \frac{\delta' + \delta}{2} \tag{5.2a}$$

$$\delta' - \delta = N \cos \frac{\alpha' + \alpha}{2} \tag{5.2b}$$

where $\alpha =$ right ascension using the mean equinox at the initial epoch t

$\alpha' =$ right ascension using the mean equinox at the precessed epoch t'

$\delta =$ declination using the mean equinox at the initial epoch t

$\delta' =$ declination using the mean equinox at the precessed epoch t'

$M =$ general precession in right ascension

$= m_m(t - t')$, t and t' in years

$m_m =$ value of m midway between t and t', where

$m =$ $3^s.07234 + 0^s.00186\ T$ (T measured in tropical centuries from 1900.0)

$N =$ general precession in declination $= n_m(t - t')$, t and t' in years

$n_m =$ value of n midway between t and t', where

$n =$ $20''.0468 - 0''.0085\ T$ (T measured in tropical centuries from 1900.0).

These equations are iterated until convergence to the desired accuracy is achieved. Another example of this method is provided by Woolard and Clemence (1966, p. 262) that expresses $\Delta\alpha$ and $\Delta\delta$ in terms of a simple power series in time. This method uses no iteration and no trigonometric functions and thus is computationally more efficient.

Equations (5.2) should be used wherever extremely high accuracy is not required, since they are the simplest to program, and take less computer time than other approaches. However, these and other similar approaches suffer from the disadvantage of using coordinate systems in which the various precession motions are not orthogonal. This means that the constants in these equations are valid only over a relatively short period of time.

Kaplan (1981) cites a more accurate algorithm that yields an accuracy of at least 0.001 seconds in right ascension and $0''.01$ in declination. It uses epoch J2000.0, the new epoch for the *Astronomical Almanac* starting in 1984. This new epoch used a system of time called Barycentric Dynamical Time (TDB), later replaced by IAU resolutions in 1991 with Barycentric Coordinate Time (TCB). Since precession occurs slowly, the difference between TCB and Coordinated Universal Time (UTC), or civil time, obtained from radio station WWV or Global Positioning System (GPS) receivers is negligible in precession calculations for pointing ground-based telescopes, and you can substitute UTC time for TCB in the equations given below.

For computing precession, assume that J_s is the Julian day of the starting epoch and J_e is the Julian day of the ending epoch. Note that Julian *dates* are ordinary dates in the Julian calendar system, while Julian *days* are the number of days (possibly including fractions of a day) from Greenwich mean *noon* (12 hours Universal Time) of January 1, -4712.

$$\zeta_o = (2306''.2181 + 1''.39656\,T - 0''.000139\,T^2)\,t + (0''.30188 - 0''.000344\,T)\,t^2 + 0''.017998\,t^3 \tag{5.3a}$$

$$z_A = (2306''.2181 + 1''.39656\,T - 0''.000139\,T^2)t + (1''.09468 + 0''.000066\,T)t^2 + 0''.018203\,t^3 \tag{5.3b}$$

$$\Theta = (2004''.3109 - 0''.85330\,T - 0''.000217\,T^2)\,t - (0''.42665 + 0''.000217T)\,t^2 - 0''.041833\,t^3 \tag{5.3c}$$

$$\alpha' - \alpha = \zeta_o + z_A + \arctan\frac{q\sin(\alpha+\zeta_o)}{1-q\cos(\alpha+\zeta_o)} \tag{5.3d}$$

$$\delta' - \delta = \Theta[\cos(\alpha+\zeta_o) - \sin(\alpha+\zeta_o)\tan(\alpha-\alpha'-\zeta_o-z_A)] \tag{5.3e}$$

$$q = \sin\Theta\left(\tan\delta + \tan\frac{\Theta}{2}\cos(\alpha+\zeta_o)\right) \tag{5.3f}$$

where $(90^\circ - \zeta_o) =$ the right ascension of the ascending node of the equator at epoch J_e on the equator of J_s reckoned from the equinox of J_s

$(90^\circ + z_A) =$ the right ascension of the node reckoned from the equinox of J_e

$\Theta =$ the inclination of the equator of J_e to the equator of J_s

$T =$ the time interval between J2000.0 and J_s

$= (J_s - 2451545.0)/36525$ in Julian centuries (5.3g)

(This definition of T is used throughout this chapter.)

$t =$ the time interval between the catalog and current epochs

$= (J_e - J_s)/36525$ in Julian centuries. (5.3h)

Meeus (1991, p. 59) gives the following method for computing Julian days (for J_e and J_s):

For the year Y, month M, and day of month D (where D can have a fractional day part), if $M > 2$, leave Y and M unchanged. If $M = 1$ or $M = 2$, set $Y = Y - 1$, and $M = M + 12$. If the date is in the Gregorian calendar (most likely for telescope pointing), set

$$A = \mathrm{INT}(Y/100) \tag{5.3i}$$

and

$$B = 2 - A + \mathrm{INT}(A/4) \tag{5.3j}$$

where the INT function represents the truncation of a floating point number to an integer. If the date is in the Julian calendar, set B = 0. The Julian Day (JD) is given by

$$\mathrm{JD} = \mathrm{INT}(365.25(Y+4716)) + \mathrm{INT}(30.6001(M+1)) + D + B - 1524.5. \tag{5.3k}$$

Although the conversion from the Julian calendar to the Gregorian calendar is usually given as the year 1582 (Julian date October 4 is followed by Gregorian date October 15), official adoption of the Gregorian calendar did not occur in England until 1752 and in Turkey until 1927, so care must be exercised in using this algorithm. Also remember that this system used by astronomers has a year 0, but there is no year 0 in the A.D./B.C. system. So year 585 B.C. is year −584 in the JD system.

The second method of computing precession is to rotate the geocentric rectangular equatorial coordinates from one epoch to another. This can be done in the following sequence:

1. Transform from Right Ascension and declination to geocentric rectangular equatorial coordinates.
2. Rotate the geocentric rectangular equatorial coordinates.
3. Transform the precessed geocentric rectangular equatorial coordinates back to precessed Right Ascension and declination.

Step 1 can be performed using the following equations (Woolard and Clemence, 1966, p. 280):

$$\begin{aligned} x &= \cos\alpha\cos\delta \\ y &= \sin\alpha\cos\delta \\ z &= \sin\delta. \end{aligned} \tag{5.4}$$

Step 2 is to perform the matrix operation.

$$\begin{bmatrix} x' \\ y' \\ z' \end{bmatrix} = \begin{bmatrix} x \\ y \\ z \end{bmatrix} \times \begin{bmatrix} \text{rotation} \\ \text{matrix} \end{bmatrix}.$$

The rotation matrix consists of the direction cosines of the apparent angle produced by precession. Let A, B, and C be the angles in the rectangular coordinate system (x,y,z) between the axes and a radius from the origin to the point in the sky being measured, and A', B', and C' be the angles between the axes of the same coordinate system and a radius to a different position in the sky. This simulates the apparent motion of an object within the coordinate system due to precession. The object appears to have moved through an angle D given by

$$\cos D = \cos A\cos A' + \cos B\cos B' + \cos C\cos C'.$$

The rotation matrix consists of the cosines between the various angles:

$$\begin{bmatrix} x' \\ y' \\ z' \end{bmatrix} = \begin{bmatrix} x \\ y \\ z \end{bmatrix} \times \begin{bmatrix} \cos(x',x) & \cos(y',x) & \cos(z',x) \\ \cos(x',y) & \cos(y',y) & \cos(z',y) \\ \cos(x',z) & \cos(y',z) & \cos(z',z) \end{bmatrix}. \tag{5.5a}$$

The rotation matrix is given by Seidelmann (1992, p. 103) as

$$\begin{bmatrix} \cos z_A \cos\Theta \cos\zeta_o & -\cos z_A \cos\Theta \sin\zeta_o & -\cos z_A \sin\Theta \\ -\sin z_A \sin\zeta_o & -\sin z_A \cos\zeta_o & \\ & & \\ \sin z_A \cos\Theta \cos\zeta_o & -\sin z_A \cos\Theta \sin\zeta_o & -\sin z_A \sin\Theta \\ +\cos z_A \sin\zeta_o & +\cos z_A \cos\zeta_o & \\ & & \\ \sin\Theta \cos\zeta_o & -\sin\Theta \sin\zeta_o & \cos\Theta \end{bmatrix}. \tag{5.5b}$$

When the matrix multiplication is performed, the results are as follows:

$$\begin{aligned} x' = \quad & x - 2\left(\sin^2(\zeta_o + z_A) + \cos\zeta_o \cos z_A \sin^2\frac{\Theta}{2}\right)x \\ & + 2\left(\sin\zeta_o \cos z_A \sin^2\frac{\Theta}{2} - \sin(\zeta_o + z_A)\right)y + (\cos z_A \sin\Theta)z \end{aligned} \tag{5.6a}$$

$$\begin{aligned} y' = \quad & y + \left(\sin(\zeta_o + z_A) - 2\cos\zeta_o \sin z_A \sin^2\frac{\Theta}{2}\right)x \\ & - 2\left(\sin^2\frac{\zeta_o + z_A}{2} - \sin\zeta_o \sin z_A \sin^2\frac{\Theta}{2}\right)y - (\sin z_A \sin\Theta)z \end{aligned} \tag{5.6b}$$

$$z' = z + (\cos\zeta_o \sin\Theta)x - (\sin\zeta_o \sin\Theta)y - 2\sin^2\frac{\Theta}{2}z \tag{5.6c}$$

where ζ_o, z_A, and Θ are the same as in Equations (5.3). Step 3 is to transform the resulting (x',y',z') to (α',δ').

As seen from the above discussion, there are basically two methods of deriving accurate precessed coordinates from catalog coordinates. The first method uses the spherical trigonometric Equations (5.3). The second method uses the matrix Equations (5.6). To help you decide which approach to use, it might be useful to understand whether one approach is computationally more efficient. If you are using a very powerful PC, it makes little difference. But if you are using an older PC with a slower processor, it makes sense to use the more computationally efficient algorithm.

The table below summarizes the number of trigonometric and arithmetic operations that are required for each method. Since sine, cosine, and tangent functions and their respective inverse functions all require roughly the same amount of time to compute, they are grouped together under the "Trig" heading. Numbers raised to an integer power n are counted as n

multiplies. The results are as follows:

Table 5-2

Item	Equation	Trig	+	−	×	/
T	(5.3g)	0	0	1	0	1
t	(5.3h)	0	0	1	0	1
ζ_o	(5.3a)	0	3	2	10	0
z_A	(5.3b)	0	4	1	10	0
Θ	(5.3c)	0	1	4	10	0
α'	(5.3d)	3	4	1	3	1
δ'	(5.3e)	1	2	4	2	0
q	(5.3f)	3	2	0	2	1
Apply	(5.3)	0	2	0	0	0
x,y,z	(5.4)	4	0	0	2	0
x'	(5.6a)	7	5	2	13	1
y'	(5.6b)	1	3	4	13	1
z'	(5.6c)	0	1	2	7	0
α', δ'	(5.4) inverse	4	0	0	2	0

If a value is computed in a prior equation, it is not counted again when it appears later on in other equations used in the same method. For example, $\sin(\alpha + \zeta_o)$ was counted for Equation (5.3d), so it is not counted again in Equation (5.3e). Method 1 uses (5.3g) and (5.3h), (5.3a)-(5.3c), and (5.3d)-(5.3f), then the resulting corrections are applied to the original (α,δ). Method 2 uses (5.3g) and (5.3h), (5.3a)-(5.3c), and equations like (5.4) to convert (x',y',z') to (α',δ'). This is summarized as follows:

Table 5-3

	Trig	+	−	×	/
Method 1	7	18	14	37	4
Method 2	16	17	17	67	4

The relative times for performing various calculations depends on the computer hardware and the algorithms used to perform trigonometric function calculations. To provide a basis for comparing different computing methods, we have assumed that "trig" functions are evaluated to eight-decimal accuracy using Chebyshev polynomials. These functions would probably be evaluated using eight adds and seven multiplies per function. To allow one to compare different types of arithmetic calculations, the author has defined an add and subtract to be one computation unit, a multiply to be two computation units, and a divide to be three computation units. These values are based on the following instruction times for 32-bit floating point operations for the DEC LSI-11.

Using these assumptions, the first method requires 272 computation units, and the second requires 532 units. It seems clear that the second method requires considerably more time than the first, but one should not infer that it would take about twice as long, since various kinds of overhead have not been included in this comparison. Furthermore, if other corrections are handled as matrix rotations, then the conversion from (α, δ) to (x, y, z) and back again need be done only once for many corrections done in a series.

Comparing the speed of modern pipelined Reduced Instruction Set Computer (RISC) chips to microcomputers of the mid-1980s is difficult because modern chips are so complex. RISC chip instruction pipelining and caching make it difficult to predict in advance how long it will take to execute a floating point instruction. Nevertheless, at the risk of oversimplification the Intel Pentium takes one clock cycle (at best) to execute an add, one for a subtract, one for a multiplying, and up to 39 for a full-precision divide. The speed of the Intel Pentium 60 MHz processor has been benchmarked at 5 MFLOPS using the LINPAK benchmark (Dongerra, 1994). This is considerably faster than the IBM PC with an 8087 co-processor that scored 0.0069 MFLOPS on the same test. This is more than enough processing power for the most complex control algorithm which will take no more than a few percent of the processor. At this point, the capabilities of the operating system for prioritizing tasks are more important than the speed of the processor. The analysis of algorithm speed is of benefit only to those with slower processors.

5.2 Nutation

5.2.1 The Physical Basis of Nutation

The action of the Sun and the Moon on the Earth produces the motion of the mean pole of the Earth described above as luni-solar precession. These bodies also produce a motion of the true pole of the Earth about the mean pole, which is nutation. As shown in Figure 5-4, this elliptical motion has a dominant period of roughly 19 years with an amplitude of about $17''$ in longitude and $9''$ in obliquity. The principal terms of the series that describe nutation depend on the regression of the node of the Moon's orbit. Other terms depend on the mean longitudes and mean anomalies of the Sun and the Moon, and on their combinations with the longitude of the Moon's node.

The motion of the true pole about the mean pole is described in terms of corrections to the longitude, $\Delta\psi$, and to the mean obliquity, $\Delta\epsilon$. These quantities are expressed in two series with 106 terms each, and include all terms of at least $0\overset{''}{.}0001$ in magnitude (Kaplan, 1981, pp. A4-A6). Terms depending on the Moon's longitude have periods of less than about 60 days, and are considered to be the short period terms. The remaining terms are longer period, and typically are interpolated in 10-day intervals in tables of apparent places of fundamental stars. Expressions for $\Delta\psi$ and $\Delta\epsilon$ given below contain all terms, both long and short period.

5.2.2 Magnitude of the Nutation Corrections

As stated above, the amplitude of the orbit of the true pole about the mean pole is about a dozen arc-seconds. Computerized systems not requiring this level of accuracy may ignore the nutation correction.

5.2.3 Computing Nutation

The methods of computing nutation parallel those for computing precession. All methods of computing nutation depend on first computing $\Delta\psi$ and $\Delta\epsilon$.

The tables of terms in Kaplan (1981) give several different coefficients for sine or cosine terms of the same argument. These have been grouped together to yield the following equations, which are accurate to $0\overset{''}{.}0001$:

$$\begin{aligned} \Delta\psi = \ & -(17\overset{''}{.}2289 + 0\overset{''}{.}01745\,T)\sin\Omega - (1\overset{''}{.}4006 + 0\overset{''}{.}00008\,T)\sin 2\Omega \\ & + (0\overset{''}{.}0242 + 0\overset{''}{.}00003\,T)\sin l + 0\overset{''}{.}0031\sin 2l \\ & + (0\overset{''}{.}0707 - 0\overset{''}{.}00017\,T)\sin l' + 0\overset{''}{.}0005\sin 2l' \\ & - (1\overset{''}{.}6341 + 0\overset{''}{.}00013\,T)\sin 2F + 0\overset{''}{.}0002\sin 4F \\ & + 0\overset{''}{.}0008\sin D + (1\overset{''}{.}3346 + 0\overset{''}{.}00007\,T)\sin 2D \\ & - 0\overset{''}{.}0002\sin 4D \end{aligned} \tag{5.7a}$$

$$\begin{aligned} \Delta\epsilon = \ & +(9\overset{''}{.}2179 + 0\overset{''}{.}00089\,T)\cos\Omega \\ & + (0\overset{''}{.}6076 - 0\overset{''}{.}00035\,T)\cos 2\Omega + (0\overset{''}{.}0105 - 0\overset{''}{.}00001\,T)\cos l \\ & - 0\overset{''}{.}0010\cos 2l + 0\overset{''}{.}0001\cos 3l + (0\overset{''}{.}0200 - 0\overset{''}{.}00004\,T)\cos l' \\ & + 0\overset{''}{.}0008\cos 2l' + (0\overset{''}{.}7104 - 0\overset{''}{.}00040\,T)\cos 2F \\ & - 0\overset{''}{.}0001\cos D + (0\overset{''}{.}5852 - 0\overset{''}{.}00034\,T)\cos 2D + 0\overset{''}{.}0002\cos 4D \end{aligned} \tag{5.7b}$$

where the fundamental arguments at J_s are

$\Omega =$ the longitude of the ascending node of the Moon's mean orbit on the ecliptic, measured from the mean equinox of date

$$= 450160\overset{''}{.}280 - (5r + 482890\overset{''}{.}539)\,T + 7\overset{''}{.}455\,T^2 + 0\overset{''}{.}008\,T^3 \tag{5.8a}$$

$l =$ the mean anomaly of the Moon

$$= 485866\overset{''}{.}733 + (1325r + 715922\overset{''}{.}633)\,T + 31\overset{''}{.}310\,T^2 + 0\overset{''}{.}064\,T^3 \tag{5.8b}$$

$l' =$ the mean anomaly of the Sun (Earth)

$$= 1287099''.804 + (99r + 1292581''.224)\,T - 0''.577\,T^2 - 0''.012\,T^3 \quad (5.8c)$$

$F =$ the difference $L - \Omega$, where L is the mean longitude of the Moon

$$= 335778''.877 + (1342r + 295263''.137)\,T - 13''.257\,T^2 + 0''.011\,T^3 \quad (5.8d)$$

$D =$ the mean elongation of the Moon from the Sun

$$= 1072261''.307 + (1236r + 1105601''.328)\,T - 6''.891\,T^2 + 0''.019\,T^3 \quad (5.8e)$$

$r =$ one revolution $= 360° = 1296000''$.

Equations (5.7) and (5.8) yield an accuracy far higher ($0''.0001$) than that needed for telescope control, so terms in these equations that are not needed should be dropped. To build a system accurate to one arcsecond with roughly a dozen sources of error, each source of error should be computed with about $0''.1$ accuracy. If there are ten terms in the equation to be dropped, then terms less than $0''.01$ can be dropped. Actually, since random errors propagate in quadrature, each of 12 terms need only be accurate to $1''/\sqrt{12}$, or about $0''.3$. However, the equations that result in dropping terms less than $0''.01$ are so improved in efficiency over those containing all terms that little additional gain in efficiency can be made by taking this effect into account here.

Using $T = 1$ (one century from the year 2000.0), when terms less than $0''.01$ are dropped, Equations (5.7) and (5.8) become

$$\Delta\psi = -(17''.2289 + 0''.01745T)\sin\Omega - 1''.4006\sin 2\Omega + 0''.0242\sin l + 0''.0707\sin l' - 1''.6341\sin 2F + 1''.3346\sin 2D \quad (5.9a)$$

$$\Delta\epsilon = +9''.2179\cos\Omega + 0''.6076\cos 2\Omega + 0''.0105\cos l + 0''.0200\cos l' + 0''.7104\cos 2F + 0''.5852\cos 2D. \quad (5.9b)$$

Note that by eliminating the superfluous terms, eight trigonometric function evaluations are eliminated, at a savings of 22 computation units per function, though nutation changes so slowly that Equations (5.7) and (5.8) could be computed once at the start of the evening.

Ghedini (1982, p. 9) uses a similar set of equations to compute $\Delta\psi$, but gives none to compute $\Delta\epsilon$. He claims an accuracy of $0''.075$ for his equations. Meeus (1982, p. 70) also employs an analogous set of equations but uses the Sun's mean longitude and the Moon's mean longitude instead of F and D, so his equations contain different constants.

Once $\Delta\psi$ and $\Delta\epsilon$ have been computed, $\Delta\alpha$ and $\Delta\delta$ can be found. To evaluate the different methods for doing this in terms of computation speed, the number of arithmetic operations of each type must be counted, using the same method that was used for precession. The table below lists these counts.

Table 5-4

Equation Trig	+	−	×	/	Computation	Units
(5.8)	0	14	5	30	0	79
(5.7)	22	21	12	48	0	613
(5.8) + (5.7)	22	35	17	78	0	692
(5.9)	12	9	3	19	0	314
(5.8) + (5.9)	12	23	8	49	0	393

It has been assumed that T was computed at the start of the evening. Equations (5.8) must be evaluated, then used in either the highest accuracy Equations (5.7) or the lower accuracy Equations (5.9).

Using the same conversion to computation units that was used previously, Equations (5.8) require 79 units, (5.7) require 613 units, and (5.9) require 314 units. Thus the high accuracy equations require a total of 692 units, while the equations with lower, but acceptable, accuracy require 393 units. To compare methods of obtaining (α,δ) from $\Delta\psi$ and $\Delta\epsilon$, Equations (5.8) and (5.9) will be used.

Once $\Delta\psi$ and $\Delta\epsilon$ have been computed, simple transformations can be computed to convert these deltas of longitude and obliquity into deltas of right ascension and declination.

The first method of computing general precession that was discussed in the previous section used equations in closed form to evaluate the changes in right ascension and declination. The analogous equations for computing these changes to the precessed coordinates (α',δ') to the precessed and nutated coordinates (α'',δ'') are (to first order):

$$\alpha'' - \alpha' = (\cos\epsilon + \sin\epsilon\sin\alpha\tan\delta')\Delta\psi - \cos\alpha'\tan\delta'\Delta\epsilon \tag{5.10a}$$

$$\delta'' - \delta' = \sin\epsilon\cos\alpha'\Delta\psi + \sin\alpha'\Delta\epsilon. \tag{5.10b}$$

The second method of computing general precession used matrix multiplications to rotate rectangular equatorial coordinates. Before transforming back to the precessed and nutated (α'',δ''), the nutation correction can be applied by adding the following changes to the rotated coordinates:

$$\Delta x = -(y\cos\epsilon + z\sin\epsilon)\Delta\psi \tag{5.11a}$$

$$\Delta y = +x\cos\epsilon\,\Delta\psi - z\Delta\epsilon \tag{5.11b}$$

$$\Delta z = +x \sin\epsilon\, \Delta\psi + y\Delta\epsilon. \tag{5.11c}$$

Second-order terms are neglected, as they can be at most one unit in the eighth figure.

The nutation correction can also be applied by multiplying the precession rotation matrix in (5.5) by the following matrix before rotating the coordinates ($\Delta\psi$ and $\Delta\epsilon$ are in radians):

$$\begin{bmatrix} 1 & -\Delta\psi\cos\epsilon & -\Delta\psi\sin\epsilon \\ +\Delta\psi\cos\epsilon & 1 & -\Delta\epsilon \\ +\Delta\psi\sin\epsilon & +\Delta\epsilon & 1 \end{bmatrix}. \tag{5.12}$$

A more accurate rotation matrix is given in Seidelmann (1992, p. 115) but using the one above produces an error in the eighth decimal place, too small to worry about. The two rotation matrices must be multiplied to handle significant second-order terms properly. Thus merely adding the two matrices element by element may not produce results of the desired accuracy.

Since all these methods use the same method of computing $\Delta\psi$ and $\Delta\epsilon$ and they all yield roughly the same accuracy, those with older computers might want to make execution speed the criterion for choosing one as the best method. Those with modern PCs or workstations may find the following discussion of academic interest only. The number of operations of each type are tabulated below:

Table 5-5

Equation	Trig	+	−	×	/	Computation Units
(5.10)	5	2	1	8	0	129
(5.11)	2	2	2	9	0	66
(5.12)	2	0	3	2	0	51

These figures assume that once a term involving a trig quantity is evaluated, it can be used elsewhere without the need to evaluate it again.

Using the same method of converting these operations into computation units that was used in the precession algorithm evaluation, the method of Equations (5.10) requires 129 computation units, (5.11) requires 66 units, and (5.12) requires 51 units. This is misleading, however, as (5.11) and (5.12) are useful only when combined with the second method of computing general precession. Results obtained using (5.10) are simply added to (α',δ') of the first method, while the results from (5.11) are added to the results of the rotation in the second method, and results from (5.12) are matrix multiplied with the rotation matrix. The matrix multiplication requires three multiplies and three adds per matrix element, for a total of 81 units.

The total number of computation units for computing precession, computing nutation (without counting the units for computing $\Delta\psi$ and $\Delta\epsilon$), computing $\Delta\psi$ and $\Delta\epsilon$, and combining the results are

Table 5-6

Equation	Computation Units
(5.3),(5.8),(5.9),(5.10)	272 + 393 + 129 + 2 = 796
(5.4) - (5.6),(5.8),(5.9),(5.11)	532 + 393 + 66 + 3 = 994
(5.4) - (5.6),(5.8),(5.9),(5.12)	532 + 393 + 51 + 81 = 1057

This clearly shows that when both precession and nutation are computed and the results are combined, the direct evaluation method of using equations in closed form for both precession and nutation is faster, with the matrix method taking about 33% longer. This is an interesting result, in light of the fact that many systems in current use favor the more mathematically elegant, but more computationally intensive, matrix approach (C.f., Huguenin and McCord, 1975).

However, if several other corrections are to be applied, the vector/matrix method could turn out to be faster. Given modern processor speeds, it probably makes little difference. For the Keck 10-m telescopes, Wallace (1988) proposed using matrices for nearly all corrections. He divided calculations into low frequency (those performed once every 5 minutes), medium frequency (every 5 seconds), and high frequency (20 times per second). By performing a calculation only as often as needed, his control computations used only 7% of a MicroVAX II minicomputer. These calculations are only part of the entire control system software, and this figure does not include operating system overhead. But since the Intel 60 MHz Pentium has about 10 times the floating point execution speed of the MicroVAX II, whether to use matrices or equations in closed form is no longer an issue for modern computers.

5.3 Polar Motion

Minor periodic perturbations not taken into account in the theories of precession and nutation, and random realignments of the distribution of the Earth's mass caused by tectonic plate drift and earthquakes, cause the true pole of the Earth to change position by a random amount in a random direction at random times, although the motion tends to make a spiral pattern when plotted over several decades. This motion is depicted in Figure 5-7, and usually amounts to no more than 50 meters per year, which, at the polar radius of the Earth, is usually less than $0''.1$ in a year. Amounts this small are difficult to measure, and the motion cannot be predicted

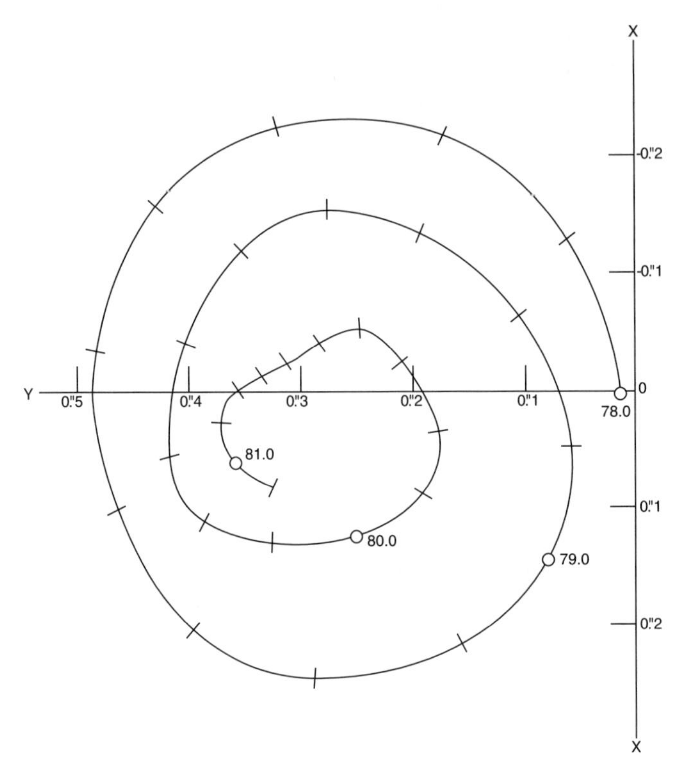

Fig. 5-7 *Motion of the Earth's Rotational Pole Over Several Decades. Annual Report 1980, Federation of Astronomical and Geophysical Sciences.*

in advance, only evaluated as an effect on the coordinate system after the motion has occurred. Therefore, there is no theory of polar motion in closed mathematical form.

Considering the magnitude of the correction involved, there is no need to attempt to correct for polar motion in most telescope control systems. The exception might be large telescopes that strive to achieve near diffraction-limited performance using active or adaptive optics. These telescopes can have very stringent pointing and tracking requirements, e.g., to achieve very high resolutions in spectrographs using very narrow (0.″2) slits. For example, Wallace (1988) quotes the pointing requirement for the Keck 10-meter telescopes as 1″ RMS at night, and 0.″1 following a 1° offset. To achieve this kind of accuracy, data from polar motion observatories must be gathered and entered into the control system or an acquisition and guiding system must be present to close the outer control loop.

5.4 Sidereal Time

5.4.1 About Time

The corrections discussed earlier tend to require months or years to accumulate significant changes. Many of the corrections discussed in the rest of this chapter and in the next depend on knowing time very accurately, so now it's about time to talk about time.

There are two ways to measure time relevant to telescope control systems and the practical means of delivering accurate time to them. The first is by using atomic clocks to provide a precise and extremely regular series of beats or pulses to keep time from a particular epoch. The second method of timekeeping is to use the motion of celestial bodies.

Atomic time is based on nuclear transitions of cesium-133 that occur at microwave frequencies (about 9 GHz). International Atomic Time (*Temps Atomique International*, or TAI) is derived from atomic clocks after correcting for known effects. The time used by the Global Positioning System (GPS) satellites to time-stamp their position messages is GPS time, which uses the same time interval (length of a second) as TAI and differs from TAI by exactly 19 seconds. GPS time coincided with UTC (see below) at the GPS standard epoch 1980 January 6.0. GPS receivers compute GPS time as part of determining the location of the receiver. They also receive a correction to compute UTC, so most GPS receivers display UTC although some permit using a jumper or software to select either UTC or GPS time.

The time needed most for telescope control is Dynamical Time, based on the motion of the Earth. Our calendar uses the motion of the Earth in its orbit about the Sun to define a year. Solar time is based on the time between transits of the Sun across a meridian, which depends on both the rate of rotation of the Earth on its axis and the speed of the Earth in its orbit around the Sun. The epoch of solar time is determined by the transits of the Sun across the Greenwich meridian. We subdivide a solar day into 24 hours each containing 60 minutes, with 60 seconds in each minute. In the time it takes the Earth to go around the Sun once in its orbit (one year), the Sun transits the meridian approximately 365.25 times, hence the number of days in a year.

Since the Earth's orbit around the Sun is not a perfect circle, and the Earth's rotation axis is tipped with respect to the ecliptic plane, apparent solar time is not a uniform time scale, which means that the time interval between successive apparent noons is not the same throughout the year. Universal Time (UT0) takes these two effects into account through the Equation of Time that applies corrections of up to about 16 minutes to apparent solar time. Polar motion corrections of about 0.3 second applied to UT0 result in UT1. As astronomers monitored the rotation of the Earth with respect to UT1, they discovered predictable deviations of unknown origin with periods of 6 months and 1 year. These deviations of about 0.3 second, applied to UT1, result in UT2. UT2 has little practical application, and is not in widespread use.

Sidereal time is also based on the Earth's rotation, and measures the time between meridian transits of distant celestial objects. If the Earth did not rotate on its axis but continued as usual in its orbit, we would see one day per year (since we would go around the Sun once, so the Sun appears to go around the Earth once), but no change in sidereal time, since very distant stars with negligible parallax would not appear to move. Thus there are approximately 364.25 sidereal days per year, one fewer than the number of solar days per year. Sidereal time is the time used most frequently in telescope control, since it is tied very closely to the Earth's rotation rate, and therefore the rate at which distant celestial objects move.

Coordinated Universal Time (UTC), the time we use for civil time (with adjustments for time zones), is a compromise between UT1 and TAI. UTC seconds are identical to TAI seconds in length. The difference between UTC and TAI is an integer number of seconds to keep UTC within 0.90 second of UT1. The difference UT1 – UTC is known as dUT1, and is broadcast on WWV/WWVH shortwave and WWVB very low frequency signals.

5.4.2 Magnitude of Time Corrections

There are 86,400 UT1 (solar) seconds in a UT1 day, and according to Seidelmann (1992, p. 52) the number of sidereal seconds in a UT1 day is given by

$$s = 86636\overset{s}{.}55536790872 + 5.098097 \times 10^{-6}\, T - 5.09 \times 10^{-10} T^2$$

where T = the number of Julian centuries since JD 2451545.0 (J2000.0).

The difference is about 4 minutes per day out of 1440, or about 0.3%. Computer clocks run at the atomic TAI/GPS/UTC rate, which is not the rate at which the Earth turns and the stars move. This means you must convert UTC to sidereal time and make sure when computing drive rates that you remember that stars appear to move 15 arc seconds per second of *sidereal* time.

5.4.3 Computing Sidereal Time

One of the quantities you must compute to control just about any telescope is the hour angle (h) of an object, which is defined as:

$$h = LAST - \alpha \qquad (5.13a)$$

where LAST = local apparent sidereal time and

$\alpha =$ right ascension of the object.

Another way of stating the same thing,

$$h = \text{GAST} - L_o - \alpha \tag{5.13b}$$

where GAST = Greenwich apparent sidereal time, and

$L_o =$ the observer's longitude (positive west, negative east of Greenwich), as determined from United States Geological Survey charts or GPS.

To compute sidereal time, first compute the local *mean* sidereal time at Greenwich at 0^{h}, GMST_0, as follows, according to Meeus (1991, p. 83):

$$\begin{aligned} \text{GMST}_0 = \; & 6^{\text{h}}41^{\text{m}}50\overset{\text{s}}{.}54841 + 8640184\overset{\text{s}}{.}812866\, T \\ & +0\overset{\text{s}}{.}093104\, T^2 - 0\overset{\text{s}}{.}0000062\, T^3 \end{aligned} \tag{5.14a}$$

where T is given by (5.3g). In degrees and decimals, Meeus (1991, p. 83) uses

$$\begin{aligned} \text{GMST}_0 = \; & 100.46061837 + 36000.770053608\, T \\ & +0.000387933\, T^2 - T^3/38710000. \end{aligned} \tag{5.14b}$$

Then compute GAST_0 by adding the Equation of the Equinoxes (nutation in right ascension):

$$\text{GAST}_0 = \frac{\text{GMST}_0 + \Delta\psi \cos\epsilon}{15} \tag{5.14c}$$

where $\Delta\psi$ is from (5.9a) and ϵ is from (5.16k) in the next section. The 15 converts seconds of arc to seconds of time. To convert to sidereal time at some time other than 0^{h},

$$\text{GMST} = 1.0027379093\ \text{UT1} + \text{GMST}_0 \tag{5.14d}$$

and

$$\text{GAST} = 1.0027379093\ \text{UT1} + \text{GAST}_0. \tag{5.14e}$$

Local apparent sidereal time (LAST) is

$$\text{LAST} = \text{GAST} - L_o. \tag{5.14f}$$

The accuracy of these equations depends, in practical terms, on the accuracy with which you can determine UT1. UTC is at most $0\overset{\text{s}}{.}9$ (13.5 arc seconds) from UT1, so if you need only $1'$ accuracy, you can substitute UTC for UT1. If you need substantially higher accuracy, then you will need to obtain a source of UT1 or dUT1 (UT1 – UTC). GPS satellites broadcast a navigation message that includes GMST, but this signal is not available to the user in most GPS receivers. You might convince your GPS vendor to modify the firmware in your receiver to make GMST available, but that could be very expensive.

Another approach is to obtain dUT1 from WWV/WWVH or WWVB. If you listen to WWV or WWVH, you will hear a 100 Hz bass note or hum following each tick. The length of this hum varies, from 0.470 second for a logical 1 to 0.170 second for a logical 0. It is possible to build an electronic circuit to decode WWV and WWVH signals to obtain this value and send it to your computer. See Figure 5-8 for a graphical representation of each minute's data frame from WWV and WWVH.

A similar code is available on the Very Low Frequency (60 kHz) station WWVB. The length of the seconds carrier drop is 0.2 second for a logical 0 and 0.5 second for a logical 1. Figure 5-9 represents a one-minute data frame from WWVB. The U.S. Naval Observatory publishes values for dUT1 in IERS Bulletin A, which is available over the World Wide Web at `http://maia.usno.navy.mil`.

5.5 Aberration

5.5.1 The Physical Basis of Aberration

When an observed object moves with respect to a fixed observer, the light travelling from the moving object to the fixed observer takes a finite amount of time to reach the observer. During this time, the object moves a certain amount. The observer receives the light from the direction of the object when it emitted its light, not from the direction of the object at the time the observation is made. The angle between the two directions is a form of aberration called the correction for light time.

This form of aberration is, for stars and other objects outside our solar system, a constant for each star but is unknown, so it is ignored. The concern of one who is designing a telescope control system is not where the star is at the time of the observation, but where to point the telescope to see the star's light. Light time correction is important only in the solar system, in which the distances that light travels from various objects to the Earth are of the same order of magnitude as the distances between various observing positions, whether caused by observing an object at different places in the Earth's orbit about the Sun, or by sending deep space probes to various parts of the solar system. This makes the light time correction for solar system objects essential. Lack of light time corrections in the ephemerides for Comet Kohoutek caused large pointing errors when radio telescopes

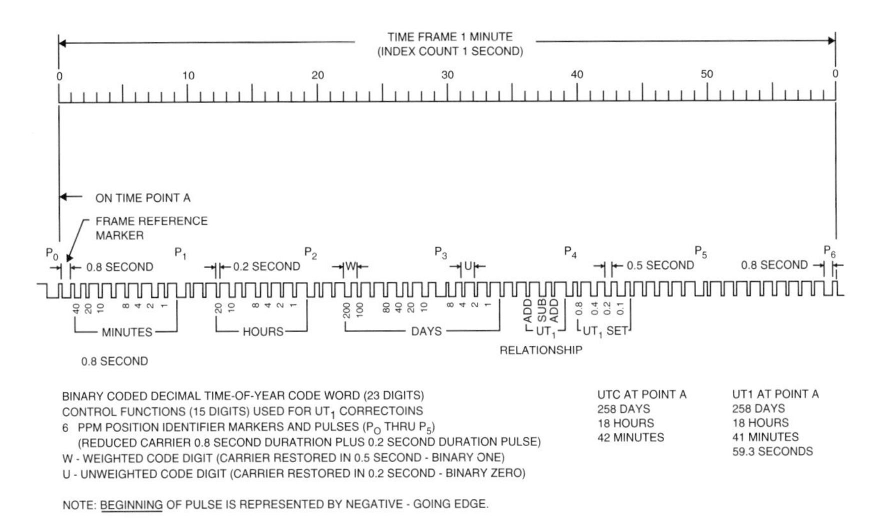

Fig. 5-8 *WWV/WWVH Time Code Format from NBS Technical Note 695, Time and Frequency User's Manual.*

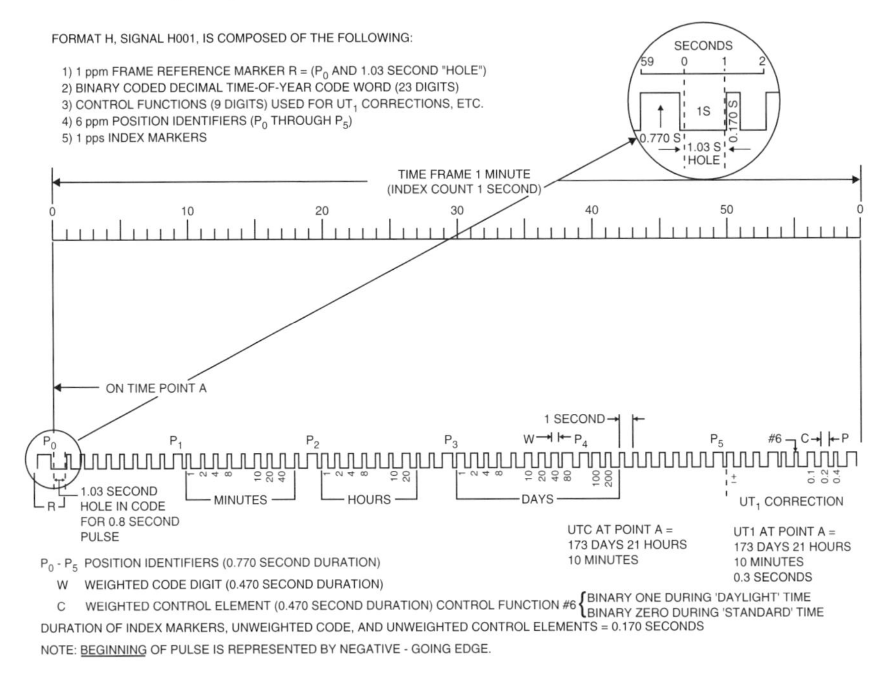

Fig. 5-9 *WWVB Time Code Format from NBS Technical Note 695, Time and Frequency User's Manual.*

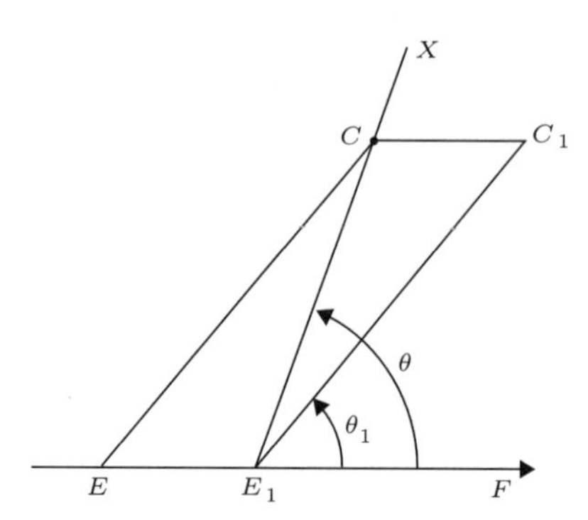

Fig. 5-10 *Geometric explanation of aberration. Courtesy Cambridge University Press.*

were used to observe the comet. The IAU standard for ephemerides of solar system objects after 1979 is to include the light time corrections in the ephemeris position.

The case where a fixed object (e.g., a star) transmits light to a moving observer is called stellar aberration. In this case, the aberration is due to making an observation in a moving coordinate system. As seen in Figure 5-10, if an observer at the eyepiece E is observing a star whose light enters the telescope at C, the motion of the observer (due to the Earth's rotation and motion in its orbit about the sun) causes motion of the eyepiece to E1 during the time it takes light to travel the length of the telescope tube. Although the light travels from C to E1, it appears to the observer to have followed a path from C1 to E1. This stellar aberration has three components, corresponding to the three main motions of the Earth: *diurnal aberration*, caused by the rotation of the Earth; *annual aberration*, caused by the motion of the Earth in its orbit about the Sun; and *secular aberration*, caused by the motion of the solar system through space. The stars and the solar system may each be considered to be moving in uniform rectilinear motion, so one cannot distinguish between the correction for light time and secular aberration. Therefore, secular aberration can be ignored.

Planetary aberration is the sum of stellar aberration and the correction for light time, and is applied only to members of the solar system.

5.5.2 Magnitude of Stellar Aberration Corrections

If one is observing an object at some elevation angle θ, then stellar aberration will cause a displacement $\Delta\theta$ of the object toward the apex of the motion of the observer. Gurnette and Woolley (1961, p. 47) give the magnitude of the correction as

$$\sin(\Delta\theta) = \frac{V}{c}\sin(\theta - \Delta\theta) \tag{5.15a}$$

or, expanding in powers of V/c,

$$\Delta\theta = \frac{V}{c}\sin\theta - \frac{1}{2}\left(\frac{V}{c}\right)^2 \sin 2\theta + \cdots \tag{5.15b}$$

where V = the velocity of the observer, and c = the velocity of light.

Since the average velocity of the Earth about the Sun is about 18 miles per second, V/c is about 0.0001, which corresponds to about $20''$. The second-order term in (5.15b) is about $0''.001$ and may be neglected.

5.5.3 Computing Annual Aberration

Smart (1979, p. 184) computes annual aberration corrections to be added to the uncorrected (α, δ) to obtain corrected (α', δ') in the following manner (expressions for k, L_t, and ϵ have been updated to J2000.0):

$$\alpha' - \alpha = Cc + Dd \tag{5.16a}$$

$$\delta' - \delta = Cc' + Dd' \tag{5.16b}$$

where

$$C = -k\cos\epsilon\cos L_t \tag{5.16c}$$

$$D = -k\sin L_t \tag{5.16d}$$

$$c = \cos\alpha\sec\delta \tag{5.16e}$$

$$d = \sin\alpha\sec\delta \tag{5.16f}$$

$$c' = \tan\epsilon\cos\delta - \sin\alpha\sin\delta \tag{5.16g}$$

$$d' = \cos\alpha\sin\delta \tag{5.16h}$$

$$k = \text{the constant of aberration} = 20''.49552 \tag{5.16i}$$

$$L_t = \text{the Sun's true geocentric longitude} \tag{5.16j}$$
(Computed as explained below.)

$$\begin{aligned}\epsilon &= \text{the mean obliquity of the ecliptic}\\ &= 84381''.448 - 46''.8150\,T - 0''.00059\,T^2 + 0''.001813\,T^3\ldots\end{aligned} \tag{5.16k}$$
Kaplan (1981, p. A3)

It is assumed α is converted to degrees or radians to find $\sin\alpha$ and $\cos\alpha$.

Meeus (1982, p. 83) computes the Sun's true longitude L_t from its geometric mean longitude as follows:

$$L_t = L_m + \nu - l' \tag{5.16l}$$

where $L_m =$ the geometric mean longitude of the Sun, referred to the mean equinox of date

$= 280°27'57''.850 + (100\,r + 0°46'11''.270)\,T + 1''.089\,T^2$ (5.16m)

$l' =$ the mean anomaly of the Sun [see (5.8c)].

To compute the true anomaly ν, Meeus (1982, p. 122) suggests using the following (the expression for e has been updated to J2000.0):

$$\nu = 2 \arctan\left[\sqrt{\frac{1+e}{1-e}} \tan\frac{E}{2}\right] \tag{5.16n}$$

where $e =$ the eccentricity of the Earth's orbit

$= 0.016708320 - 0.000042229\,T - 0.000000126\,T^2$ (5.16o)

$E =$ the eccentric anomaly of the Earth in its orbit

$= l' + e \sin E.$ (5.16p)

Equation (5.16p) is computed by iteration until convergence. Equations (5.16l) – (5.16p) can be computed at the beginning of the evening.

This procedure neglects terms, known as the E-terms, which depend on the eccentricity and longitude of perihelion of the Earth's orbit. These E-terms are about 0''.34 and are constant throughout the year for a particular star, changing very slowly over a period of centuries. The E-terms are evaluated by

$$\text{(in longitude) } \Delta\lambda = ke \sec\beta \cos(\omega - \lambda) \tag{5.17a}$$

$$\text{(in latitude) } \Delta\beta = ke \sin\beta \sin(\omega - \lambda) \tag{5.17b}$$

where $k =$ the constant of aberration $= 20''.49552$ (5.17c)

$e =$ the eccentricity of the Earth's orbit [see (5.16o)]

$\omega =$ the longitude of perihelion of the Earth's orbit

$= 102°56'18''.046 + 1°43'10''.046\,T.$ (5.17d)

Meeus (1991, p. 88) computes ecliptic longitude (λ) and latitude (β) as follows:

$$\tan\lambda = \frac{\sin\alpha\cos\epsilon + \tan\delta\sin\epsilon}{\cos\alpha} \tag{5.17e}$$

$$\sin\beta = \sin\delta\cos\epsilon - \cos\delta\sin\epsilon\sin\alpha. \tag{5.17f}$$

The effect of the E-terms on (α,δ) can be computed as follows:

$$\alpha' - \alpha = c\Delta C + d\Delta D \tag{5.17g}$$

$$\delta' - \delta = c'\Delta C + d'\Delta D \tag{5.17h}$$

where c and d are given by (5.16e and 5.16f) and

$$\Delta C = +ke\cos\omega\cos\epsilon \tag{5.17i}$$

$$\Delta D = +ke\sin\omega. \tag{5.17j}$$

These quantities are added to α and δ, respectively, to make the correction for E-terms.

By convention, catalog mean places of stars already contain the E-terms, until 1984 (see Section 5.11 below). Theoretically, one should remove the E-terms from the catalog position, compute precession, nutation, and proper motion, then recompute the E-terms using the new equinox and epoch, and apply them to the new coordinates. Since the E-terms are relatively constant over short periods of time (decades), the maximum error per century created by treating the E-terms as constant instead of following the above procedure is given by Gurnette and Woolley (1961, p. 145) as $0^{s}.0001$ in $\Delta\alpha\cos\delta$ and $0''.002$ in $\Delta\delta$. These errors are so small that the E-terms can be treated as being constant in ephemerides published before 1984, unless exceptional accuracy is required.

5.5.4 Computing Diurnal Aberration

Gurnette and Woolley (1961, p. 49) express the velocity of an observer on the surface of the Earth as $vr\cos L_c$, where $v(= 0.46$ km/sec) is the equatorial rotational velocity of the surface of the Earth, r is the geocentric radius of the Earth at the observer's site expressed in units of the Earth's equatorial radius, and L_c is the geocentric latitude of the observer. This corresponds to a constant of diurnal aberration of

$$\frac{v}{c} r\cos L_c = 0''.320\, r\cos L_c = 0^{s}.02133\, r\cos L_c.$$

This can be resolved into the following corrections to (α,δ) to obtain corrected (α',δ'):

$$\alpha' - \alpha = 0\overset{s}{.}0213\, r \cos L_c \cos h \sec \delta \tag{5.18a}$$

$$\delta' - \delta = 0\overset{''}{.}320\, r \cos L_c \sin h \sin \delta \tag{5.18b}$$

where

$$h = \text{the hour angle} \;=\; \text{local apparent sidereal time} - \alpha. \tag{5.18c}$$

Since the equatorial radius of the Earth (6378140 m) differs from the polar radius (6356755 m) by only about two parts in 600, and any Earth-based observer's geocentric radius is likely to fall in between the two, r can be taken to be 1 independent of the observer's position, and still keep the correction for diurnal aberration accurate to $0\overset{''}{.}1$. This also means the distinctions among geocentric, geodetic (geographic), and astronomical latitude can be ignored in most cases. For those requiring higher accuracy, Gurnette and Wooley (1961, p. 489) give corrections for obtaining r as

$$r = \frac{21.4 \sin^2 \phi + a}{a} \tag{5.18d}$$

where ϕ = geodetic latitude
a = equatorial radius
= 6378.140 km.

5.5.5 General Relativistic Effects

The aberrations discussed above are apparent bending of light based on physical phenomena analyzed assuming light travels in a straight line. Einstein's theory of general relativity shows how gravity can change the metric of space-time to make light appear to bend near massive objects. This phenomenon was observed by Eddington at a solar eclipse when stars very close to the Sun appeared to be even closer to the Sun (by about 1″) than when observed at a time when the Sun was not in the star field. Green (1985, p. xi) states "...general relativity must be introduced into astrometry at a fundamental level." This is true when working at the milliarcsecond level, such as in tracking solar system space probes. Such effects can be ignored in optical telescope control systems. The interested reader should consult Green (1985, pp. 75–79).

5.6 Parallax

5.6.1 The Physical Basis of Parallax

Since the positions of the Sun, Moon, planets, and stars are published in geocentric or heliocentric coordinates, and since the observer is not at the position of reference for the published coordinates, these bodies usually appear to be displaced in the sky. When viewed by an observer on the Earth's surface, the displacement depends directly on the distance of the observer from the reference point and inversely on the distance of the body from the reference point. The parallax of a close star caused by the orbital motion of the Earth about the Sun is depicted in Figure 5-11.

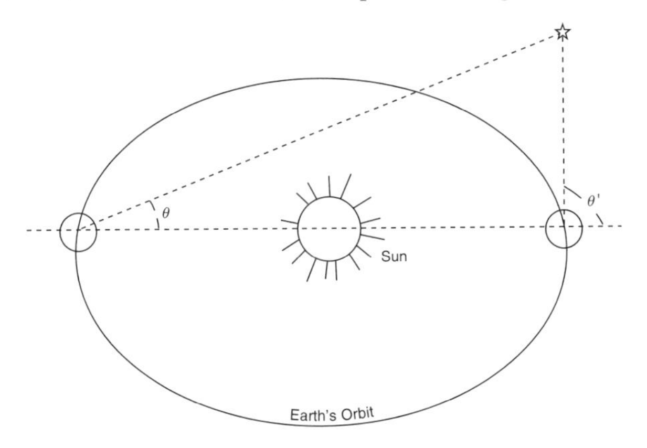

Fig. 5-11 *Heliocentric parallax.*

5.6.2 Magnitude of Parallax Corrections

Stellar positions are typically referred to the Sun, but the heliocentric parallax caused by the Earth's distance from the Sun is less than $0\overset{\prime\prime}{.}1$ for all but the closest stars, which have a parallax of about one arc second. The geocentric parallax, caused by the observer's being separated from the center of the Earth by its radius, is negligible for stars.

For the Sun and planets, both the heliocentric and geocentric parallax corrections are on the order of a few arc seconds, and may be approximated by first order theory with sufficient accuracy. The close proximity of the Moon to the Earth requires the use of exact formulae to compute the Moon's geocentric parallax, and the much closer proximity of artificial satellites requires that their positions be computed exactly from a known geocentric ephemeris. Because most astronomers have no interest in pointing a telescope accurately at the Moon or artificial satellites, these cases are ignored in the discussion below. Furthermore, most astronomers have little interest in those few close stars that would require a stellar parallax correction, so in most cases, it, too, can be ignored. The discussion of stellar parallax is presented for completeness.

5.6.3 Computing Stellar Parallax

Gurnette and Woolley (1961, p. 64) give the corrections to the barycentric place (α,δ) to the geocentric place (α', δ') of a star due to heliocentric parallax as follows:

$$\alpha' - \alpha = \pi \left(\frac{Y \cos \alpha - X \sin \alpha}{\cos \delta} \right) \tag{5.19a}$$

$$\delta' - \delta = \pi (Z \cos \delta - X \cos \alpha \sin \delta - Y \sin \alpha \sin \delta) \tag{5.19b}$$

where $(X, Y, Z) =$ the geocentric coordinates of the Sun

$\pi =$ the annual parallax of a star

$=$ $1''/D$, where D is the heliocentric distance of the star in parsecs (1 parsec = 3.26 light years).

Often, neither the heliocentric distance of a star nor its annual parallax are known, so the correction for parallax cannot be made. This usually occurs, however, only when the parallax is too small to matter. Many star catalogs contain values of heliocentric parallax when it is of the order of $0''.1$ or larger.

Meeus (1982, p. 85) uses the following method to compute (X,Y,Z):

$$X = R \cos L_t \tag{5.20a}$$

$$Y = R \sin L_t \cos \epsilon \tag{5.20b}$$

$$Z = R \sin L_t \sin \epsilon \tag{5.20c}$$

where $R =$ the distance from the Sun to the Earth in A.U.

$L_t =$ the Sun's true longitude referred to the mean equinox of date [see Equation (5.16l)]

$\epsilon =$ the mean obliquity of the ecliptic for that date [see Equation (5.16k)].

Meeus (1982, p. 83) computes R as follows:

$$R = \frac{1.0000002\,(1 - e^2)}{1 + e\cos\nu} \tag{5.20d}$$

where $e =$ the eccentricity of the Earth's orbit [see (5.16o)]
$\nu =$ the true anomaly of the Earth in its orbit [see (5.16n)].

As an alternative method, Gurnette and Woolley (1961, p. 64) suggest using the star constants c, d, c', and d' as follows:

$$\alpha' - \alpha = \pi(Yc - Xd) \tag{5.21a}$$

$$\delta' - \delta = \pi(Yc' - Xd') \tag{5.21b}$$

where c, d, c', and d' are given by Equations (5.16e)–(5.16h). This permits corrections for annual parallax to be included with the annual aberration terms of the reduction from mean to apparent place, as follows:

$$\alpha' - \alpha = (C + \pi Y)c + (D - \pi X)d \tag{5.22a}$$

$$\delta' - \delta = (C + \pi Y)c' + (D - \pi X)d'. \tag{5.22b}$$

These equations can be simplified if the annual parallax is small enough, as follows:

$$\alpha' - \alpha = C(c + d\pi k_1) + D(d - c\pi k_2) \tag{5.23a}$$

$$\delta' - \delta = C(c' + d'\pi k_1) + D(d' - c'\pi k_2) \tag{5.23b}$$

where

$$k_1 = \frac{R\sec\epsilon}{20''\!.49552} \tag{5.23c}$$

$$k_2 = \frac{R\cos\epsilon}{20''\!.49552} \tag{5.23d}$$

in which $20''\!.49552$ is the constant of aberration (J2000.0). This method does not require that X, Y, or Z be computed.

Although the method using (5.23), by itself, is slightly more efficient than the method using (5.19), when the other arithmetic operations in (5.23a,b) are taken into account, the real savings accrue when both annual aberration and annual parallax are computed, as is usually the case when computing annual parallax. Thus the preferred method of computing both annual aberration and annual parallax is to combine them using the latter method.

5.6.4 Computing Solar or Planetary Parallax

For bodies such as the Sun, the planets, or comets, published catalog positions are referred to the center of the Earth, so there is no need to compute heliocentric parallax.

Gurnette and Woolley (1961, p. 63) give equations similar to the following for computing geocentric parallax, updated to J2000.0 and reordered to give corrections to geocentric coordinates (α,δ) to obtain corrected coordinates (α',δ'):

$$\alpha' - \alpha = -\pi(r \cos\phi \sin h \sec\delta) \tag{5.24a}$$

$$\delta' - \delta = \pi(r \cos\phi \cos h \sin\delta - r \sin\phi \cos\delta) \tag{5.24b}$$

where

$\pi =$ the horizontal parallax of the object $= 8\overset{''}{.}794148/D$ where D is the geocentric distance of the object in A.U.

$r =$ the geocentric radius of the Earth at the observer's position in units of the Earth's equatorial radius (see Equation [5.18d])

$\phi =$ the geocentric latitude

$h =$ the topocentric hour angle of the object

$\delta =$ the topocentric declination of the object.

Gurnette and Woolley (1961, p. 57) give equations to compute the geocentric latitude if the geodetic latitude is known. This kind of conversion is necessary only for objects less than about 1/4 A.U. from the Earth, such as the Moon and nearby Earth-crossing minor planets.

5.7 Refraction

5.7.1 The Physical Basis of Refraction

As light leaves interplanetary space, considered to be a vacuum with index of refraction $\mu = 1$, and penetrates the Earth's atmosphere on its way to a telescope, it passes through regions of the atmosphere with different μ. This is depicted in Figure 5-12. To compute the effective refraction of a distant body for a plane-parallel atmosphere, only the index of refraction at the Earth's surface is required. Since the Earth's atmosphere is not strictly plane-parallel, the resulting errors at low altitudes can be on the order of minutes of arc. This, combined with other factors, such as temperature inversions, make it impossible to compute refraction corrections from fundamental considerations of the physical theory of refraction. This is in

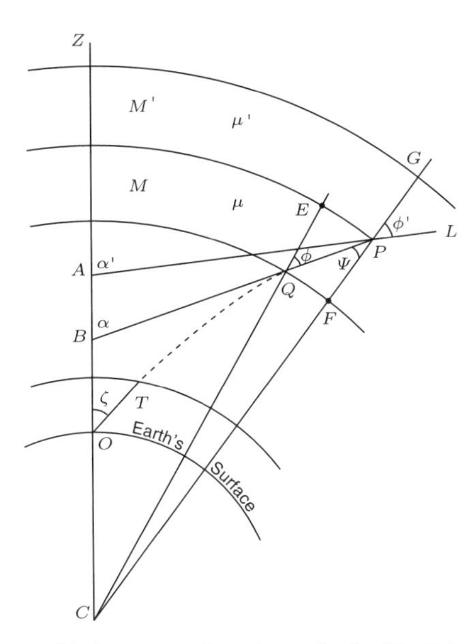

Fig. 5-12 *Refraction of light in passing through the Earth's atmosphere. Courtesy Cambridge University Press.*

contrast to precession, nutation, aberration, and other phenomena that can be predicted accurately on the basis of theoretical considerations.

Instead, models of atmospheric refraction are usually based on extensive observations. These observations are often published in the form of a set of tables, the most widely-used being those of Pulkovo Observatory (Orlov, 1956). These tables are quite lengthy, so when building a computerized control system, it is not realistic to try to place the tables on a mass storage device and access them in the required amount of time. Some mathematical models of refraction are described below. These models differ widely in structural, mathematical, and computational complexity, with the most computationally intensive model being most accurate.

The theory of refraction is complicated by the fact that different colors exhibit different amounts of refraction, so star images are actually tiny vertical spectra. In Table 5-7 Woolard and Clemence (1966, p. 84) list the differences in refraction angle between red and blue light for varying zenith angles z.

Refraction is also complicated by the fact that turbulence in the atmosphere changes the refraction of a given volume of air with time. This effect makes the apparent position of a star change so rapidly that it smears the star's point-like image into a "seeing disk" for the naked eye and all instruments except those with very high time resolution. This "seeing"

Table 5-7

z (degrees)	Delta R
30	$0\farcs35$
45	$0\farcs60$
60	$1\farcs04$
75	$2\farcs24$

effect is ignored in the discussion below, since there is nothing one can do about it without a very expensive adaptive optics system. Another effect is differential refraction, which tends to distort extended objects. This effect can be overcome using a pair of rotating prisms, but is usually ignored in all but large telescopes.

5.7.2 Magnitude of Refraction Corrections

Gurnette and Woolley (1960, p. 55) give values of $J = (\mu \sin(Z - R))/\sin Z$ for an ideal spherical atmosphere, where μ is the index of refraction of the atmosphere at sea level, Z is the true zenith distance of a star, and R is the angle through which the star's light is refracted, for various angles of Z. The values of J were derived using mean observed values of R and μ. Using $\mu = 1.00029241$ (at a wavelength of 5780Å, from Garfinkel, 1967, p. 248), we have computed the corresponding refraction angle R. These values are reproduced in Table 5-8. Note that the values for the angle of refraction R in Table 5-8 are in arc-minutes. It is clear that even for relatively small zenith distances, refraction can be quite large, compared to other sources of error. Therefore, except for systems of modest accuracy, refraction corrections are essential.

Table 5-8

Z (degrees)	J	R
0	1.000000	-
30	1.000000	-
60	1.000001	$1\farcm7$
70	1.000003	$2\farcm7$
75	1.000005	$3\farcm7$
80	1.000011	$5\farcm5$
82	1.000016	$6\farcm7$
84	1.000026	$8\farcm6$
86	1.000046	$11\farcm8$
88	1.000095	$18\farcm1$
89	1.000148	$23\farcm7$
90	1.000242	$34\farcm5$

5.7.3 Computing Refraction

Smart (1979, p. 73) gives the *approximate* effect of refraction for small zenith distances on the true coordinates (α,δ) of a star based on the refraction law $R = k \tan Z$ (where k is a constant), producing a corrected place (α', δ') given by

$$\alpha' - \alpha = -R \frac{\sec^2 \delta \sin h}{\tan z(\tan \delta \tan \phi + \cos h)} \tag{5.25a}$$

$$\delta' - \delta = -R \frac{\tan \phi - \tan \delta \cos h}{\tan z(\tan \delta \tan \phi + \cos h)} \tag{5.25b}$$

where $R =$ the angle of refraction $= Z - z$
$h =$ the hour angle of the observed body
$z =$ the apparent zenith distance
$Z =$ the true zenith distance
$\phi =$ the observer's latitude.

The angular units of $\Delta\alpha = \alpha' - \alpha$ and $\Delta\delta = \delta' - \delta$ are the same as those of R, computed using one of the methods given below.

If a more accurate refraction law is used, the correction from first principles is given by Wasserman (1986) as:

$$\delta' - \delta = R \cos \eta \tag{5.26a}$$

$$\alpha' - \alpha = h' - h = -R \sec \delta' \sin \eta \tag{5.26b}$$

where η is the parallactic angle and h is the hour angle. Wasserman computes $\cos \eta$ and $\sin \eta$ as follows:

$$\cos \eta = \frac{\sin \phi - \cos z \sin \delta'}{\sin z \cos \delta'} \tag{5.26c}$$

$$\sin \eta = \frac{\cos \phi \cos h'}{\sin z} \tag{5.26d}$$

where ϕ is the observer's latitude. Note that to compute (5.26a) you need to first compute (5.26c) to obtain η, but (5.26c) requires η from (5.26a). A similar situation pertains to z, which is what you are trying to compute if you are controlling an alt-az mount. To escape this circle of prerequisites, first compute (5.26c) using δ instead of δ' and Z instead of z, then iterate between (5.26a) and (5.26c) until the change in δ' is less than some small number, for example $1''$.

Many astronomers use a single simple equation to compute the refraction angle R over the entire range of zenith distances $0 \leq z \leq 90°$. Such methods typically do not offer high accuracy at large zenith distances, but are quite adequate for most applications. For example, the *Almanac for Computers* (p. B13) lists the following:

$$R = \frac{P}{273 + T_C} \{3.430289[z - \arcsin(0.9986047 \sin 0.9967614z)] - 0.01115929z\}$$

where $R =$ the refraction correction in minutes of arc
$z =$ the corrected zenith distance in degrees
$T_C =$ the temperature in °C
$P =$ the atmospheric pressure in millibars.

This equation is accurate to 0.′1 for altitudes greater than 15°, 1.′0 above 3°, and 3′ near the horizon. To obtain more accuracy, Meeus (1991, p. 102) computes R in minutes of arc as follows:

$$R = \frac{283P}{1010(273 + T_C)} \left[\frac{1.02}{\tan\left\{(90° - Z) + \frac{10.3}{(90° - Z) + 5.11} + 0.0019279\right\}} \right] \tag{5.27}$$

where $R =$ the refraction correction in minutes of arc
$Z =$ the true zenith distance in degrees
$T_C =$ the temperature in °C
$P =$ the atmospheric pressure in millibars.

This equation is accurate to about 4″. Eisele and Shannon (1975) fit the equation

$$R = A \tan Z + B \tan^3 Z$$

at the points $Z = 0°$, 45°, and 85°, where Z is the true zenith distance.

For $85° \leq Z \leq 90.°6$, the equation

$$R = A\left[e^{-B(90° - Z)} + e^{\frac{-B(90° - Z)}{5}}\right]$$

was used. The resulting equations are as follows:

$$R = \frac{17P}{460 + T_F}(57.626039 \tan Z - 0.05813517 \tan^3 Z) \tag{5.28a}$$

for $0 \leq Z \leq 85°$, and

$$R = \frac{17P}{460 + T_F} 871.94412 \left[e^{-0.53520501(90° - Z)} + e^{-0.107041(90° - Z)} \right] \quad (5.28b)$$

for $85° \leq Z \leq 90.°6$,

where $R =$ the refraction angle $(Z - z)$ in arc seconds

$Z =$ the true zenith distance

$P =$ the atmospheric pressure in inches of Hg measured at the Earth's surface

$T_F =$ the atmospheric temperature in °F measured at the Earth's surface.

Equations (5.27a) and (5.27b) are identical at $z = 85°$. Overall accuracy of this set of equations over the range $0 \leq z \leq 90°$ is given as $3.''7$.

Although this level of accuracy does not compare with the $0.''1$ accuracy that is possible in the other astronomical corrections used to compute corrected place from mean place, it is adequate for all but the most demanding of real-time telescope control applications. Garfinkel (1967) describes a polytropic model of the atmosphere and an algorithm for computing refraction to high accuracy. He claims overall error of less than one arc-second, with most computations being in agreement with the standard tables within $0.''1$, even for zenith distances greater than 90°. Garfinkel's computer program contains roughly 300 lines of FORTRAN statements, and performs several iterations within many different loops. Its author estimates that it would take on the order of ten seconds to run this program on a typical minicomputer (Rodin, 1982) made in the early 1980s. This would translate to about a second on modern PCs programmed in a high level language using floating point hardware, and roughly 10 to 20 times that if done in interpretive BASIC. Since refraction must be re-computed with changing altitude as the object of interest rises and sets, to obtain $0.''1$ accuracy, refraction should be computed at least once per second when observing near the horizon. Unless you want to dedicate a PC to this task, the computation time makes Garfinkel's algorithm impractical.

The problem in using Garfinkle's algorithm in a real-time control system is not simply the execution speed of the control computer. Equation (5.28b) can be used to estimate the changes in P or T which would change R by $0.''1$. Using $z = 90°$ as the worst case, (5.28b) reduces to

$$R = \frac{17P}{460 + T_F} \times 871.94412 \times 2$$

with $T_F = 50°$F and $P = 30$ inches of Hg, $dR/dP = 0.000279$ radians per inch of Hg, $= 57.''6$ per inch of Hg, so R changes by $0.''1$ when P changes by 0.0017 inches of Hg. This kind of pressure change can easily happen within a minute or so. Similarly, under the same conditions, $dR/d\,T_F = -0.0000164$ radians per 1°F (at $T_F = 50°$F) $= -3.''9$ per 1°F, so R changes by $0.''1$ when T_F changes by 0.03°F. Again, changes of this magnitude can happen very quickly. Since either P or T_F or both can change in a period of time comparable to that required to compute refraction directly from Garfinkel's algorithm, this method is not satisfactory for field use.

This also brings to light another problem. Atmospheric pressure cannot be measured with an error less than about 0.01 inches of Hg, and temperature cannot be measured with an error less than about 0.25°F without very expensive instrumentation. This means that even if Garfinkel's algorithm is capable of computing refraction to $0.''1$ accuracy, field instrumentation will probably limit the accuracy of refraction calculations to 1–5 arc seconds. Thus the equations of Meeus (5.27) or Eisele and Shannon (5.28) are well-suited to the real-time telescope control environment because of their minimal computation requirements and the accuracy of typical temperature and pressure field instruments.

If a means could be found to compute Garfinkel's algorithm quickly and to measure P and T very accurately, telescope pointing accuracy under normal field conditions could be improved. Wellnitz (1983) claims that changes in atmospheric pressure at high altitudes are always accompanied by the appropriate changes on the ground, since the pressure measured at the Earth's surface is simply the hydrostatic pressure of the air mass above the sensor. However, a given temperature measured at the Earth's surface does not imply a particular atmospheric temperature distribution along the line of sight, since temperature inversions and inhomogeneities in azimuth temperature change rapidly. Despite this, Wellnitz claims that the error introduced by these effects is only about $0.''05$ near the zenith, growing to about $1''$ RMS between 10° and 40° elevation. This means that refraction algorithm accuracies of about $0.''5$ would be useful under most observing conditions. Wellnitz also confirms the limits imposed by temperature and pressure field instrumentation.

Seidelmann (1992, pp. 141–143) gives a method of computing R using numerical integration. Not having tried this algorithm, the authors do not know how accurate it is, nor how much time it takes to compute. It appears this algorithm may be more accurate than Eisele and Shannon's but less accurate than Garfinkel's, and will probably take less time to compute than Garfinkel's for convergence criteria of $1''$ or larger. We leave it as an exercise to the reader to determine if these speculations are accurate.

5.7.4 Parallactic Refraction

The equations for atmospheric refraction given above assume the light is coming from a "distant" object, such as a star. Taff (1991, p. 93) states that when computing refraction for artificial satellites and sometimes the Moon, the equations for R should include a term r where $R+r$ is the total refraction correction, and r to lowest order is given by:

$$r = -(421''/D) secZ \tan Z \tag{5.29}$$

where $Z =$ the uncorrected zenith distance in degrees

$D =$ the distance from the object to the observer in kilometers.

Assuming you are controlling an astronomical telescope and not a satellite tracker, you must be observing the Moon at $Z > 89°$ for this effect to be as large as $1''$. It can be ignored for all other celestial bodies unless extremely high accuracy is needed.

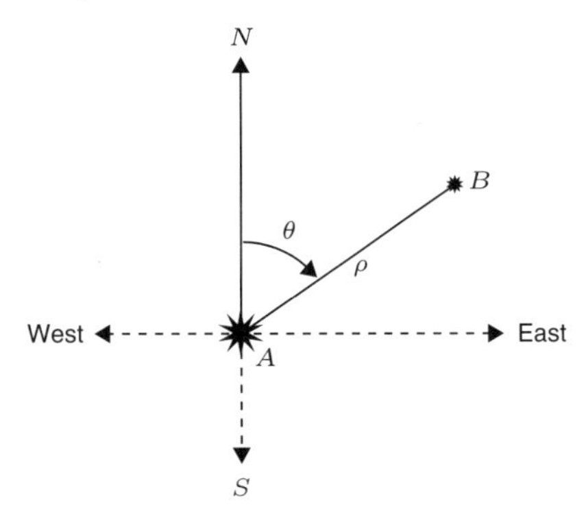

Fig. 5-13 *Changing star orientations due to orbital motion. Courtesy Cambridge University Press.*

5.8 Orbital Motion

The motion of two stars in a binary star system causes their observed positions in the sky to change with time, as shown in Figure 5-13. If the stars are relatively far apart, the changes in the stars' positions will be large enough to be observable. The rate of change in corrected position depends upon the separation between the stars and the distance of the binary system from the observer.

Most binary systems are at a distance sufficient to keep the changes in positions of the two stars less than $0\overset{''}{.}1$ in any appreciable time. This is because to move rapidly in an orbit, the stars must be so close together that the angle the orbit subtends from the Earth is small. Conversely, for the orbit to subtend an appreciable angle on the sky, the stars must be so widely separated that they move very slowly in their orbits.

In either case, once the desired star is located (by using a recent catalog), the effects of its orbital motion can usually be ignored. In those few cases where it cannot be ignored, orbital motion can be considered to produce a position change that varies linearly over time in the course of a year, and is usually noted in the star catalogs in a form that is readily entered into the computer for a trivial correction calculation.

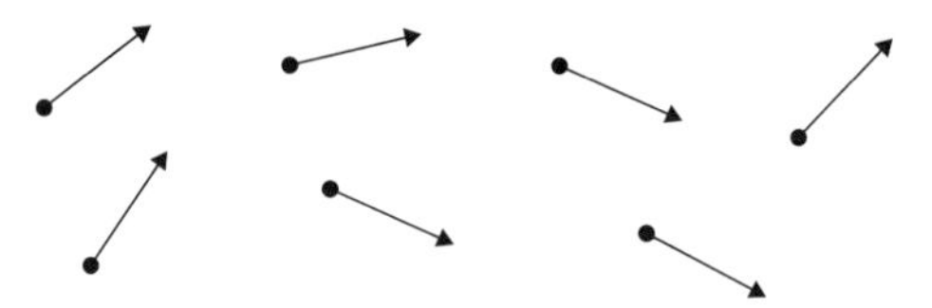

Fig. 5-14 *Proper motions of stars.*

5.9 Proper Motion

The proper motion of a heavenly body is its motion relative to the "fixed stars" (objects known not to move significantly with respect to each other over long periods of time). Many star catalogs give values for $\Delta\alpha$ and $\Delta\delta$ due to proper motion over some period of time, such as a year or a century, for those few stars with known proper motions. Such motion is depicted in Figure 5-14. Most stellar proper motions of significant size are known, with the largest being on the order of about $10''$. Most of the less than 5,000 known proper motions are less than $1''$. For those stars listed in fundamental catalogs, or in catalogs based on the fundamental catalogs, the mean place includes a correction for proper motion.

Green (1985, p. 264) gives the correction to (α, δ) for proper motion as

$$\alpha' - \alpha = \left(\mu_\alpha + \frac{1}{2}t\,\frac{d\mu_\alpha}{dt}\right)t \tag{5.30a}$$

$$\delta' - \delta = \left(\mu_\delta + \frac{1}{2}t\,\frac{d\mu_\delta}{dt}\right)t \tag{5.30b}$$

where $\mu_\alpha =$ the annual proper motion in α given in the star catalog, in arc seconds per year or century

$\mu_\delta =$ the annual proper motion in δ given in the star catalog, in arc seconds per year or century

$t =$ the time interval in years or centuries (depending on the units for μ_α and μ_δ) between the catalog and current epochs

$$d\mu_\alpha/dt = -0.422 V_r \pi \mu_\alpha \sin 1'' + 2\mu_\alpha \mu_\delta \tan\delta \sin 1'' \qquad (5.30c)$$

$$d\mu_\delta/dt = -0.422 V_r \pi \mu_\delta \sin 1'' - 225\mu_\alpha^2 \sin\delta \cos\delta \sin 1'' \qquad (5.30d)$$

where $V_r =$ the radial velocity of the star in km/s
$\pi =$ the annual parallax of the star in arc seconds.

If the radial velocity or annual parallax of the star is not known, the second terms in the parentheses in (5.30a,b) can be set to zero.

In vector notation, the vector μ representing the proper motion correction in arc seconds to the unit position vector (x, y, z) is

$$\begin{aligned} \mu_x &= -15 \sin\alpha \cos\delta \mu_\alpha - \cos\alpha \sin\delta \mu_\delta \\ \mu_y &= 15 \cos\alpha \cos\delta \mu_\alpha - \sin\alpha \sin\delta \mu_\delta . \\ \mu_z &= \cos\delta \mu_\delta \end{aligned} \qquad (5.31a)$$

The corrected unit vector (x', y', z') is given by

$$\left| \begin{matrix} x' \\ y' \\ z' \end{matrix} \right| = \left| \begin{matrix} x \\ y \\ z \end{matrix} \right| \left(1 + V_r \frac{\pi t}{4.74} \sin 1'' \right) + \left| \begin{matrix} \mu_x \\ \mu_y \\ \mu_z \end{matrix} \right| . \, t \sin 1''. \qquad (5.31b)$$

Green (1985, p. 265) gives corrections to μ_α and μ_δ due to precession. Since proper motions are themselves small, corrections to them for precession are very small, usually less than a few arc seconds per century.

The proper motion of other (closer) bodies, such as planets and comets, is the basic phenomenon of interest to those preparing ephemerides for these objects, and thus proper motion is included in ephemerides of these bodies. Therefore, the designer of a computerized telescope control system need not be concerned with corrections for proper motion of solar system objects.

5.10 Reduction from Mean to Topocentric Place

The preceding sections have described methods of computing corrections to the mean (catalog) position of an object to obtain its topocentric position as one would see it from a point on the Earth's surface. Proper motion, orbital motion, stellar aberration, refraction, and parallax cause changes in the direction in which a star is actually observed, but do not affect the frame of reference. Precession, nutation, and polar motion cause changes to the fundamental coordinate system, but do not affect the direction in which a star is observed. The corrections for precession, nutation, and annual aberration are large enough to cause significant errors if the corrections are not applied in the proper order and the cross terms are neglected.

Gurnette and Woolley (1961, p. 150) quote an analysis by Porter and Sadler indicating that the most accurate method of reducing mean to topocentric place is to apply the correction for aberration to the mean place, then to use the resulting coordinates for the precession correction, then to apply the nutation correction to the precessed coordinates. The corrections for the heliocentric parallax of stars, proper motion, and orbital motion are small enough to be applied at any convenient point. The correction for refraction is applied last. Table 5-9 summarizes the process of reducing a mean place, represented by $(\alpha,\delta)_m$, to a topocentric place, represented by $(\alpha,\delta)_t$.

In Table 5-9 each correction is treated as if it were a computer software subroutine, in which a set of input coordinates is transformed into a set of output coordinates by computing a correction and applying the correction to the input coordinates. The input coordinates for Step 1 are the mean place right ascension and declination, and after applying all the corrections, the output coordinates of Step 8 are for the topocentric place of the object. It is these latter coordinates that are used to point the telescope. Polar motion has been ignored in this table because of its negligible effect and our inability to model it. Because it was found above to be computationally more efficient to combine the computation of heliocentric stellar parallax

Table 5-9

Step	Correction	Input	Output
1a	Annual aberration	$(\alpha,\delta)_m$	$(\alpha,\delta)1$
1b	Stellar parallax (included as part of Step 1a)		
2	Precession	$(\alpha,\delta)1$	$(\alpha,\delta)2$
3	Nutation	$(\alpha,\delta)2$	$(\alpha,\delta)3$
4*	Orbital motion	$(\alpha,\delta)3$	$(\alpha,\delta)4$
5*	Proper motion	$(\alpha,\delta)4$	$(\alpha,\delta)5$
6	Diurnal aberration	$(\alpha,\delta)5$	$(\alpha,\delta)6$
7	Planetary parallax	$(\alpha,\delta)6$	$(\alpha,\delta)7$
8	Refraction	$(\alpha,\delta)7$	$(\alpha,\delta)_t$

* Usually not needed

and annual aberration, these corrections are listed together.

In a computerized control system, those corrections that need be computed only once per evening should be at the top of the list, then the other corrections should be ordered by their frequency of computation, with the lowest frequency corrections being closer to the top of the list. This is because the output of each step is the input to the succeeding step. This constraint places diurnal aberration near the bottom of the list, instead of at the top, as recommended by Porter and Sadler. However, the error resulting from this is typically negligible in most telescope pointing applications.

Note that the process depicted in this table is *serial,* with each correction being a sequential step in which the result of the previous step undergoes a transformation to obtain the input for the succeeding step. A different approach is the one used before electronic digital computers became available, and which is still used today because of this tradition. This latter approach is to compute Besselian day numbers, which, when used with star constants computed using the star's mean place coordinates, form a series of eight terms in right ascension and seven terms in declination. This *parallel* approach is popular because the Besselian day numbers depend only on the time an observation is made, not on either the object's or the observer's coordinates. Thus the Besselian day numbers can be computed in advance for the entire year, and published in an ephemeris. The only calculations the observer must do are to compute the star constants, then apply the Besselian day numbers for the observation date, and compute the two series for right ascension and declination.

This approach is still useful for one doing the calculations by hand or using a pocket calculator, but it has no advantage over the serial approach when used in a small computer to control a telescope in real-time. Furthermore, Besselian day numbers are used to compute *apparent* place only, not *topocentric* place. Therefore corrections for geocentric parallax, diurnal aberration, and refraction are not included in the Besselian approach.

Another approach along the same lines is to use independent day numbers, which are used in a different series for position. Those interested in Besselian day numbers are referred to Taff (1991, Ch. 5) and to Green (1985, pp. 288–295).

5.11 Changes in the 1984 Ephemerides

The method of preparing the catalogs that astronomers use as the sources for the coordinates they enter into telescope control systems changed in 1984. These changes include new constants for computing some of the corrections given in previous sections, but the equations given in this chapter already reflect these changes, so there is no need to bring them up to date. The 1984 changes can affect very high accuracy calculations of precession from a catalog using an epoch before 1984. For most applications, these changes will not contribute a significant error.

Kaplan (1981) and Seidelmann and Kaplan (1982) list changes in the method of preparing the Astronomical Ephemeris starting with the ephemeris for 1984. These were brought about by resolutions adopted by the International Astronomical Union in 1976, 1979, and 1982, and are summarized below.

1. The new fundamental epoch is J2000.0, corresponding to January 1.5, 2000, or Julian Day number 2,451,545.0. The previous fundamental epoch (1950.0) was 1950 January 0.5. A Julian century of 36525 ephemeris days will continue to be used as the time unit for most calculations.

2. The new constant of general precession in longitude, per Julian century, is $5029''.0966$. The previous precession constant for epoch 1900.0 was $5025''.64$.

3. The equinox of the *Fifth Fundamental Catalog* (FK5) has replaced the FK4 equinox as the origin of Right Ascension. This produced a shift of $0^s.06$ in 1984 plus a correction that is time dependent.

4. The new constant of aberration is $20''.49552$. The previous constant was $20''.496$. The E-terms of aberration are no longer included in catalog mean places of stars.

5. The new solar parallax constant is $8''.794148$, compared to the previous value of $8''.794$. The new equatorial radius of the Earth is 6378140 m (the previous value was 6378160 m), with a new flattening factor of $0.00335281 = 1/298.257$, compared to the previous value of 0.00335285.

6. The new constant of nutation is $9''.2025$. The constant at epoch 1900.0 was $9''.2235$. A new theory of nutation, based on a non-rigid model of the Earth, has been adopted. The new model uses two 106-term series to evaluate nutation.

7. Other changes, including the definition of new time scales and new astronomical constants, do not affect telescope pointing algorithms, but do affect the way in which ephemerides will be generated from 1984 onward.

The corrections described in this chapter are used to convert mean catalog positions to topocentric position in the sky. In the following chapter, corrections for telescope characteristics that affect pointing accuracy are discussed. Both sets of corrections must be made for high accuracy telescope control.

Chapter 6

Systematic Errors II — Mechanical Corrections

The previous chapter dealt with the problem of where in the sky an object appears to be placed and how to compute this corrected position from the coordinates given in a catalog or ephemeris. If a perfect telescope were set to the corrected position coordinates, the object would appear in the center of the field. Any pointing errors seen in the eyepiece would be due only to errors in computing the corrected position. The computed corrections are based on factors external to the particular telescope being used. They are computed in the same way for all observers at the same place and time observing the same object.

This chapter treats the problem of how to point an imperfect telescope accurately. These corrections are computed in a manner that can be used with any telescope of a particular mounting type, but the values of the error parameters are unique to each individual telescope, since they are characteristics of that telescope.

In a typical telescope control system, the corrected position is what is presented to the control computer as the desired position. In closed-loop systems, a position feedback sensor produces a number corresponding to the angular position of the axis. This number is converted to a meaningful coordinate, such as declination, by applying a calibration algorithm to the raw encoder readout. The corrections discussed in this chapter are incorporated into the calibration algorithm to produce numbers representing the true position of the telescope. The control computer then compares the corrected position of the target object with the position of the telescope (determined from the encoder readings), and generates commands to the motors to minimize the difference between the object and telescope positions.

In open-loop systems, the control software keeps track of where it "thinks" the telescope is pointing, so there is no position comparison to make. Since there are no encoder readings in open-loop systems, the control computer uses lookup tables or models of telescope errors using coefficients derived from special tests to adjust the desired position of the telescope to compensate for the telescope's mechanical errors.

The mechanical corrections of interest are as follows:

1. Zero offset (the 0° point in hour angle or declination)
2. Polar axis misalignment (equatorial mounts)
3. Azimuth axis misalignment (alt-az mounts)
4. North-South axis misalignment (alt-alt mounts)
5. Non-perpendicularity of the axes
6. Collimation errors
7. Tube flexure
8. Mount flexure
9. Servo lag
10. Gearing errors
11. Bearing errors
12. Drive train torsion errors.

Since the size of each correction depends so much on each individual telescope, each type of correction can be any size, from fractions of an arc second to several degrees.

6.1 Telescope Mount Designs

Before discussing how to correct errors in each type of telescope mount, we will briefly review the major mount types and their advantages and disadvantages. This will help you select the best mount design for your type of observing.

By far the most common type of mount for both large and small telescopes is the *equatorial* mount. This mount has one axis (the *polar* or *right ascension* axis) aligned parallel to the Earth's rotation axis, while

Fig. 6-1 *The 40-inch Yerkes Refractor uses a German equatorial mount. Courtesy Perry Remaklus.*

Fig. 6-2 *The Kitt Peak 4-m telescope uses a fork equatorial mount. Courtesy National Optical Astronomy Observatories.*

Fig. 6-3 *The 2.5-m Hooker telescope on Mt. Wilson uses an English yoke equatorial mount. Courtesy Dennis diCicco.*

the *declination* axis is perpendicular to the polar axis. Because of the simplified tracking that equatorial mounts offer, these mounts have been popular for over a century and consequently, there are many variations. The German equatorial offsets the optical tube from the polar axis, simplifying the design (see Figure 6-1). The fork mount usually aligns the center of the optical tube with the polar axis, as shown in Figure 6-2. This design reduces the amount of dead weight (e.g., the counter weight) the motors have to move and the bearings have to support. Other interesting variations include the English yoke that suspends the telescope between North and South piers for extra stability (see Figure 6-3), the cross axis mount that also uses two piers (see Figure 6-4), and the Springfield mount, that directs the beam through the polar axis and permits the viewer to remain stationary no matter where the telescope is pointing.

The equatorial mount has the advantage of ease of tracking celestial objects, since once the telescope is pointing at an object, tracking involves moving only the polar axis at a constant rate. In addition, there is no rotation of the image, since the field is always aligned in a constant (α, δ) system.

Disadvantages include a blind spot at the celestial pole (not a problem for most observers), and the primary mirror is placed in many

Fig. 6-4 *The Cerro Tololo Interamerican Observatory Curtiss Schmidt uses a cross axis equatorial mount. Courtesy National Optical Astronomy Observatories.*

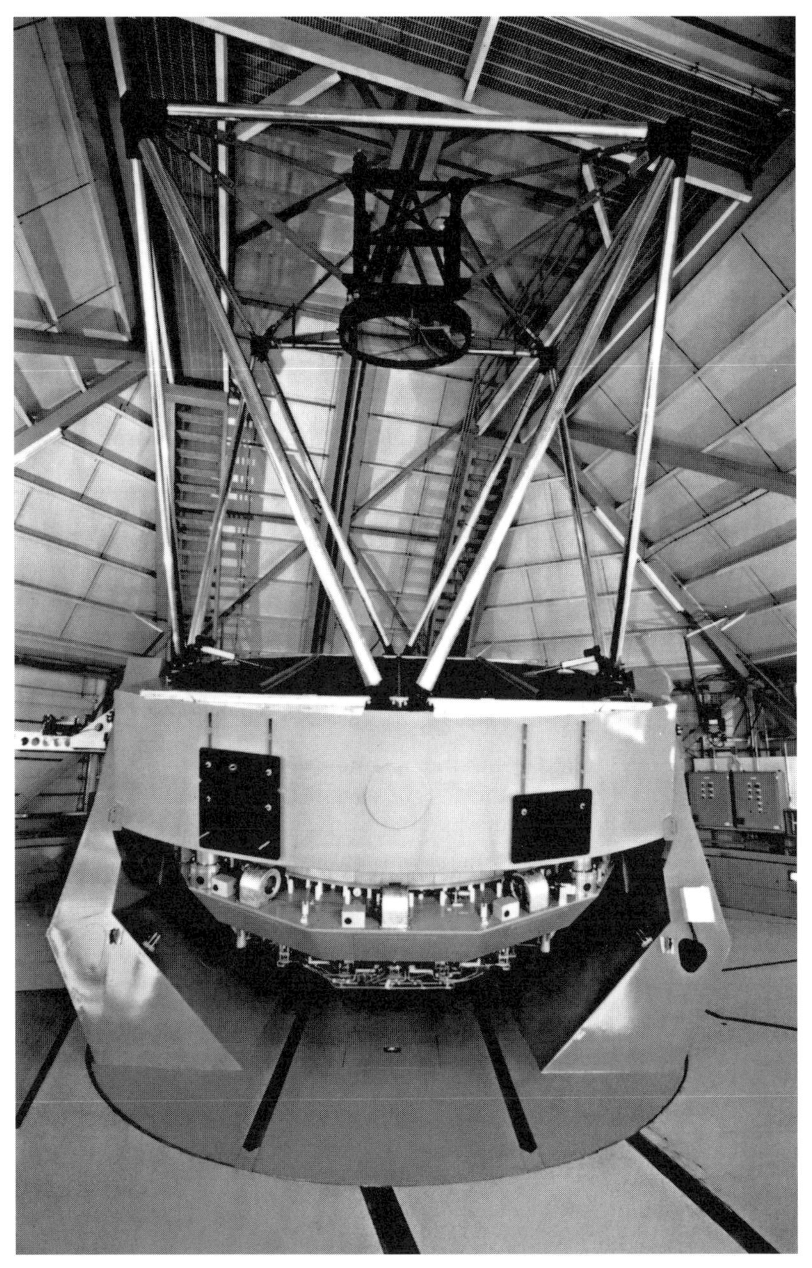

Fig. 6-5 *The Kitt Peak 3.5-m Wisconsin-Indiana-Yale-NOAO (WIYN) telescope uses a fork alt-az mount. Courtesy National Optical Astronomy Observatories.*

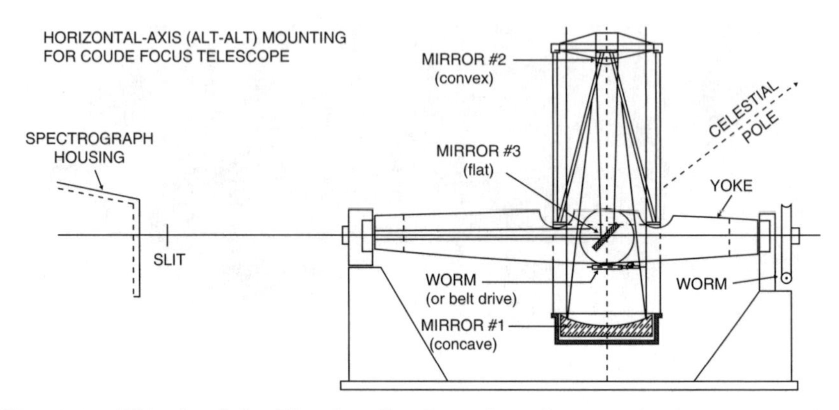

Fig. 6-6a *This sketch by Theodore Dunham shows how an alt-alt mount can be envisioned as an English yoke mount with both piers having the same height.*

different orientations with respect to gravity, which makes designing the mirror cell more difficult, especially for larger telescopes. The German equatorial often must be moved to place the optical tube on the other side of the polar axis when tracking at high declinations to avoid banging the tube into the mount. Equatorial mounts also have a high center of gravity. This is not a problem in permanently-located telescopes, but mobile telescopes need to be more compact and have a lower center of gravity to make transport safe and convenient.

The electronic digital computer has made it possible for telescope designers to explore other mount designs that make tracking celestial objects accurately more difficult, but have offsetting advantages. The *altitude-azimuth* or alt-az mount has one axis pointing to the local zenith that rotates in azimuth, and the other axis perpendicular to the first. An example is shown in Figure 6-5.

The alt-az mount has several advantages. It is very compact, enabling the construction of larger telescopes at lower cost, since a certain size and weight of the primary mirror requires less mass (and cost) in the mount. If the telescope is smaller, its dome or enclosure can also be smaller, and less expensive. The mirrors in an alt-az mount tip in only one direction with respect to gravity, simplifying the design of the mirror cell, which is further simplified if the optics tube is not permitted to go through the zenith. Alt-az telescopes also have a low center of gravity, which makes them the best mount for mobile applications.

These advantages have a high price. To track a celestial object, both axes of an alt-az mount have to be moved at varying rates that depend on the location of the object in the sky. The range of rates is very wide, putting severe demands on the drive motors. The focal plane rotates at varying

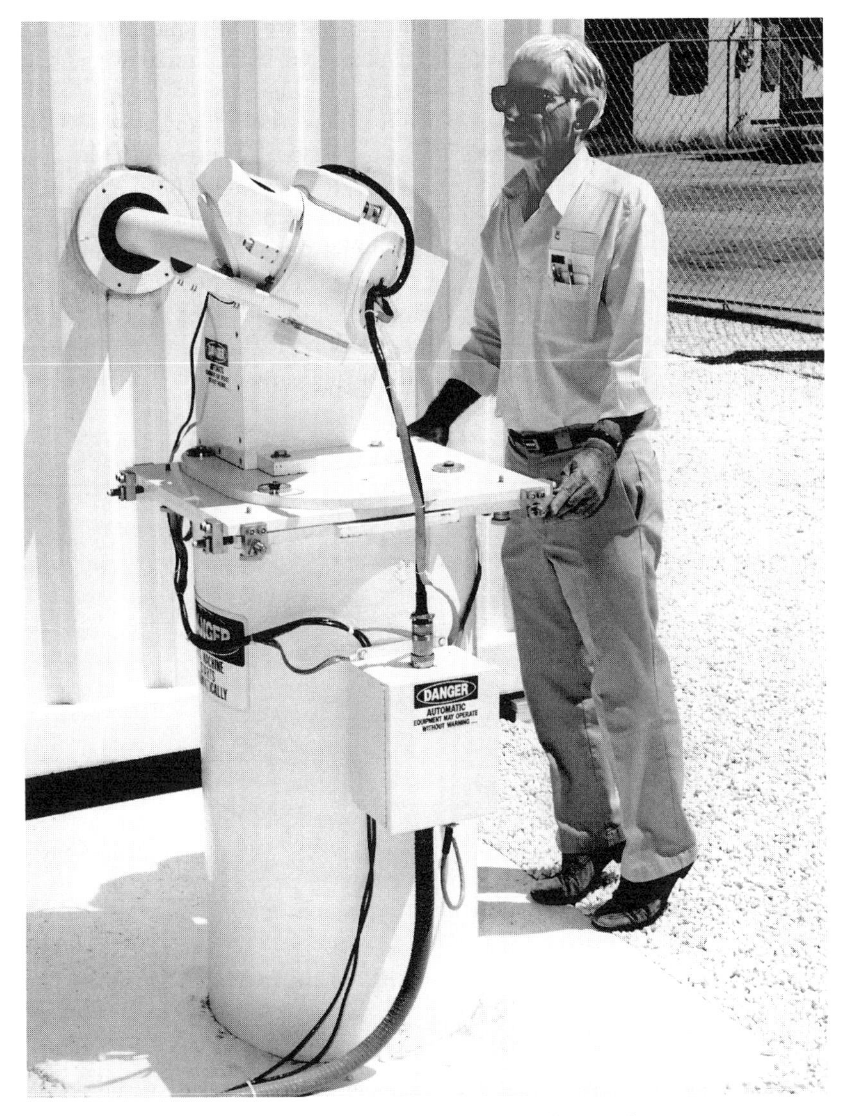

Fig. 6-6b *The Global Oscillation Network group (GONG) solar telescopes use an alt-alt mount. Courtesy National Optical Astronomy Observatories.*

rates, so imaging requires despinning the focal plane. Alt-az mounts have a small cone around the zenith where tracking speeds approach infinite.

A little-known type of mount is the *altitude-altitude*, or alt-alt. A type of alt-alt mount is shown in Figure 6-6a and 6-6b. You can think of this mount as an English yoke in which both piers are the same height. This mount is attractive to those who need access to both the zenith and the celestial pole, or who need other features of this type of mount. Using a third (rotating) mirror, it can place an image at a fixed place, as does the Springfield equatorial mount. The alt-alt mount shares with the alt-az mount the disadvantages of a wide range of drive rates for both axes, and a rotating focal plane. It has tracking and pointing blind spots at the horizon (where few observations are made) in the directions of its major supports.

This is not an exhaustive list of telescope mount designs. There are three- and four-axis designs for tracking satellites or other special targets, and many interesting telescope mounts intended for unusual or unique applications. The interested reader should consult *Unusual Telescopes* (Manly, 1991).

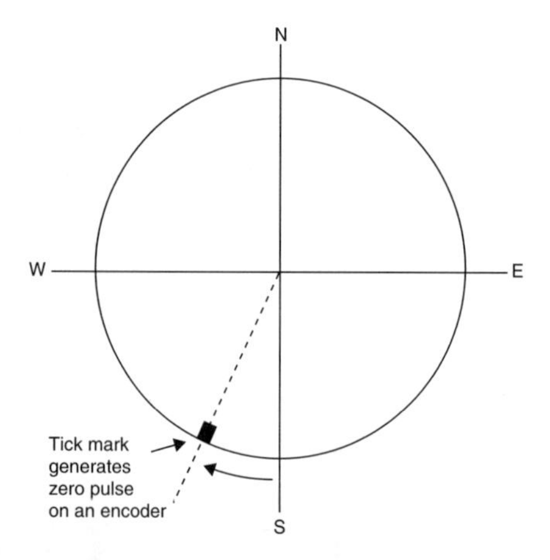

Fig. 6-7 *Zero offset error in azimuth.*

6.2 Telescope Pointing Corrections—Equatorial Mount

6.2.1 Zero Offset

Closed-loop systems using shaft encoders must determine the encoder reading that corresponds to the zero point of that axis. The simplest form of the encoder calibration algorithm is a linear model, in which the hour angle (or declination) reading is obtained by multiplying the number produced by the encoder by a fixed factor that is the number of arc seconds per least significant bit of the encoder count, then adding a constant. Included in this constant is the offset of the zero position of the encoder relative to the direction of zero hour angle. This correction, shown in Figure 6-7, is unnecessary in open-loop systems, which are oriented by pointing at a star and entering its coordinates, or by using limit switches and a clock to determine the telescope's initial or home (h,δ) position.

Associated with zero offset is the scale factor for converting from encoder counts to an angle. This factor can be measured accurately with an autocollimator, or by sighting a distant (stationary) object, rotating the telescope through 360°, and sighting the object again. The scale factor can be found by noting how many encoder counts occur during the 360° rotation.

6.2.2 Polar Axis Alignment

Careful measurements of star positions with any equatorially-mounted telescope will indicate that the polar axis of the telescope is not exactly parallel to the Earth's rotation axis. The angle between these two axes cannot be expressed as a constant correction to hour angle or declination, since the corrections to the position (h,δ) depend on h and δ. However, this angle, once measured, can be resolved into components of the elevation error (M_{el}) and azimuth error (M_{az}) of the polar axis, as shown in Figure 6-8. Wallace (1975, p. 284) uses the following equations to compute corrections Δh and $\Delta\delta$ to the telescope position (h,δ):

$$\Delta h = \tan\delta(M_{\mathrm{el}}\sin h - M_{\mathrm{az}}\cos h) \tag{6.1a}$$

$$\Delta\delta = M_{\mathrm{el}}\cos h + M_{\mathrm{az}}\sin h. \tag{6.1b}$$

These equations were derived using a small angle approximation for M_{el} and M_{az}.

6.2.3 Driving Rates

If an equatorial telescope is perfectly constructed, to track an object the telescope need not move at all in declination, and must move 15″ per sidereal second in right ascension. To obtain the correct tracking rate, one could compute the right ascension drive rate directly. Seidelmann (1992, p. 52) gives the rate of rotation of the Earth as 15.″04106717866910 per second of UT1 time. Since the difference between UT1 and UTC rates produces a difference in the two times of no more than 1 second every 6 months (1 part in 16 million), you can safely substitute the UTC rate for the UT1 rate for most applications.

To correct the drive rates of both axes properly to track an object accurately, you must recompute sidereal time using Equations (5.14) and the alignment, mount, and other corrections in this chapter each time through your software servo loop, preferably several times per second. In effect, you

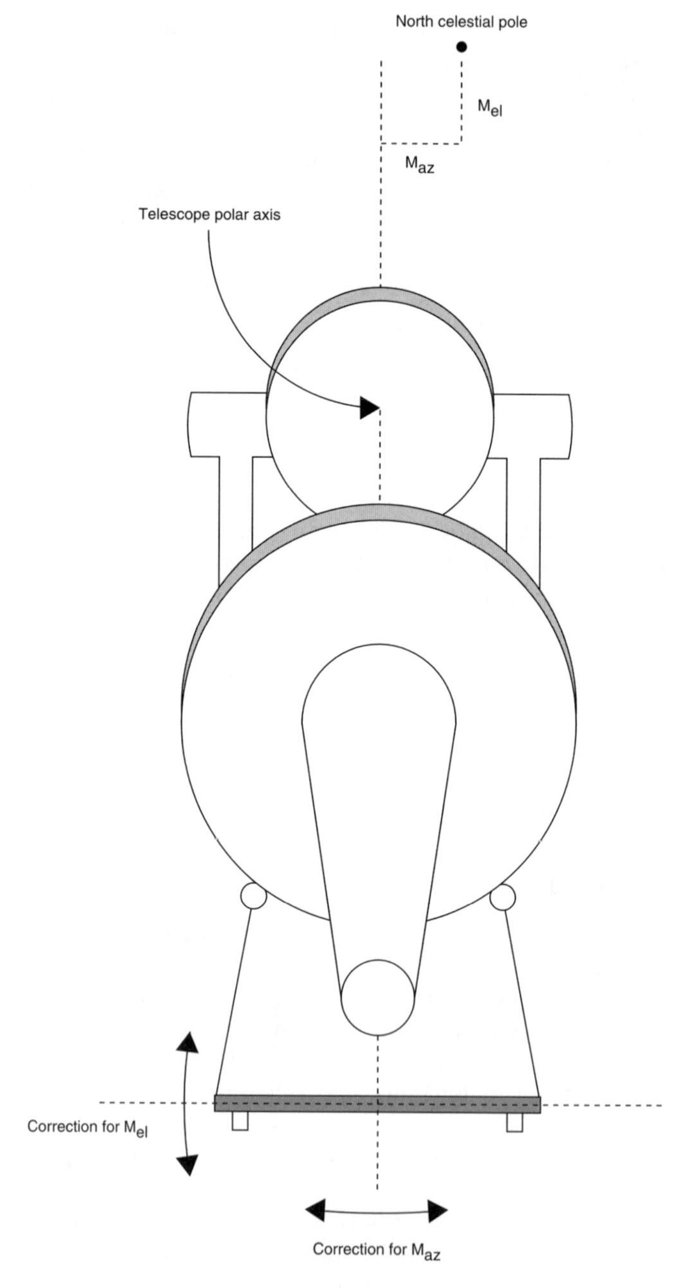

Fig. 6-8 *Corrections for polar axis misalignments.*

compute a new "corrected" object position in each axis, then update the motor speeds accordingly. This places a considerable load on the control computer and is appropriate only when good accuracy is required.

6.3 Telescope Pointing Corrections—Alt-Az Mount

6.3.1 Zero Offset

In closed-loop systems, the zero offsets of the encoders on an alt-az mount are included as constant terms in linear encoder calibration models. This is exactly analogous to the equatorial mount zero offset, and therefore is not necessary in open-loop control systems.

6.3.2 Azimuth Axis Alignment

The alt-az mount analogy to polar axis misalignment is zenith misalignment. This is shown in Figure 6-9. Imagine the rectangular coordinate system (x,y,z) has its origin at the intersection of the telescope's elevation and azimuth axes, and its x-y plane is parallel to the horizon such that the x-axis points North, the y-axis points west, and the z-axis points to the zenith. A telescope azimuth plane that is tilted can be represented by the coordinate system (x'',y'',z'') that has been rotated through angle **a** in elevation about the x-axis to form coordinate system (x',y',z'), and then rotated through angle **b** in azimuth about the z'-axis. Typically, angle **a** is small, but angle **b** can have any value. The matrix representing the first rotation is

$$\begin{bmatrix} 1 & 0 & 0 \\ 0 & \cos a & \sin a \\ 0 & -\sin a & \cos a \end{bmatrix}$$

while the matrix representing the second rotation is

$$\begin{bmatrix} \cos b & \sin b & 0 \\ -\sin b & \cos b & 0 \\ 0 & 0 & 1 \end{bmatrix}$$

with the resulting matrix product representing the complete set of rotations on (x,y,z) to obtain (x'',y'',z'') as follows:

$$\begin{bmatrix} \cos b & \cos a \sin b & \sin a \sin b \\ -\sin b & \cos a \cos b & \sin a \cos b \\ 0 & -\sin a & \cos a \end{bmatrix}. \tag{6.2}$$

When this resulting matrix is applied to the coordinate system (x,y,z), the expressions for the coordinates (x'',y'',z'') are as follows:

$$x'' = x \cos b + y \cos a \sin b + z \sin a \sin b \tag{6.3a}$$

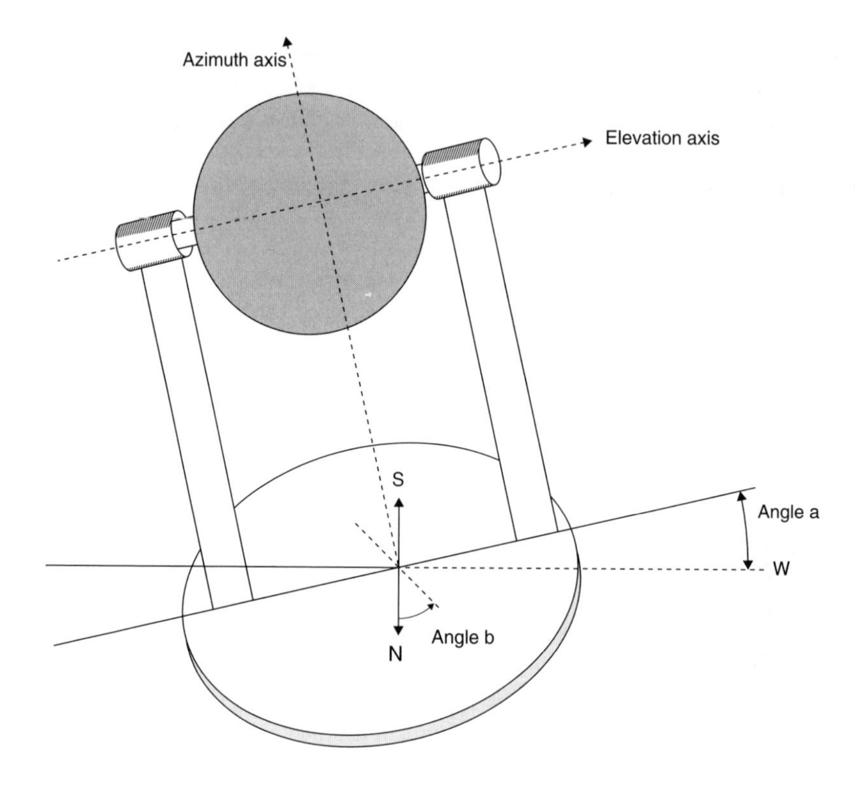

Fig. 6-9 *Corrections for azimuth axis misalignments.*

$$y'' = -x \sin b + y \cos a \cos b + z \sin a \cos b \tag{6.3b}$$

$$z'' = -y \sin a + z \cos a. \tag{6.3c}$$

Using the small angle approximations $\sin a = a$ and $\cos a = 1$, Equations (6.3) become

$$x'' = x \cos b + (y + za) \sin b \tag{6.4a}$$

$$y'' = -x \sin b + (y + za) \cos b \tag{6.4b}$$

$$z'' = -ya + z. \tag{6.4c}$$

If the azimuth A and zenith distance Z are the computed place for a star, but the star is observed at azimuth A'' and zenith distance Z'' with all other error sources taken into account, angles a and b can be found by re-arranging (6.4) into the form

$$a = \frac{z - z''}{y} \tag{6.5a}$$

$$\sin b = \frac{x''(y+za) - xy''}{x^2 + (y+za)^2} \tag{6.5b}$$

and by using the spherical-to-rectangular coordinate system transformations

$$x = \sin Z \cos A \qquad x'' = \sin Z'' \cos A'' \tag{6.6a}$$

$$y = \sin Z \sin A \qquad y'' = \sin Z'' \sin A'' \tag{6.6b}$$

$$z = \cos Z \qquad z'' = \cos Z''. \tag{6.6c}$$

By sighting only a single star, computing its topocentric position (A,Z) from its catalog mean coordinates, and measuring its position at (A'',Z''), angles a and b can be found. Additional measurements can be used to improve the accuracy of the values measured for a and b.

Once a and b are found, the true (A,Z) of the telescope is found from the position (A'',Z'') maintained in the program (open-loop) or read from the shaft angle encoders (closed-loop). The matrix (6.2) may be rearranged to give

$$x = x'' \cos b - y'' \sin b$$

$$y = x'' \cos a \, \sin b + y'' \cos a \, \cos b - z'' \sin a$$

$$z = z'' \sin a \, \sin b + y'' \sin a \, \cos b + z'' \cos a.$$

When the same small angle approximation is applied, the result is

$$x = x'' \cos b - y'' \sin b \tag{6.7a}$$

$$y = x'' \sin b + y'' \cos b - z'' a \tag{6.7b}$$

$$z = x'' a \, \sin b + y'' a \, \cos b + z''. \tag{6.7c}$$

Knowing a and b from previous measurements and (A'',Z'') from the telescope axis angle encoders, one can compute the (A,Z) where the telescope is actually pointing using Equations (6.6) and (6.7). Once Equations (6.7) are evaluated, Z is obtained from $z = \cos Z$, then A is found from $x = \sin Z \cos A$. To expedite computing these equations, values for $\sin b$ and $\cos b$ can be computed once and stored when a and b are found from measurements.

Another approach to alignment errors for any telescope (not just ones with alt-az mounts) is to define a telescope coordinate system based on how the scope happens to be aligned, then find a rotation matrix (similar to that used for precession in Equation (5.5)) describing the transformation to (α,δ). Although this simplifies many of the alignment error computations, it does not take into account axis non-perpendicularity and other mechanical corrections. It also complicates monitoring mechanical correction coefficients for trends indicating imminent bearing failure or other problems.

6.3.3 Equatorial to Alt-Az Conversion

Catalogs and ephemerides generally give positions in right ascension and declination coordinates. The encoders attached to the axes of equatorial mounts give readings in terms of these coordinates, so they are the most convenient coordinates for a computerized system controlling an equatorially-mounted telescope.

A control system for an alt-az mount, however, computes motor commands in the altitude (elevation)-azimuth coordinate system. Depending on the mechanical design of the mount and tube or truss and on how various error sources are modeled, one might choose to work in either alt-az or equatorial coordinates. Thus, a conversion from one set of coordinates to the other is usually required at some point.

6.3.3.1 Conversion Equations

Smart (1979, p. 35) gives the following equations for transforming (α,δ) into alt-az coordinates:

$$\cos Z = \sin\phi \sin\delta + \cos\phi \cos\delta \cos h \tag{6.8a}$$

$$\cos A = \frac{\sin\phi \cos Z - \sin\delta}{\cos\phi \sin Z} \tag{6.8b}$$

where $A =$ azimuth, measured westward from the South
$h =$ local hour angle
$\phi =$ observer's latitude
$\delta =$ declination
$Z =$ zenith distance $= 90°-$ altitude.

The numerator of (6.8b) has the numerator terms reversed in sign from Smart, who measures azimuth westward from North. Meeus (1982, p. 44) gives a mathematically identical form for azimuth:

$$\tan A = \frac{\sin h}{\cos h \sin\phi - \tan\delta \cos\phi}. \tag{6.8c}$$

Note that A can have a value anywhere in a complete circle (0°–360°), and that using either Equation (6.8b) or Equation (6.8c) by itself will result in confining the telescope to one quadrant of the circle. This is because to find A, the control program must compute the arccos or arctan, both of which have quadrant ambiguities. For example, if $\cos A$ has a value of 0.707, A can be 45° or 315° (−45°). If $\cos A = -0.707$, then A can be 135° or

225°(−135°). Meeus's expression using $\tan A$ has the same problem, with the additional problem of "blowing up" at 90° and 270°.

This problem of quadrant ambiguity can be solved by noticing that $\cos A$ and $\tan A$ have quadrant ambiguities shifted by 90°. That is, if both $\cos A$ and $\tan A$ are computed, the correct angle for A can be deduced from the following table:

		$\cos A$ +	$\cos A$ −
$\tan A$	+	$0° <= A <= 90°$	$180° <= A <= 270°$
	−	$270° <= A <= 360°$	$90° <= A <= 180°$

With Equation (6.8c) blowing up at $A = 90°$ and $A = 270°$, the computer program implementing this algorithm must check for overflows and underflows, and when one is detected, use this information to deduce the quadrant for the result from Eqn. 6.8b.

Another approach is to use direction cosines, which eliminates the quadrant ambiguity problem altogether. Taff (1991, pp. 2 and 13) converts the unit vector to an object in the sky (x, y, z) in (α,δ) coordinates to the same unit vector (x', y', z') pointed at the same point in the sky in (A, Z) coordinates using

$$\begin{bmatrix} x' \\ y' \\ z' \end{bmatrix} = \begin{bmatrix} \cos(90° - \phi) & 0 & \sin(90° - \phi) \\ 0 & 1 & 0 \\ -\sin(90° - \phi) & 0 & \cos(90° - \phi) \end{bmatrix} \times [x\ y\ z].$$

Substituting for the two unit vectors and simplifying the trig functions in the rotation matrix gives

$$\begin{bmatrix} \sin Z \cos A \\ \sin Z \sin A \\ \cos Z \end{bmatrix} = \begin{bmatrix} \sin\phi & 0 & -\cos\phi \\ 0 & 1 & 0 \\ \cos\phi & 0 & \sin\phi \end{bmatrix} \times [\cos\delta\cos\alpha \quad \cos\delta\sin\alpha \quad \sin\delta] \tag{6.8d}$$

where ϕ is the observer's latitude. The value of z' is used to find $\cos^{-1} Z$, then either x' or y' is used to find A. Care must be taken with (6.8d) in the Southern Hemisphere, where ϕ is negative.

6.3.3.2 Driving Rates

In contrast to the equatorial mount, which is driven in one axis only at a constant rate (excluding minor corrections) to track celestial objects, both axes of an alt-az mount must be driven at rates that vary over a wide dynamic range. Differentiating Equations (6.8a,b,c) with respect to h, then multiplying by $dh/dt = 15$ arc seconds per second of sidereal time gives

$$dZ/dt = 15 \sin A \cos \phi \tag{6.9a}$$

$$dA/dt = 15(\sin \phi + \cot Z \cos A \cos \phi) \tag{6.9b}$$

or

$$dA/dt = -15 \sin A \cos A (\cot h + \sin \phi \tan A) \tag{6.9c}$$

where Z, h, A, and ϕ are as before and the rates dZ/dt and dA/dt are in units of arc seconds/sidereal second. These equations give the rates for the topocentric place in the sky and a perfectly aligned telescope. As for the equatorial mount, you must first correct the telescope position Z and A given by the encoders for alignment, mount, and other errors in this chapter, then use the current and predicted values of Z and A to determine the amount of motor motion to perform in the time before the next pass through the software servo loop.

As an extreme example, suppose the azimuth axis is tipped through angle ϕ (your latitude) and aligned with the North pole so that you have an equatorial mount. This is as if ϕ is now 90°, so $\cos \phi = 0$, making $dZ/dt = 0$ and $dA/dt = 15''$ per second of sidereal time. But ϕ is a constant, and it appears to the control software that Z and A are changing at unusual rates. This will be discussed in greater detail in Chapter 13.

Note that as Z approaches zero, $\cot Z$ in Equation (6.9b) grows without bound, forcing dA/dt to do the same. Therefore, there is a cone around the zenith of angular size determined by the maximum azimuth drive rate within which celestial bodies cannot be tracked. Figure 6-10 is a plot of dZ/dt versus h and Figure 6-11 is a plot of dA/dt versus h for bodies of different declinations. From these plots it is seen that the drive rates needed to cover a useful fraction of the total sky vary considerably. This places severe requirements on the motors used in the drive, since they must be capable of delivering the required torque over a wide range of speeds. This is in contrast to equatorial mounts, in which motor speed requirements are far less severe, since motor speeds during tracking are relatively constant. To estimate the size of the "cone of exclusion" at the zenith, assume $\cos A$ and $\cos \phi$ are both 1 in Equation (6.9b). Notice that when $\cot Z$ gets large, $\sin \phi$ can be ignored. Equation (6.9d) then gives the motor's maximum speed (in units of arc seconds per sidereal second) as

$$\left. \frac{dA}{dt} \right|_{\max} = 15 \cot Z_{\min} \tag{6.9d}$$

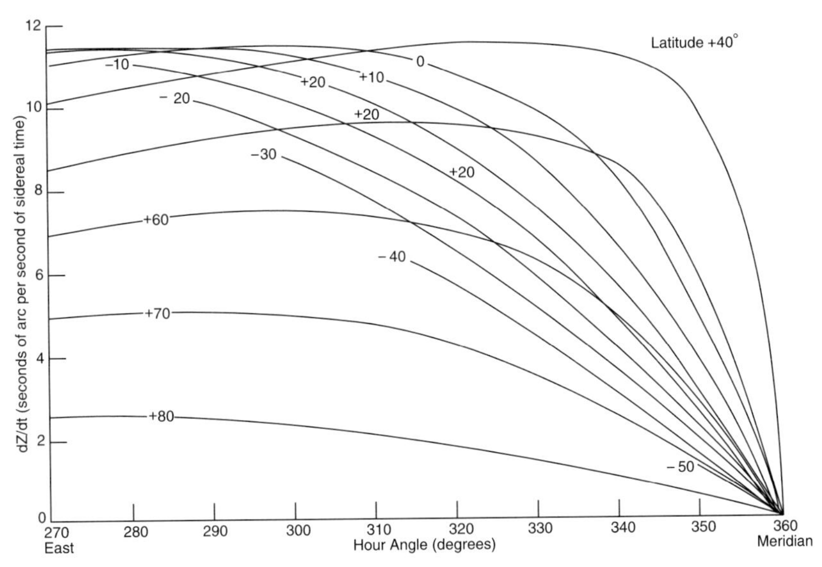

Fig. 6-10 *dZ/dt versus hour angle for various declinations. Courtesy Vincent A. Sempronio.*

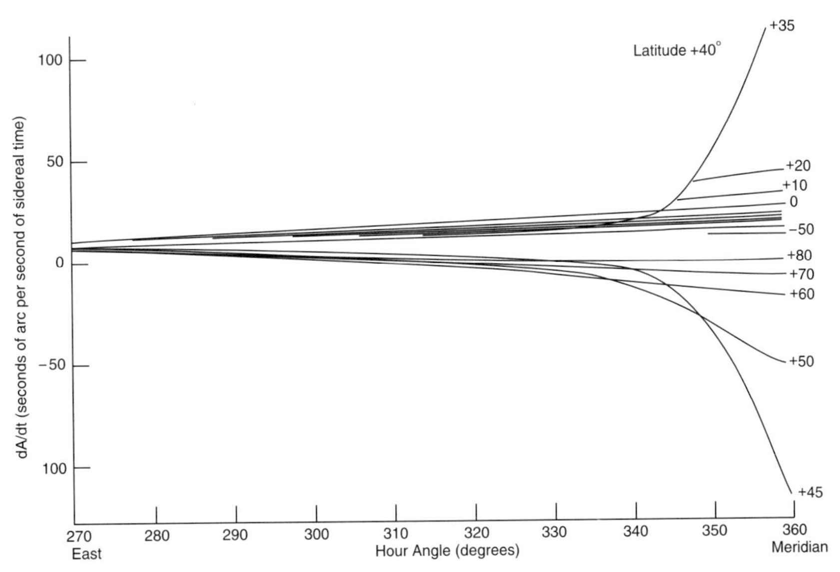

Fig. 6-11 *dA/dt versus hour angle for various declinations. Courtesy Vincent A. Sempronio.*

where $Z_{\min}$ is the angular radius of the cone.

6.3.3.3 Field Rotation Corrections

One final aspect of alt-az mounts is that the field of view rotates in the image plane as the telescope tracks a celestial body. This is because the image plane of an alt-az telescope stays aligned with the local vertical, whereas the image plane of an equatorial telescope stays aligned with the hour circle through the object being tracked. Since the stars in the field of an equatorial telescope are constant in right ascension, their relative hour circle alignments will not change. But the field as a whole in an alt-az telescope appears to rotate at the rate of change of the parallactic angle.

As shown in Figure 6-12, Smart (1979, p. 34) describes a spherical triangle whose apexes are the celestial pole (P), the zenith (Z), and the star (X). Angle XPZ is the hour angle h, angle XZP is the azimuth A (measured using Smart's method), and the remaining angle is the parallactic angle, the angle of interest (PXZ). Side PZ is the co-latitude ($90° - \phi$, where ϕ is the observer's latitude), side PX is the co-declination ($90° - \delta$), and side ZX is the zenith distance of the star. From the law of cosines,

$$\cos PXZ = -\cos XPZ \, \cos XZP + \sin XPZ \, \sin XZP \, \cos PZ.$$

Let η denote the parallactic angle PXZ. Then

$$\cos \eta = \cos h \, \cos A + \sin h \, \sin A \, \sin \phi. \tag{6.10}$$

The minus sign disappears from the first term when our convention of measuring A west from South is used.

Differentiating both sides with respect to h,

$$\begin{aligned} \frac{d\eta}{dt} = & \frac{15}{\sin \eta} \left[\sin h \cos A \left(1 - \sin \phi \, \frac{dA}{dh} \right) \right. \\ & \left. + \cos h \sin A \left(\frac{dA}{dh} - \sin \phi \right) \right] \end{aligned} \tag{6.11a}$$

where $d\eta/dt$ is in units of $''$/sidereal second. This reduces to

$$\frac{d\eta}{dt} = \frac{15 \cos \phi \cos A}{\sin z} \tag{6.11b}$$

with the convention that a positive value represents a *clockwise* rotation of the field despinner when looking at the Cassegrain focus for telescopes in the Northern Hemisphere. You must reverse the sign for telescopes in the Southern Hemisphere.

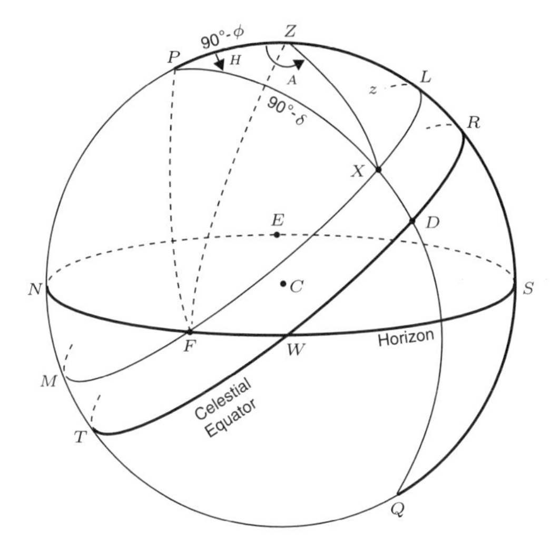

Fig. 6-12 *The geometry of the parallactic angle (PXZ). Courtesy Cambridge University Press.*

If one is doing simple photometry on a star centered in the field, field rotation can be ignored. However, if one is photographing the entire star field, a separate field rotation motor must be used to rotate the camera in synch with the rotation of the image. The rate for driving this motor, $d\eta/dt$, must be recomputed frequently enough to prevent field stars from smearing significantly on the plate. If the image field is 4 inches in diameter and stars must be kept stationary to within 0.0001 inches (2.5μ) on the edge of the field, the field cannot rotate more than about $10''$. The rates at which the image plane must be driven to counteract the field rotation are plotted in Figure 6-13.

6.4 Telescope Pointing Corrections—Alt-Alt Mount

6.4.1 Zero Offset

As with the other mount types, closed-loop systems using encoders must establish a zero offset for the conversion from encoder units to axis units for each axis.

6.4.2 North-South Axis Alignment

Alt-alt mounts can be aligned in any direction in azimuth, but to simplify the equations transforming (α, δ) to alt-alt coordinates, the axis supported by the piers is usually aligned along the local meridian in a North-

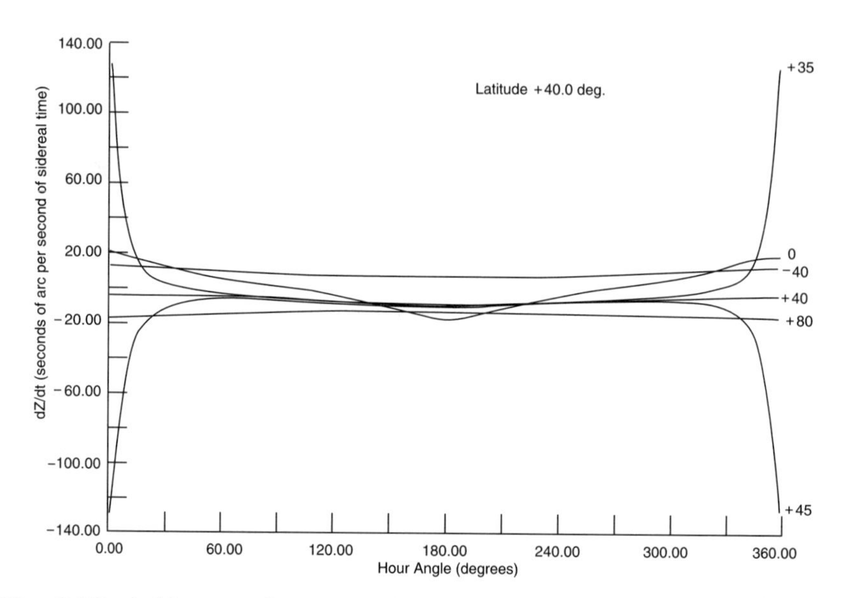

Fig. 6-13 *$d\eta/dt$ versus hour angle for various declinations. Courtesy Ronald McDaniel and Soft-Tec Systems.*

South direction (we use ξ to represent rotation angles of this axis) and parallel to the local horizontal, while the other axis is the "East-West" axis (χ represents angles in this axis), perpendicular to the N-S axis.

This situation is similar to the alt-az mount, so again referring to Figure 6-9 and assuming the two fork arms are aligned N-S, the N-S axis can be misaligned in two directions:

1. tipped away from horizontal through an angle a, and

2. rotated through angle b about an axis midway between the North and South piers perpendicular to the N-S axis.

To understand these effects, we define the Cartesian coordinate system (x, y, z) in which the origin is the intersection of the N-S and E-W axes. Coordinate x points East, y points North, and z points to the zenith in a properly aligned system. Misalignments can be represented as a rotation in x through angle a to yield coordinates (x', y', z'), then a rotation in z through angle b. These rotations can be represented as follows:

$$\begin{bmatrix} x' \\ y' \\ z' \end{bmatrix} = \begin{bmatrix} 1 & 0 & 0 \\ 0 & \cos a & \sin a \\ 0 & -\sin a & \cos a \end{bmatrix} \times [x\ y\ z]$$

and

$$\begin{bmatrix} x'' \\ y'' \\ z'' \end{bmatrix} = \begin{bmatrix} \cos b & \sin b & 0 \\ -\sin b & \cos b & 0 \\ 0 & 0 & 1 \end{bmatrix} \times [x' \; y' \; z']$$

for a combined rotation matrix of

$$\begin{bmatrix} \cos b & \cos a \sin b & \sin a \sin b \\ -\sin b & \cos a \cos b & \sin a \cos b \\ 0 & -\sin a & \cos a \end{bmatrix}$$

which is the same as Equation (6.2). Note that the (x, y, z) coordinate system for the alt-alt mount is different from that for the alt-az mount. For the alt-alt mount, the (x, y, z) coordinate system is:

$$\begin{aligned} x &= \sin\xi \cos\chi \\ y &= \sin\xi \sin\chi \\ z &= \cos\xi \end{aligned} \tag{6.12}$$

In a manner similar to the alt-az mount, you find angles a and b by working backwards using the residuals from multiple observations, then you apply a and b when pointing to new objects.

6.4.3 Equatorial to Alt-Alt Conversion

In a manner analogous to the alt-az mount, one must convert $(\alpha,\ \delta)$ to $(\xi,\ \chi)$ coordinates, determine correct drives rates for both axes, and use the correct rate to de-spin the image plane.

6.4.3.1 Conversion Equations

Dunham (1978) gives the following equations for transforming $(\alpha,\ \delta)$ into alt-alt coordinates:

$$\sin\xi = \sin\delta\cos\phi - \cos\delta\sin\phi\cos h \tag{6.13a}$$

$$\tan\chi = \frac{\cos\delta\sin h}{\cos\delta\cos h\cos\phi + \sin\delta\sin\phi} \tag{6.13b}$$

where

$\xi =$ rotation in the N-S axis measured westward from the East horizon

$\chi =$ rotation in the E-W axis measured from the “middle” position where northward is positive, and southward is negative h = hour angle

$\delta =$ declination

$\phi =$ latitude.

Pay attention to quadrant errors, just as with the alt-az mount.

6.4.3.2 Driving Rates

As with the alt-az mount, the drive rates for the alt-alt mount vary over a wide range. Dunham (1978) gives the following rates in units of arc seconds per sidereal second:

$$\frac{d\xi}{dt} = 15 \sin\phi \sin\chi \tag{6.14a}$$

$$\frac{d\chi}{dt} = \frac{15(\tan\phi \tan\delta \cos h + 1)\cos\phi \cos^2\delta}{\cos^2\xi}. \tag{6.14b}$$

6.4.3.3 Field Rotation Corrections

As with the alt-az mount, the field rotates at a changing rate that depends on where the telescope is pointing at the moment. The rotation angle deciphered from a GONG software listing provided by Harvey (1994) is:

$$\sin p = \sin\phi \cos\delta \cos h - \cos\phi \sin\delta$$

from which we derive

$$\frac{dp}{dh} = -\frac{\sin\phi \cos\delta \sin h}{\cos p}. \tag{6.15a}$$

Similarly,

$$\sin r = \frac{\cos\delta \sin h}{\cos p}$$

$$\frac{dr}{dh} = \frac{\cos\delta}{\cos r \cos^3 p}(\cos^2 p \cos h - \cos\delta \sin^2 h \sin\phi \sin p) \tag{6.15b}$$

and

$$\sin g = \frac{\sin r \sin\phi}{\cos\delta}$$

$$\frac{dg}{dh} = \frac{\sin\phi \cos\delta \cos r \dfrac{dr}{dh}}{\cos g \cos^2\delta}. \tag{6.15c}$$

The angle between the local meridian and the vertical in the field of the alt-alt telescope is given by

$$\rho = r + p - g$$

so the rate at which this angle changes in arc seconds per sidereal second is

$$\frac{d\rho}{dt} = 15\left(\frac{dr}{dh} + \frac{dp}{dh} - \frac{dg}{dh}\right). \tag{6.15d}$$

We leave it as an exercise for the reader to rederive these equations based on an arbitrary orientation of the axes, instead of the usual N-S/E-W orientation. These equations would be useful to someone building a trailer-mounted alt-alt telescope to align the telescope to get the best view of a particular target, instead of aligning the major axis along a N-S line.

6.5 Intrinsic Telescope Corrections

The previous two sections dealt with the orientation of the telescope mount on the face of the Earth. The corrections in this section are concerned with intrinsic features of the mount itself. Each correction is analyzed in the coordinate system which is most natural. Transformations from the coordinate system chosen for the analysis of the correction to the equatorial, alt-az, or alt-alt coordinate systems are not detailed for each correction, since these transformations have been discussed above.

6.5.1 Non-Perpendicular Axis Alignment

The alignment errors discussed above concerned the polar axis in an equatorial mount, the azimuth axis in an alt-az mount, and the N-S axis in an alt-alt mount. If the complimentary axis in each case is not exactly perpendicular, errors will result that require correction.

In an equatorial mount, assume $90° + p$ is the angle between the polar and declination axes, as shown in Figure 6-14. In Figure 6-22, if the telescope's optical axis crosses the equator at point C, the star is transiting the meridian at B, and point A has the same true declination as B and lies on the arc described by the declination axis, then angle ACB is p, angle ABC is a right angle, the side BC is the declination δ, and side AB is the hour angle correction Δh. Using Napier's rule,

$$\sin BC = \tan AB \tan(90° - ACB)$$

$$\sin \delta = \tan \Delta h \cot p$$

$$\tan \Delta h = \tan p \sin \delta.$$

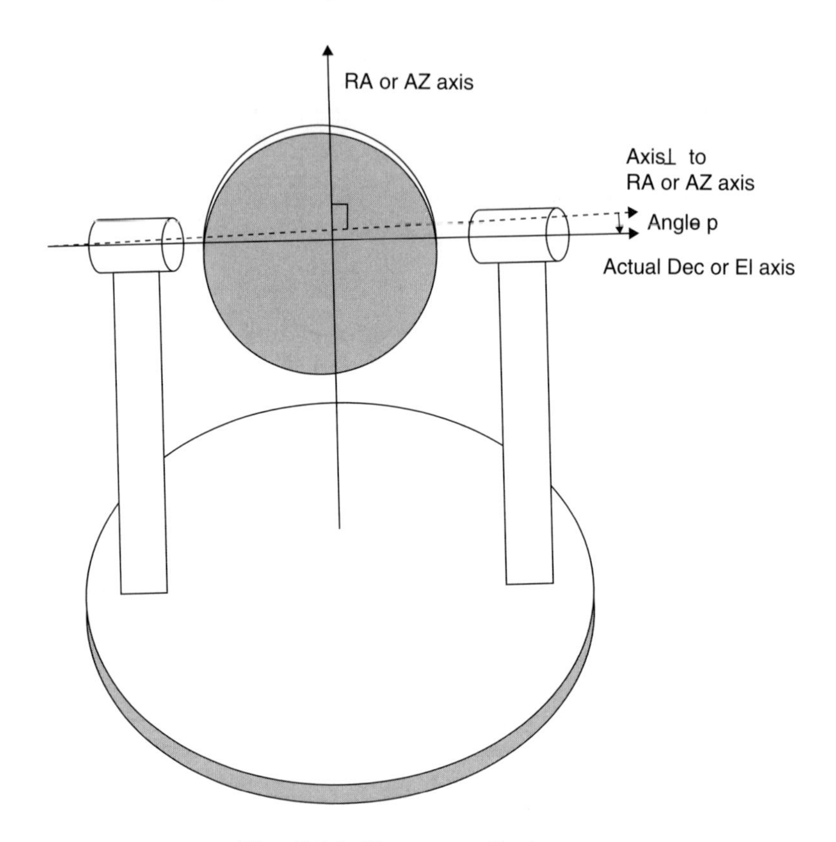

Fig. 6-14 *Non-perpendicular axes.*

Since Δh and p are small angles, hour angle positions can be corrected using

$$\Delta h = p \sin \delta. \tag{6.16a}$$

Since side $AC >$ side BC, there exists a point D on arc AC where DC is the declination δ. Again, using Napier's rule

$$\sin \Delta h = \cot p \, \tan \Delta\delta$$

$$\tan \Delta\delta = \tan p \, \sin \Delta h.$$

Since $\Delta\delta$, p, and Δh are all small angles,

$$\Delta\delta = p \, \Delta h. \tag{6.16b}$$

By analogy, for the alt-az mount

$$\Delta A = p \sin \text{(altitude)} = p \cos Z \tag{6.17a}$$

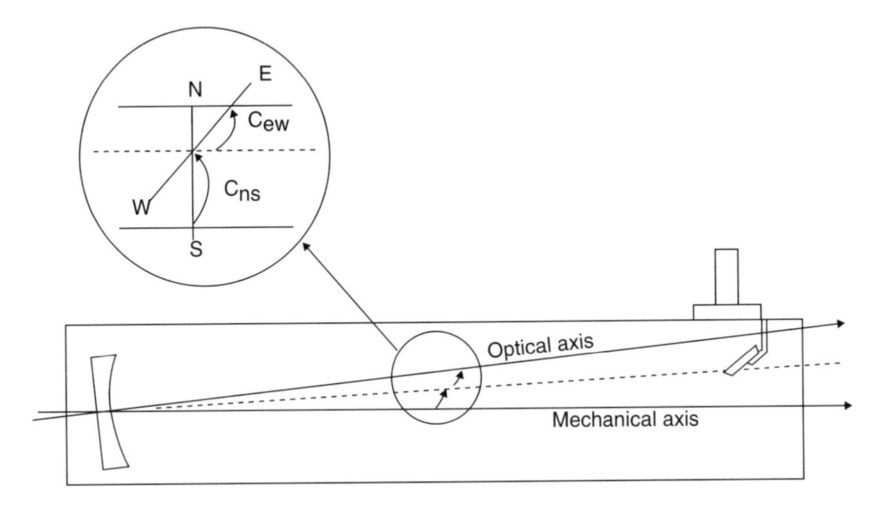

Fig. 6-15 *Collimation errors.*

$$\Delta Z = p\Delta A, \tag{6.17b}$$

and for the alt-alt mount,

$$\Delta\xi = p\,\sin\chi \tag{6.18a}$$

$$\Delta\chi = p\,\Delta\xi. \tag{6.18b}$$

6.5.2 Non-Alignment of Mechanical and Optical Axes

Collimation errors between the mechanical and optical axes have two sources:

1. a static component, due to simple constant offsets built in when the telescope was constructed and the optics aligned, and
2. dynamic components, which arise from flexure (droop) of the tube or truss, and of the mount.

The dynamic collimation errors are treated in the following two sections.

Static collimation errors are small constant angles affecting the hour angle and declination in an equatorial mount. As shown in Figure 6-15, to correct an East-West collimation error angle of $C_{\rm ew}$ measured on the celestial equator, Wallace (1975, p. 295) uses

$$\Delta h = C_{\rm ew}\sec\delta. \tag{6.19}$$

A North-South collimation error angle of $C_{\rm ns}$ is the correction to declination.

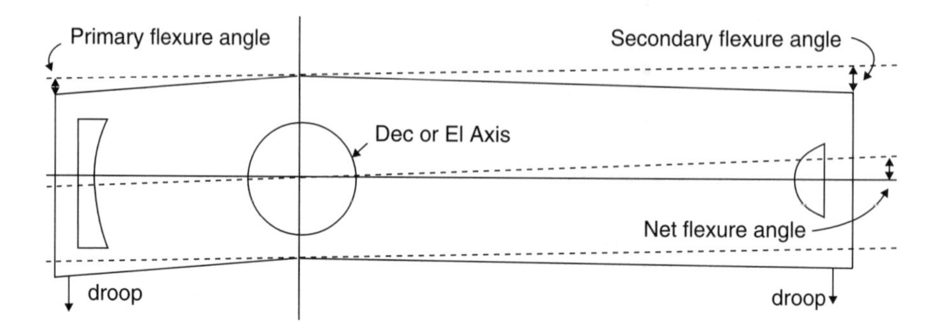

Fig. 6-16 *Tube flexure.*

In an alt-az mount, if C_{ew} is measured at the horizon,

$$\Delta A = C_{\mathrm{ew}} \csc Z. \tag{6.20}$$

A "North-South" (actually, zenith distance) collimation error angle of C_{ns} is the correction to zenith distance. The corrections for the alt-alt mount are analogous.

The formulation of these corrections is based on the assumption that the optical components of the telescope are in good alignment among themselves, and that the alignment of the optical axis as defined by the optical system as a whole is what accounts for the collimation error with the mechanical axis, which is defined by the bearings. If the optical components are not aligned among themselves, the result is usually visible as unduly large coma or other optical anomalies, which can be detected, diagnosed, and corrected relatively easily.

6.5.3 Tube Flexure

The cylindrical tube (on small telescopes) or the truss (on larger telescopes) that supports the optical components bends or droops due to gravity by amounts in zenith distance that vary with the zenith distance. The actual amount of the flexure depends on the masses of the truss members and optical components, the stiffness of the truss tubing material, the tubing diameter and wall thickness, and the geometrical design of the truss. Such a structure could be modeled crudely in real-time on a PC. For example, the model of the KPNO 4-meter Serrurier truss requires three seconds to execute on a CDC 6400 computer (Abdel-Gawad, 1969, p. A-1), and modern Pentium processors should not take much longer. But simpler equations exist that are entirely adequate.

Tube flexure is depicted in Figure 6-16. The table below (based on Abdel-Gawad, 1969, p. 33) shows the flexure of the upper truss (F_u), an empirical approximation of the upper truss flexure using

$$\Delta Z = F_u(90°)\sin Z, \tag{6.21}$$

the net flexure of both the upper truss (holding the secondary mirror) and the lower truss (holding the primary mirror) (F_n), and an approximation of the net flexure using (6.21) for F_n. All angular deflections are in arc seconds.

Z(deg)	F_u	$F_u(90°)\sin Z$	F_n	$F_n(90°)\sin Z$
0	0.000	0.000	0.000	0.000
15	1.190	1.188	0.159	0.096
30	2.390	2.295	0.210	0.185
45	3.240	3.246	0.260	0.262
60	3.980	3.975	0.320	0.320
75	4.410	4.434	0.330	0.357
90	4.590	4.590	0.370	0.370

This table shows that Equation (6.21) is useful in modeling net tube or truss flexure. Abdel-Gawad (1969, p. A-2) gives a listing of a FORTRAN program useful in predicting the value of $F_n(90°)$ for a Serrurier truss. Wallace (1975, p. 295) uses the same $\sin Z$ model, but quotes the secondary support flexure at $f/8$ in the Anglo-Australian telescope as $23''$ at the horizon. The large difference between the KPNO 4-meter deflections quoted in the table above and the AAT deflection indicates that the value of $F_n(90°)$ should be found for each telescope. This can be done using the program listed by Abdel-Gawad, although a direct measurement of tube flexure probably would be more accurate. To reduce the effects of residual errors from the calibration of the control system, $F_n(90°)$ should be minimized first by using good design and construction techniques, then the remaining flexure can be modeled. According to Abdel-Gawad (1975, p. 71), this is more accurate than relying on the computer to compensate for a bad design dynamically because the residual error varies with the size of the total error.

The effect of flexure is easy to compute for an alt-az mount. To compute the change in (h,δ) in an equatorial mount due to flexure,

$$\Delta h = \Delta Z \sin\eta \tag{6.22a}$$

$$\Delta\delta = \Delta Z \cos\eta \tag{6.22b}$$

where η is the parallactic angle found in (6.10). Therefore, Equations (6.8)–(6.10) and (6.21) must be evaluated for both equatorial and alt-az mounts, unless tube flexure is small enough to ignore. Flexure for alt-alt mounts is analogous.

6.5.4 Mount Flexure

When equipment of large mass is attached to the telescope mount and is not counterbalanced, or if the mount is not stiff enough to support the telescope mass properly, the mount will flex by a noticeable amount. Examples of mount flexure are shown in Figure 6-17. The amount and direction of flexure varies with the design and construction of the mount, and must be measured empirically for each telescope. This can be done by recording pointing residuals during observing operations, then analyzing them for variation with hour angle or declination.

Wallace (1975, p. 295) has determined by such an empirical method that the horseshoe flexure in hour angle h of the AAT is given by

$$\Delta h = -18''\!.9 \sin h \sec \delta. \tag{6.23}$$

Since the AAT is capable of at least $2''\!.5$ RMS pointing accuracy, the mount flexure must be modeled to obtain this level of accuracy.

Mount flexure is more of a problem in equatorial and alt-alt mounts, in which the force vectors due to the weight of the telescope tube assembly shift with changing hour angle. In alt-az mounts, the weight borne by each part of the mount does not change with telescope position, so mount flexure does not change. Any mount flexure is, therefore, embedded automatically in the shaft encoder calibration. The mount flexure should be modeled in equatorial and alt-alt mounts when high pointing accuracy is required, but it can be ignored in alt-az mounts.

6.5.5 Servo Lag Errors

In a linear one-term servo, the motor speed command on either axis is directly proportional to the error signal calculated for that axis. If a damping term is added (e.g., by adding velocity feedback) and the servo is over-damped, as most telescope control systems are, or if it is critically damped, the telescope approaches, but never equals, the desired position. This lag can be determined by measuring the servo performance (error signal) in response to a step function. The time constant τ of the servo can be measured, then used in the equation for an underdamped servo,

$$\text{lag error } = I\, e^{-t/\tau}$$

where I is the amount the motor speed changes at the instant the step function is applied, time t is measured from the time the new error signal is computed, and τ is the measured time constant. The constant exponential term to be used in the servo lag model is determined by using for t the

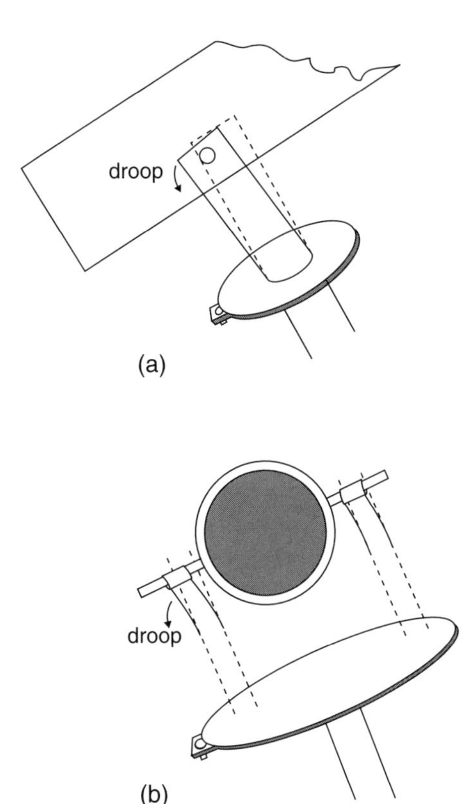

Fig. 6-17 *Mount flexure.*

time between error signal calculations (which are, presumably, performed at fixed time intervals), so $\tau' = e^{-t/\tau}$ is a constant. The servo lag is then

$$\text{lag} = I\tau' \tag{6.24}$$

where I is the motor speed correction before the servo lag error correction is applied.

6.5.6 Position Encoder Errors

The encoders used most frequently for high accuracy shaft angle position feedback in telescopes are optical or Inductosyns. Errors in the master pattern of either encoder type, or errors in depositing the pattern onto the encoder substrate, can cause random "tooth-to-tooth" errors of up to about 10% of the angular size of one encoder count.

These errors are usually small enough to be ignored. This is because optical encoders always give the correct integral number of counts per revolution. If the encoder has the intrinsic resolution to be connected directly to the telescope shaft without gear reductions, errors of up to 10% of the desired resolution are not important. If the encoder is connected to the telescope shaft through a reducing gear train, the size of the pattern errors is reduced by the same gear reduction factor as the angular size of an encoder count, so again, the error can be ignored.

Small sensor misalignments inside optical encoders typically do not produce errors. Absolute encoders use gray codes to prevent false readings. Incremental encoders often use two tracks for both count and direction sensing. The logic circuitry prevents false counts. If a sensor is in the wrong position, the encoder will not function. Therefore, if an optical encoder works, sensor misalignments can be neglected.

Other types of encoders may have a non-linear response curve (e.g., a differential capacitor or a resolver), or some other kind of error. The error modeling that is required depends on the device. If you have an angle position encoder that is not optical, find someone with an encoder more accurate than yours, and calibrate yours against his. A power spectrum analysis (a one-dimensional Fourier transform) of the resulting curve can be used to find the periodic dependence (if any) to use in modeling the encoder errors. A much simpler approach is to use encoders specified at or better than the required accuracy, to avoid the need to model the encoder errors.

6.5.7 Gearing Errors

There are many different types of gear errors, as briefly discussed below. The terms and equations are from the Gemini Project Cassegrain Rotator description dated 8 June 1994. All dimensions are in inches unless stated otherwise.

Center distance variation is a variation in the distance from the center of the gear to the outermost part of the tooth, that is, a variation in the radius of the gear (otherwise known as axial runout). If the gear teeth are involute (the space between gear teeth is the same form as the teeth, only inverted), then center distance variations do not affect indexing (position) accuracy of gears but they do affect backlash.

Tooth thickness variation is a variation in the width of each tooth compared to its neighbors. The indexing error in arc minutes is given by

$$e_{\text{tooth}} = 3428 \left(\frac{\Delta B_o - \Delta B_i}{D} \right) \tag{6.25a}$$

where ΔB_o and ΔB_i are the tooth thickness of adjacent teeth and D is the pitch diameter of the gear.

Pitch error is the difference in tooth spacing along the pitch line. The indexing error in arc minutes is given by

$$e_{\text{pe}} = 6875 \left(\frac{\Delta p_o - \Delta p_i}{D} \right) \tag{6.25b}$$

where Δp_o and Δp_i are pitch errors of adjacent gear teeth and D is the pitch diameter of the gear.

Involute profile error is the deviation of the tooth form from the true involute. The indexing error in arc minutes is given by

$$e_{\text{ip}} = 6875 \left(\frac{\Delta a_1 - \Delta a_2}{D \cos \varphi} \right) \tag{6.25c}$$

where Δa_1 and Δa_2 are the profile errors of adjacent gear teeth, D is the pitch diameter, and φ is the pressure angle (obtained from the gear's manufacturer).

Pitch diameter eccentricity is eccentricity of the pitch line of the gear. Its indexing error in arc minutes is given by

$$e_{\text{pd}} = 3438 \left(\frac{e \sin \Theta_e}{R} \right) \tag{6.25d}$$

where e is the eccentricity, R is the pitch circle radius, and Θ_e is the rotation angle about the rotation axis.

Lead angle error is a tilt in the face of the tooth away from being perpendicular to the gear plane (or nominal face angle). This error can cause an indexing error in arc minutes given by

$$e_{\text{la}} = 6875 \left(\frac{F \tan \lambda_e}{D} \right) \tag{6.25e}$$

where F is the face width of the gear, λ_e is the lead angle error, and D is the pitch diameter.

Lateral runout is caused by misalignment of the gear teeth to the axis of rotation. The indexing error in arc minutes is given by

$$e_{\text{lr}} = 6875 \left(\frac{F \sin \lambda}{D} \right) \tag{6.25f}$$

where F is the face width of the gear, λ is the tilt angle of the tooth, and D is the pitch diameter.

Figure 6-18 shows these gear shape errors. In addition, when a drive train consists of multiple gears, there is backlash when the drive gear reverses direction. Backlash can be reduced by using precision-cut split gears that are spring loaded. The other errors can be modeled if very high accuracy is needed.

In a closed loop system, if a shaft encoder is placed in the gear train between the motor and the telescope axis, any errors due to gears between the motor and the encoder can be ignored, but errors caused by gears between the encoder and the telescope axis appear in the graph of telescope axis position (as measured by calibration stars corrected to topocentric position) versus encoder reading. This is the encoder calibration curve. Since any errors that appear in this curve must be modeled in some manner, it is best to place the smallest possible number of gears between an encoder and the telescope axis, and to ensure there are no clutches or other means to detach the encoder from the telescope axis. The encoder gears should be marked before assembly and calibration so that upon subsequent disassembly, the calibration is not lost.

Periodic gear errors appear in the encoder calibration curve as sine terms, whose amplitudes are found empirically. If there are several gears in the drive train, a power spectrum analysis of the curve will show an amplitude for each gear's periodic error. To avoid complicating the gear error model, the number of gears in the drive train should be kept as small as possible. This approach also reduces the total backlash.

To model small high frequency, regularly occurring gear errors, one can keep track of gear positions and use the residuals in the calibration curve to detect the tooth-to-tooth gear errors. As in the case of tube flexure, the modeling accuracy depends on the size of the systematic error being modeled. Rather than try to implement software corrections for sloppy gears, a better approach is to use precision gears, especially in the stages of the drive gearbox closest to the driven telescope axis, since the errors in gears closer to the motor are divided down by the larger gear ratio between a given gear and the telescope axis. If the power spectrum of error residuals still shows peaks at the gear rotation frequencies, a simple sine wave gear error model can be implemented.

The best approach is to avoid gear errors altogether by using a "toothless" reduction, that is, a friction drive. Such drives are still subject to periodic errors, but they are small and of low frequency, because the high frequency errors (e.g., small bumps on the roller or the driven segment) tend to be ground down over the years with use. Friction drives have been used successfully on telescopes up to 3.5 meters in aperture, but may lack the stiffness required in some applications.

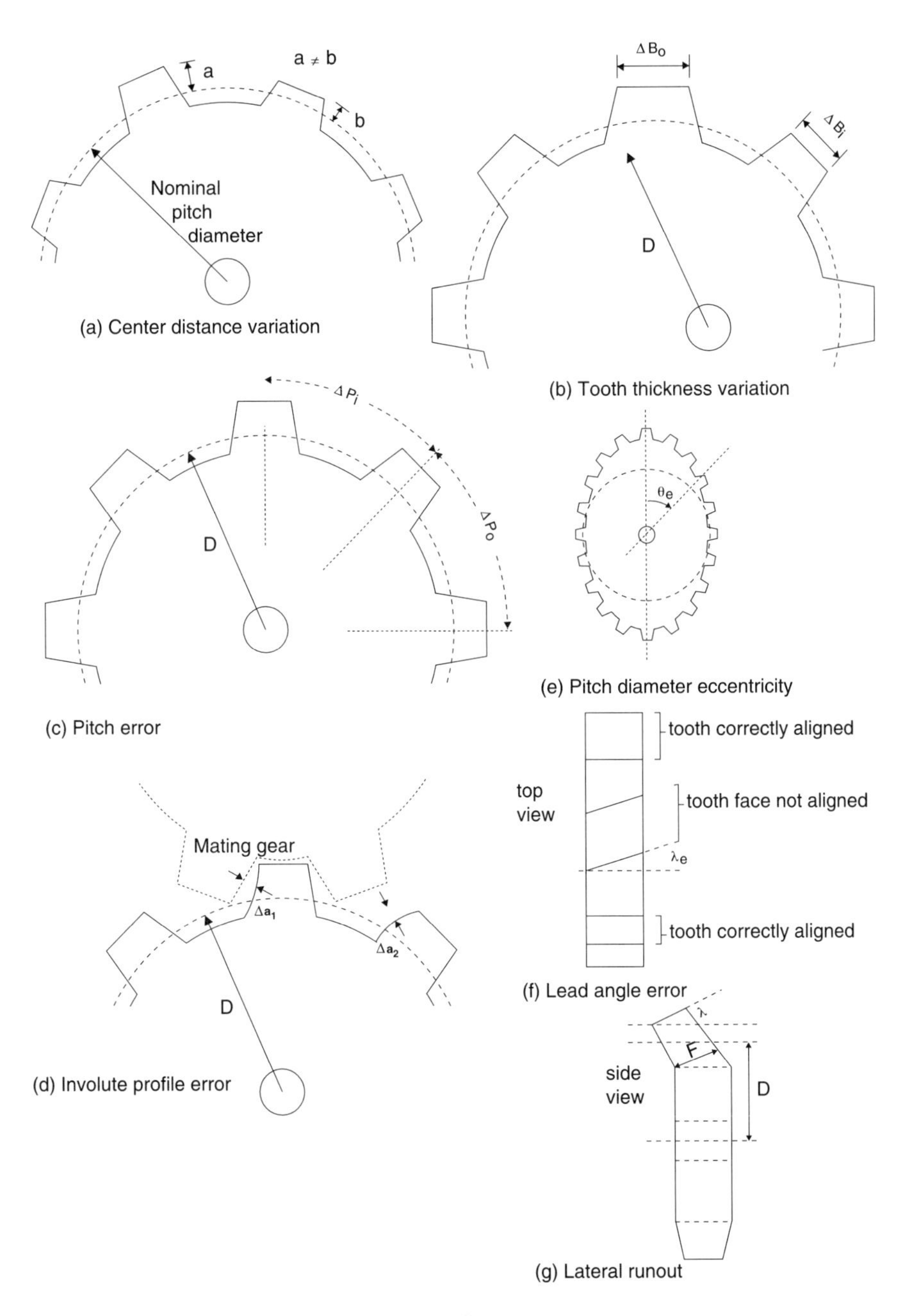

Fig. 6-18 *Gear errors.*

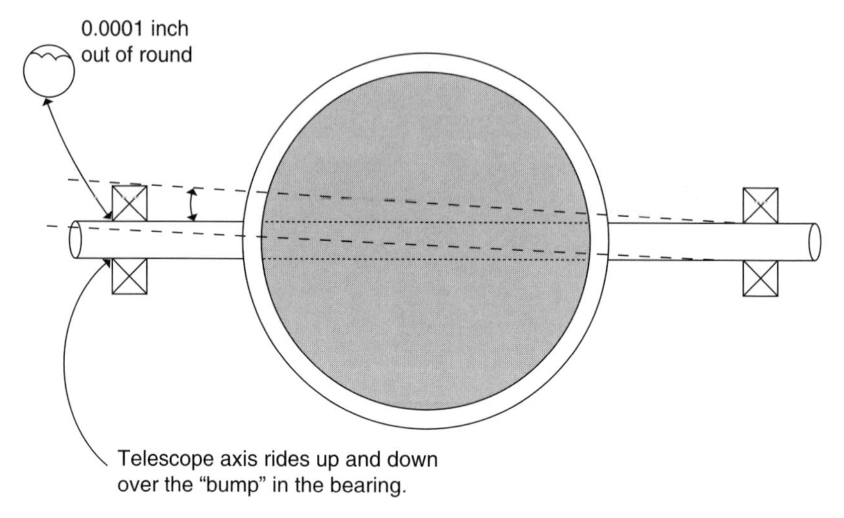

Fig. 6-19 *Bearing errors.*

6.5.8 Bearing Errors

Bearing runout has the effect of periodically changing the gear reduction slightly between the motor and the driven shaft. The tolerance on ball or roller bearing race runout on good bearings is typically held to about 0.0001 inches or less. As shown in Figure 6-19, if one altitude bearing in an alt-az mount has a runout of 0.0001 inch and the other bearing has the same runout in the opposite direction, and the two bearings are separated by 40 inches (e.g., a 30-inch telescope), the runout produces an apparent non-perpendicularity between the altitude and azimuth axes of about $1''$ that can appear to change direction as the shaft is rotated. From Equations (6.17), this results in a pointing error that grows larger at small zenith distances. If the runout is in a direction parallel to the azimuth axis, it is indistinguishable from non-perpendicularity between the axes. If the runout is in a direction perpendicular to the azimuth axis, it is indistinguishable from a zero offset in azimuth. Runout in any other direction can be resolved into components perpendicular to and parallel to the azimuth axis. Either way, the altitude bearing runout errors are a periodic form of non-perpendicularity and zero offset errors.

Runout in the declination bearing of an equatorial mount is treated much the same as in the altitude bearing, except that Equations (6.16) are used in modeling the runout that looks like axis non-perpendicularity. Runout in the right ascension bearing is a periodic form of polar axis misalignment.

Only the runout in the bearings supporting the principal telescope axes has any large consequence. Bearing errors in the drive train produce very small telescope pointing effects.

Bearing errors are dynamic, in that the axis running through the bearing exhibits the error only when a ball or roller is travelling over the "hill" in the race that has the runout. These errors are small and difficult to isolate, so they probably aren't worth worrying about until larger errors have been modeled successfully. As in the case of gearing errors, a power spectrum analysis will indicate what periodic terms and coefficients to include in the model. An accurate dial indicator will help in determining what fraction of the residuals is due to bearing errors.

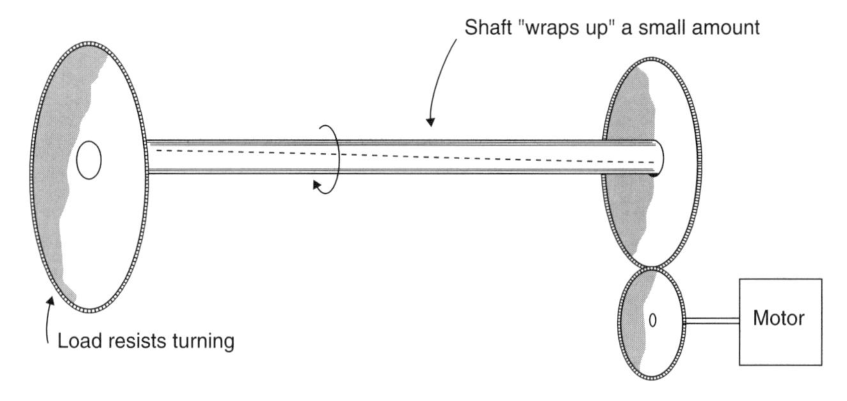

Fig. 6-20 *Drive train torsion errors.*

6.5.9 Drive Train Torsion Errors

If one end of a rod of length L is held fixed and a torque T is applied to the other end with the direction of the vector T along the longitudinal axis of the rod, the rod will twist through angle Θ according to the formula

$$T = KL\Theta \tag{6.26}$$

where the coefficient K accounts for the stiffness, cross sectional area, and diameter of the rod. This is shown in Figure 6-20. Equation (6.26) is the expression for a Hook's law relationship, provided the torsion angles remain within the rod's elastic limit. Since the gears in a telescope's drive train amplify the motor torque to very large values, there is a (variable) difference between a gearbox shaft angle position and the telescope axis position produced by the mass of the telescope and the torque of the motor, as amplified by the gearbox. The amount of twist in the drive train gear shafts will vary, especially in alt-az mounts, in which the motor speeds vary considerably. This is because a motor's torque decreases rapidly with increased speed, and the motor acceleration varies with time.

This source of error can be reduced to tolerable levels simply by using short thick drive shafts with very high K coefficients and by placing an encoder as near to the telescope axis as is possible. This error can also be controlled by using a separate gear system to drive the encoder from the telescope axis. This latter approach also has the advantage of allowing one to choose convenient gear ratios for both driving the telescope axis with the motor and reading the telescope axis position using the encoder.

6.6 Reducing the Effects of Systematic Errors

The easiest way to eliminate some of the errors discussed in the preceding sections is to provide adjustments on the telescope to counteract them. This works for static errors, such as position alignment errors (e.g., polar axis alignment on the refracted pole), axis perpendicularity errors, and collimation errors, but you cannot compensate for dynamic errors, such as flexure, in this manner. For smaller telescopes and in applications not requiring high accuracy, compensating for errors with mechanical adjustments is the best approach. There is no point in burdening a small computer with software corrections that can be made unnecessary with a few simple adjustments. However, it is difficult to provide adjustments in larger telescopes that do not detract from the overall stiffness of the mount, and it is easier to measure an error to a certain accuracy than it is to make a compensating mechanical adjustment to the same accuracy. Therefore, in larger telescopes, and in applications requiring high accuracy (say $30''$ pointing accuracy or better), the preferred approach is to model the errors in software.

6.6.1 Mechanical Adjustments

The first mechanical adjustment is to align the major axis (e.g., the polar axis in equatorial mounts). Until this is done, measurements of the other errors do not make sense. One way of aligning the polar axis is to use the "drift" test. In this procedure, a star near the meridian and celestial equator is centered and tracked, with only RA tracking adjustments being allowed. The eventual movement in Dec suggests which way the azimuth of the mount should be changed. A similar procedure can be applied to a star also on the celestial equator, but near the eastern horizon instead of the meridian for an indication of mount elevation adjustment. For highly accurate setting, this procedure is a bit slow, as it requires long drifts to achieve high accuracy. Figure 6-21 depicts polar axis alignment errors.

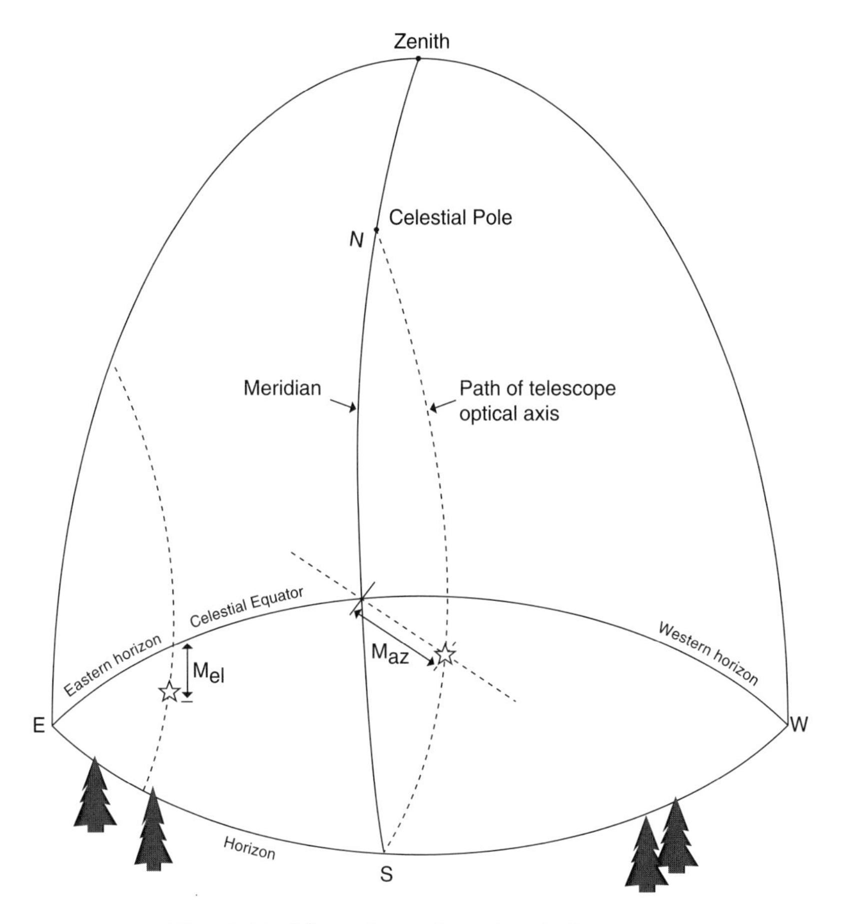

Fig. 6-21 *Measuring polar axis misalignments.*

For computer controlled telescopes, a modification of the drift test can be used that is much faster and quite accurate. This method was used to align the DFM Engineering mount on the Automatic Photoelectric Telescope at Fairborn Observatory East. The mount is first rough aligned using whatever convenient methods are at hand. The scale factors relating steps or counts to movement in RA and Dec are then determined. A star of known position located near the celestial equator and meridian is centered on the crosshairs and its position (current epoch) is entered. The telescope is then commanded by the computer to move to a star near the pole and on the meridian (within 20° of the pole and a few degrees of the meridian is adequate). When the telescope stops, presumably somewhat away from the star, it is centered using the Dec control on the paddle and the azimuth adjustment of the mount. This is then repeated using a star on the equator near the eastern horizon, and the star near the pole is chosen with a similar RA. After moving to the star near the pole, the star is centered using the Dec paddle control and the elevation adjustment of the mount. This can be repeated a time or two if desired.

The errors that remain after polar alignment tend to be mixed together. If pointing residuals are carefully analyzed, errors proportional to RA or Dec, or a trigonometric function of RA or Dec, can be found. For example, Figure 6-22 shows how to measure pointing errors related to non-perpendicularity of the RA and Dec axes. As given in Equation (6.16a), this causes errors in RA proportional to the sine of the declination. In Chapter 13, Trueblood discusses his method for error correction in the example system.

6.6.2 Compensating for Mechanical Behavior in Software

Although mechanical adjustments can be made in small telescopes to correct for static errors, in those applications requiring dynamic error compensation, or in large telescopes that do not permit mechanical adjustments, error correction can be performed in software. Kibrick (1984) describes a program at Lick Observatory to use software to compensate for a periodic error in the RA worm gear, and to damp low-frequency telescope vibrations of the 120-inch telescope at Lick. Figure 6-23 shows the error in the RA worm gear appearing every two minutes (of time). An optical encoder was attached to the large worm gear using a friction drive. When the encoder count read by the control computer indicates that the spot on the worm that has the error is about to enter the tooth mesh with the worm gear, the software sends commands to the RA motor to change speed to compensate for the error.

Figure 6-24(a) shows the other problem—excessive ringing of the truss structure due to sudden starts or stops, or wind gusts. Although better mechanical design could have reduced this problem to a tolerable level, a new truss is too expensive on a telescope this large. A less expensive solution is to compensate for the low-frequency ringing in software. The pattern of ringing was carefully recorded and analyzed, to ensure that an accurate model of it could be embedded in the software. When a command is given that would produce mechanical ringing, the software issues commands to the motors to produce motions to counteract the ringing, according to the software model. The model is stored as a table of constants that are used in an equation (developed empirically) to compute the correction. Results

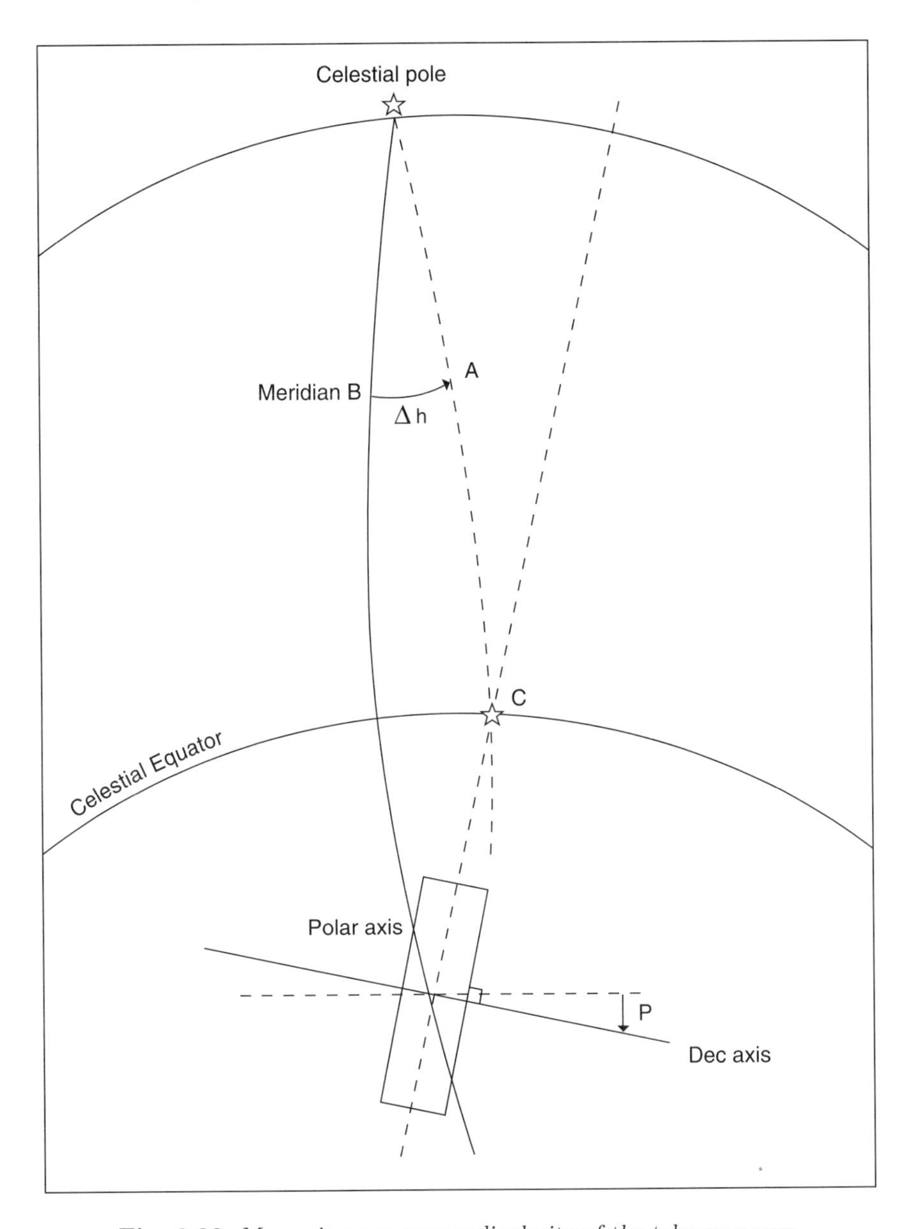

Fig. 6-22 *Measuring non-perpendicularity of the telescope axes.*

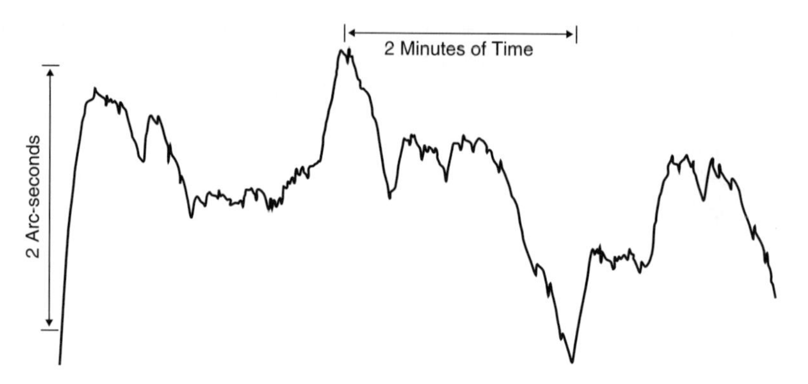

Fig. 6-23 *Tracking error vs. time in the Lick 120-inch telescope. Courtesy Lick Observatory.*

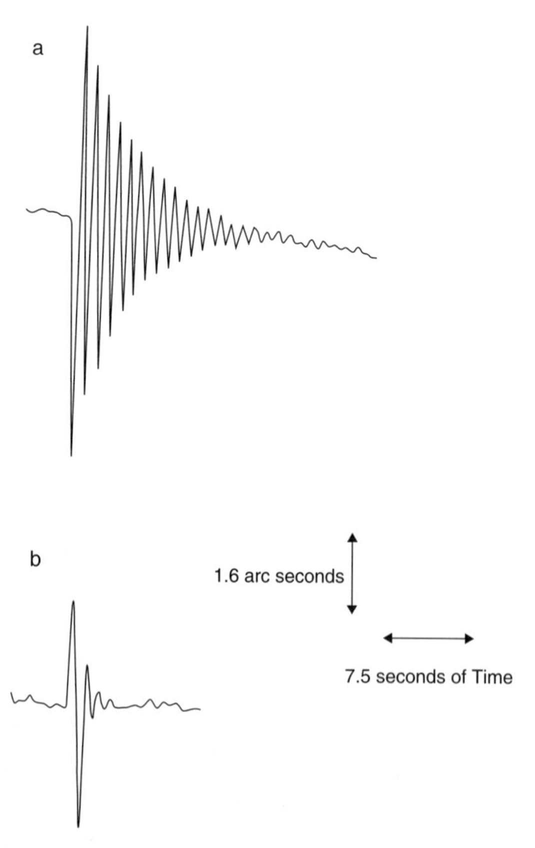

Fig. 6-24 *Telecope vibration compensation in software. Courtesy Lick Observatory.*

of this effort are shown in Figure 6-24(b). Both the initial excursion and the time it takes for vibrations to damp are reduced considerably, making the telescope more compliant to commands issued by the computer, and better able to handle a series of short quick motions.

This concludes our discussion of the theoretical aspects of telescope control system design. The next chapter is devoted to good design techniques you should use to make life easier for you.

Chapter 7

Practical Design Considerations

In this chapter, some of the practical considerations involved in the approach taken at the Fairborn Observatory to telescope control are described. These are matters that ended up being of concern to us, and we pass them along to you as things worth considering. We must admit that in many cases, we only became concerned about these after we had learned the hard way about the serious consequences of unconcern! Many of the ideas developed in this chapter also appeared in a *Byte* magazine article on real-time control (Genet, Boyd, and Sauer, 1984). We thank the editors of *Byte* for permission to exerpt from the original material.

7.1 Operator Convenience

Although some computer-controlled telescopes are intended for only fully automatic operation, most control systems will be operated by an astronomer through the computer. A good control system should be easily operated by an average astronomer without specialized training. It is tempting to have lots of dials, knobs, and pretty red lights, but simplicity is preferable. It is also tempting to use specialized computer terms and jargon, but plain language oriented to the operation of the telescope is more appropriate. It is not easy to develop a good astronomer/telescope control system interface, so if the real intent of a project is total automation, the equipment and project can be simplified somewhat if no attempt is even made to provide for a human observer.

The first data-logging and photometer-control systems at the Fairborn Observatory (Genet, 1980, 1982) used a hand-held hexadecimal keypad and remote video monitor for all communications between the astronomer and computer. All instructions and requests were displayed on the monitor in plain English, and the operator did not even need to know that a computer was involved. While a telescope can be controlled from a hexadecimal keypad, it is so unlike the traditional telescope control paddle that significant relearning must take place. Also, it is questionable if stars can be moved into position and centered as quickly with a hexadecimal keypad as they can be with a more traditional control paddle. For these reasons, a normal control paddle should be used in computerized telescope control systems. As a control paddle is not useful for entering numeric data, a separate numeric keypad may also be appropriate.

In some cases, it would be possible to place the computer and its terminal/keyboard inside the observatory dome for the astronomer to enter responses directly. However, not only is the observatory not a hospitable environment for electronic equipment, trying to look through a telescope while cradling a computer in your arms (or running across the room to enter each response) is not conducive to productive observing! The way to make it most convenient for the operator is to have a fully automatic control system that operates the telescope while the astronomer sleeps. For the "part time" astronomers that have a regular job during the day, a telescope that totally operates itself is convenient indeed!

7.2 Hardware/Software Tradeoffs

There is a natural, but often expensive and occasionally disastrous tendency to make the hardware as simple and inexpensive as possible, and to place the burden on the software. The hardware costs are both immediate and obvious, but the software development and life cycle costs are off in the future and less obvious—especially to the optimists among us. Many real-time control projects have been abandoned in failure when it finally became clear that the software programming would take an order of magnitude more time and money than originally anticipated.

Removing the burden from software development and placing it on the interface hardware is what standardized buses with plug-in slots are all about. Such bus-based computers can initially cost more than those without standardized slots for expansion. However, this investment can open up the world of ready-to-go, plug-in interface cards. These cards come fully assembled and debugged. The better ones have excellent documentation and have been "burned in" (tested with full power applied for several hours to weed out bad components). Cards for popular computer buses often come with device drivers or software libraries ready to install with whatever operating system is most popular with that bus.

Many of these interace cards are highly intelligent, with their own built-in processors and machine language programs in EPROM. Some are programmable by your PC using software that comes with the card. The software work has been done by the card manufacturer, and its cost spread out over hundreds or thousands of users. Smart interface cards are often used by control system designers to do the fast, repetitious, and mundane tasks, such as controlling stepper motors and reading shaft angle encoders. By delegating these standard tasks to the off-the-shelf smart cards, it is often possible to program most or all of the "custom" software in a convenient high-level language such as BASIC, and in some systems, the complexities of interrupts may be reduced, or the need for interrupts may even be eliminated entirely. By picking the right bus and making a careful selection from the off-the-shelf interface cards available for this bus, the cost and time spent in custom software development can often be reduced by an order of magnitude. This can be the difference between the success and failure of an entire project.

As chips have become more capable over time, it has been possible to pack an increasing number of functions in a given amount of space. A number of manufacturers realized that by placing an entire computer on a board along with the chips needed to implement most real-time control functions, that some very low-cost control systems would be possible. If one of these single board control systems will do the job for you, and future flexibility is not critical, then there will not be a cheaper solution. The second generation Fairborn Observatory APT used such a single board computer and non-bus interface. It is almost an unbelievably simple system from the hardware viewpoint, suggesting that under some circumstances, a non-bus approach truly has merit.

For an example of hardware/software tradeoffs, consider the computer control of stepper motors. At one extreme, the computer can calculate the bit pattern needed to drive each individual phase of each of the stepper motors. Typically, this function would be implemented in machine language and embedded in a larger BASIC or C program. To slew the telescope, it is necessary to "ramp" the steppers up to a high speed, and then ramp them down again. A chip or two added to the hardware can simplify the software considerably. For instance, Hurst Mfg. makes a one-chip stepper controller/driver, with TTL pulses for stepper motion and direction on the input, and a direct connection to small steppers on the output. Sigma Instruments makes a two-chip set (which is used with a few discrete components) that takes a pulse stream of some length as the input, and provides a ramped pulse stream as the output.

At the other smart extreme, there are intelligent stepper controllers, of which the Whedco STD card is a prime example. From a high-level language such as BASIC (or from a low-level language), the Whedco stepper controller can be told the start speed for a stepper, the ramping acceleration, top velocity, and initial position. Anytime thereafter, a simple command, such as to move 1,235,476 steps CCW, will be executed exactly, including the ramp up, slew, and ramp down—bringing the stepper to a smooth stop exactly where desired. Both hardware and software flags are available to signal "movement complete." An onboard register keeps accurate track of the total motion, and hence the current position. Unramped moves and single steps can be commanded. In-motion stops, either ramped or immediate, can be commanded at any time. While the Whedco controller has an onboard Z-80 and its own program in EPROM, all this is totally transparent to the user, and the controller just appears as an output (and input) port to the user to which the high-level commands are issued.

The Fairborn Observatory APT's use two steppers, one each on the right ascension and declination axes. By allowing the telescope to go in only one of eight directions at any one time, and breaking up the move between two stars into two separate segments (with the telescope coming to a momentary stop between the segments), it is possible to get by with somewhat simpler software. Here is a case where the overall concept (telescope movement) was made slightly more complex, and the total system efficiency was reduced (imperceptibly) by adding a "waypoint" stop in every move—all to simplify the hardware and software.

In the end, your tradeoffs should reflect your talents and capabilities. If your strength is in hardware and you are weak in software, then you ought to try to minimize the software task as much as possible. If you are more at home developing software, try to simplify the hardware task by using commercial electronics as much as possible, and use software to "glue" the system together. Design your system to highlight your strengths and to minimize your weaknesses.

7.3 Single Board Computers and Buses

Selecting the right bus-based or single board computer for any given real-time control task is relatively straightforward if prejudices not related to real-time control can be set aside. If the task is reasonably well defined and will not be changing in the future, if a number of essentially identical systems are to be made at low cost, and if a single board computer will do the job, then use it.

The single board computer used in early versions of the APT is based on the Motorola 6809 and is made by Peripheral Technology in Marietta, Georgia. Fred Brown, president of this small company, says that his philosophy is to put everything needed on one board and to make it at the lowest cost. What this board has is a 6809E processor, 56KB of RAM and 4KB of EPROM, a floppy disk controller, two serial and two parallel ports, and a real-time clock. If this meets your needs, then for less than $300 you can get a fully assembled and tested system, but if you need more, then you might consider a bus-based system with plug-in slots. For convenience, buses will be discussed in three wide categories: hobby, personal computers, and industrial control. Hobby buses would certainly include the venerable and ever popular S-100, which has done much to make the 8080 and Z-80 processors well known, and the SS-50 bus which remains the favorite of the 68XX fans. Categorization as a "hobby bus" simply means that besides the fine complement of ready to go commercial systems and cards, these buses are patronized by those hardy individualists, the home brewers. Personal computers with a wide selection of plug-in cards for their slots are really limited to the Macintosh and the IBM-PC and their very close clones. If the real-time control task at hand is not overly difficult, or if one of the systems mentioned above is available and adequate to the task, then it makes a lot of sense to use it. Honeycutt (1977) developed one of the first computer control systems based on the S-100 bus, while Tomer controls his mobile telescope with a PC. Skillman (1981) developed an automatic telescope that was controlled by a combination of Apple II and KIM computers.

If the real-time control job is a tough one, or if one desires to tap the full potential of what is available off-the-shelf for real-time control, then it is necessary to turn to the buses used by industry in their control systems. Good examples of industrial buses are the STD Bus, VMEbus, and Multibus.

The STD Bus has a large number of different manufacturers (about 100) and a wide selection of different cards (nearing 1000). The card selection is particularly rich in the area of such real-time control tasks as smart stepper controllers, optical encoder interfaces, and DC servo motor interfaces. While there are 16-bit STD systems, this bus is primarily an 8-bit bus that favors the Z-80 microprocessor and the CP/M operating system. For greater crunching power, the VMEbus and Multibus are good choices. The VMEbus, an open architecture first developed by Motorola and now supported by hundreds of manufacturers, is used extensively in professional observatories for a wide variety of control applications.

7.4 Hardware Approaches

There are essentially three approaches that can be taken to the hardware. First, commercial off-the-shelf electronics can be used in all or most of the system. This usually implies that a bus approach be used—preferably a bus with a good selection of real-time control cards (such as the PC/AT or ISA bus, STD, or VMEbus). This approach has a number of advantages. First, fully developed, tested, and functioning cards can be used in the system. Second, commercial cards that have been in use for some time tend to have their "bugs" worked out, and they are more reliable than almost any new "first time" cards. This is especially so if the cards are from a well-known supplier of cards to industry (cards built for the hobby market are not necessarily as reliable). Third, such cards usually come with rather complete documentation—obviating the need to write your own documentation. Fourth, for "smart" off-the-shelf cards, the software is already written and stored in ROM—thus additional effort need not be invested to program the intelligent card. Finally, using off-the-shelf commercial cards can often save time, money, and grief. It is difficult to design, build, test, debug, and document a board cheaper than one can be bought. This is due, of course, to the economies of scale inherent in commercial production.

The second approach that can be taken to hardware is to fabricate to an established design. This saves design time, and can reduce, but not eliminate, the debugging and documentation time. This approach is reasonable when no commercially available card exists for a critical function. Sometimes, however, one is forced to do the best with what is available, and fabricating to an established design makes some sense.

The third and final hardware approach is designing, building, testing, debugging, and documenting cards entirely from scratch. When, for some critical function, no commercial card is available, and no design is available, then one has no choice but to go through the entire process. Plenty of time and money should be set aside to do a good development job, as otherwise the performance and reliability of the total system will be jeopardized because of one poorly done card. For non-bus systems, making custom cards is likely to be the rule. If the design is done very carefully (such as that done by Boyd, Genet, and Slonaker), then this approach can work well, as exemplified by the current Fairborn Observatory APT control system, which resides on a single printed circuit (pc) card. If not thought out well, it quickly becomes a nightmare.

7.5 Interface Software

The complexity of software in telescope control systems makes the approach to software development of critical importance. Methods of reducing the software complexity with good system design and the use of appropriate hardware have already been discussed. By selecting the right development and run-time environments, the effort of building and debugging the software that remains in the system can be minimized.

Assembly language has its uses. An excellent use in real-time control is by the manufacturers that put it in the EPROMs on the smart interface cards. On rare occasions, a high-level language has a task to do that is simply beyond its speed capabilities, though that is rarely a problem with modern computers. The primary use for assembly language today is in coding device drivers, parts of dynamic link libraries, or new operating system services that must become part of the operating system instead of being a regularly scheduled task.

Really good high-level languages are a pleasure to use. The Fairborn Observatory has gravitated towards the use of two BASIC languages. On our Z-80 based CP/M systems, we ran Microsoft MBASIC in both its interpretive and compiled versions. Programming in the interpretive version was convenient, and if greater speed was needed, we could always use the compiled version. MBASIC and all its look-alikes are among the most widely used computer languages in the world, and thus need no description here.

Our current 68000-based systems use Microware's "Basic09" BASIC interpreter running under the OS9 operating system. For the true high-level language control system aficionado, this language is such a delight to use, that a short digression to mention its origins and salient features is in order. When Joel Boney and Terry Riter at Motorola started the design of the 6809 microprocessor, they contracted with Ken Kaplin at Microware to develop a language and operating system that would fully utilize the many unique and valuable features of this most advanced of the 8-bit microprocessors. What Microware came up with was a language that supported structured programming (sort of a cross between BASIC and Pascal), that was completely modular (allowing independent programs to work together), and that included an interactive compiler with very high-level debugging functions. The ability to develop the software for a complex control system as a number of independent programs has proven to be very valuable.

Having selected system and hardware approaches and an operating system and language, software development can begin once sufficient hardware is on hand. Just as an evolutionary approach to the hardware makes sense—starting as simple as possible and adding the frills later—so it makes sense to start with the simple and move to the more complex with the software. Breaking the program into little modules, where each module accomplishes a specific function, makes the entire task much easier. Each function can be checked out while the program is fresh in one's mind and the bugs worked out right away. It is sometimes helpful to write a small, highly simplified program on the side to check out the major features of some function, e.g., to see if the equipment does what you expect it to. Unlike normal, non-real-time programming, programming a control system usually requires that the control system itself be in operation, that a logic probe and circuit schematic diagrams be close at hand, and the expected effects of the software on the hardware be constantly checked.

Sometimes it is possible to check out some specific function with just a small part of the system in operation, or even with a special piece of equipment used just for checking software. For instance, in the Fairborn Observatory 0.4-meter telescope control system, two small steppers and short connection cables were used to check out the software as it was written. For example, a command in declination, instead of moving the telescope, could turn a little stepper right beside the terminal.

Besides having special cables and peripheral equipment just for software development, it is quite helpful if a second and identical computer can be set aside for development of the software and hardware while the primary computer stays hooked up to the telescope and is used to control it. Later on, the second computer can be used for data analysis, word processing, and as a backup to the primary computer.

7.6 Adaptability

No matter how carefully you plan out your system on paper (and it should be most carefully planned!), there will be changes—usually quite a few of them—before you get the system operational. No matter what the configuration of the system is when it first becomes operational, it will be modified later (sometimes extensively so). This suggests that the prudent thing to do is to plan on change from the very beginning so that changes can be made as gracefully as possible.

One approach to having an adaptable system is to use a standardized bus. If some of the hardware doesn't work, only a card's worth needs to be tossed. As suggested earlier, if you are making your own card and a commercial one is available, then going with a bus allows a fallback position to the commercial card. If, as sometimes happens to the best of projects, the system should be scrapped to make way for a new one, then much of the electronics can be salvaged for use on the new system if both systems were on the same standardized bus. If the hardware is very simple, as it is in the non-bus APT system described later, evolution can occur by designing and building a new interface card.

Just as the interchangeability of cards on a bus increases adaptability, so having interchangeable program "software modules" helps the adaptability of the software. When things don't work, only modules, not entire programs, have to be scrapped. Contrary to popular belief, modularized programs do not take significantly more code or memory, and execute almost as quickly as in-line code. Even if this were not true, memory is dirt cheap these days, and by using smart interface cards, there is no requirement for raw speed.

At first it might seem that having smart cards with unalterable programs stored in their ROMs would reduce adaptability, but this is not the case. The fixed tasks performed by these smart cards are not likely to be in need of change, but by relieving the main CPU of the time-critical tasks and allowing a high-level language to be used, changes in the overall control program can be made much more easily, because the structure of these programs is greatly simplified by the use of these cards. High-level language programs, particularly if they are well documented, are changed with relative ease compared to a program written in assembler, or a language designed to implement lower level hardware dependent primitives, such as FORTH.

Careful modularization of the hardware and cables along functional lines (just as the software was modularized) improves adaptability. If a function has to be changed, it does not require a rework of all the electronics, half the cables, and the entire back panel—just a single card, one cable, and a modular strip on the back panel. Real thought needs to be given to dividing a system along functional lines and minimizing interactions. Having slightly more hardware or extra cables is sometimes worthwhile if this cleans up things along functional lines. Building in spare capability is usually worthwhile. Extra wires in a cable and extra pins on a connector cost very little and may save making up an entirely new cable later. Terminal strips a little larger than needed often are helpful. Extra reserve in power supplies will almost certainly be used as time goes by. Extra slots in the card cage are always useful.

7.7 Reliability

Many astronomers rightly fear that a computerized control system will fail on them at some crucial moment. Such fears are often based on past experience with early computer systems which indeed were unreliable. However, if careful attention is paid to reliability throughout the design and construction process, a highly reliable system will most likely result. It might be noted that an increasing number of industrial processes are being computerized without reliability difficulties.

For complex systems, reliability often is enhanced by putting the electronics on a standardized bus. Having all the electronics plug into a single card cage minimizes the number of interconnections to be made, and reduces the susceptibility of the system to electromagnetic interference (EMI). The importance of reducing interconnections and EMI cannot be stressed too strongly, and a sad but true story is appropriate here.

The first computer at the Fairborn Observatory was a Radio Shack TRS-80 Model I. It served faithfully for a number of years as a data-logger for the photometer, and as an analysis machine. It was then applied with confidence to controlling the 0.4-meter telescope. While it worked in this capacity, it did not work well. Random "crashes" of the system were frequent and extremely annoying, and there were several connector and cable problems. Because of these problems, this approach was abandoned in favor of the STD bus. The STD system that replaced the TRS-80 never crashed, and its reliability was outstanding.

Going to a bus will not, by itself, eliminate all connector and EMI problems. Good grounding and shielding practices must be followed throughout the entire system. It costs very little extra to run shielded instead of unshielded cables out to the telescope, and the card cage, power supplies, etc., should all be well shielded. All equipment and circuits should be properly grounded, and a single ground point for each ground circuit should be used to avoid "ground loops." Separate ground circuits should be used for power, signals, and chassis connections, and these three should never be tied together.

Connectors should be of high quality; clamp on the cable, and lock it to the device being connected, so connections will not jiggle loose. Their contacts should be gold plated to reduce resistance, and should use wiping action to assure a positive connection. The screw-on type A/N connectors are best, although quite expensive. However, reliability is usually a good investment.

An even better approach is to eliminate as many connectors from the system as is practical. In the single pc card version of the second generation Fairborn Observatory APT control system, all the electronics except the computer and power supply have been placed on a single card to eliminate connectors. This not only saves money, but it significantly improves reliability.

The quality and reliability of computers and boards varies widely. For some customers, low cost is important, and for them, cheap board materials are used, holes are not plated through, and chips are marginal seconds. This may be perfectly acceptable to someone playing games on his personal computer, where a failure would only be slightly annoying. However, a failure in a telescope control system can not only result in lost observations, but a failure can place the telescope drive in a mode that causes damage to the telescope itself. Therefore, only the highest quality glass epoxy boards, top quality chips, and the best construction techniques are acceptable. Cards are usually burned-in and environmentally stressed before they are sold. It is rare when a card doesn't work the first time it is turned on and keep on working without any failure for years. Reliability costs more to begin with, but it is worth it in the long run.

Even when using well designed equipment of high quality construction, it is prudent not to expose electronics to environmental stresses unnecessarily. Temperature cycling, moisture condensation, overheating, and dust all take their toll and reduce reliability. The floor of an observatory is a particularly inhospitable environment for electronics, and it pays to keep all or almost all of the electronics in an environmentally controlled room. Astronomers also like to operate the telescope in shirt-sleeve comfort! If this is not possible, thought should be given to a small enclosure with a thermostatically controlled heater to provide some protection to the electronics.

In spite of all precautions, an occasional failure will occur in almost any complex system. When this happens, maintainability becomes important.

7.8 Maintainability

Systems always fail at the worst possible moment. They fail when they are needed most and when it would be most difficult to fix them. The good design inherent in a control system (or lack of it) shows up when something fails at 2AM in the middle of a "chance in a lifetime" observing run. Documentation is the most critical aspect of maintenance. If the system is well documented, then troubles can be found in a logical, organized, and expeditious manner. Complete documentation is important. Every card, cable, and wire should be documented—right down to the color of the wires in the cables. The documentation should all be assembled together in one place—preferably in a single bound volume so that nothing gets lost. There should be at least two copies of the documentation. As mentioned earlier, one of the advantages of buying off-the-shelf commercial cards is that most of them are thoroughly documented. Not only do commercial cards rarely fail, but they are relatively easy to fix when they do.

Hand-in-hand with documentation goes labeling. Connectors, terminal strips, boards, etc. all need to be labeled, and the labels should agree with the identification on the documentation. When in doubt, label.

Ease of access is important for maintenance. If things are hard to get at, not only are they hard to fix, but other things can be damaged in the process, not to mention frayed nerves and strained eyes. Large removable covers are helpful. Wire-wrapped cards are generally more difficult to repair than pc cards. In repairing a wire-wrap card, it is not unusual to induce another fault, and in trying to repair it, sometimes yet another.

The special little cables, test devices, etc., used to make software development easier are also useful in isolating troubles. A well-rounded selection of modern electronic test equipment is really a necessity. If possible, this should include a good logic probe, digital voltmeter, and wide-band oscilloscope.

Isolating a fault and bringing a system back up on line is much easier if spare boards are available, or even better yet, a completely spare system. Spare pieces and parts should also be kept. Keeping a spare system and spare boards is not cheap, but the cost can be reduced if emphasis is placed on commonalty and standardization.

7.9 Safety

Safety is the most important design consideration of all. A poorly designed control system that does not take safety into consideration can kill the operator, other observers, the telescope itself, or other equipment. Safety must be built into a control system, and on no account should a known hazard be allowed to exist just because it is difficult, time consuming, or costly to correct.

Usually, it is easy to build safety into your control system. For example, make sure there are two limit switches for each motion that has a travel limit. The first is a software limit that tells the control system software that the limit has been reached. The second limit switch should have adequate current-carrying capacity to interrupt the flow of current to the motor. The purpose of the second switch is to prevent a disaster in case of a hardware or software failure in the control computer.

Another way to build safety into your system is to use software timers to check on motion. When you program the software to perform a move, you should also calculate the amount of time it will take at the speed you are commanding to hit a limit switch, even if you don't expect to move as far as the limit switch. Add a bit of margin, say a second or two (depending on the commanded speed), then tell the operating system to post an interrupt from the system timer if the timer is not cancelled by the time it goes off. When the motor driver software indicates the move is complete or a limit switch is reached, then the application software should cancel the timer. Otherwise, if the timer interrupt goes off, it should generate an alarm and stop the motor.

Other measures can contribute to safety, including using the key switch on your computer (or installing one, if it doesn't have one) to prevent unauthorized use; ensuring that high voltage power supplies for photomultiplier tubes or avalanche photodiodes are properly insulated to prevent electric shock; building into the software some limits on slew speeds, to reduce the risk of banging an unwary observer with the telescope at a public night; placing guards or using sensors to ensure that fingers are out of the way of the dome or roll-off roof on public nights; and following all building codes.

Sometimes safety relates more to the equipment than to humans. What can be a nuisance to humans can be threatening to equipment. For example, if you suffer a power failure in the middle of the night just as a thunderstorm approaches (power failures correlate highly with thunderstorms), could you close your roll-off roof or dome shutter without power, in time to prevent rain damage to your telescope and other equipment? Having manual backups to essential computer-controlled functions is good design practice.

We can't enumerate all the things you should do to make your observatory, telescope, and control system safe for you, your visitors, and your equipment but we do urge you to design your system to work with safety as the primary objective. A careful reading of local and national building codes, talks with inspectors and design engineers and other institutions with similar equipment can be very productive.

7.10 Conclusions

Many different and sometimes conflicting issues must be resolved during the design and development of any complex real-time control system. There are many ways to go wrong, but only a few ways to go right. Of course most observatories have small budgets and there is constant financial pressure to cut corners. However, the greatest expense is having to completely redo a system, as a system that really never works quite right just is not useful in the demanding world of observational astronomy. Throughout the history of systems development, there is never enough time to do the job right the first time, and plenty of time to do it over. Don't make this classic mistake!

Evolutionary development of a bus-based system using proven commercial cards can result in a highly reliable, easily maintained, and easily modified system. If great design care is taken on a "hardware simple" system, such as the Fairborn Observatory APTs, excellent results can be achieved with just a single board computer and a few fairly simple custom boards.

Part III

Telescope Control System Components

Chapter 8

Motors and Motor Controllers

The previous part of the book described the things you should keep in mind when designing a control system and the problems your design should solve to meet your performance requirements. This part introduces the components you have at your disposal to use in your design. We focus on motors and sensors on the hardware side, and the operator interface and operating systems on the software side. For each type of component, we describe how it works, what kinds of performance you can expect from it, and approximately how much it will cost.

The principles discussed in the previous sections will not change much with time, but the components available to telescope control system designers change constantly. This part is not intended to be a comprehensive tutorial on what is available, because it will be obsolete soon after it is published. Instead, this is intended to help you ferret out the components that will be available at the time you are designing your system by showing you what was available at a particular moment in time. You can use this snapshot as a starting point for exploring the current commercial market.

8.1 A Standard Telescope Problem

In this chapter, we describe various kinds of motors and motor controllers that can be interfaced to computers. Although this kind of technical information is interesting and useful, it is not the only point of interest.

New motors are continually being introduced with improvements over earlier designs, and specifications of current offerings are changing as the technology advances. Appendix B lists the names and addresses of several manufacturers of motors and controllers appropriate to telescope control. We recommend that you contact the manufacturers listed there and discuss your specific applications with their engineers. One long distance telephone call may save hundreds of dollars and many hours of grief.

Another aim of this chapter is to demonstrate how to evaluate various motor options. Once the decision has been made to build a particular type of control system, a particular motor, gear, and motor controller combination can be selected that meets the system performance requirements. To help compare the performance of motor and gear combinations, we have devised a hypothetical model of a telescope. The rotational inertia of this model is computed first. This defines the load that each motor and gear combination must move. In succeeding sections, each type of motor is described, then its applicability to the hypothetical telescope is assessed. In this way, a motor and gear combination can be selected. An example of this process is presented in Chapter 13.

When retro-fitting a telescope with a computerized control system, one might consider using the existing drive motors and controls, by simply connecting these controls to the computer. When this is too difficult or is otherwise inappropriate, or when designing a new telescope, motors and controllers should be selected with a view to how easily they can be integrated into the computerized system. This is as important as their ability to provide the needed torque and slew, set, guide, and track rates. To do this, one should compare different control servo design proposals with numerical calculations using actual motor and load characteristic data.

A hypothetical alt-az telescope is depicted in Figure 8-1, with the numbers in the figure identifying the mass elements described below. The mass elements define the standard load on the azimuth motor, using the moment of inertia equations in Figure 8-2 to compute the moment about the azimuth axis. U.S. units (pounds, feet, etc.) are used. Consequently, since the pound is a unit of force, the slug (one pound/foot/sec/sec) is the unit of mass used in the table, computed by dividing the force in pounds by the gravitational acceleration constant of 32 ft/sec^2. In those cases in which two moment arms are involved, both are given. The total weight of 1005 lbs. is distributed in such a way as to yield a moment of inertia through the vertical azimuth axis of 174.1 slug-ft^2 or 5571.2 lb-ft-sec^2 (1,069,670.4 oz-in-sec^2).

The following paragraphs describe various types of motors, how they would be integrated into a computerized control system, and how their characteristics affect telescope control performance, based on this hypothetical model.

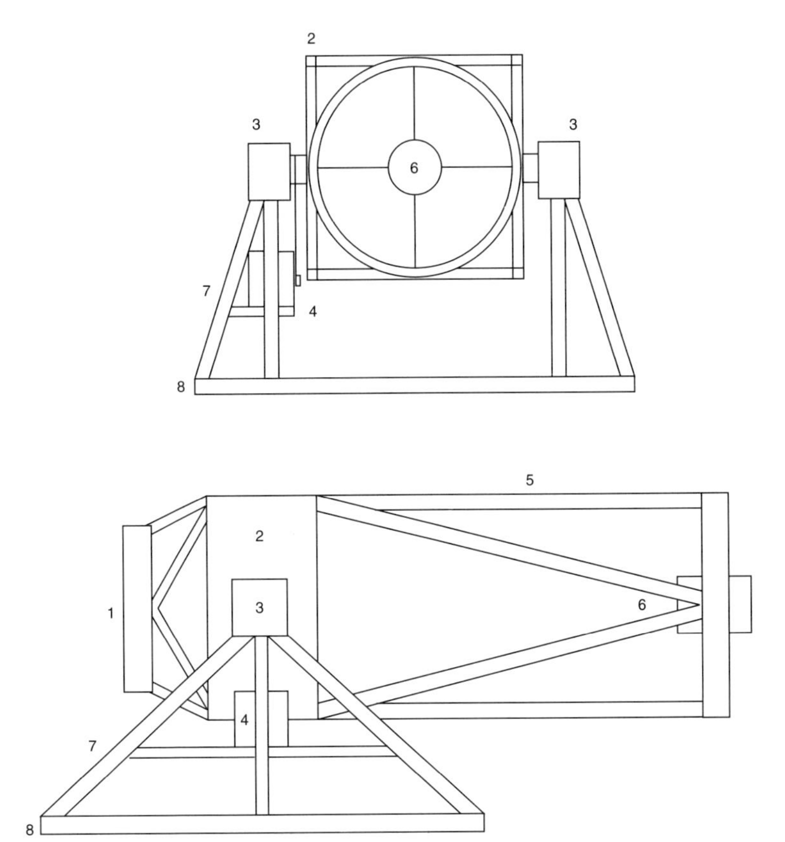

Fig. 8-1 *Hypothetical telescope for comparing motor drive designs.*

Mass Element	Weight (pounds)	Moment Arm(s) (feet)	Moment of Inertia (slug-ft^2)
1. Mirror cell and supports	250	1.5	17.6
2. Altitude axis support box	100	1.0, 1.46	4.8
3. Altitude axis bearing housings	100	1.67	8.6
4. Altitude axis drive	100	1.67	8.6
5. Tube truss	160	3.0, 1.25	22.8
6. Secondary mirror and motor	75	6.0	84.2
7. Altitude axis support beams	120	2.0	15.0
8. Azimuth bearing ring	100	2.0	12.5
Total weight = 1005 lbs.			
Total moment of inertia about the azimuth axis = 174.1 slug-ft^2			

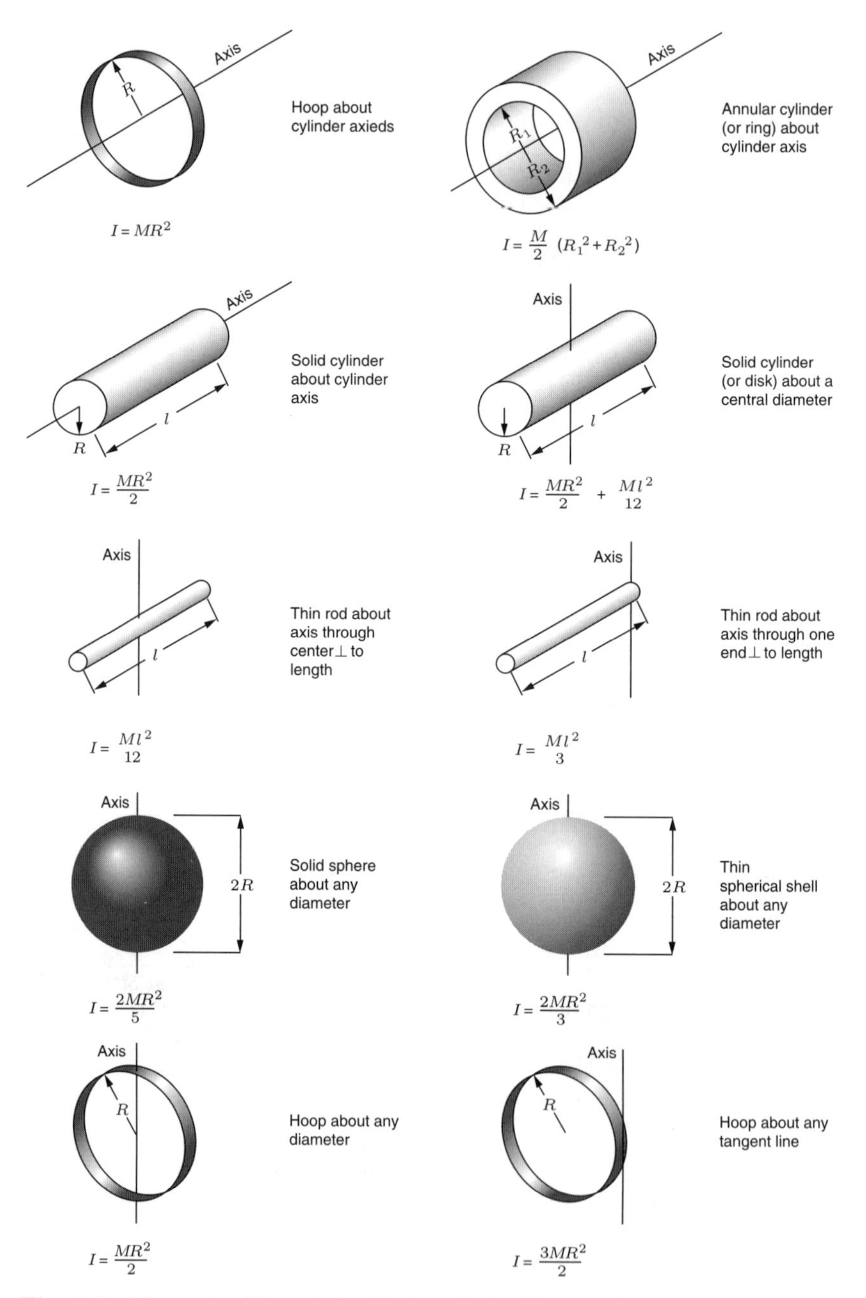

Fig. 8-2 *Moments of Inertia for Various Body Shapes. Courtesy John Wiley & Sons, Inc.*

8.2 Large DC Torque Motors

The first motor configuration we consider is a large direct current (DC) torque motor mounted directly on the driven axis. This configuration is used on a few modern mid-sized telescopes of 0.5 to 1.5 meters in aperture designed for special applications (low Earth orbit satellite tracking) that require very high speed, optimum servo performance. Most telescopes used for ordinary astronomical research do not use this approach, for reasons that are discussed below.

Torque motors are DC servo motors that use a permanent magnet stator. They are often frameless, which means that the usually large stator housing is bolted to the telescope mount, and the rotor has a hole through it so it can be press-fit mounted onto the telescope axis shaft. The telescope shaft bearings maintain correct rotor alignment inside the stator. These motors are intended for applications requiring high torque to inertia ratios. Although pulse width modulation control can be used with torque motors, linear proportional control is used in the high performance telescope control applications requiring DC torquers. Figures 8-3 and 8-4 show typical torque motors. An example of the largest practical motor is the Inland Model T-12008, which is the motor used in the calculations below.

Fig. 8-3 *A typical brushless DC torque motor. Courtesy Inland Motor, Specialty Products Division.*

Fig. 8-4 *Open frame DC torque motor. Courtesy Inland Motor, Specialty Products Division.*

One advantage of a direct drive torque motor is that gearing errors are eliminated, since the telescope axis is driven directly by the motor. However, when DC torque motors are used in a simple analog servo system, they can suffer from an even larger error—the motor dead band, which is the error signal voltage (corresponding to an error angle) to which the motor just fails to respond. For example, to achieve a 1″ dead band, assuming a peak voltage V_s of 49.8 volts and a servo amplifier gain K_a of 20 volts/volt, the control gain K_s, which is $2V_s/K_a$, must be 4.98 volts per arc second. To avoid exceeding the maximum applied motor voltage of 49.8 volts, the error signal cannot represent more than 10 arc seconds. This now poses problems in generating the error signal, given that the error can be as much as 180° if the previous object you were tracking is in one area of the sky and the next object is in the opposite part of the sky. Conversely, if a servo gain K_s of 4.98 volts per 180° were used so as never to exceed V_s, and assuming a servo amplifier deadband (input bias current, converted to a voltage through the input resistor) of 15μV (typical of high power servo amplifiers), the motor deadband is $(15\mu V/4.98\ V) \times 180°$, or 0.0005426° (703″, or 11′.7). At the 15μV error signal level, the current is only 20V/V $\times$ 15μV/0.970 ohms (winding resistance), or 309μA, which produces only

0.0012 lb-ft of torque. This is well below the 1.0 lb-ft motor friction, which is due mainly to the brushes. To produce 1.0 lb-ft to just overcome the friction, the error signal must be 0.0124 V. This produces a dead band of 0.45°, which is too large for most applications.

The deadband is a problem only in simple analog servos, where a potentiometer or a similar device is used to generate the reference voltage representing the desired position. In more complex analog servos, or in digital servos, the control gain K_s is easily adjusted according to the size of the error. In a digital servo, the computer can apply the maximum voltage to the motor until it senses the error is less than 10″ (the error signal value which has acceptable dead band). At this point, the program can revert to the linear mode until the desired position is reached. This would not be practical, since with a maximum error of 10″ in the linear mode, the inertia of the telescope would cause significant overshoot at slew speeds. A more practical control approach is to vary K_s in several fixed ranges, according to the current value of the error signal. This adjustment of the control gain is another example of the use of a forward path integral.

Aside from the higher cost of the large torque motor ($6,000) and servo amplifier ($1,500 or more), the power requirements for DC torque motors can be high. The Inland Model T-12008 develops 68 lb-ft of torque, but requires a 5,200 watt servo amplifier (assuming 50% efficiency). Dissipating 5,200 watts near the telescope optical path can cause air currents that degrade the image. There is usually no need to supply this much power when a good friction drive or even gears can be used to increase the effective motor torque, which permits the use of a smaller motor. This can reduce motor power consumption to 100 watts or less. Furthermore, most telescope mounts could not bear the weight of the T-12008 (190 lbs.) on each axis without sacrificing some instrument weight capacity.

Despite these problems, the acceleration of the torque motor is excellent. Even assuming the motor delivers only its continuous torque (which is lower than its peak torque) throughout a slew, the torque of 68 lb-ft (13,056 oz-in) accelerating a mass moment of 1,069,670.4 oz-in-sec^2 gives an angular acceleration of 0.012 rad/sec^2. That is, it can move the telescope through 180° in less than 20 seconds (counting ramp up and ramp down times)! Although this is impressive performance, that much mass moving that quickly could rip tangled cables to shreds or severely injure an unwary bystander before the observer could intervene. Furthermore, the telescope would be difficult to point accurately without a tachometer generator to provide velocity feedback to damp the system.

Thus from the standpoint of cost, power, weight, and ease of control, direct coupled DC torque motors are inappropriate to the typical telescope control problem (although they are appropriate in those few instances in which extremely rapid slewing or tracking performance is required). A DC torque motor smaller than the T-12008 would have less weight, power consumption, and cost. However, its dead band would be about the same, and small variations in the load due to friction or wind gusts would cause large over- or undershoots when the driven shaft is near its desired position, since a smaller motor would not have adequate torque to compensate for these variations.

8.3 Servo Motors

Torque motors are but one member of the general class of DC servo motors. Other examples include wound-field, permanent magnet (PM), moving coil, and printed circuit types. Figure 8-5 shows a typical servo motor and controllers that might be used with it. Figure 8-6 shows a high torque servo motor suitable for a wide range of telescope control applications.

Wound-field servo motors use coils in both the armature (rotor) and the field (stator). By controlling the current in either the armature or the field, the speed of the motor is controlled. It is better to use armature control in telescope control applications, since the requirement for larger driving signals is usually offset by greater ease of control.

PM motors use a permanent magnet in the stator instead of field coils. They usually have two wires instead of the four found in a wound-field type. Advantages include higher starting torque, a linear speed-torque relationship, smaller size, and lighter weight. PM motors usually use brushes for commutation, but brushless types are now available. This is the type of servo motor used most frequently in modern telescopes.

The moving coil motor is a PM motor with a different type of construction. Rather than using a conventional cylindrical armature, they use either a flat disk for the armature, or a shell armature that has no iron; epoxy or fiberglass holds the copper conductors in the armature. A variation is the printed circuit motor, which uses copper foil on printed circuit board material to form a flat disk armature. Although moving coil motors have low inertia, high acceleration, and other good qualities, they are more expensive than the conventional PM motors, and therefore are rarely used in telescope control applications.

In general, DC servo motors have relatively flat torque curves, so that torque does not fall off too rapidly or fluctuate wildly with higher speed. This feature, along with the availability of controllers that interface easily to computers, makes servo motors good candidates for telescope control.

The second motor configuration to be tested against the standard telescope model is a small PM instrument DC servo motor geared to each

Fig. 8-5 *A Modern Servo Motor With Internal Resolver and Typical Controllers. Courtesy Compumotor Division, Parker Hannifin.*

telescope axis using precision gears. A typical motor to use in this case is the Inland Motor Model T-1814, which has 85 oz-in of stall torque.

Assuming that a 359:1 reduction (a common ratio for RA worm gears used to approximate sidereal rate from UTC) is used to gear down the motor, the load moment of inertia is reduced by $^{1}/_{(359)}{}^{2}$ to 8.30 oz-in-sec^2. When combined with the rotor inertia, this yields a total inertia of only 8.32 oz-in-sec^2 with a ratio of load inertia to rotor inertia of 415, a bit high for this motor and application. If the continuous stall torque were applied when the load is at rest, the acceleration would be 85 oz-in/8.32 oz-in-sec^2, or 10.2 rad/sec^2, which, when divided by the gear ratio, accelerates the telescope from rest at a rate of 0.03 rad/sec^2. Although this acceleration figure is higher than that for the DC torquer, the top speed of the torquer is significantly higher, since there are no gears between the torquer and the telescope axis to reduce the speed.

Fig. 8-6 *High Torque Servo Motor With Internal Encoder and Controller. Courtesy Compumotor Division, Parker Hannifin.*

To compute the overall gain K of the servo, assume the control gain, K_s, is 4.98 V/π radians, or 1.585 V/radian; the servo amplifier gain, K_a, is 20 V/V; the torque sensitivity, K_v, is 36.8 oz-in/A; and the gear reduction, N, is 359. Therefore, K is 1.585 V/radian $\times$ 20 V/V $\times$ 36.8 oz-in/A $\times$ 359/(4.07 ohms), or 102,910 oz-in/rad, which gives a damping constant for critical damping, F_c, of 368.6 oz-in/rad/s. This makes the servo heavily overdamped, which eliminates the possibility of overshoot, but increases the time required to settle to the new position. This is not a disadvantage, since most telescope control systems need to be overdamped to the point where it may take a half minute to slew 180°. Assuming the same servo amplifier characteristics as in the previous example, the dead band is reduced by the amount of the gear reduction from 703″ in the previous example to about 2″, an acceptable level for many applications.

The motor example worked out above using a linear amplifier is representative of the traditional approach to the servo motor design problem. This traditional approach is still available from a variety of manufacturers. Figure 8-5 shows a modern approach sold by Parker Compumotor. Called the Compumotor Plus Series, these PM motors use an additional magnet section of the rotor and coils at the rear of the motor to form a resolver (see the next chapter). The resolver tells the motor controller (the black boxes in the photo) the angular position of the motor at any moment, and the controller adjusts the current in the motor's windings to bring the motor to where it should be at that moment. The accuracy of the resolver and controller is about 12 arc minutes. These motors are available with torques of 130, 400, or 1,100 oz-in with rotor inertias of 1.64, 9.23, or 51.72 oz-in^2 respectively, so in most cases a gear reduction is needed, not only to match the load and load inertia to the motor torque and rotor inertia, but to reduce the position error to a reasonable value. These motors (with controllers) range in price from about $2,000 to about $3,000 each. Though expensive, they are very high quality and rugged enough to provide years of trouble-free service.

The next step up in price and performance is the Parker Compumotor Z Series. These servo motors also have internal resolvers and offer speeds to 7,000 rpm. There are 11 models with torques in the range 346 oz-in to over 9,000 oz-in. They range in price (with controller) from about $4,900 to over $10,000 each, and have a rotor position accuracy of 8 arc minutes, with a repeatability of 0.088° unloaded, so these motors also must be geared to a telescope axis to reduce the position error. Motor and controller combinations similar to the Compumotor Plus and Z Series are also sold by Whedco and Galil Motion Control (see Appendix B for addresses).

Most of the servo motors described to this point require a substantial gear reduction ratio between the motor and the driven telescope axis. The problem is that the more gear reduction stages there are in the system, the larger the backlash and periodic errors. If you limit the number of reduction stages to one (using a powerful and accurate motor), you can more easily control your gear errors. If you use a direct friction drive (a small roller 1–4 inches in diameter pressure loaded against the edge of a much larger disk) with carefully machined parts, you can virtually eliminate backlash and bring periodic errors under control. Such a drive is used on the WIYN telescope in Figure 6-5.

A servo motor with the torque of a large DC torquer but the ease of control and lack of dead band of the (geared down) stepper motor, coupled with exceptional accuracy, is shown in Figure 8-6. The Compumotor Dynaserve Direct Drive brushless servo motors come in 25 models ranging in torque from 6 ft-lbs (1,152 oz-in) to 370 ft-lbs (71,040 oz-in). Their high rotor inertia (up to 25,160 oz-in^2) permits them to drive a friction roller even on large telescopes. With such low gear ratios (friction drives typically have reductions of 10:1 to 50:1) it is critical to match the motor's rotor inertia to the rotational inertia of the driven axis. Compumotor recommends a ratio of telescope axis rotational inertia (divided by the friction drive ratio squared) to motor rotor inertia in the range 50-200, depending on the top speed of the driven axis. Since they use accurate resolvers or high accuracy optical encoders for feedback, these motors can obtain up to 25 arc second accuracy and 2 arc second repeatability under load.

Using the standard problem for comparing motors, let's assume a Parker Compumotor Dynaserv DM1015B with 11 ft-lbs (2,112 oz-in) of torque and a rotor inertia of 660 oz-in^2 drives a 2-inch diameter friction roller against the edge of the 48-inch diameter azimuth bearing. The moment of inertia of the telescope about the azimuth axis is 1,069,670.4 oz-in-sec^2. The load inertia on the motor (reduced by the friction drive) is this number divided by $(48/2)^2$, or 1,857 oz-in-sec^2. The rotor inertia of 660 oz-in^2 is converted to equivalent units by dividing by 32ft/sec^2 or 384 inches/sec^2, giving a rotor inertia of 1.7 oz-in-sec^2. The ratio of telescope axis to rotor inertia is 1,080, so a smaller roller, additional gear reduction, or a larger motor is required. A suitable motor is the Model 1050A with a rotor inertia of 5,250 oz-in^2 or 13.7 oz-in-sec^2. This gives a ratio of load inertia to rotor inertia of 135.8, within the recommended range of 50–200. The Model 1050A has a peak torque of 37 ft-lbs (7,104 oz-in). The total inertia is $1,857 + 13.7 = 1,871$ oz-in-sec^2.

If the peak torque of the Model 1050A were applied to the azimuth axis when it is at rest, the axis would accelerate at a rate of 7,104 oz-in/1,871 oz-in-sec^2 = 3.8 rad/sec^2 which, when divided by the gear ratio of 24, gives 0.16 rad/sec^2 (9.06°/sec^2), which is very respectable performance. This motor/controller combination has 1,024,000 steps/revolution, so the reduction of 24:1 makes one step equal to 0.053 arc second. This is a good design, since a typical tracking rate of 15″/sec means the motor would step at the rate of 284 steps per second, well above the mount resonant frequency that is likely to be in the range of 5-15 vibrations per second (Hz). The controller has a top speed of 1,600,000 steps per second, or over 23°/sec. That is an extremely fast slew speed, probably too dangerous for star parties. This motor is fully capable of meeting all reasonable performance requirements for the hypothetical telescope problem.

Unfortunately, this level of high performance is expensive, ranging from about \$3,700 to almost \$14,000 per single motor/controller combination. But from many points of view, this is the ideal motor for driving most telescopes' axes.

The performance of the geared servo motor makes it entirely acceptable as a telescope drive, with the only disadvantage being that modern motors and controllers make such performance expensive.

8.4 Servo Motor Controllers and Computer Interfaces

Servo motors require a linear power amplifier to drive them that is not too different from a good quality audio amplifier for a stereo sound system.

For a DC motor, the amplifier must accept a DC voltage from the computer and drive the motor at a speed proportional to this input (control) voltage. Usually, a tachometer attached to the motor generates a feedback voltage proportional to the motor's speed. This serves as a second input to the servo amplifier, which adjusts the current in the motor coils to keep the motor speed proportional to the control voltage.

Desirable servo amplifier characteristics are low input bias voltage and current (the voltage or current that just barely produces no output), and very linear output voltage or current (without velocity feedback) or linear motor speed (with velocity feedback) response to the control voltage. Good servo performance is expensive. For example, the Torque Systems C0401 servo controller, a popular servo amplifier used in older telescope control systems by DFM Engineering, Inc., costs about $460. With a Torque Systems "snapper" servo motor and tachometer combination unit, which sells for about $220, the motor and controller total price is about $700.

A variation from the traditional linear servo amplifier that is less expensive and easier to interface to a computer is the technique of pulse width modulation (PWM). There are integrated circuits available that convert a DC control voltage to a series of pulses with widths that are proportional to the control voltage. Such chips can drive small servo motors directly, or large DC torquers with external power transistors. Often they can also accept tachometer inputs for speed control. If this approach looks promising for your application, a discussion of your motor performance requirements with the manufacturer of the telescope drive motor will aid in your assessment of the ability of PWM to meet these requirements. For a wide variety of uses, PWM can provide adequate control of DC servo motors inexpensively. In those applications requiring better performance with larger motors, linear proportional control can give more precise control, but at a higher price.

Motion Science offers a servo monitor/PWM amplifier combination using either an STD Bus or RS-232 interface. The continuous torque of their motors ranges from 200 to 650 oz-in., with speeds to 3,000 rpm. The system is expensive, up to $2,000 per axis.

There are two basic ways to attach DC servo motors to computers. The first is to use an interface card with the servo amplifiers built-in. This is exemplified by the CyberResearch CSVS 931A single axis PWM servo card for the PC/AT (Industry Standard Architecture, or ISA) bus and by the nuLogic 1–4 axis servo cards also for the ISA bus. The CyberResearch PWM single-axis system (including controller, amplifier, power supply cables, and a 40 oz-in peak torque motor) costs $1,395, while their CSVS 843A three-axis linear proportional servo system (including similar accessories and three 126 oz-in peak torque motors) costs $3,695. The nuLogic boards (without motors) start at $995 for a single-axis system and go to $3,200 for a rack mounted three-axis system. The nuLogic products are available from Texas Micro Express.

The other approach is to use a card employing a digital-to-analog converter (DAC) to apply a control voltage to a separate servo amplifier. The computer loads a number into a bus register that the card converts to a corresponding voltage. Such cards are available for a large number of computers, including the IBM PC and STD bus. For example, the Cyber-Research CYDAC 02 has two 12-bit DACs and sells for $199. Note that the resulting tracking resolution is one part in 4,096, or roughly 0.02%, which may not be enough. Boards with higher resolution (16 bits) are available from other manufacturers. In this approach, the computer must be capable of all the control loop calculations normally provided by servo motor controllers. These calculations can be quite complex, as they consume the greater part of the 68040 processors used in servo motor controllers.

A closed-loop DC servo motor control system can be built very inexpensively. Such a system would not be very accurate, but it would place the object within the field of the finder. This is all that some applications require. The system can be built with power operational amplifiers (about $10 for 100 watts) driving common DC motors. Conductive plastic potentiometers can provide the angle position feedback. The computer interface would consist of a 14-bit Burr-Brown D/A converter to drive the op amp, and a 14-bit Burr-Brown A/D converter to read the voltage on the wiper of the potentiometer. This drive system would be used for slewing only, then a normal synchronous motor and drive corrector would be used for tracking. The AC drive corrector (with a set speed) would also be used to center the object in the field of the main telescope. An estimate of parts costs is given below.

Item		Cost (new)
Conductive plastic potentiometers (2)		$ 60
DC motors (2)		20
Inexpensive gears (2 sets)		60
DC power op amps (2)		20
A/Ds and D/As		80
Miscellaneous interface chips, etc.		80
	Total	$320

Costs could be reduced to about $100 using parts bought at radio amateur "ham fests" and surplus electronics stores. In this approach, the computer reads the potentiometers, corrects them for non-linearities using a calibration curve determined from star measurements, and adjusts the motor control voltage based on the potentiometer readings. The computer program slows down the motors as the desired positions are approached, to avoid overshoot. The only correction such a system has to compute is precession. A refraction correction could be added if observations are to be made near the horizon. If higher accuracy in pointing and tracking are required, the system would cost more, due primarily to the higher costs of more accurate shaft angle encoders and their computer interfaces.

In the mid-1980s, Galil Motion Control and Hewlett-Packard helped to develop a new concept in interfacing servo motors to computers. The problem was twofold:

1. system designers did not always know enough control theory to adjust the servo parameters (including bandwidth, phase shifts, and gain) correctly in the computer software, and

2. smaller computers then available spent most of their time on computing the servo control algorithm, without leaving enough computer power for the application.

To solve these problems, both companies introduced circuits that accepted step and direction inputs from inexpensive stepper motor computer interfaces and inputs from a shaft angle encoder attached to the servo motor. The circuit output is a voltage to a servo motor. The step and direction digital inputs tell the servo controller circuit where the motor should be, and the encoder on the motor shaft tells the controller where the motor actually is. The controller determines the difference between where the motor is and where it should be, then adjusts its output voltage up or down to bring the motor to where it should be. This approach makes the servo motor look like a stepper motor to the computer, without some of the torque and other performance problems of steppers mentioned in the next section.

Galil Motion Control now offers a complete line of servo motors, complete integrated amplifier/controllers and separate amplifiers, power supplies, and controllers, as well as software to aid in designing your servo system.

The Parker Compumotor motors in the Compumotor Plus series have a resolver yielding a system resolution of 12,800 steps per revolution of the rotor shaft. The Compumotor Plus controllers pictured in Figure 8-5 both use the Proportional-Integral-Derivative (PID) control algorithm that measures motor errors with the motor installed in your system experiencing actual load conditions, and adjusts control parameters for both a position loop and a velocity loop to provide the best system performance. They use high speed 16-bit and 32-bit arithmetic to update the control signal every 300μs. These controllers allow you to program the number of steps per revolution of the motor from 200 to 25,600 steps per revolution. They have an integrated power supply, so you simply plug them into 120 VAC power, and they also have an RS-232 serial interface that permits you to program servo parameters. The "X" version of the controller uses Compumotor's control language over the serial port, relieving your computer of computing acceleration ramps, keeping track of the number of steps moved, and other functions.

The Parker Compumotor controller for the Compumotor DM Series Dynaserv motors shown in Figure 8-6 works basically the same way, using outputs from an optical encoder attached to the motor shaft to provide either 655,360 or 1,024,000 steps per revolution (depending on motor model) with step rates up to 1,600,000 steps per second. It also uses an adaptive control algorithm to adjust control parameters to actual load conditions. The controller permits running the motor in a position, velocity, or torque mode. In the position mode, it accepts step and direction digital signals, while in the velocity or torque modes, the controller accepts ± 10 VDC analog control voltages. The controller uses 125/230 VAC single phase power.

Modern controllers make it easy to interface servo motors to computers, making servo motors a very attractive choice for telescope control.

8.5 Stepper Motors

The third motor configuration to be considered is a stepper motor geared to each telescope axis using precision gears. A stepper is typically a DC motor with a permanent magnet armature consisting of a dozen or more magnet faces and four or more field windings. Examples of small angle stepper motors and a typical controller are shown in Figure 8-7.

When a current is generated in one of the field windings, the armature is forced to move through an angle determined by the number of armature magnet faces and field windings, then it comes to rest. This action is a "step." By turning successive field windings on and off in a set pattern, the motor can be made to step repeatedly until the armature shaft rotates through 360°, as shown in Figure 8-8. Steppers are available in sizes from a fraction of an oz-inch of torque (for quartz "analog dial" wristwatches) up to 3,000 oz-inches or more, and from 2 steps per revolution up to thousands of steps per revolution. Typically, a special drive card is needed to convert a train of TTL digital logic pulses into the correct sequence of field winding currents, although Cybernetic Micro Systems makes integrated circuits that interface easily to a computer bus and convert commands sent in standard ASCII character codes to stepper motor coil currents. A typical motor for telescope control is the Superior Electric M092 Slo-Syn stepper with 200 steps per revolution.

Fig. 8-7 *Typical PM DC Stepper Motors and Controller. Courtesy Compumotor Division, Parker Hannifin.*

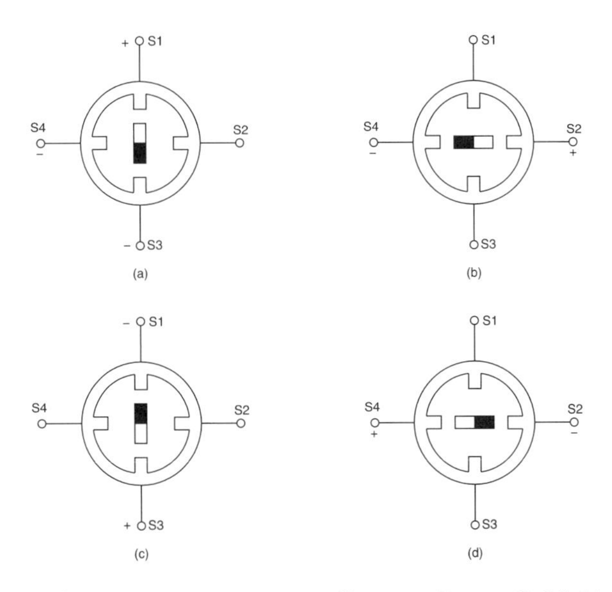

Fig. 8-8 *Stepper motor operation. Courtesy Reston Publishing Co.*

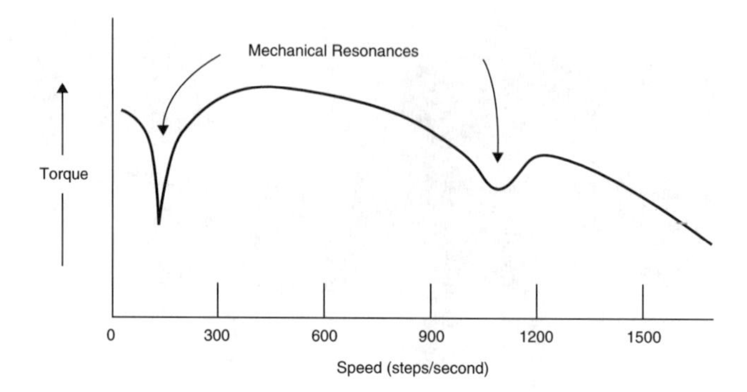

Fig. 8-9 *Typical stepper motor speed vs. torque curve.*

The maximum speed of the motor is determined by the maximum number of pulses per second it can translate into steps. The servo and stepper motors share the property that their torques decrease rapidly with increasing angular velocity. However, the two motors differ, in that the servo motor torque versus speed curve is essentially linear, while the stepper motor torque curve is quite non-linear. The characteristics of the circuit used to drive the stepper motor are as influential as any other factor in determining a stepper's torque versus speed curve. Circuits that compensate for the dynamic impedances of the motor coils enable the motor to deliver higher torque at high speeds, but still do not make the torque versus speed curve linear. A typical stepper motor torque curve is shown in Figure 8-9.

Because of the divergence of the running torque and start-up torque versus speed curves at even moderate step rates (with start-up torque being lower), the controller should ramp a stepper up to speed rather than accelerating instantaneously from a standstill. Even if ramped, steppers may suddenly seize at some point in the ramp when the load is such that the rate of stepping causes the rotor to oscillate at a harmonic of the mechanical natural frequency. This problem can be reduced somewhat by using the telescope and rotor inertia matching design practices discussed earlier. Mechanical rotational dashpots are sometimes added to steppers to increase damping, which decreases the per-step settling time. Ramping a stepper permits it to achieve higher step rates, and is necessary in worm gear drives to prevent gear damage, and to reduce telescope oscillations when a slew is completed. The automatic photoelectric telescope described in Part V initially ran without ramping, but was modified to include this feature. In that system, a step size of 1″6 is used with a tracking rate of 11.5 steps/second and a slewing speed of about 5,000 steps/second.

In evaluating stepper performance, note that the stepper's effective dead band is essentially its step size. For motors with 200 steps/revolution, each step is 1.8°. Gear reductions should be chosen to make each step an appropriate size. In systems with separate slew and track motors in which the track motor is a stepper, the step size should be considerably smaller than the seeing disk, and small enough to avoid any noticeable vibration, say $1/2$ to $1/20$ arc second. In systems where there is only one stepper motor per axis, then the step size and slew rate must be balanced against each other. Having high maximum step rates eases this tradeoff, but it is possible that one may have to live with a large step size or a slower slew speed than one might have otherwise wanted. In some applications, small step size is only important for tracking in RA. In these cases, the step size can be large (1″ or more) in Dec, and in RA while slewing. A higher gear reduction ratio is usually needed for a stepper motor than that typically used with a servo motor because of the torque curve characteristics.

Despite the need for care in design, ramping, and additional gearing, steppers can give performance almost as good as servo motors, and at reasonable cost. Servo motor and stepper motor costs are roughly equal when the motor, controller, power supply, computer interface, and gearing are all taken into account (except for the Dynaserv and other high-performance and costly servo motors). All of the items needed to build a telescope axis drive using either a servo motor or a stepper motor can be purchased for about $1,000 per axis—considerably less than for a large DC torque motor. If you have a sharp eye at the surplus electronics stores, the patience to wait for a bargain, and the willingness to build some of the electronics yourself, you can cut the price to $300 per axis, or even less.

Stepper motors have several significant advantages over other types of motors which make them often the best choice for a low-cost computerized telescope control system. First, they can be used without position or velocity feedback in simple open-loop systems of moderate accuracy (up to a few arc minutes). This eliminates the cost of shaft angle encoders and their computer interfaces, if very high pointing and tracking accuracy are not required. The computer can determine the number of pulses to be sent to the stepper motor, and using the (geared down) step size, the software can estimate the telescope's position. This is not possible using either the DC torquer or the smaller DC servo motor, except for the expensive servo motors with modern controllers.

The second advantage of the stepper motor is that it is easy to interface to a computer. All that is needed is a source of pulses. Dozens of computer interface boards are available that generate pulses and a direction bit, or that generate the motor winding currents directly. Although some DC servo motor controllers also use the step/direction interface, they tend to be quite expensive.

One disadvantage of both steppers and small servo motors is the need for gear reduction. Aside from introducing some errors, which can be reduced to a reasonable level using precision gears, the use of gear reduction requires that these motors be capable of a high maximum speed, which is often difficult to achieve. To solve this problem, and because motors are not capable of full torque at high speeds, a separate slew motor is often used, especially on older telescopes larger than one meter in aperture. Although it requires an additional computer interface per axis and the slew motor must have higher torque than the tracking motor, small telescope slew motors often can be operated at a single speed, so it may be connected to line voltage with a simple on/off switch under computer control (e.g., an Opto-22 module). Most slew motors on larger telescopes are ramped, however, to prevent gear damage.

One form of mechanical reduction that is quite useful is precision belting, such as that available from Winfred M. Berg, Inc. A good toothed belt cushions stepper pulses slightly, which helps to prevent stalling due to mechanical resonances. Toothed belts have negligible backlash, and the slight amount of elasticity in the belt tends to average out tooth-to-tooth belt errors. Belts are most appropriate as the first stage in the drive train right after the stepper, so their errors can be reduced by later gear stages.

The use of two motors per axis, one for tracking and one for slewing, is expensive. To avoid this expense, a high maximum step rate is needed so a single stepper per axis may be used. One method of obtaining this high step rate is to employ a higher than normal voltage to drive the stepper when it is slewing the telescope. The field windings have the electrical characteristics of a pure resistor in series with a pure inductor. When the normal operating voltage of the stepper motor, say 6 volts, is applied to the field winding, the resistor limits the total current flow, and the inductor uses the inrushing current to build up a magnetic field about its windings. It takes a relatively small, but finite amount of time to build up the magnetic field in the windings. This time is only small when compared to the length of time between steps at low step rates; at high step rates, it can be a significant fraction of the time between steps. This reduces the torque for high step rates, which effectively limits the maximum step rate for most steppers to about 500 to 1,000 steps per second.

By using an over voltage of a few times the normal operating voltage to slew the motor, say 24 volts for a 6 volt motor, the time it takes to build up the field is reduced, since the initial current is larger. This produces higher torque at higher step rates. By shortening the drive pulse "on" time, the energy stored in the field windings is delivered at a faster rate, so the power (at any given step rate) is increased. This technique can be used only for high step rates, since when the motor is stopped or stepping slowly, the amount of time the drive voltage is switched to any given winding is long enough for large currents to build up and overheat the winding. A variation of this approach, which can be used for all motor speeds, is to apply the higher voltage only for a short time at the beginning of a step, then to reduce the drive voltage to the rated voltage. Circuits to implement this "bi-level" approach can be found in the literature of some of the stepper motor manufacturers including Anaheim Automation. The concept is illustrated in Figure 8-10.

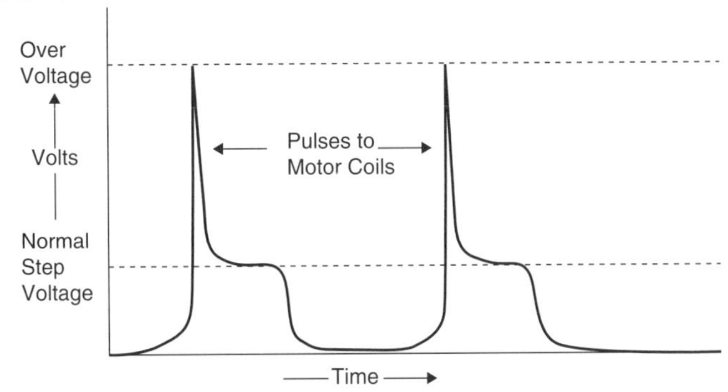

Fig. 8-10 *Bi-level stepper driver characteristics.*

Fig. 8-11 *Single-Motor and Three-Motor Bi-Level Stepper Motor Drivers. Courtesy Anaheim Automation.*

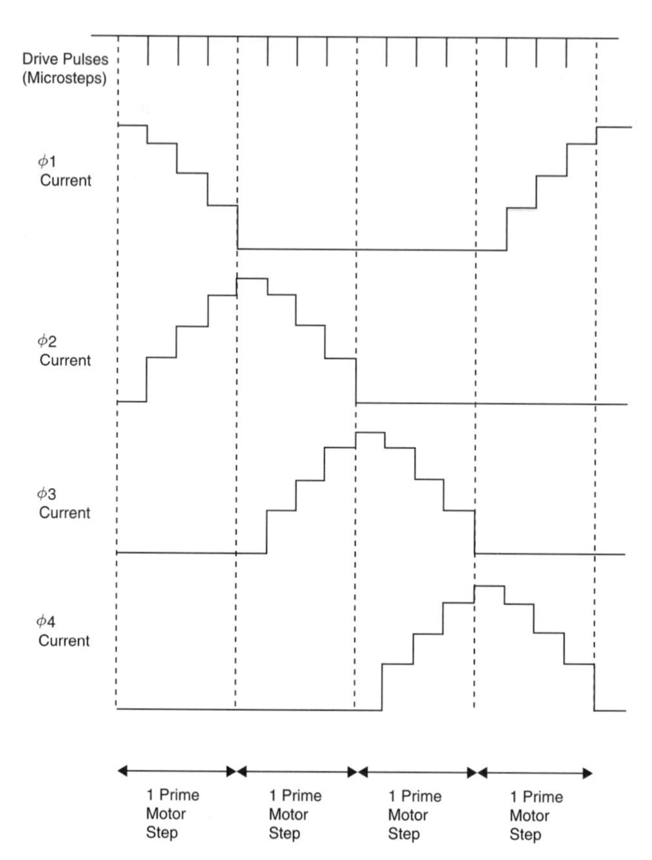

Fig. 8-12 *Microstepping timing chart. Courtesy Mesur-Matic Electronics Corp.*

Another method of improving the speed range of a stepper motor is to drive the motor in half steps instead of full steps. Half steps are achieved by driving two field windings simultaneously, rather than just one at a time. This produces 400 steps in a 200 step per revolution motor, which allows more freedom in choosing mechanical reduction ratios. It also produces smoother steps, which means the motor is affected less by mechanical resonances. Superior Electric, Anaheim Automation, Parker Compumotor, and other manufacturers make stepper motor controllers capable of half step operation. The use of half steps (or smaller) is recommended over the use of full steps, which disturb both the motor and the entire drive train with large "kicks" of torque that tend to induce vibrations in the telescope's image.

The half step concept has been extended by Parker Compumotor, Anaheim Automation, and others to microstepping. Microstepping can be achieved by placing several small current steps per full motor step on pairs of field windings, or by a pulse width modulation technique. The net effect of both methods is to divide the current flow between the pairs of windings. The Compumotor LN Series has 16 resolutions from 200 to 101,600 steps per revolution. The technique used by Mesur-Matic to divide each full step of a 200 step per revolution motor into 4 microsteps is illustrated in Figure 8-12. Step rates up to 2 million steps per second are possible with microstepped motors. At the other end of the speed range, microstepping allows motors to be operated at about one-tenth the speed achievable using conventional stepper drivers without introducing noticeable vibration. This provides the wide speed range required for drives using one motor per axis, especially for alt-az and alt-alt mounts. One skilled in electronics could also solve these problems by building a stepper driver that can be switched between two or more step sizes to help solve the step size/slew rate tradeoff problem mentioned earlier. Several manufacturers, including Parker Compumotor, make stepper controllers that can change step sizes under software control.

The speed range of any motor is necessarily limited. Generally, DC servo motors have a wider speed range than stepper motors. Among the steppers, those that employ microstepping have the widest speed range. For example, Anaheim Automation steppers can rotate at speeds up to 50 revolutions per second. If separate slew motors are used, then speed range will not be a problem. However, if one is willing to make the necessary compromises, then the conventional stepper is the least expensive single-motor-per-axis approach. If better performance is desired, then a microstepper or DC servo motor approach is preferable.

One characteristic of steppers that should be taken into account is the motor's absolute accuracy specification. When a torque load (as opposed to a pure friction load) is placed on a stepper at rest then the stepper is powered up, the holding torque of the stepper is offset by the load, producing a shaft position error. This effect is depicted in Figure 8-13. The following table gives a rough idea of shaft error (deviation from the no load position) as a function of load torque (expressed as a percentage of the motor's holding torque) for a typical 200 step per revolution motor.

Load Torque (% of holding torque)	Error (degrees)
25	0.3
50	0.6
100	1.7

In a dynamic situation with no load on the motor initially, when the motor starts to turn at some constant rate, a constant load torque will

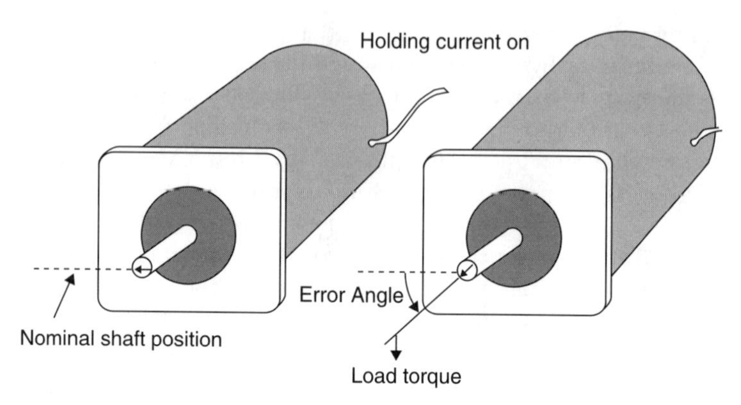

Fig. 8-13 *Stepper error induced by load torque.*

introduce a *phase error* (shaft position error) into the servo loop of an amount given roughly by the table above. This error will remain constant as long as the torque load causing the error remains constant, and will appear to be servo lag position error. However, as the torque load changes with changing telescope pointing angles or wind loads, the motor shaft position error will vary by amounts similar to those given in the table for the static case. For the dynamic case, the motor torque of importance is the pull-in torque, not the holding torque. Gearing should be chosen so that the change in torque load on the motor for all telescope pointing angles and wind loads is a small fraction of the motor's pull-in torque. Gear reduction not only amplifies the motor's torque, it also reduces the effect of a given motor shaft position error on the driven telescope axis.

As an example of how a manufacturer specifies a stepper's absolute accuracy, Parker Compumotor specifies the unloaded absolute accuracy of its LN Series microstepping motors as $\pm 5'$. The loaded error varies with the load, and is specified as being approximately equal to the unloaded error plus a factor proportional to the residual friction loading when stopped. This additional error is given as $1'$ for a friction load of 1% of the motor torque. This is roughly linear, so a 50% load would produce a dynamic shaft position error of $50'$. When added to the unloaded error of $5'$, this would produce a total motor shaft error of $55'$, or almost 1° (with no gear reduction). Each motor step in a motor with 25,000 steps per revolution is only $52''$, so the total shaft error can be as large as 60 times the step size if the motor's torque is much too small to handle load variations.

Another effect, unique to microstepped motors, is torque ripple. The motor is microstepped by applying different currents to two windings simultaneously. When one of the motor's permanent magnet poles is nearly aligned with a field winding, almost all of the current is in the one field

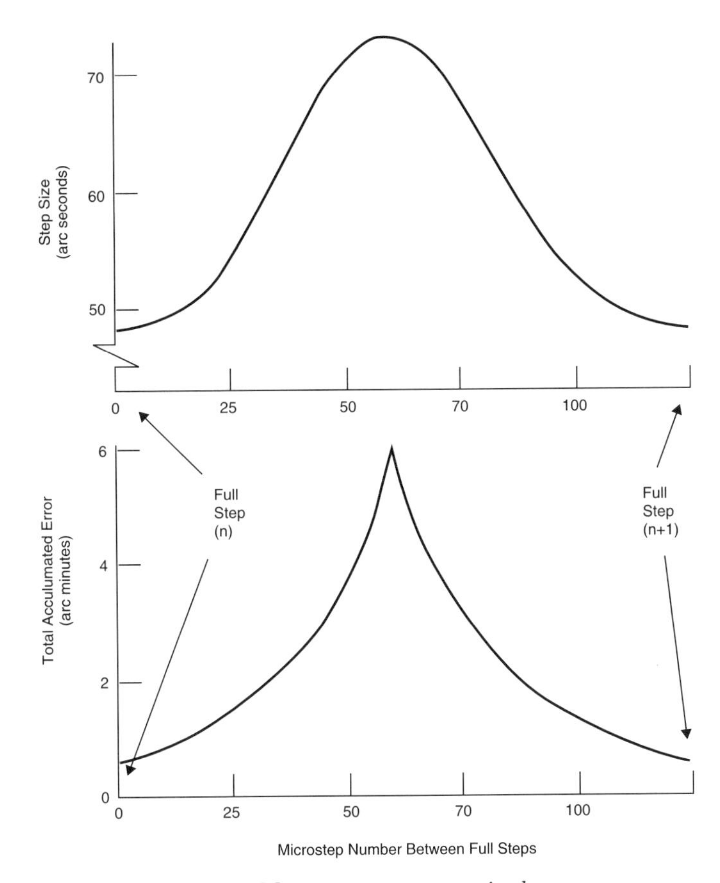

Fig. 8-14 *Microstepper torque ripple error.*

winding nearest the magnet pole. This produces a slight dip in the torque at this rotor position. With a constant load, this reduction in torque shortens the step size slightly. Conversely, when the magnet pole is in the middle between two field windings, both windings are driven by roughly equal currents. This raises the torque slightly, which lengthens the step size. In a stepper with 25,000 steps per revolution, this effect causes the step size to vary sinusoidally between $50''$ and $70''$. The error does not accumulate over a complete $360°$ rotation of the rotor, but it does accumulate temporarily between the motor poles to a maximum value of $\pm 5'$, as shown in Figure 8-14. Since the individual step error (as much as $20''$) is a large fraction of the nominal microstep size ($52''$), and the maximum error between motor poles ($5'$) is so large, the motor should be geared to the telescope axis in a manner that reduces this effect to a tolerable level.

Fig. 8-15 *A stepper with integral gearhead reducer. Courtesy Vernitron Corp.*

When dealing with such small step sizes, step-to-step errors due to small variations in the spacing and shape of the rotor magnet poles and the placement of field windings (both of which vary from motor to motor and from step to step within a particular motor) become a significant effect. Compumotor corrects for these effects by measuring each unloaded step size at the factory, and storing correction current values in a PROM that is placed in the motor controller. This feature customizes the field winding currents for each microstep to correct for motor geometry variations, but it cannot eliminate the torque ripple errors.

Another error source in microstepped motors is hysteresis. If the motor is stepped some number of steps in one direction, then the same number of steps in the opposite direction, the shaft does not end up at the original position. As much as $3'$ of the $5'$ total unloaded shaft error in the Compumotor LN Series motors can be due to hysteresis (White, 1984).

The effects of dynamic load torque and torque ripple errors can be reduced through gear reductions between the motor and the driven shaft. By choosing the gear reduction and motor torque carefully, a microstepper can be a very effective telescope drive motor.

Steppers are available with many options. Figure 8-15 shows a stepper with an integral gearhead reducer, and Figure 8-16 shows a stepper with an integral lead screw, which would be useful for moving a Cassegrain secondary for focus adjustment.

8.6 Stepper Motor Controllers and Computer Interfaces

Unlike servo motors, steppers require a controller that turns the winding currents on and off in a precisely defined sequence. This can be done directly

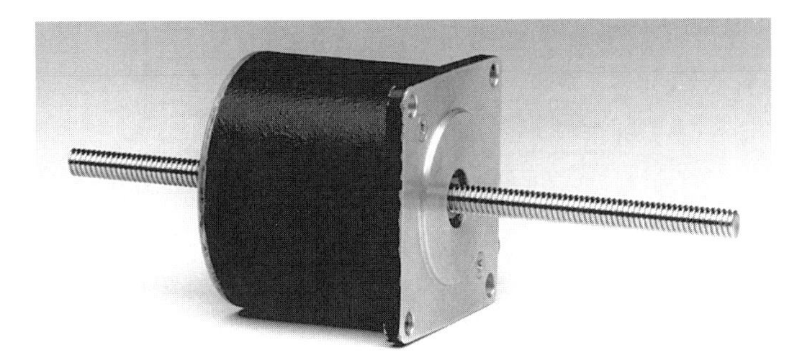

Fig. 8-16 *A stepper with integral lead screw. Courtesy Anaheim Automation.*

by the computer at low cost. Eight single bit outputs (four per motor) can be toggled by the computer in the proper drive sequence for two motors. These outputs drive power transistors that feed coil currents to the steppers. This requires a sizeable number of the available CPU cycles of an 8-bit microcomputer, but if other computations do not consume very much of the CPU, this is probably the least expensive way to drive a telescope with a computer, especially since steppers are readily available at many surplus electronics stores.

If you use your computer in this way, you will need to build a board to drive the coils with the proper current. This is not recommended, for the reasons given in Chapter 7. But if you insist, a schematic diagram for a bi-level driver board is given in Figure 8-17, and Figure 8-18 shows how three such circuits looked when assembled.

The next step up in price is to use a commercial stepper controller. This approach reduces the load on the CPU at the cost of adding more hardware to the system. Anaheim Automation and other manufacturers offer low-cost controllers capable of either full or half steps. They usually accept TTL-level inputs on two separate input lines. These can be in the form of either pulse rate and direction, or up-pulses and down-pulses (pulse one line for CW rotation, pulse the other line for CCW rotation).

Although some inexpensive stepper controllers can accept pulse rates up to 1 MHz, they typically do not compensate for motor characteristics that change with speed. This usually limits the top speed of the motors driven by this driver to under 1,000 full steps per second, since the motor torque falls off rapidly after about 500 full steps per second. The computer could be used to toggle a single output bit at the desired step rate. This output would serve as the signal to the driver card. A second bit could be used to indicate motor direction to the controller card. This is the approach used in an early version of the APT system described in Part V. The APT stepper

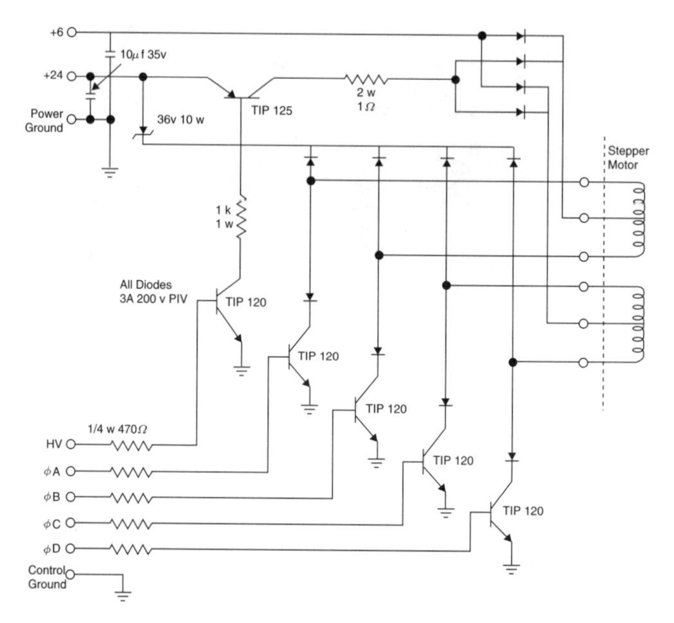

Fig. 8-17 *A Bi-Level Stepper Driver Schematic Diagram.*

driver is capable (with ramping) of up to 5,000 half steps per second with ordinary Slo-Syn steppers. Most of the parts for this driver can be obtained at Radio Shack for about $30. More sophisticated driver cards are offered by Superior Electric (manufacturer of the Slo-Syn steppers) and others.

If the microcomputer generates the pulse train, a good deal of the CPU can be consumed in this activity. In the APT, the CPU has little else to do, so here CPU loading is not an issue. However, in those applications where the CPU must perform other tasks, an interface card can be used to generate the pulses that are translated by a stepper controller into motor steps. Pulse output cards designed to drive steppers are available for many different computers.

One example is the Oregon Microsystems Model PC58-8 shown in Figure 8-19. This card has eight output channels for driving up to eight motors. Each channel can send different pulse rates from 0 to 1,044,000 steps per second with separate lines for speed (pulse train) and direction. This can be used with motor driver cards for either steppers or servo motors that accept step and direction inputs and generate the proper coil currents for the motors they drive. The card also has ±10 VDC or PWM outputs for directly driving servo amplifiers.

The computer determines the pulse output rate by using a simple ASCII language and software that comes with the card to help you load commands.

Fig. 8-18 *Three Assembled Bi-Level Stepper Driver Boards.*

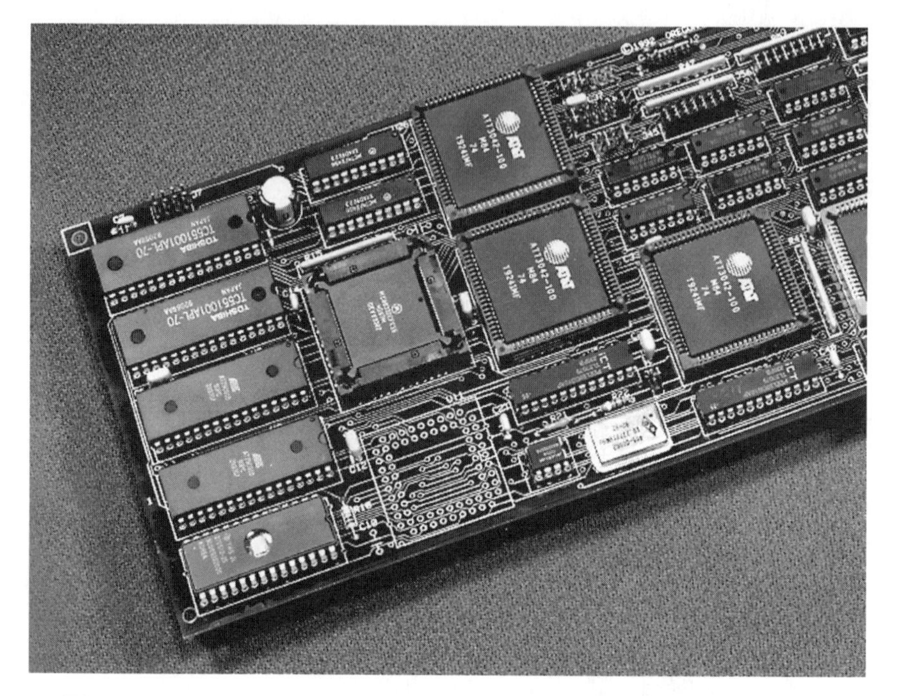

Fig. 8-19 *8-Motor PC Interface Card. Courtesy Oregon Micro Systems.*

The card provides commands for setting maximum speeds and accelerations for ramping up and down, setting a particular speed, setting a multiplier so commands can be in user units, reading back the position of the motor shaft based on the net number of pulses sent (taking into account the direction of each command), and dozens of other functions. To use this card, you set up the various parameters (e.g., maximum speed and acceleration), then send the card the speed command. From that point on, the card sends out pulses at the commanded rate without further reliance on the computer until the computer sends a new command. Besides the pulse and direction output lines, this card has independent home and limit switch inputs for each axis, 22 bits of user-definable I/O, and a position range of 67 million pulses on each axis.

Anaheim Automation makes a series of boards similar to the Oregon Microsystems EISA/ISA boards. The CLCI500 Series comes in one- or two-motor versions, and can step at rates up to 24,444 steps/second with step counts up to 65,535 steps per index (position move command). The CLCI2000 Series comes in versions for 1–4 motors, with step rates up to 2,500,000 per second and step counts up to 16,777,215 per index.

Another approach is the Superior Electric stepper driver that uses the

General Purpose Interface Bus (GPIB), otherwise known as the Hewlett-Packard Interface Bus (HPIB) or the IEEE-488 standard. Since many computers are available with GPIB interfaces, this driver or one using a similar approach could become popular.

We do not recommend designing and building your own computer interface card, but for hobbyists with lots of time and very little cash, one approach is to obtain a prototyping board for your computer's bus, and add presettable divide-by-N counters and a crystal oscillator. To set a speed, you load a binary number into the card's registers that represents the crystal frequency divided by the number of steps per second you want to generate, rounded to the nearest integer. If the crystal is about 20 Mhz and you use at least three bytes (24 bits) of divide-by-N counters, you can generate a very wide range of drive rates with very high speed resolution. This approach would be suitable for driving an alt-az or alt-alt telescope, but you would need to write software to control ramping and other motor control functions done by commercial boards. You would also spend a great deal of time debugging the board.

One final point to bear in mind when designing the drive system is how the motor/gear combination affects the stability of the control servo. For example, direct-drive systems using a large DC torquer can be difficult to fine-tune and usually require both position and velocity servo loops, and some even have acceleration loops (Klim and Ziebell, 1985). This is because structural resonances are reflected directly into the velocity loop, and the inertia of the motor is small compared to the inertia of the load. This usually requires a separate compensation (filter) network for each servo loop and can easily lead to stability problems. This problem is solved in modern servo controllers using a powerful CPU chip and a smart algorithm to tune servo parameters to the actual load conditions, but these features are expensive.

A smaller motor geared to the telescope has a much larger inertia as "seen" by the load (larger by the factor n^2, where n is the gear reduction), so the structural resonances are decoupled from the motor frequencies above the locked rotor resonance frequency. The bandwidth of the velocity loop can therefore be high without major stability problems.

Although servo motors have better torque curves, better performance in general, and can be as easy to interface to computers as steppers, the latter do offer a real advantage when used in secondary mirror focusing mechanisms. Normally, both servo motors and stepper motors have some current in their coils even when stopped. This current can generate heat that can cause bad seeing from convection currents inside the telescope tube. Anaheim Automation and others offer stepper controllers that can be commanded by a computer to turn off all current when stopped. If the stepper moves a non-backdrivable gear (e.g., a precision lead screw), then it can be turned off without moving the secondary mirror. Usually, this cannot be done with servo motors, so steppers have a clear advantage in this application.

Stepper motors also have a clear advantage when cost is a major factor, and when the best performance is not needed. Examples of applications not needing outstanding performance are moving filter wheels, flip mirrors, field derotators, and other optical parts of telescopes and instruments.

This is only a small sample of the wide range of motors and controllers that are available. Having discussed in some detail how these components can be used in telescope control systems, we now turn our attention to methods of providing position feedback from the telescope axes and other information to a computer, so the control system can "sense" where the telescope is actually pointing and environmental factors affecting telescope operation and performance.

Chapter 9

Sensors

Any computerized control system needs information from the environment in which it is controlling something. In a simple open-loop telescope control system, this could be the coordinates of a bright star centered in the eyepiece that the observer enters on a keyboard. More sophisticated telescope control systems might use angle and linear position sensors to detect the position of the telescope axes and the movable secondary mirror (for focusing). They might also receive time signal inputs, and perhaps temperature and barometric pressure information for computing refraction. These are but a few of the sensors that are useful to telescope control. In this chapter, we review some of the more popular types of sensors in use and briefly explain how they function.

In a classical closed-loop telescope control system, one or more position feedback sensors may be placed anywhere between the motor shaft and the telescope focal plane. Once a position feedback sensor is calibrated so that its output can be converted to topocentric position in the sky, the sensor will automatically correct, through its calibration, for any systematic errors (e.g., gear errors) from the motor shaft up to the position feedback sensor. The position sensor cannot be used to correct for errors between it and the focal plane. These "down stream" errors (mount flexure, tube flexure, collimation errors, etc.) may be modeled by the error and command computer, which then determines a new position feedback sensor reading corrected for the modeled errors. By placing the position feedback sensor as close to the focal plane as possible, one can use the sensor calibration to remove systematic errors directly, rather than increasing the computational burden of the error and command computer in computing these error models.

Many methods are available for providing shaft angle feedback for closed-loop servo telescope control, with a wide range of accuracies and prices. Most telescopes using control systems with position feedback employ optical shaft angle encoders, so particular attention should be paid to this portion of the chapter. Inductosyns and synchros are also used in larger telescope applications. We would expect most modest-sized telescope control systems to be open-loop, a few to use precision potentiometers for low cost position feedback, and most of the rest to use incremental optical encoders of moderate accuracy. Other types of sensors are covered to stimulate interest and ideas for improving the accuracy of low-cost position feedback devices. Appendix C lists the names and addresses of several manufacturers of position angle feedback devices, while vendors of other types of sensors are listed in Appendix E.

9.1 Precision Potentiometers

One frequently-used form of shaft angle encoder is the *precision potentiometer.* Two examples are shown in Figure 9-1. This device is typically used with a well-regulated DC power supply to provide a DC voltage proportional to the potentiometer's shaft angle. In analog servos, it can be used for both position control input and for actual position feedback. A precision op-amp can be used to compute the error signal, which is then amplified to drive a DC servo motor. This very simple system was described in the previous chapter.

The characteristic that determines a potentiometer's suitability for position encoding is its conformity to the desired function of resistance versus shaft angle. The function most often desired in a telescope control system is a linear relationship between resistance and shaft angle. Any deviation from linearity is a characteristic of the potentiometer, and cannot be offset by adjusting other system parameters. Wire-wound potentiometers are quite common, and can be obtained on the surplus market with a linearity of 0.1% (0.36° if used directly on the measured axis) for roughly $15–$25. The accuracy of these potentiometers is often limited by the potentiometer's resolution. For example, when the pot is directly coupled to the telescope axis, the resolution is 360° divided by the number of turns of wire used to make the winding. Since 1000 turns or more of wire is typical for these potentiometers, their resolution roughly equals their linearity, or 0.1%.

The other parameter that influences the accuracy of wire-wound potentiometers is temperature stability, which may be on the order of 0.05% per degree C. This limits the overall accuracy to about one-half degree of arc for typical observing conditions. This may be adequate for those who wish to find only Messier objects. A way around the temperature stability problem is to use the ratio of resistances (or voltages) of the two legs of the pot instead of the absolute resistance (or voltage) of just one leg. This can be done in an analog circuit by placing one leg between the inverting input of

Fig. 9-1 *Precision potentiometers. Courtesy Vernitron Corp.*

an op-amp and ground, and the other leg between the inverting input and the output of the op-amp. In a microcomputer-based system, both legs of the pot can be connected to the computer using analog-to-digital (A/D) converters, and the ratio of voltages can be computed directly. However, power supply voltage ripple and random noise on the supply voltage to the pot will still introduce errors into the readings of the voltages on the two legs, and limit the overall accuracy to about 5–10 arc minutes.

Several suppliers manufacture so-called "infinite resolution" potentiometers that use a thick film of conductive plastic as the resistive element. The resolution of such potentiometers is not limited by the pot itself. In analog systems, the overall system resolution of "infinite resolution" pots is determined by the bias voltage and drift of the DC amplifier used in the servo. In digital systems, the sample-and-hold amplifier of the A/D converter and the resolution of the A/D converter itself determine the pot's overall resolution in the system. These factors, plus the voltage regulation of the DC supply feeding current to the pot limit the accuracy of these pots to about 0.1° of arc. Conductive plastic potentiometers may be obtained on the surplus market for about \$10–\$30.

Potentiometers are packaged in many different shapes, sizes, and configurations. Both wire-wound and conductive plastic pots are available in

Fig. 9-2 *Multi-turn conductive plastic potentiometer. Courtesy Vernitron Corp.*

multi-turn versions that can be geared to each axis to take advantage of the increased resolution and accuracy, since many of the pot errors are decreased by the gear ratio. This arrangement does introduce gear errors, but if high precision anti-backlash gears are used, the gear errors on a multi-turn pot should be less than the intrinsic errors of a single-turn pot. Figure 9-2 shows a 10-turn conductive plastic pot. Figure 9-3 shows two linear potentiometers that could be used for determining the position of a Cassegrain secondary when its separation from the primary is used to determine the location of the focus. In systems employing a multi-turn pot, it is important to "unwrap" the pot from time to time instead of always rotating the telescope axis in the same direction. This prevents the pot from hitting its stop and being damaged by additional telescope axis travel in the same direction.

When using a pot in a control system, it is important to ensure that the voltage response of each leg of the pot is linear. This means the impedance of the computer interface that senses the voltage must be orders of magnitude greater than that of the pot. But if one makes the total resistance of the pot too low (in an attempt to make the sensing impedance relatively high), the larger resulting currents produce self-heating of the pot, which can cause hot spots in the resistive element of the pot that lower the pot's accuracy. Taking the ratio of the voltages of the two legs of the pot does

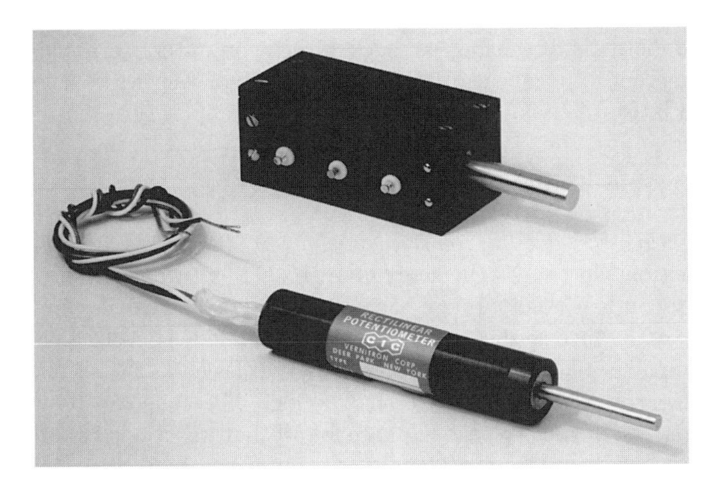

Fig. 9-3 *Linear potentiometers. Courtesy Vernitron Corp.*

not compensate for these hot spots. This means that the pot should be of high enough resistance to avoid significant self-heating, and the input impedance of the circuit that senses the voltage should be very high (10 to 100 megohms).

There are two common methods for measuring a voltage with a computer. The first is to use an analog-to-digital converter (A/D), which is a module or chip designed to convert a voltage input directly to a digital value. Typical A/D converters come in resolutions from 6 to 22 bits, with 8-, 10-, 12-, and 16-bit converters being the most popular. The resolution of the A/D converter determines how many different voltage values can be sensed. For example, a 12-bit A/D converter divides the input range into 2^{12}, or 4096, different possible output values. For an input range of 0–10 volts DC, each "count" of the A/D converter is 10/4096 volts, or about 2.44 millivolts.

The second way for a computer to sense a voltage is to apply the voltage to the input of a voltage-to-frequency (V/F) converter, then count the pulses from the V/F converter for a fixed period of time (e.g., one second). V/F converters are available with a frequency range of 0 to 1 MHz over a voltage range of 0 to 10 volts DC. This gives much greater resolution than most ordinary A/D converters, and V/F converters are typically as linear as A/D converters. A simple counter circuit, which is very easy to interface to a computer, can count the pulses from the V/F converter. Several companies make both A/D converter cards and pulse counting cards for popular standard computer buses, including Intelligent Instruments,

Data Translation, CyberResearch, and Computerboards (see Appendix D).

These methods of sensing voltages with a computer are not limited to potentiometers, but can also be used for many of the other sensor types discussed later, such as barometric pressure or temperature sensors.

9.2 Variable Reluctance Transformers

The *rotary variable differential transformer* (RVDT) is another type of angle position sensor. As the shaft of an RVDT is turned, an iron core or secondary coil is moved through the field of the primary coil, changing the magnetic reluctance of the transformer. A regulated AC voltage is applied to the primary coil, and the output is an AC voltage on the secondary coil that is linearly dependent on the shaft angle. The coils can be wound to provide the AC output voltage in phase for all shaft positions (a V response) or 180° out of phase for shaft angle positions less than a predefined zero (straight line response). RVDTs usually use a ferrite core with a cardioid shape to linearize the response. While most RVDTs allow a full 360° of rotation, the output is linear over only about 80–120° of rotation. Later designs have extended this to 340° of rotation, but this could still present some difficulty in calibration when a full 360° of rotation is needed, such as for the azimuth axis on an alt-az mount.

In an analog servo, two RVDTs may be connected in a master/slave circuit with a servo amplifier and motor, so that the motor shaft angle always agrees with that of the master RVDT, whose shaft is turned by the observer to indicate desired position. This is analogous to the simple servo using pots as both input and feedback devices. In a digital servo, the RVDT output would be converted to a DC voltage using a precision true RMS filter, then fed to an A/D converter connected to the control computer.

RVDTs are more accurate than precision potentiometers, with overall accuracy of a few arc minutes possible with precision units. The temperature stability is better than potentiometers, but they are expensive ($200 on up) and are rarely seen on the surplus electronics market. They typically are not used for telescope position feedback because of their cost, the expense of the required support circuitry, and the difficulty of building an interface to connect them to a computer.

Microsyns are also variable reluctance transformers, but are typically constructed with a cylindrical ferrite armature that rotates inside a stator with four field poles, two each for the input and output coils. Microsyns are used when the total angular motion is only a few degrees, and thus are unsuited for telescope axis angle feedback.

Fig. 9-4 *Three high-accuracy resolvers. Courtesy Singer Co., Kearfott Division.*

9.3 Resolvers

A *resolver* is a transformer with (usually) two or more windings, with at least one on the stator and one on the rotor, that are coupled so that the output voltages are proportional to the sine and cosine of the rotary shaft angular position. In this way, it differs from an RVDT, which has a linear output. Figure 9-4 shows some examples of resolvers. Most resolvers available today use two stator windings and two rotor windings to enable the calculation of complex (four quadrant) trigonometric functions in an analog system. Figure 9-5 shows the internal construction of a typical resolver.

Some resolvers are capable of very high resolution and accuracy—18 bits (262,144 counts per revolution) or greater. They are also very expensive. The Northern Precision Laboratories Model 801397-2 with 18-bit resolution costs $11,500. This is somewhat less than a new optical shaft angle encoder of the same resolution and accuracy, but such high resolution resolvers are often a bit larger and heavier, require more power, and require more support electronics than optical shaft angle encoders.

Very high resolution and accuracy can be obtained using two resolvers, one tied directly to the axis of the telescope, and the other geared to the axis such that it rotates several times with one turn of the main shaft. This type of resolver is called "two-speed," and uses two lower resolution resolvers, each turning at a different speed. Figure 9-6 is a block diagram of such a two-speed resolver configuration, and Figure 9-7 shows the kind of electronics board used to combine the outputs of, for example, a 14-bit and an 11-bit

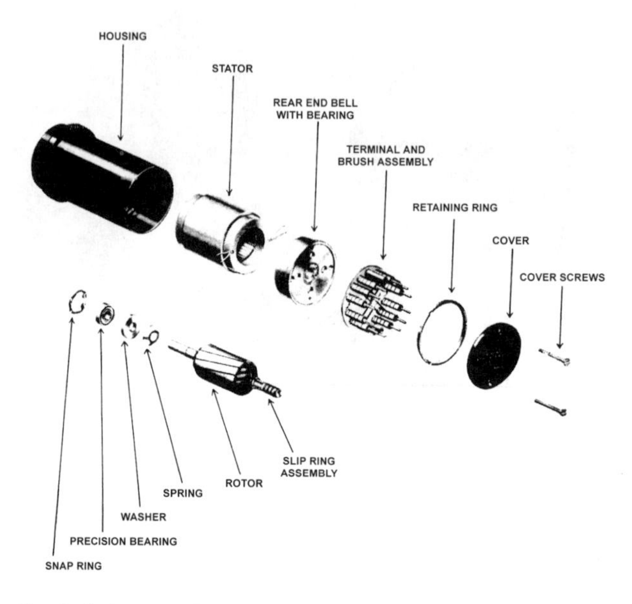

Fig. 9-5 *Exploded view of a 4-coil resolver. Courtesy Reston Publishing Co.*

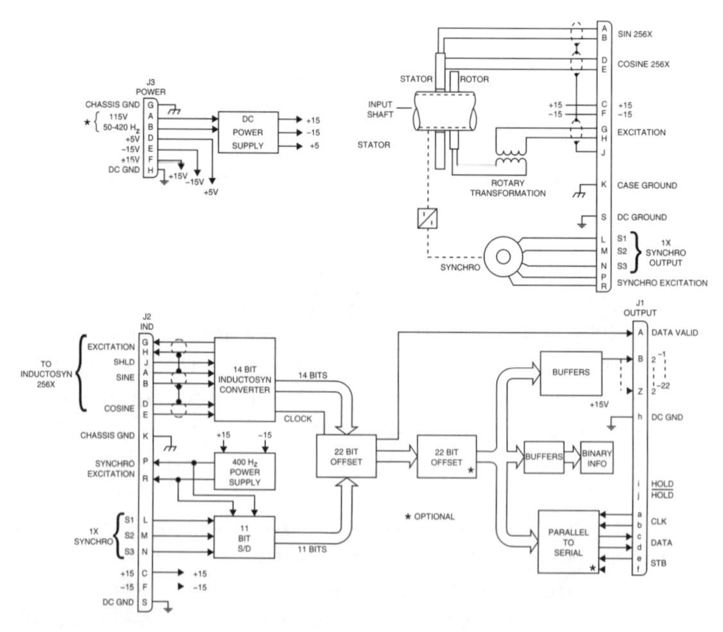

Fig. 9-6 *Electrical and mechanical layout of a two-speed resolver. Courtesy Northern Precision Laboratories, Inc.*

Fig. 9-7 *Digital circuit board with 800 Hz oscillator and VCO. Courtesy Northern Precision Laboratories, Inc.*

unit to obtain 22-bit (roughly 0″.3) resolution. The 256:1 gearing gives three overlapping bits between the two units, which the electronics board uses to sort out gearing errors and intrinsic resolver errors. The result is a unit with an RMS error of about 1″. Figure 9-8 shows a two-speed resolver with very high resolution and accuracy.

Northern Precision Laboratories sells circuit boards that provide the interface between their resolvers and computers using the Multibus. Adapter boards are available to connect a Multibus to a few of the more popular personal computers. Analog Devices offers a series of modules that make it easy and inexpensive to build a computer interface for common types of resolvers.

If one is able to obtain a high resolution resolver inexpensively on the surplus market, it would be worth considering as an angle position feedback

Fig. 9-8 *Two-speed, multi-turn absolute resolver. Courtesy Northern Precision Laboratories, Inc.*

sensor, but surplus optical shaft encoders of higher resolution are usually more readily available for less money. Therefore, resolvers are not expected to see much use in telescope control applications, except, perhaps, at larger professional observatories.

9.4 Synchros

Synchros are transformers that are used in pairs to transmit shaft angle position over large distances without a direct mechanical linkage. The synchros are connected together in an electrical circuit such that when one shaft is turned, the other shaft also turns to the same angle. The name comes from the fact that synchros are "self-synchronous." Other names are *selsyn*, *autosyn*, or *teletorque*. Synchros are also used for torque transmission or for voltage indication. Synchros could be used as a pair to form an open-loop analog control system, in which one synchro is the input device for giving desired position, and the other is directly attached to the corresponding telescope axis, and drives the telescope to the desired position. Such a scheme, though possible, would be impractical, since the second servo would need the characteristics of a large DC torque motor.

A variation of the synchro that would be useful as an angle position feedback sensor is the synchro/resolver. This is a single synchro whose output, rather than being wired to another synchro, is electronically decoded into a useful form. A function module is available from Computer Conversions Corp. for about $500 that converts the three-phase synchro outputs

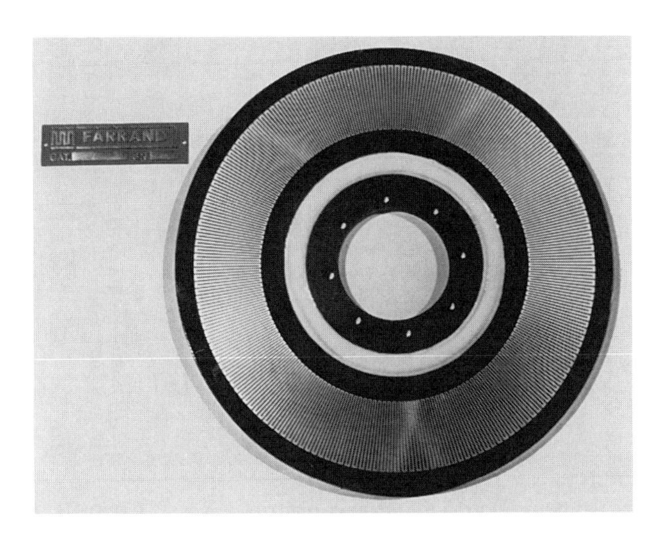

Fig. 9-9 *Inductosyn rotor. Courtesy Farrand Controls, Inc.*

to linear DC with ±6 arc minute accuracy. A similar unit from Vernitron is shown in Figure 9-12. A Scott T transformer is available from Magnetics, Inc. for $180 that converts the three-phase synchro outputs to sine and cosine voltages with 10 arc second accuracy. These could be fed to A/D converters, which are then read by the control computer and converted to an angle.

Analog Devices offers 14-bit synchro-to-digital converter modules costing about $200 that utilize a Scott T transformer and a quadrant selector, sine and cosine multipliers, and a subtractor to compute a binary number representing the angle. Two synchros can be linked together with gears and two-speed logic in a manner analogous to the two-speed resolvers. The two-speed logic is available for about $1,500.

Synchros are available for about $25–$50 each on the surplus electronics market, and the synchro-to-digital converter modules are available at reasonable prices. One could use just the Scott T transformer and an A/D converter, and do the sine/cosine calculations in the microcomputer, but that increases the computation load on the computer. Analog Devices offers all the modules needed for about $300–400 per axis that yield roughly 10 arc minute accuracy. They have written a 191-page book on the subject of synchros, resolvers, and Inductosyns that is an excellent introduction to these subjects.

Fig. 9-10 *Inductosyn stator. Courtesy Farrand Controls, Inc.*

9.5 Inductosyns

A variation of resolvers used extensively in telescope control applications is the Inductosyn (registered trademark of Farrand Controls, Inc.), a kind of resolver with optical encoder accuracy. One factor limiting the accuracy of a resolver is that wires are wound in a coil. The placement of the wire turns in the coil varies slightly from turn to turn, introducing non-linearities that are difficult to model. The Inductosyn is made using a photoresist technique to deposit a metal pattern on a stable dielectric substrate.

The Inductosyns made by Farrand consist of two aluminum or hot rolled steel disks, each containing a pattern of loops in metal foil on a dielectric substrate. The disks are mounted coaxially with the sides containing the foil pattern facing each other. One disk is the stator (fixed to the case) and the other is the rotor (rotates with the axis whose position is being sensed). The rotor and its foil pattern are shown in Figure 9-9, while the stator is shown in Figure 9-10. These patterns are deposited with an accuracy of a few arc seconds or better.

The device works by passing an AC signal of 200 Hz to 200 kHz (nominally 10 kHz) through the foil pattern of the rotor, then detecting the AC signal that is generated on the stator by inductive coupling across a small air gap between 0.003 and 0.008 inches wide. This is shown in Figure 9-11(a). The induced voltage is a maximum when the loops are facing. As shown in Figure 9-11(b), the induced AC voltage amplitude is zero when the loops are oriented midway between each other, then goes to a minimum when aligned with turns opposite to each other. With a foil pattern pitch of P

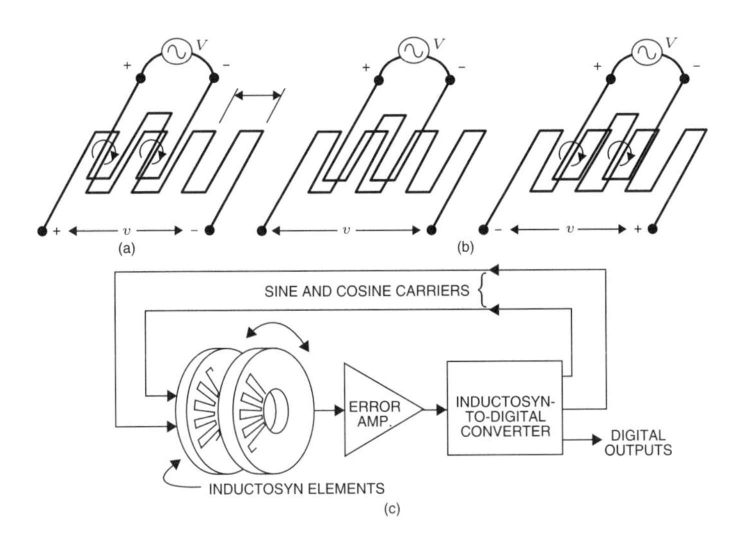

Fig. 9-11 *Principles of Inductosyn operation. Courtesy Farrand Controls, Inc.*

and a rotor input voltage amplitude V, the relation between the induced voltage V_a and displacement x from perfect alignment is given by

$$V_a = kV \cos \frac{2\pi x}{P}.$$

To provide a reference for comparing the voltage amplitudes, the stator has a second pattern laid down 90° out of (spatial) phase, or $P/4$, with the first. This produces a signal

$$V_b = kV \sin \frac{2\pi x}{P}$$

where k is a coupling ratio that depends on the size of the air gap and other factors. Both signals are in phase with the excitation AC signal in this mode of operation. However, by exciting the stator quadrature coils with signals phase shifted in time, a phase shifting operating mode can be achieved, which undergoes a shift of 360° for each displacement of one cycle length.

The rotor can have from 18 to 2048 poles (the units in Figs. 9-9 and 9-10 have 360 poles). When used in normal (non-phase shifting) mode, the stator output voltages act as a resolver of 18 to 2048 speed. The two outputs can be subdivided electronically to provide very high resolution. Electronics from Farrand convert the quadrature signal into digital pulses to make the

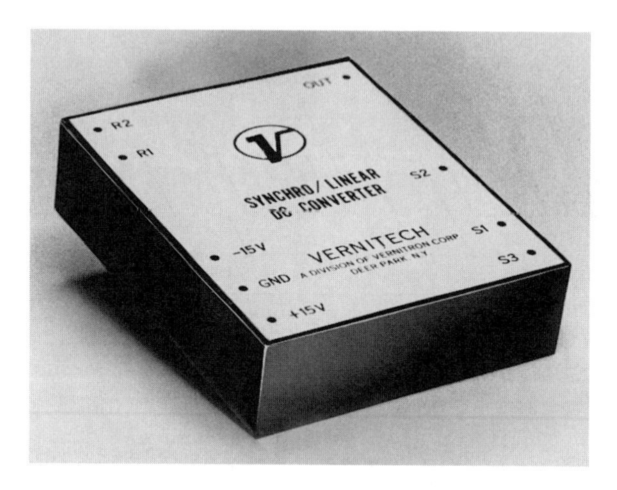

Fig. 9-12 *A synchro to DC volts converter. Courtesy Vernitron Corp.*

unit act similar to an incremental optical encoder. This package divides the sine (or cosine) wave from each pole into 1000 or 2000 parts. Other divisors are available as options. A standard 12-inch Inductosyn with 2000 poles and electronics using a divisor of 2000 has 4,000,000 counts (pulses) per 360° shaft rotation, or $0''.324$ resolution. The accuracy is quite high, since the signal is averaged around all the poles of the unit, so small random pattern errors tend to be averaged out. Production accuracies are typically $2''$–$8''$, but units with an absolute accuracy of $1''$ can be specially selected. The most common Inductosyn configuration is shown in Figure 9-11(c).

Because the pattern is much simpler to generate and replicate than that of an optical encoder, for very high resolution, (18 bits or more), the Inductosyn can cost less than an optical encoder of equivalent resolution and accuracy. For example, a very high accuracy optical encoder can cost over $11,000. The Inductosyn plates can be purchased new from Farrand Controls for about $1,800 a pair, with an extra $1,400 for the excitation and conversion electronics, and a $575 charge for selecting a pair of plates with $1''$ accuracy. Electronics from Analog Devices are less expensive (about $700) and offer even higher resolution.

Farrand also makes the Inductosyn pattern on straight substrates for linear positioning applications. Such a device would be useful for encoding the position of a Cassegrain secondary mirror. Farrand can select units with 0.000080-inch accuracy, or supply standard units with 0.0002-inch accuracy. This level of accuracy is not needed for encoding focusing positions on many telescopes, since temperature changes from night to night will change the absolute position of the focus by that amount or more. Many professional

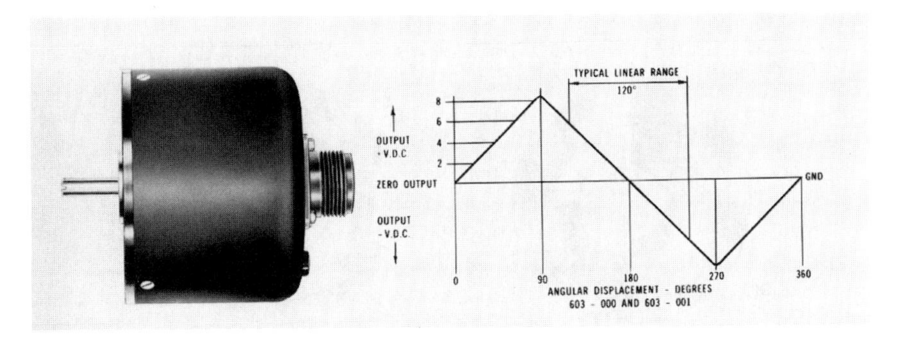

Fig. 9-13 *Trans-Tek Series 600 rotary differential capacitor. Courtesy Trans-Trek, Inc.*

telescopes automatically adjust the focus to compensate for temperature, and such telescopes could use this level of accuracy or even higher. Most telescopes could use the level of repeatability offered by linear Inductosyns, so that throughout the course of a single evening, one can shift among several foci and return predictably to the desired focus.

9.6 Rotary Differential Capacitors

An interesting angular displacement transducer manufactured by Trans-Tek, Inc. is the Series 600 variable differential capacitor, which is shown in Figure 9-13. The unit contains a precision variable capacitor in an LRC oscillator circuit. As the shaft is rotated, a semi-circular plate attached to the rotor moves with respect to a similar stationary plate, changing the capacitance of the unit, and, therefore, the frequency of the oscillator. The oscillator output is demodulated and converted to a DC signal of 100 mV per degree of rotation. Electronics inside the unit use 12–16 volts DC external power to generate the AC oscillator carrier and perform all signal conversion. The unit has 0.05% of full scale linearity, and theoretically has infinite resolution. With a full scale of 160°, the linearity is 4.8′. Practically, the unit has 3.6 arc second resolution, which corresponds to a 100 nV signal, close to the limit one would want to try to detect reliably in an environment of EMI/RFI caused by motors and computers.

The unit has one drawback—its linear response region is limited to at most 160° of rotation. Thus it cannot be used on some types of telescope mounts that require a full 360°, or even 180°, of rotation. If the telescope is a German equatorial and has a bad horizon, or will always be used at elevations of at least 10° above the horizon, the Trans-Tek unit is a viable choice for an angle position sensor. It would also be useful in an alt-az

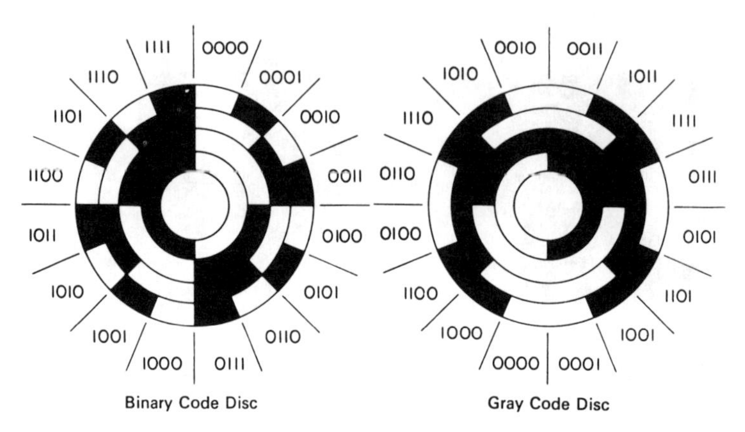

Fig. 9-14 *Two absolute optical encoder disk patterns. Courtesy Reston Publishing Co.*

mount if motion in one axis is restricted, as it usually is in elevation to a range of 90°. The unit itself costs about $350 new (the authors have not seen them on the surplus electronics market), and support electronics, including a power supply and an analog-to-digital converter card for a PC, could be purchased for roughly $150–$200, which places it within the budget of a small system user. To obtain coverage over a full 360°, three units can be used, but this approach is usually more expensive than using some types of optical shaft angle encoders. We are not aware of any telescope control systems using this device.

9.7 Optical Shaft Angle Encoders

Optical shaft encoders are devices that consist of a disk, usually glass, on which is deposited by photographic means one or more patterns of alternating transparent and opaque slots or lines which, when rotated through the path of a light source and photodetector, generates a series of pulses. The number of transparent and opaque line pairs around the disk determines the number of pulses or counts per turn of the disk, which is the resolution of the encoder. Examples of two absolute encoder patterns in common use are shown in Figure 9-14. Figure 9-15 shows a few of the wide variety of optical encoders that are available. Encoders may give an absolute position using a unique output code for each resolved shaft angle (*absolute encoder*), or they may send out a pulse for each incremental resolved shaft angle step (*incremental encoder*). An absolute encoder typically has a resolution of 2^N, in which there are N concentric rings of transparent and opaque line

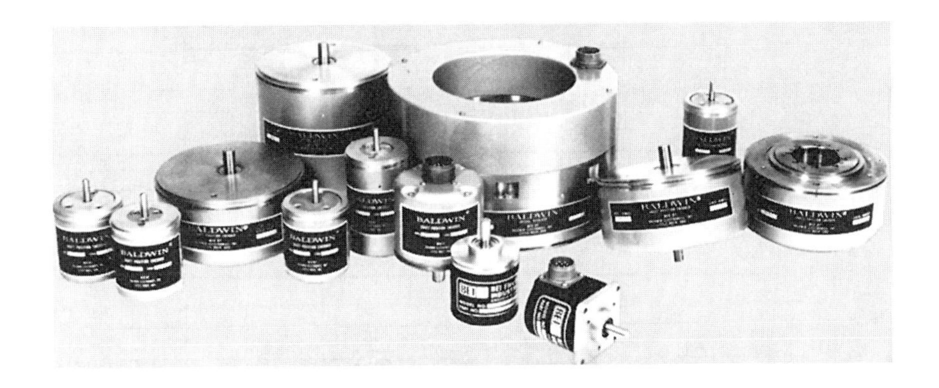

Fig. 9-15 *Some optical shaft angle encoders. Courtesy BEI Electronics, Inc.*

pairs, with each successive ring having twice the number of line pairs as the ring inside it, beginning with the innermost ring, which is half transparent and half opaque. A light source and photodetector placed on opposite sides of the disk for each of the N rings give N bits of a binary number representing the shaft angle. To prevent small pattern misalignments from causing false readings taken when several bits are changing state from one count to the next, a Gray code is often used instead of straight binary, since in a Gray code, only one bit changes state from one count to the next. This is shown in the following table (from Spaulding, 1965, p. 10):

Decimal	Binary Code	Gray Code
0	00000	00000
1	00001	00001
2	00010	00011
3	00011	00010
4	00100	00110
5	00101	00111
6	00110	00101
7	00111	00100
8	01000	01100
9	01001	01101
10	01010	01111
11	01011	01110
12	01100	01010
13	01101	01011
14	01110	01001
15	01111	01000
16	10000	11000
.	.	.
.	.	.

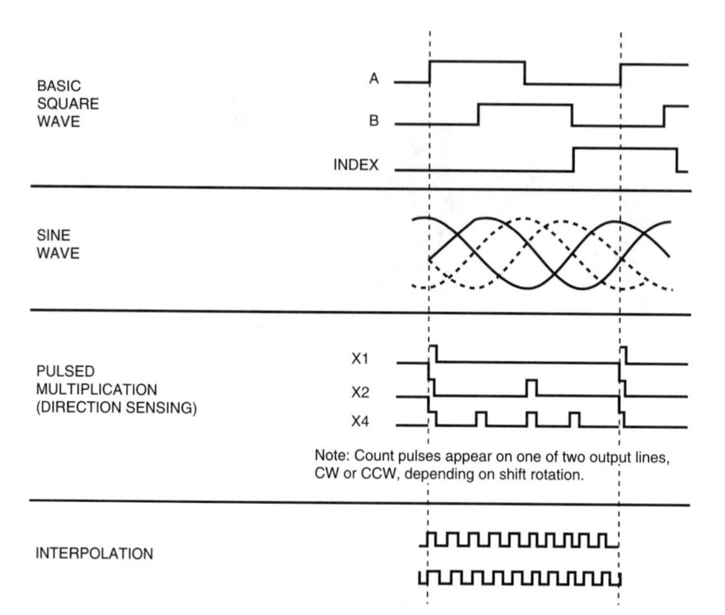

Fig. 9-16 *Characteristic waveforms of incremental encoder options. Courtesy BEI Electronics, Inc.*

The Gray code is easily converted into a binary number with either hardware or software, using the following rules (from Spaulding, 1965, p. 10):

1. The most significant natural binary bit is identical to the most significant Gray bit.

2. If the nth natural binary bit is 0, the $(n-1)$th natural binary bit is identical to the $(n-1)$th Gray bit.

3. If the nth natural binary bit is 1, the $(n-1)$th natural binary bit is the opposite of the $(n-1)$th Gray bit.

This is expressed in the formula

$$\mathrm{B}(n-1) = (\mathrm{G}(n-1).\mathrm{AND}.(.\mathrm{NOT}.\mathrm{B}(n))).\mathrm{OR}.((.\mathrm{NOT}.\mathrm{G}(n-1)).\mathrm{AND}.\mathrm{B}(n))$$

where $\mathrm{B}(n)$ is the nth natural binary bit, and $\mathrm{G}(n)$ is the nth Gray code bit.

Other anti-ambiguity techniques are in common use, such as window codes, cyclic decimal, U-scan, V-scan, and 2-out-of-5 (a kind of parity code).

While an absolute encoder uses N rings for N bits of resolution, an incremental encoder with the same resolution uses two concentric rings of

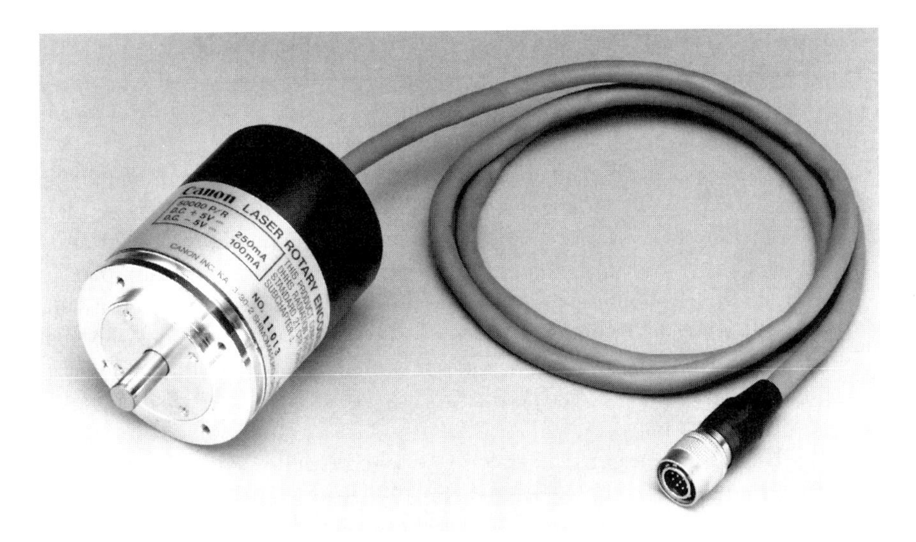

Fig. 9-17 *Canon Laser Rotary Encoder. Courtesy Canon U.S.A., Inc., Components Division.*

2^N line pairs. The two ring patterns are in spatial quadrature with each other. Logic chips can be used to count and store the current shaft position, and the direction of the counter can be determined from the relative phases of the pulses received from the two rings. Thus the shaft angle is determined electronically rather than optically. There is often a third ring with a single mark indicating the zero position. A pulse from this track is used to zero the counting logic.

Either absolute or incremental encoders can be used for telescope control. For resolutions up to about 16 bits, incremental encoders are far less expensive than absolute types, by a factor of as much as 10. A 16-bit absolute encoder has 2^{16} (65,536) counts per revolution, which gives it a resolution of 1,296,000″ per circle divided by 65,536, or about 20″ per count. An absolute encoder with this resolution costs about $2,000, while an incremental encoder with almost the same resolution sells for $450. Resolutions up to 24 bits and custom mechanical designs are available on special order. For example, a stock 20-bit absolute encoder costs about $10,000, while a custom 21-bit encoder with a 6-inch hollow shaft hole for direct mounting on a telescope axis costs about $20,000 on special order. Multi-turn encoders of either the incremental or absolute type are available that use a lower-resolution (and therefore less expensive) encoder disk geared to the input shaft of the encoder assembly. These units can offer the high resolution (but not necessarily the accuracy) of single-turn units at significantly lower cost.

Optical shaft encoders are available on the surplus electronics market if one pursues them and is patient. For example, Trueblood was given a 13-bit absolute encoder in 1971, bought two 15-bit absolute encoders in 1977 for $125 each, and bought two 16-bit absolute encoders in 1981 for $25 each. Ads for optical shaft encoders appear in *Sky & Telescope* magazine at odd intervals. Such encoders can be geared to each telescope axis to yield high resolution (1″ or better). Measurements can be made, and algorithms can be developed and implemented in the control software to minimize the errors introduced by the gears to a few arc seconds if a great deal of time is spent taking measurements, reducing the data, and studying the gear error sources. One should be wary of shaft angle encoders obtained on the surplus market, since often the electronics are bad, and occasionally the encoder disk is damaged.

The BEI H-25 incremental encoder is a good choice for those who wish to avoid the problems that often accompany surplus optical encoders. The basic unit is small, inexpensive compared to other types of optical encoders (about $450), and has 2540 counts per turn. As shown in Figure 9-16, interpolation logic multiplies the pulses per turn on each output phase by a factor of 5, up to 12,700. The pulses from the quadrature outputs can then be put through optional "steering" circuitry, producing 50,800 pulses per revolution. These pulses are sent out on one of two lines, depending on whether the encoder shaft is being turned CW or CCW. These two lines can be connected to the "up" and "down" inputs of a series of 74LS193 counters to obtain an integrated position count. Both interpolation and pulse steering circuitry can be placed inside the encoder by BEI at very little additional cost (about $60 in 1983).

You need to investigate TTL output and pullup resistor options for the H-25 that will increase the noise immunity—a must when there are motors and computers radiating interference. This is crucial with incremental encoders, since stray pulses on the line will give inaccurate position readings that accumulate throughout the night.

Instead of designing custom logic to count the pulses from the encoder, this circuitry can be part of the computer interface. An example is the Oregon Microsystems PC38-4E, with four axes of stepper motor control and four encoder inputs.

Another low cost incremental encoder to consider is the Litton Model 81 1000 count-per-turn (roughly 10-bit) encoder, which sells new for under $200. Similar units for even less may be available by now. For those interested in experimenting with making their own low resolution optical encoder, Opal (1980) describes an encoder yielding 50, 100, or 200 counts per turn. He includes a circuit for converting the two quadrature outputs to pulse and direction outputs.

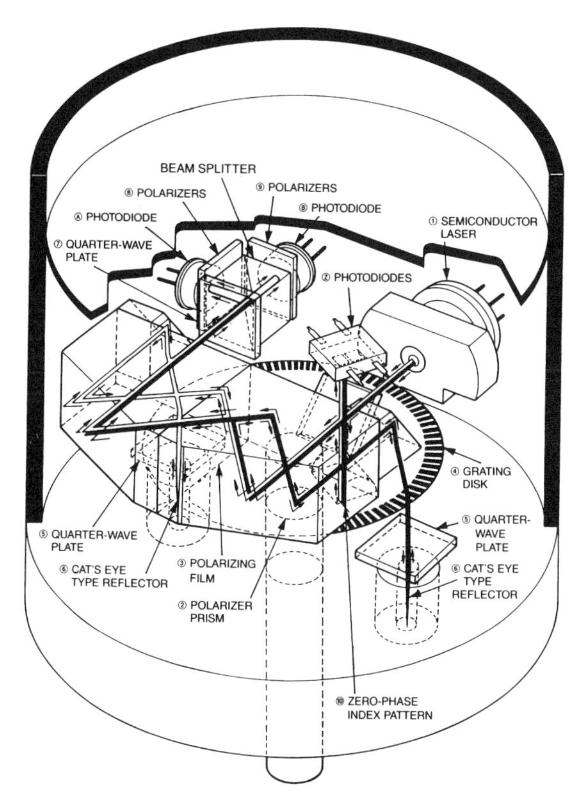

Fig. 9-18 *Principles of Operation of the Canon Laser Rotary Encoder. Courtesy Canon U.S.A., Inc., Components Division.*

In the late 1980s, Canon developed rotary and linear encoders based on laser interferometry. The Model X-1 rotary encoder has a basic resolution of 225,000 pulses per revolution, that can be increased by interpolation to 18,000,000 pulses per revolution ($0\rlap{.}''072$ per pulse). This unit sells for \$11,890 while the Model K-1 that sells for \$2,100 has a basic resolution of 81,000 pulses per revolution that interpolation increases to 1,296,000 ($1''$ per pulse). These units are very accurate even with interpolation, since interpolated accuracies are typically 1.5 times the resolution.

The operation of the Canon rotary encoder is shown in Figure 9-18. A semiconductor laser generates coherent light with a wavelength of 780 nm that is split into two beams by a polarizer prism. Both beams pass through the grating disk, which generates plus and minus first order diffracted beams that are modulated during one pitch rotation of the grating disk. Each beam 180° apart from the other passes through the same

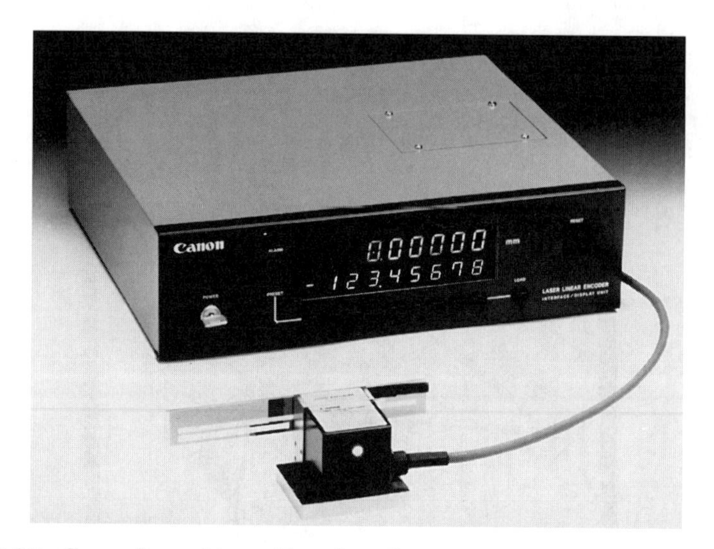

Fig. 9-19 *Canon Laser Linear Encoder. Courtesy Canon U.S.A. Inc., Components Division.*

slit in the disk in both the forward and reverse directions, causing four phase modulations. By interfering the two first order diffracted beams, four cycles of sinusoidal light intensity change are obtained from each change of disk rotation by one pitch. Electronics are used to interpolate each sine wave for higher resolution.

Canon's linear encoders work in a similar fashion, and have travels of 50, 100, or 150 mm (roughly 2, 4, and 6 inches), with resolutions from 0.1μm (0.0000039 inch) to 0.01μm (0.00000039 inch). The accuracy is equal to the resolution at constant temperature. Another scale below the first on the same substrate generates a home pulse in the middle of the scale, providing a convenient reference point. The Model L-104 has a slit pitch of 1.6μm, while the L-108 uses a slit pitch of 3.2μm. Both models have scales with a coefficient of expansion with temperature of 5.2×10^{-7} °C, so the temperature-induced error is small. This level of performance makes them ideal for encoding the position of the secondary mirror for controlling the focus of a Cassegrain telescope. They tend to be expensive, close to $4,000 with an interpolator and readout unit. A slightly less expensive unit (about $3,000) without a home pulse is sold by Opti-Cal, while Heidenhain offers a similar unit.

Sony Magnescale America, Inc. also offers a full line of linear laser encoders with 0.01μm accuracy and resolution. Their open units (Model B575A-NS) consisting of a scale and a separate sensor vary in length from 30–210 mm and start at $4,100, while units with 0.1μm start at $2,250.

9.8 Other Encoder Types

A variation on the optical encoder is the mechanical encoder, which is made by the same process used to make etched printed circuit cards. A pattern is photographed onto photoresist over a layer of copper foil on glass epoxy material. The resist is developed, the pattern is etched, then a set of brushes is used in place of lights and photodiodes to make and break one circuit per bit (on an absolute encoder), with the foil and brush acting as a switch. Single-turn mechanical encoders are usually limited to 10 bits or less, since the brush contact width needed to ensure good electrical connection becomes larger than the line widths at resolutions of about 10 bits. Multi-turn mechanical encoders, which use gears to speed up the rotation of the 10-bit disk with respect to the encoder shaft, are available inexpensively with up to 21-bit resolution. Examples of mechanical encoders are shown in Figs. 9-20 and 9-21.

Sony Magnescale America, Inc. makes a full line of both low and high resolution rotary angle encoders based on their magnetic sensing technology used in their machine tool linear scales. Their RE series is available in four resolutions from 1000 to 2048 pulses per revolution and cost about $170. Their MSE hollow shaft encoder has an outer diameter of 57 mm, a bore of 24 mm, ranges in resolution from 18,000 pulses per revolution (ppr) to 360,000 ppr, and costs about $3,500 with an MD-20 sensor head. An enclosed Model MSS-201B with a range of 43,000 ppr to 860,000 ppr sells for $2,045. Sony also makes a larger unit for sending a beam through the encoder. The Model RS1 is 132 mm in outer diameter with a bore of 121 mm. A unit with a single detection head with resolutions in the range of 72,000–1.4 million ppr costs $4,190 while a double head unit with resolutions in the range of 144,000–2.8 million ppr sells for $5,636.

Another encoder package is the motor/encoder combination that typically uses a DC torquer and a very high resolution encoder to form a single compact unit. An early example is shown in Figure 9-22. Such motor/encoder integrated units are the precursors to the digital servo motors developed by Galil and Parker Compumotor. Canon offers a similar unit.

Those with Cassegrain telescopes or folded Newtonians often control the focus position by moving the secondary mirror. For a Cassegrain telescope, if the secondary mirror multiplies the primary mirror $f/$-ratio by a factor N, then moving the secondary mirror a distance d moves the focal plane a distance $(N^2 + 1)d$. N is usually in the range 2–6, so for a secondary mirror with a factor $N = 3$, moving the secondary mirror a distance d moves the focal plane a distance $10d$. Some applications require controlling the focal plane to within 0.0001 inch, which (for $N = 3$) requires controlling the secondary mirror position to within 0.00001 inch. Field curvature

Fig. 9-20 *Two-speed mechanical encoder. Courtesy Northern Precision Laboratories, Inc.*

Fig. 9-21 *A mechanical encoder and encoder disks. Courtesy Vernitron Corp.*

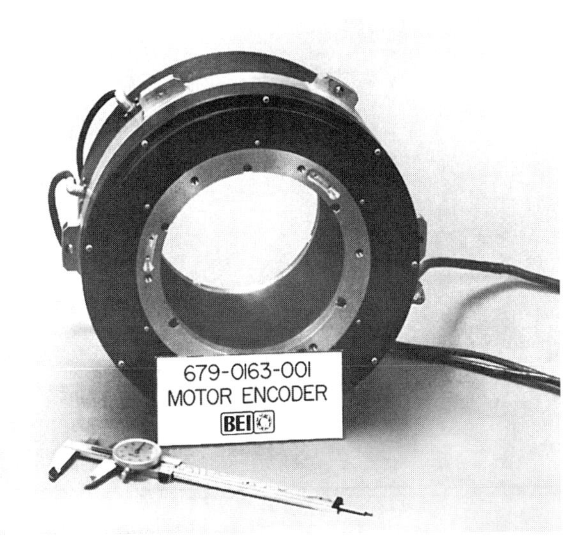

Fig. 9-22 *A single-piece motor/encoder unit. Courtesy BEI Electronics, Inc.*

across a small-format CCD can be almost 0.001 inch, which determines the resolution needed. Thus for most telescopes using small-format CCDs, the secondary mirror position should be controlled to within 0.0001 inch. Note that this is the *resolution* and *repeatability* needed, not the absolute *accuracy*.

A linear encoder with this level of resolution and repeatability is the Temposonics II, made by MTS Systems Corporation. The unit is shown in Figure 9-23, while the principle of operation is shown in Figure 9-24. The Temposonics II consists of an electronics head attached to a metal tube in which is suspended a wire. A permanent magnet doughnut attached to the moving part encircles the tube. The head (fixed to the spider assembly) sends out a current pulse in the wire that generates a magnetic field pulse. This magnetic pulse interacts with the magnetic field of the permanent magnet attached to the moving secondary mirror, generating a torsional strain pulse in the tube that travels back to the electronics head at a known speed. This twist in the tube is sensed by a pickup coil in the head, which determines the elapsed time between the current pulse and the received strain pulse.

This time is converted to an absolute distance along the tube with a resolution and repeatability of 0.0001 inch and an absolute accuracy of 0.002 inch. The temperature coefficient of the rod is 3 parts per million/inch of stroke/°F, low enough for most focusing applications. With an external

Fig. 9-23 *MTS Temposonics II magnetostrictive linear sensor. Courtesy MTS Systems Corporation.*

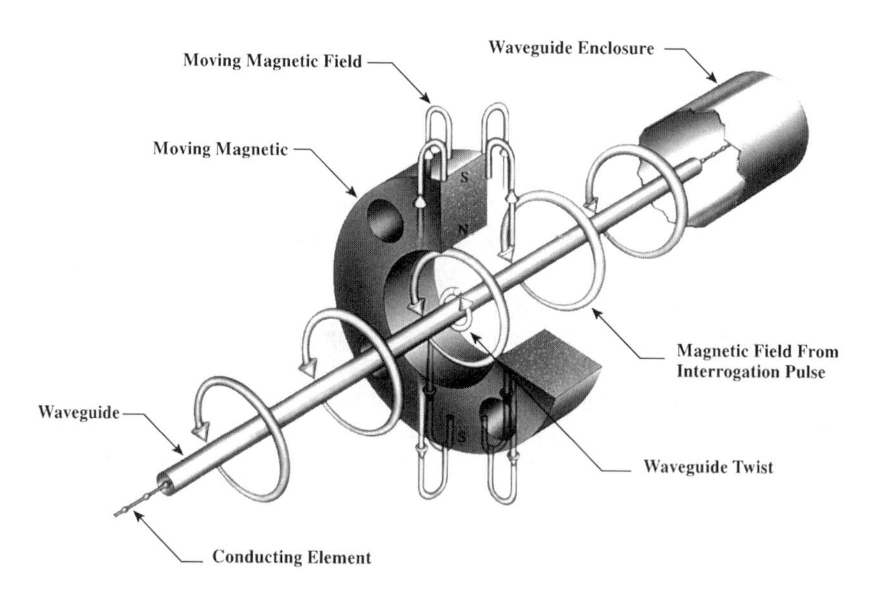

Fig. 9-24 *Principles of Operation of the MTS Temposonics II magnetostrictive linear sensor. Courtesy MTS Systems Corporation.*

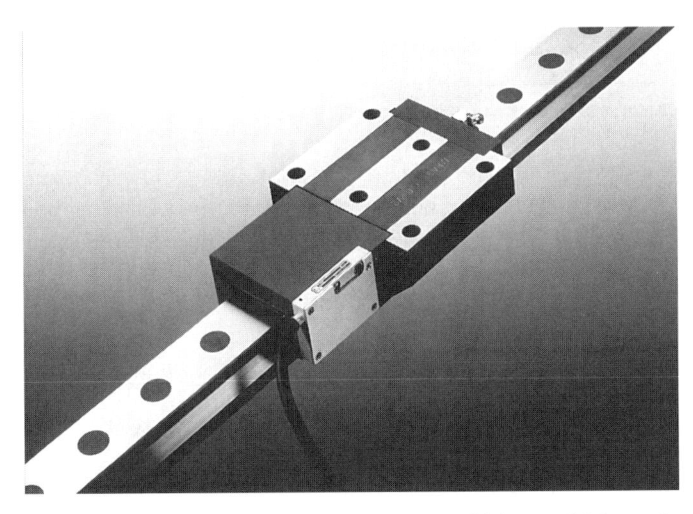

Fig. 9-25 *Schneeberger Monorail linear bearing and Monorail Measuring System linear encoder.*

interface card, the Temposonics II costs about $1,000. Note that the Temposonics II gives absolute position, which means you do not have to move the secondary mirror to a home position to reset a counter, then count incremental encoder pulses to your focus position, the way you must with the Canon and Opti-Cal linear encoders. This means you can use a slower, less expensive stepper motor for your focus motor.

Schneeberger combines a linear monorail bearing suitable for most small telescope secondary mirror assemblies with a magnetic linear encoder similar to the Sony Magnescale. The monorail bearing has a solid, tight feel that ensures minimum runout when moving the secondary mirror. Most Cassegrain systems are very sensitive to tilts and decenters, so a few micrometers can make a difference. Schneeberger offers six carriage/rail size combinations each with four tolerance classes and three preload classes, allowing you to choose the size and accuracy bearing that best fits your application.

Figure 9-25 shows the Schneeberger Monorail bearing with the Monorail Measuring System (MMS) attached. Schneeberger places magnetic fields on a strip applied to the monorail, and the sensor is attached to one end of the bearing carriage making a very compact unit that is easily integrated into a small telescope's secondary mirror assembly. The MMS must be ordered with the monorail—an existing monorail cannot be upgraded later, since the dimensions of the rail and the carriage are different depending on whether the MMS is present.

The magnetic North-South "grating" has a flux reversal every 200μm along the monorail for the basic position encoding, and another reference mark every 50 mm that helps in determining the absolute position of the carriage along the rail. A cable carries the signals to a small electronics box that interpolates the signals 50-fold to an encoder quadrature signal that can be further interpolated to a resolution as small as 1μm. This is not as small as the Canon linear encoder (0.01μm), but at about $1,320 it is about one-third as expensive as the Canon unit. Unless you have an unusual optical design, 1μm should be small enough for most applications.

One final note—judging from most catalogs for mechanical or electronic equipment, manufacturers generally expect their customers to have a working knowledge of their product. Trueblood finds that most product catalogs are nearly indecipherable. In stark contrast, Schneeberger's catalog is a model for how a vendor should design a catalog. It explains the various uses for their products, goes into how to select the right product for your application, explains every available option and how to calculate the correct length of monorail for the model and options you have selected, and even explains how to lubricate and care for the unit once it arrives.

9.9 Time Code Receivers

To control a telescope, the control computer must have an accurate source of time. An error of one second of time translates to about $15''$ on the sky, too large for some observers. For those with modest pointing requirements (e.g., $1'$), a simple short-wave radio located in North America can receive time signals from the National Institute of Standards and Technology (NIST, formerly the National Bureau of Standards) radio stations WWV and WWVH on frequencies 2.5, 5, 10, 15, and 20 MHz or from Canadian radio station CHU at 3.330, 7.335, and 14.670 Mhz. You listen to the radio for the upcoming minute's time, type that time into your telescope control computer, then hit `Return` when you hear the minute signal. Your reaction time will usually be in the range 0.1–0.3 second, which can introduce an error of as large as $5''$ on the sky.

During the 1960s, those who needed more accurate time used NIST station WWVB that broadcasts on the Very Low Frequency of 60 kHz. WWVB receivers were rare and expensive, and the loop antennas are large and bulky. During the 1970s, this was supplemented by obtaining time from line 10 of any U.S. 525-line color television broadcast signal or by using a TrueTime GOES weather satellite receiver. Loran C also provides very accurate time. Some of these time sources are available only in limited areas.

Fig. 9-26 *Datum Bancomm Model bc627AT GPS Receiver. Courtesy Datum Inc., Bancomm Division.*

With the introduction of the U.S. Air Force Global Positioning System (GPS) in the late 1970s, accurate time and position became available to all, anywhere on the surface of the Earth. Hand-held GPS receivers sell for under $500, often including a serial interface to a computer. Civilian users can obtain position accurate to 100 meters and time accurate to 0.5μs. This is more than accurate enough time for all telescope control applications.

Odetics, Motorola, TrueTime, and several other manufacturers make rack mounted or smaller units with interfaces to computers (see Appendix E for a list of GPS manufacturers). Several manufacturers make units with parallel time outputs that avoid the delays of serial interfaces. These units can be connected to a computer using parallel interfaces from National Instruments, Computer Boards, CyberResearch, or other manufacturers. Almost all GPS receivers have a *disciplined* oscillator (a timing source with a frequency that is synchronized to GPS) to provide timing when GPS signals may be weak or unavailable. Since GPS satellite radio signals use a short wavelength (a few centimeters), GPS antennas are quite small, about 4-5 inches in diameter and a few inches high, or even smaller.

TrueTime makes a Model GPS-XL module that can be incorporated into your own electronics, and Datum Bancomm makes a card that fits into the ISA slot of a PC. The Datum Bancomm Model bc627AT GPS receiver places accurate time on the ISA bus, for rapid transfer to the control software. It can also give longitude, latitude, altitude, and other data over a serial port, and capture the time of an external event and send that time over the serial port. This latter feature is very useful for time-tagging data. With its antenna, an MS-DOS device driver, and a C demo program, the bc627AT sells for about $3,700.

9.10 Other Useful Sensors

Besides the positions of your major mount axes and of your secondary mirror, there are other items of information your telescope control computer needs to do its job. If you intend to compute atmospheric refraction accurately, your computer must have sensors for ambient temperature and barometric pressure. Omega and others make accurate and reliable temperature sensors, while Setra makes a very accurate atmospheric pressure sensor, shown in Figure 9-27.

The Setra Model 270 standard accuracy is $\pm 0.05\%$ of full scale (1100 millibar) or ± 0.55 millibar, with $\pm 0.03\%$ of full scale or ± 0.33 millibar accuracy as an option. This compares with a typical field accuracy of ± 1 millibar in other units.

Another ambient parameter to measure is dew point. This is important for determining whether to activate anti-dew devices that could disturb your images if left on all the time. If dew is about to form, you should turn off high voltage devices, such as image intensifiers, photomultiplier tube high voltage power supplies, and avalanche photodiode power supplies. In an automatic observatory, you might want to turn off everything and close down for the night.

Dew point can be sensed in a number of ways. One is to detect both dry air temperature and relative humidity, the latter using a polymer sensor, then compute dew point. Relative humidity by itself is not very useful, unless you are computing atmospheric refraction using Garfinkel's algorithm or you plan to protect electronic equipment with a relative humidity specification. Relative humidity is difficult to measure accurately ($\pm 3\%$ RH is typical), which can lead to errors of a few °C.

Another way to sense dew point is to measure it directly, the most accurate means being a chilled mirror dew point sensor such as the DewTector by Protimeter. This unit reflects light off a mirror that is cooled by a Peltier thermoelectric cooler, similar to what is used to cool CCD chips in some astronomical CCD cameras. When dew forms, the light received from the mirror is less intense due to scattering by the dew. After the mirror dews, the current in the Peltier device is reversed to heat the mirror several degrees above the dew point, and the cycle is repeated. The DewTector has an internal wiper, to remove water droplets from the mirror surface. This unit, shown in Figure 9-28, has an absolute dew point temperature

Fig. 9-27 *Setra Model 270 Barometric Pressure Transducer. Courtesy Setra.*

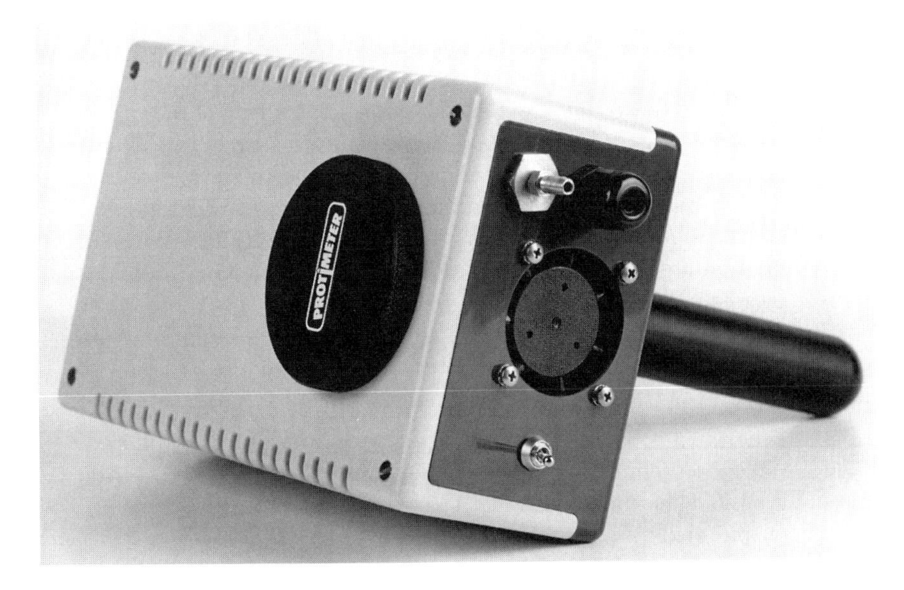

Fig. 9-28 *Protimeter DewTector Chilled Mirror Dew Point Sensor. Courtesy Protimeter, Inc.*

accuracy of $\pm 0.2°$ C, and sells for about $1,500.

An extremely useful sensor developed by Louis J. Boyd detects clouds. The following is Lou's description of his device:

> Cloud sensing at an observatory is desirable for determining when observations can be made and for giving advanced warning of precipitation. The first of these tasks is by far the more difficult. For photometric observations very thin clouds can have drastic effects on the quality of observation. There is probably no passive detector more sensitive than a stellar photometer for detecting cirrus clouds. For some spectroscopic observations thin clouds only increase integration times as a function of the extinction. The determination of what is acceptable is a major part of the problem.
>
> The second of the tasks is the one addressed here. For any astronomical observatory the protection of the instrument from rain or snow is critical for long life and quality of the mirror surfaces. Rain drop detectors are useful, but rain can occur suddenly and water spotting can occur before a roof or dome can be closed. Clouds which are capable of producing rain or snow are relatively dense and low. These clouds are capable of reflecting a significant amount of infrared radiation in the 10 micron region. This property is exploited in the cloud sensor described here.
>
> The cloud sensor must detect the radiation (or lack of radiation) from the sky and yet be insensitive to fluctuations in the ambient air temperature. Attempts were made to simply look at the sky with a thermistor and compare this to the air temperature with another thermistor. This was unsuccessful because of fluctuations of the air temperature and with tracking problems between the thermistors. A better approach was to measure the radiation from the sky and compare it to the radiation from the ground. This worked better but still had the problem of thermistor tracking due to changing ambient temperature.
>
> What was needed was a device which would make differential measurements of the radiation from the sky and from the ground. Thermocouples met this requirement but had very low output signal.

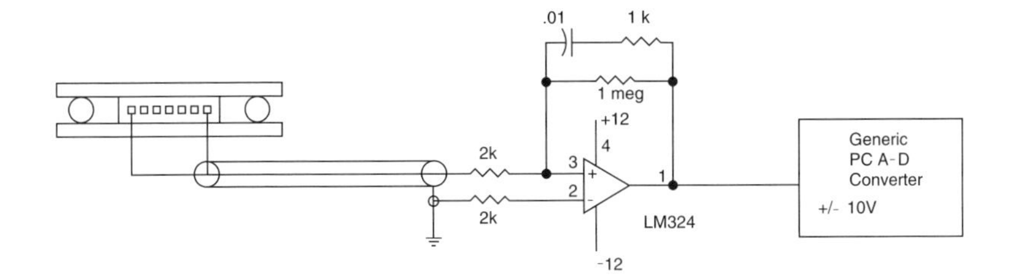

Fig. 9-29 *Louis Boyd's Cloud Sensor Schematic Diagram.*

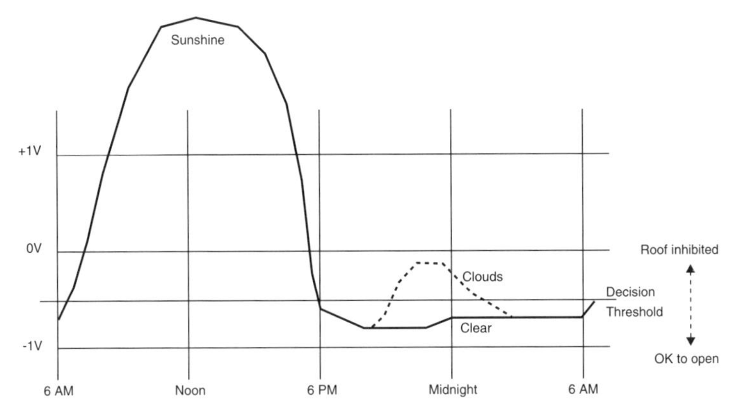

Fig. 9-30 *Plot of Output Voltage Versus Time for a Typical Day in the Life of Louis Boyd's Cloud Sensor.*

An array of thermocouples in series has a greater output. A relatively inexpensive and efficient thermopile can be obtained in the form of a Peltier device, otherwise known as a thermoelectric cooler. These devices are designed to produce a temperature difference between two surfaces when voltage is applied, but they function in reverse. They will generate a voltage whenever there is a temperature difference across them.

A bare Peltier device mounted outdoors with one side facing the sky and the other facing the ground would function as a cloud detector, but it would have little collecting area and would probably be damaged by rain. To overcome these problems the cloud detector at Fairborn Observatory was constructed of two aluminum plates with the Peltier device sandwiched between them. The plates are 8″ diameter and 0.25″ thick. There is a 6″ diameter groove 0.125″ deep in the upper plate which centers a 6″ by 0.25″ "O" ring. Short Nylon threaded rods compress the "O" ring through six equally spaced 0.25″ holes on a 7″ diameter. The Nylon rods have stainless steel acorn nuts installed (on both ends) to protect them against break down from sunlight UV. The lower plate has an aluminum conduit fitting which serves as a junction box and mount. The Peltier device is held in place by the pressure of the two plates. It is thermally connected with heat sink compound.

Bare aluminum has low emissivity at 10 microns so all outside surfaces are painted with white gloss epoxy enamel which has good emissivity at 10 microns but fairly low absorption of sunlight. The paint also prevents corrosion of the aluminum.

The output of the Peltier device is in the 10 millivolt range with a clear sky at night. This can be brought up to a more useful level with a single op amp. The op amp also gives a little protection against lightning to the A-D converter which reads the output. A typical 12-bit data acquisition board would probably be sensitive enough without the op amp.

Figure 9-30 shows the typical response over the course of 24 hours. At night the sky gives much less 10 micron radiation than the ground and heat energy passes from the lower plate to the upper through the Peltier device. When a cloud passes overhead, energy is reflected from the ground and the radiation on the upper plate increases, reducing the thermal transfer. On mountain top observatories, it is not uncommon for the cloud to reflect energy from the valley below, resulting in the sky looking warmer than the ground when clouds pass overhead. The response of the unit is quicker than the typical movement of clouds, and clouds can be detected before they are directly overhead.

If protection from precipitation is the only objective, and if only nighttime cloud cover is of interest, then a fixed level threshold is adequate to make a determination. If daytime operation is desired, then a dynamic threshold must be used which predicts what the voltage should be as a function of time with no clouds and then checks for deviation from the expected value. This has to be adjusted with the seasons.

Like most sensors, this one is sensitive to things other than clouds. When locating the detector, the area under the detector should be free of artificial heat sources and should be of high thermal mass and of high emissivity. Ordinary dirt, rocks, pavement, or concrete is adequate. It should not be in a place

> where heat from buildings, people, or vehicles can radiate onto the detector. Wind has the effect of slightly reducing the sensitivity of the detector. Snow piled on the detector slows its response time, but usually doesn't cause a problem.
>
> The cloud sensor at Fairborn Observatory has been in operation since about 1988 and has proven quite reliable. There was one failure due to corrosion when the original nylon bolts broke as the result of sun exposure. In no case has the cloud sensor given false indication of clouds when the sky was photometric, or failed to show clouds before they produced rain.

A variety of weather conditions can dictate closing after observing has begun, or not even opening an observatory at the start of the evening. An automatic observatory needs a complete weather station with wind speed, air temperature, dew point, barometric pressure, and a Boyd cloud sensor. Several vendors sell weather stations with conventional propeller or cup anemometers, but two of the most reliable and interesting weather stations with no moving parts are the TFV 4056 and the WST 7000 from MesoTech International. The WST 7000 is shown in Figure 9-31.

The method used by MesoTech to produce wind speed and direction signals is shown in Figure 9-32. A central heated core is surrounded by eight thermocouples that record the temperature of the outside of the core every 45° around the core. With no wind, all parts of the outside of the core are at the same temperature. When there is a wind, it cools part of the core surface but not the other parts on the lee side, so the different thermocouples record different temperatures. By orienting the weather station to true North and using the thermocouple inputs, the on-board microprocessor can determine wind speed in the range 0–50 m/S (meters per second, roughly 0–112 miles per hour) with a resolution of 0.1 m/S (0.22 mph) and an accuracy in the range 0.1–0.5 m/S (0.22–1.12 mph) for wind speeds up to 6 m/S (13.4 mph) and 0.5 m/S ±5% of the reading above 6 m/S. The MesoTech stations determine wind direction with a resolution of 1° and an accuracy of ±3° over the entire range of 0–360°, without suffering from the electrical gap, starting threshold, and distance constants of conventional anemometers. Values are sent by RS-232/422/485 to a serial port every 100 ms.

The TFV 4056 equipped with only wind speed and direction outputs costs $3,525. Models with air temperature, barometric pressure, a magnetometer for automatic North sensing, automatic deicing heaters (reduced

Fig. 9-31 *MesoTech Model WST 7000 Automatic Weather Station. Courtesy MesoTech International.*

temperature accuracy between −0.5°C and +0.5°C), and other options cost more. The WST 7000 standard model costs $5,400 and comes complete with wind speed and direction, air temperature, barometric pressure, relative humidity, dew point (calculated), and tipping bucket rain guage input all standard. Extra-cost options include automatic deicing heaters, magnetometer for automatic North sensing, PC-compatible data display and archive software, submersible version for marine applications, and others. Thus the base model TFV 4056 is a wind speed and direction sensor, while the base model WST 7000 is a complete weather station and offers some standard features (e.g., relative humidity/dew point and tipping bucket input) that are not even options on the TFV 4056. Either model can be the basis for a complete weather station to give information needed by an automatic observatory control computer.

Going beyond merely recording weather conditions, it is now possible to receive the information needed to predict the weather with some degree of accuracy (slightly above pure guessing). This information cannot be

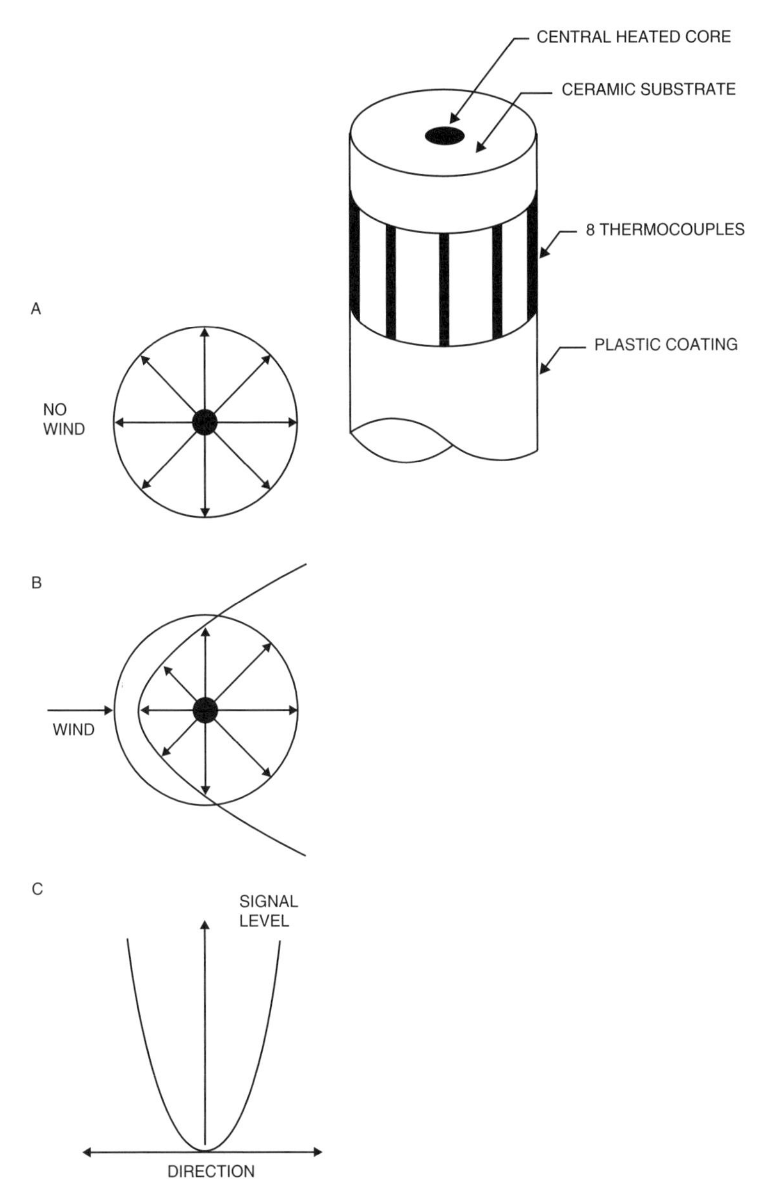

Fig. 9-32 *Principle of Operation of the MesoTech Automatic Weather Stations. Courtesy MesoTech International.*

Fig. 9-33 *Satellite Weather Image Receiving and Processing System in the WMO Van.*

Fig. 9-34 *Typical GOES Satellite Weather Image.*

processed automatically by an automatic observatory control computer, but it can help you operate your observatory more efficiently.

The National Oceanic and Atmospheric Administration, an arm of the U.S. Department of Commerce, provides two different types of satellite weather imagery. The "NOAA/TIROS" series of satellites orbit the Earth so they pass over its poles, with the plane of the orbit perpendicular to the equator. The orbits are low, only about 300–500 miles in altitude, bringing each satellite over any given location about 2 to 4 times each day. With six polar satellites operating, they provide 12–24 photos of your location per day. Since the orbits are low, you must be within about 1800 miles of the nadir point of a polar satellite to receive the signal broadcast at 137 MHz, and you need a demodulator that connects its output to your computer's serial or parallel port. The raw photos transmitted by these satellites have very high resolution, but lack the familiar political boundary lines seen on the TV weather report.

The system at the Winer Mobile Observatory (WMO), shown in Figure 9-33, is designed to receive TIROS satellite weather photos and to use them to understand weather patterns that could affect the decision whether to open up or what kind of scientific program to pursue. A PC can be used to obtain satellite weather photos from the Vanguard Electronics Labs WEPIX-2000-B receiver and MultiFAX external demodulator on the parallel port. The Vanguard receiver system consists of an APT-2 WEFAX antenna tuned to 137 MHz that also contains their low noise preamp, and the WEPIX-2000-B Environmental Satellite Receiver designed for receiving NOAA/TIROS and METEOR (Russian weather satellites) images. The receiver sends the image to the MultiFax external demodulator that supplies image data to the parallel port of the computer. The MultiFAX MFMAP PC software adds the political boundary lines and longitude/latitude marks to the TIROS images, along with many other useful features. A typical package of antenna, receiver, demodulator, and software (not counting the cost of the personal computer) is around $500–750.

The other series of NOAA satellites are the Geostationary Operational Environmental Satellite (GOES) satellites. Figure 9-34 shows a typical GOES image. The first of the GOES NEXT series, GOES I, was launched in early 1994. These satellites are in geostationary orbit along the equator, where they orbit the Earth once in 24 hours, so they stay in one place in the sky as viewed from the Earth. They image the entire disk of the Earth in one hemisphere at least once per hour in both visible and infrared bands, and they have the ability to zoom in to obtain higher resolution images of severe storms or other phenomena of interest to meteorologists. The GOES satellites transmit their images to Wallops, Virginia where they are relayed to Suitland, Maryland outside Washington, DC. There the raw images are

Fig. 9-35 *A Magneswitch Fiducial Sensor. Courtesy Sony Magnescale America, Inc.*

registered and corrected, and political boundary lines between states and countries are added. These images are then sent back to the satellites, which retransmit them in S-band microwave at 1691 MHz.

It is GOES images that you see on the TV weather report, strung together into a film loop to show the evolution of major weather systems. To receive GOES images directly, you need either a TV satellite dish antenna with a 1691 MHz feedhorn or a very large loop yagi antenna, a downconverter to a lower frequency, and a receiver for that lower frequency. A complete package including antenna, downconverter, receiver, demodulator, and software (again without the PC) costs about $1,500.

A less expensive alternative to the GOES and NOAA satellites is to receive shortwave broadcasts from the Air Force, Navy, Coast Guard, or private carriers. These broadcasts are typically on frequencies between 3 MHz and 25MHz. The broadcasts are 80% weather charts and 20% images from satellites, and it takes about 15 minutes to receive each image. If you have a PC and a good shortwave receiver, you can obtain a simple demodulator and software package for as little as $100. Depending on your budget, you can assemble a system rather inexpensively to obtain information vital to your observatory operations. More on the use of NOAA satellite imagery is given in Trueblood (1995).

One final sensor type is a type of position sensor that can convert an incremental encoder to an absolute encoder (with the help of your control computer). This is the Sony Magneswitch, made by Sony Magnescale America. These units cost about $100 each and can detect a magnet (supplied with the sensor) with a repeatability of ± 1 μm (± 0.00004 inch). This translates to one arc second on a disk only 8.25 inches in radius. For relatively low cost, a Magneswitch will give you a "home" position on each of the two major telescope axes and on the focusing secondary mechanism. Using computer software to seek this home position slowly, constantly reading the Magneswitch between each stepper pulse or series of pulses, you can find the home position then count pulses from an incremental encoder or run open-loop and count steps sent to the motor to know where you are. This makes it possible to replace absolute encoders with lower cost incremental encoders. This approach saves money, but it increases wear and tear on motors, bearings, and drive surfaces and it does slow down operations somewhat, especially recovery from a power failure. Figure 9-35 shows a Sony Magneswitch sensor.

The Magneswitch is based on linear position sensing technology used to encode and display the position of machine tool beds and other industrial applications. The sensor consists of a pickup head similar to the head in a magnetic tape recorder, and is usually mounted on the fixed part of the telescope mount. The magnet is mounted on the moving part. As the magnet sweeps by the sensor, it generates a sinesoidal output waveform. This is interpolated electronically to determine the peak voltage (where the magnet is exactly centered over the sensor), then a +12 volt DC pulse is generated within 0.1ms of the peak. To obtain ± 1 μm repeatability of the pulse output, the magnet must be moving no faster than 10 mm/s.

In this chapter, we discussed the sensors that were available in the mid-1990s to help your computer control your telescope and observatory structure. In the next chapter, we explore how to design your control system to make it easy for a person to operate it.

Chapter 10

The Operator Interface

10.1 Types of Operator Interfaces

The previous chapter described the types of sensors available to tell your telescope control computer where the telescope is pointing and about the environment in which it is operating. Part of that environment is the human operator (if one is present), and the telescope control computer must be informed of the intentions of the human operator. The operator must also know what is happening to the telescope and any instruments attached to it. This chapter describes how an observer conveys intent to a telescope control computer, how the computer conveys status to the operator, and the forms in which this communication can occur.

The operator interface is often overlooked when designing a control system, yet it is as important as the motors, computer, or any other component. If a motor is not sized or geared properly, slewing or tracking performance can suffer. If the control system is not designed properly, the operator might not know about an important status condition or make a mistake late at night that could damage the telescope or instrument. We will review the types of operator interfaces used to control telescopes, then look at how to make a good interface.

One of the earliest operator interfaces to a telescope was Galileo's pointing his telescope by grabbing ahold of it and swinging it to the object he wished to view. This works well for small telescopes and binoculars, and is used today with Dobsonian telescopes, but it is not accurate enough for CCD imaging, spectroscopy, and many other uses for a telescope.

Another operator interface is the setting circle. As a method of telling an operator where the telescope is pointed, it is simple and effective. On larger equatorial telescopes, the circles can be quite large (3 feet or more in diameter) and divided quite accurately. With vernier scales they can be read to an arc minute or better. Setting circles are still in widespread use, but are not useful as a means of telling a control computer where the telescope is pointing.

Soon after telescopes were motorized, the first hand paddle was installed on a telescope to permit an operator to control the motors. Hand paddles are still used today in a broad spectrum of telescopes, from the very largest professional telescopes (now used mostly by engineering crews) to small amateur telescopes. They are convenient, in that the operator does not have to remember any special syntax to enter on a keyboard. Instead, the operator glances at the hand paddle and presses the appropriate button. The hand paddle offers, in effect, an easily understood "menu" of choices and an obvious way to make a choice, that is, by pressing a button or rotating a knob. This paradigm is important, for it offers a clue as to how to minimize the bad effects of late-night fatigue with good interface design.

After World War II, some of the encoder technology developed for tank turrets, ship rudders, and other warfare applications made its way into telescope control systems. This technology included synchros, selsyns, and resolvers. Starting in the 1950s, larger telescopes were equipped with a dial looking somewhat like a clock face with hands driven by a selsyn to indicate RA and Dec. This dial was typically installed in a console or cabinet located slightly away from the telescope mount, making it easier for the operator to read than a setting circle located high up on a telescope mount.

The operator interface became more complex when computers began to control telescopes in the early 1970s. Early computers that were small enough and inexpensive enough for telescope control did not have sufficient processing power, graphics hardware, or memory to make graphical user interfaces possible. Consequently, the operator interface was less intuitive. Instead of glancing at a hand paddle and pressing a button, the operator stared at a screen with often cryptic messages and prompts for input, and was forced to remember the exact sequence of characters to type on a keyboard. Status information was displayed in the form of dozens of numbers on a screen, without any graphic aids to give the operator cues as to what is important information and what may be ignored. Many operator interfaces still in use today are of this type, perhaps enhanced a bit using forms on alphanumeric terminals. Although effective in controlling the telescope, such control systems make it difficult for the operator to understand the state of the telescope and instruments, and therefore to make the correct decision late at night.

Now that affordable computers have dozens or hundreds of times the processing speed needed to compute required control algorithms (such as the reduction from mean to topocentric place), more processing capability can be devoted to helping the human operator understand the status of the telescope and its instruments, make the correct choice for a course of action, and enter the correct command to effect that choice. We will now look at what constitutes a good operator interface, and how to design a modern Graphical User Interface (GUI).

10.2 Characteristics of a Good Operator Interface

The history of software and system development is littered with thousands of operator interfaces that were ill-conceived and poorly implemented. But there are many systems with good operator interfaces, so it is possible to make a good human interface in a computer system. All it takes is a good deal of thought, and knowledge of what makes a good operator interface. There are many characteristics of a good operator interface, a few of which are mentioned here.

1. **A good interface reflects how the telescope is used.** Assuming the telescope is to be operated in real time by a human operator, before designing your control system you should understand what types of instruments will be on the telescope, and what types of research programs will be performed using these instruments. What types of interactions do these instruments and research programs require? Is it mostly wide-field imaging, where you point the telescope, select a filter, and take the image without much operator interaction? Or is it spectroscopy of faint objects, in which you must take an image off the slit jaws, offset the telescope to put the object in the middle of the slit, then use a post-slit viewer to make certain the object is in the middle of the slit, then take the spectrum as a series of exposures with appropriate sky and calibration exposures? How many different things do you need to control at once, and can you do all this using one computer, one monitor, and multiple windows, or would it be better to use multiple monitors?

2. **The interface is intuitive.** This means you can look at the monitor and understand immediately the status of the telescope and instruments and how to give the desired command. A good interface does not require the operator to consult a manual before issuing a command. Sensor values and commanded rates are labeled with astronomically meaningful terms (e.g., position in degrees, minutes and seconds of declination instead of motor steps, or wind direction in degrees instead of A/D units). Commands are selected from menus or issued using dialog boxes with appropriate prompts and help features.

   More than likely, you will not be able to fit all the information you want to present to the telescope operator on a single page (window). A well-designed operator interface groups related information together on the same page, and names the pages with names that are easy to remember. The page names appear in menus, so that the operator does not need to remember a page name to display it.

3. **The system responds quickly.** The only hint that operators have about how their system is working is what they see on the screen, and how quickly the system responds to their commands. Although you could compute CPU loading and display that parameter (not a bad idea in any case!), it is also a good design practice to provide feedback as soon as possible after the operator issues a command, something to let the operator know the command was received and is being processed. Even if all you do is display the response "Working ...," at least it lets the operator know that the command was received. There are dozens of ways to provide a quick, effective response, including both visual and audio cues. For example, Andy Tomer's software (see Chapter 12) uses the standard PC audio system (no sound blasters, nothing fancy) to sound a tone or combination of tones both before and after a slew. Combined with visual cues on the screen (the selected object line on the screen changes color) these tones provide good feedback, reassuring the operator that all is operating correctly, as well as warning others nearby that the telescope is about to move.

4. **The interface is ergonomic.** Humans operate telescope control systems (at least those that require human intervention), so telescope control system design must take into account the way human sensory and motor systems work, and their strengths and weaknesses, especially under typical observing conditions. For example, at night the color sensitivity of the human eye undergoes the Purkinje Shift, in which the peak sensitivity of the eye shifts from the green towards the blue. If the telescope is used for visual observing, the observer needs to avoid seeing blue light, or even white light (which is generated by turning on blue, green, and red phosphors on a computer screen). This means the colors used on video screens must be reds, oranges, and other colors toward the red end of the spectrum. Furthermore, you must take into account how colors are generated on a computer screen using red, green, and blue phosphors. Some colors that might seem safe enough at first aren't if they use blue phosphors to generate them.

Going beyond the operation of eyes or ears by themselves, you have to consider how everything works together. A human operator can usually pay attention to only one thing at a time. If he is trying to diagnose an equipment problem and the computer detects another problem (e.g., hitting a soft limit switch), the computer should both display an alarm message and sound an audible alarm tone. Multi-sensory outputs can be more effective at getting the attention of the operator late at night, especially if things have not been working well all night, and the operator is working at high elevation. Such conditions tend to reduce the operator's effective IQ, so the control system should be designed keeping this in mind, with features that tend to compensate for the operator's reduced ability to reason.

Even if your computer does not have sound outputs, there are good practices you can follow when designing displays. One rule is to never steal a page from the operator. If the operator has called up a page of information, that information is important. The operator interface software simply cannot be smarter than the operator, so the software should not decide what page of information the operator should see.

Another rule is to get error conditions noticed by the operator. This means they can't be buried on a page that isn't being viewed by the operator. To avoid violating the previous rule, you need to find a way to display error conditions on every page, or to display a dialog box in a way that does not interfere with something else. This was handled in the Hubble Space Telescope control center by placing an alarms area on every page, with all alarms written to all pages. Furthermore, it is a good practice to require the operator to acknowledge error conditions, at least major ones. This can be done with dialog boxes, with a special mouse-activated alarms acknowledge button, or with a dedicated keyboard function key.

Modern PC applications running under Windows make use of many of these features. When designing your telescope control system operator interface, you might do well to investigate several commercial applications written for the combination of computer and operating system you plan to use, and incorporate into your system the features of the ones you find easiest to learn and to use. Borrowing good ideas from other good applications makes your control system that much easier to learn for those already familiar with the operator interfaces used in popular applications.

Even though this book is about controlling telescopes with computers, note that the general characteristics of good operator interfaces listed above apply equally well to pointing a Dobsonian telescope by hand!

10.3 Design of Graphical User Interfaces

The modern Graphical User Interface (GUI, sounds like *gooey*) was developed in the late 1970s and early 1980s in a number of places, including Xerox Corporation's Palo Alto research center and Apple Computer, and was adapted to other computers by Microsoft (PCs) and Sun Microsystems (UNIX workstations). GUIs were invented to make it easy for the general public to use computers. What made GUIs possible were (1) small computers were getting fast enough that they had some compute power left over after doing a reasonable amount of useful work at a useful rate, (2) semiconductor memory was getting dense enough and cheap enough to put a useful set of hardware for bitmapped graphics on a single small, inexpensive board that could fit into a personal computer bus, and (3) the public was interested in using computers in the home and for business, but existing human interfaces were difficult for the general public to learn to use because they were too removed from common experience.

When it came time to take advantage of excess CPU cycles to make computers more accessible, the designers of modern GUIs needed a model to make the computer more accessible. They wanted to relate this new technology to something that the general public was already familiar with, to make it easier to learn how to use computers. Although the first small computers based on a single microprocessor, such as the Altair 8800 and the Apple II, were originally aimed at the home market (homes occupied by those already familiar with and interested in computers), it did not take Apple Computer and IBM long to understand that the real market was the business world. In particular, the market for small computer hardware and software was business professionals who worked at a desk.

Therefore, the user interface designers chose as their model an office desk with file folders on it. You can lay one file folder on top of another, then bring the bottom folder up to cover the (previously) top folder, or put the two folders side by side, or move a folder off to one side out of the way. GUI designers put electronic equivalents of file folders on the computer screen, and they allowed you to use the mouse to move these folders about, as you can do on a real desk. The electronic desktop became an easily understood extension of the real desks used every day by those likely to buy desktop computers.

Before proceeding further, let's define some terms. The *monitor* is the box sitting on your desk that looks like a TV set that displays information. It has a *screen*, formed by a glass picture tube or some other display technology upon which information is displayed. In the screen's active display area, the system presents a *window*, which is a portion of the screen used by a single computer program or application, or possibly a logical portion of an application. Modern systems permit several applications to run concurrently, so as a consequence they display several windows at a time. However, you interact with only one window at a time.

To work with a particular application, you move the mouse to position the cursor within the application's window and click a mouse button to move the *focus* to that window. From that point on, all keyboard entries and mouse button clicks go into that window (the *active* window) until you move the focus to a different window. Information that is grouped together logically and displayed within a single window is a *page*. Within a single window, the operator can command one of several pre-defined pages to be displayed, but only one page can be displayed at a time.

GUIs can be fancy or plain, but they all have in common that they minimize the amount of typing on the keyboard by using a more ergonomic device, such as a mouse or trackball, to help you move a cursor on the computer screen to make a selection from a menu of choices. Wherever a finite number of possible choices exists, these choices are presented in some form of a menu, in your native written language, as opposed to requiring you to remember a command and to type it in correctly. The mouse/menu paradigm is more intuitive (provided the menu option labels are chosen to be meaningful) and more ergonomic than a keyboard.

One rule of well-designed window systems is to make interacting with a computer a *dialog*, a two-way conversation. Every time the operator does something with the mouse or keyboard, this is acknowledged on the screen. If the operator moves the mouse, the cursor moves. If he clicks a button to move focus to a new window, the borders of the old and new windows change, to reflect the change in focus. If the operator clicks on a choice in a menu bar, the selected menu pops up. As the mouse moves down the menu, each item in the menu is highlighted to indicate what choice would be selected if the mouse button were clicked. When the operator does click on a menu choice, the menu disappears and the selected action occurs. All of this happens quickly, in direct response to the operator inputs, as part of the operator/computer dialog, with the same quick response one would expect in a conversation.

Another rule is to permit more than one way of entering a command. Modern window applications allow you to issue a command by selecting it with the mouse from a menu, by entering a predefined "hot button" (function key or combination of keys reserved for that command), or by typing the command at a command prompt. The key point is to adapt to the user's way of doing things, to make it easy for the user to work in the manner most comfortable for the user, not the programmer.

Yet another windows rule is that if a menu item cannot be selected now (usually because of previous menu choices or keyboard inputs), that item is displayed in a different color or is otherwise made to look different from the valid menu choices. The item is not dropped from the menu, it maintains its usual place so that the menu looks familiar. Usually, menus have a white background with black characters for valid choices and grey characters for invalid choices. You can change these colors to help maintain night vision. This feature of menus can be used to prevent operators from issuing commands that, for example, move a spectrograph grating in the middle of an exposure. This rule forces the control system application programmer to analyze the state of the control system that results from each possible menu choice, and then determine what menu choices are possible (and which should be prohibited) from the resulting new system state.

An example of a telescope control page developed for a modern windowing system (in this case, UNIX) is shown in Figure 10-1. It is the main control page for the Wisconsin-Indiana-Yale-NOAO (WIYN) 3.5-meter telescope on Kitt Peak, which began operating in 1994. The University of Wisconsin developed the WIYN control system as part of its contribution to the project, with X-Window control pages (such as the one shown) developed by NOAO.

The telescope operator issues commands to control various functions of the telescope from this page using a mouse to move a cursor to a control *button*, a small area of the page shaded and labelled with an action, such as "(turn) off," "track," or "center." Commands that require more extensive operator input, such as "Offsets ...," "Tel info ...," or "Help ..." have ellipses ("...") after the key word to indicate that a more extensive interaction is required after selecting these buttons. Those control functions that have a very large number of possible choices, such as RA or Dec, have an area set aside for keyboard inputs. The operator moves the cursor to the line where text is to be entered, clicks a mouse key to put the text entry cursor on that line, then enters the text using the keyboard.

You might be asking yourself "This is all well and good, even attractive and interesting, but is there any meaningful difference between all this 'window dressing' and the old alphanumeric command-line interface? Why go to all this trouble?"

The answer involves the characteristics of good operator interfaces mentioned above. A good GUI reflects the type of telescope being controlled and how it is to be used (characteristic #1). By laying out the pages well and grouping on each page those commands and status information that naturally go together (according to how the telescope will be used), you avoid the nuisance of switching between pages or having lots of pages displayed simultaneously. Older style command-line systems can do this as well, but GUI systems really force the designer to think about what items should be on the same page, and how the operator will interact with them.

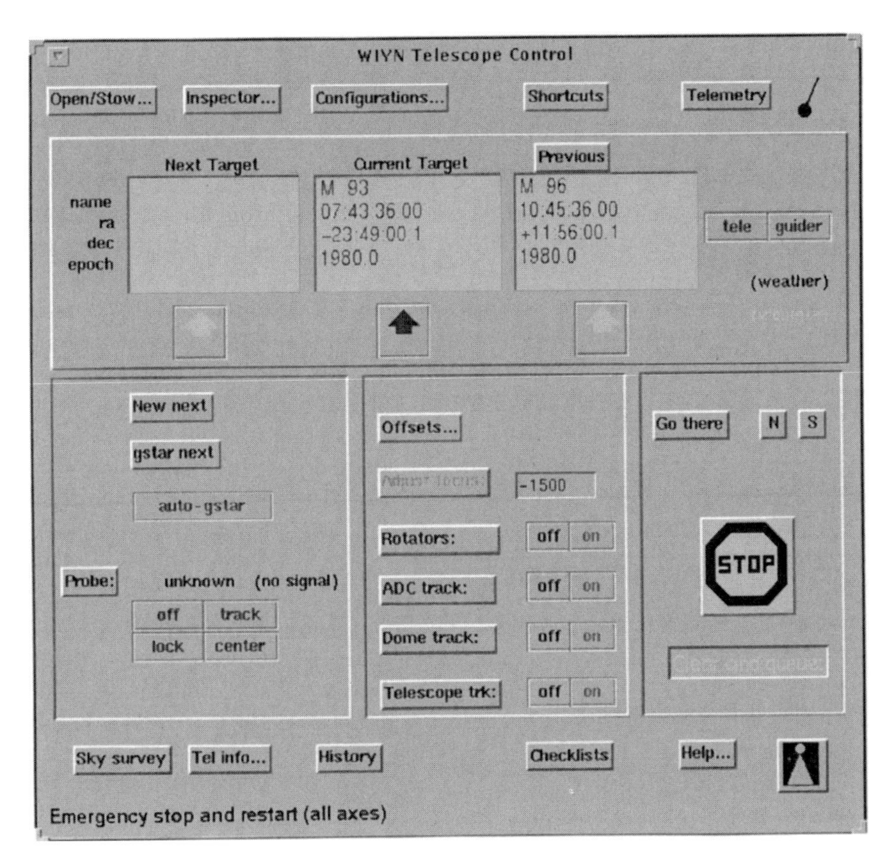

Fig. 10-1 *A WIYN telescope control page.*

Modern GUI-based operating systems (e.g., Windows NT and OS/2) are so complex and difficult to program by traditional methods that you almost have to use the development tools (see the next chapter for details). This makes developing the application much quicker with fewer errors (since the tool generates much of the low-level software), and it forces the designer to focus on the layout of each page and how each menu item works by eliminating the need to worry about the details of displaying data and receiving operator inputs. All this makes it far easier to build intuitive interfaces (characteristic #2) using modern GUI development systems, because each button on the screen is required to be labelled by the development software. The designer is not just putting labels on a computer screen, you are laying out buttons and little work areas. In effect, you are building a software version of a hand paddle, so you are forced by the development tools to think about how the hand paddle controls should be laid out in a logical way.

Although GUI systems execute no more rapidly (characteristic #3) than command line systems, on modern computers they are no slower. The advantage to GUIs here is that when you do something with the mouse or keyboard, the system responds in a dialog by highlighting a menu choice or changing a window border, and it does so quickly enough to let the operator know the CPU is not bogged down.

The biggest advantage to GUIs is that they are ergonomic (characteristic #4), making the operator interface intuitive. Humans find it easier to push buttons than to type cryptic commands on a keyboard. GUI buttons look and work just like hand paddle buttons and switches, except it is easier to enter names of objects and positions using a keyboard than it is to use a typical hard-wired hand paddle (which typically do not permit such inputs). When you click on a GUI button with a mouse, the graphical representation of the button is changed (typically three-dimensional shading is modified) to make the button look as though it is pushed in. Values of status data from sensors can be represented with analog dials with meter needles and colors, e.g., green for "OK," yellow to indicate a warning condition, and red for an alert or error condition that must be dealt with immediately. Using such graphical representations instead of a page full of numbers makes it easier for an operator to detect an error and its cause when operating under observing conditions (late nights and high altitudes) that cause mistakes.

The differences between software control panels and hard-wired hand paddles is that (1) you can have several pages, like having multiple hand paddles, one for each type of function to be performed, and (2) each page can usually be changed quite quickly and easily. Try rewiring a hand paddle to add or change something in only a few minutes! That's all it takes to change a page with modern software development tools.

Although a great deal of progress has been made in the last few years to develop affordable operating systems and software development tools to make good operator interfaces possible, there is still room for improvement. In a few years, voice input for control systems will be common, eliminating the clumsy keyboard and even the mouse. At first, voice input will be an alternative to keyboard or mouse input, so that you issue a command by speaking a menu choice instead of clicking on it. This is because voice input in affordable systems will have a limited number of words in the vocabulary that the system can recognize. Later, as speech processing improves and similar advances are made in extracting meaning from natural language sentences, you will be able to tell your computer, for example, to "Point the telescope at Spica and track it". Such control systems will be truly ergonomic, in that they will fit so naturally with human senses and motor skills that they will hardly be noticed.

The main point of this chapter is to suggest that you think carefully about the plight of the user of the system you are building, and to make it as easy as possible for that user to operate the telescope correctly and safely. Don't overlook this important system component!

Chapter 11

Computers and System Software

The final components of telescope control systems to be discussed in this part of the book are the computer and system software. Included under the generic title of system software is the operating system and all the software normally distributed with it for both the development and run-time environments. The development environment is the computer, operating system, and development tools (software design aids, editors, compilers, linkers, and debuggers) that you use to design and develop your control system software. The run-time environment is the computer, operating system, and special hardware (limit switches, motor controllers, shaft encoder interfaces, etc.) that control the observatory structure, telescope, and instruments.

Most telescope control system developers do not ignore the computer and system software to the extent that they so often ignore the operator interface, but generally not enough thought is given to them, either. This may be due in part to the tradition among astronomers of having to do most, if not all, telescope and instrument development themselves. Those astronomers specializing in observing often have a relatively strong electronics and non-real-time scientific applications software background, and tend to generalize the viewpoints of these disciplines when it comes to selecting a computer and operating system for real-time applications.

This approach does not work well. For analyzing data on your desk, it often does not matter whether you use an Apple Macintosh, a PC clone, or a UNIX workstation. Which you use is a matter of personal preference, usually based on whatever you know the best. But when it comes to real-time applications, some very popular and widespread operating systems are not designed to handle them properly, which could lead to injured observers, damaged equipment, or inoperable telescopes. This chapter explores the special requirements that real-time systems impose on computers and operating systems, and describes how to go about choosing the hardware and system software needed to build a telescope control system.

11.1 Characteristics of Real-Time Command and Control

A real-time system is one in which certain actions must be taken within some (usually small) period of time. Data analysis on your desk is not considered to be a real-time application. If you are a young faculty member trying to get tenure at a university, you have to publish some number of papers per year, but it doesn't really matter if the next paper is finished today or tomorrow. In real-time systems, it is critical that certain things happen within milliseconds of other events, or even microseconds in some cases.

In a telescope control system employing a software servo loop, certain calculations have to be done some number of times per second to achieve a design servo bandwidth and pointing or tracking accuracy. For example, if your servo design requires that you issue commands to the motors 10 times per second, then all the computer instructions to read the encoders, apply the encoder calibration to compute where the telescope is, compute where the telescope should have been when the encoders were read, compute the error and the new motor speed, and to issue the commands from the motors must happen every 100 ms per motor, or 50 ms per calculation for two motors (one for each major telescope axis). If you use 100% of the CPU, then you have 50 ms to do each calculation for two axes. As a rule of thumb, real-time systems should use less than 50% of the processor, so each calculation should take no more than 25 ms, if nothing else is happening in the computer.

If you read and took to heart the previous chapter, you know that lots of other things are happening in the computer at the same time. You should count on allocating somewhere around 10–20% of the CPU to the operator interface and other operating system functions. That reduces the time for each set of calculations to around 15 ms, even less if you have several other motors to control and several sensors to monitor.

All of a sudden, it begins to look like you have no time at all to do everything. Fortunately, modern inexpensive computers based on the Pentium and PowerPC chips are quite powerful. For example, the 60 MHz Pentium is rated by Dongarra (1994) at 5 million floating point operations per second (MFLOPS). In 15 ms, such a CPU can execute 75,000 floating point operations. Although the trigonometric functions that are so prominent in Chapters 5 and 6 each take about 20–30 floating point operations to compute, that still leaves a healthy margin. But you should examine your control algorithms closely to see if the computer you plan to use can handle them, assuming that only about 20% of the CPU will be available to execute them. Honeycutt (1992) reports that controlling a fully automatic equatorial telescope requires about 7% of a MicroVAX II, a computer that Dongarra rates at 0.12 MFLOPS. But that system does not have a modern point and click graphical user interface. Be careful not to extrapolate too far from this one data point!

Another characteristic of real-time systems, especially those used in command and control situations, is quick and reliable response to interrupts. Just about any device hooked to a computer, be it a keyboard or mouse on a serial port, a printer on a parallel port, or a motor controller card plugged into your computer's backplane, has a data path and a status path between the device and the processor or backplane. The status path consists of one or more lines (*interrupt* lines) dedicated to letting the processor know when an input or output is complete. Serial ports and parallel printer ports typically issue one interrupt for each 8-bit byte transferred, while disk drives issue an interrupt after the transfer of the last byte of a single disk block (often 512 bytes) filling a sector of one track on the disk. Direct memory access (DMA) input/output boards usually have programmable block sizes, so you set a register on the board to indicate the number of bytes or words to transfer before issuing an interrupt.

This somewhat esoteric concept of an interrupt can save your life! A telescope control system must respond to interrupts of many kinds, including those from limit switches, which are usually placed to prevent damage to equipment. But properly placed switch bars and interlocks can also prevent you from getting your head pinned between your telescope and your dome, or stop another moving part from hurting someone. If the computer hardware or software cannot guarantee an acceptable maximum response time to an interrupt, then it is not suitable for use in a real-time system.

11.2 Selecting Computer Hardware

Computers made in the 1950s and on into the 1960s polled the interrupt lines every so often. When it detected that an interrupt line had been raised, the processor would execute the appropriate software (the *interrupt service routine*) to load and send the next byte to the device, or return from the device driver and indicate that the full data transfer was complete. Interrupt polling introduces an unnecessary delay between the time a device posts an interrupt and when the CPU begins to process it.

Modern computers use vectored interrupt processing. In this scheme, each device on the input/output bus has an assigned area in the computer's main memory where the address of the interrupt service routine (ISR) for that device is stored. This memory area is the *vector*, and often contains both the address of the ISR and a processor status word to load when processing the interrupt. The processor status word contains information about preventing other interrupts from interrupting the processing of the current interrupt and other information about the mode of the CPU when processing the interrupt.

The ISR is usually part of the *device driver* for the particular device plugged into your computer's bus. Device drivers are software written to be part of the operating system that control devices for the operating system. When you plug a new device into your bus, you have to load the device driver for that device, to permit the operating system to access that device. Part of loading the device driver is to figure out where the ISR will be stored and to load that address into the vector. Those who design and build operating systems dedicate specific areas of computer memory to device drivers and vectors, then publish these reserved address areas. Others developing new boards and device drivers then use these published specifications when writing device drivers for that operating system.

Since most modern personal computers and workstations use vectored interrupts, you don't have to go to any special trouble to get the right hardware for your control system. But to make interrupts work well, both the hardware and the operating system must be designed for the task, and for each other.

Another thing to consider when choosing computer hardware is the availability of interface cards for the computer bus. As pointed out in Chapter 7, a very good approach for reducing development time and cost is to use a standard computer bus for which there are many third-party manufacturers of interface cards. Some common computer PC buses are the Industry Standard Architecture (ISA), Extended ISA (EISA), and Peripheral Component Interconnect (PCI) buses for desktop PC computers, and the Personal Computer Memory Card International Association (PCMCIA) interface for both desktop and notebook computers. Other buses include the NuBus for Apple Macintosh computers, the S-bus for Sun workstations, the Turbochannel for DEC workstations, the SCSI bus for many different makes of workstation and PC, and the VMEbus for ruggedized and industrial control. Of these, interface cards are more generally available for the ISA and VMEbus buses, especially those boards most needed for telescope control systems, including motor controllers, encoder interface cards, counter/timer boards, and digital input/output boards for limit switches, turning AC motors on and off, and other applications.

While investigating computer buses, you should look into the rate at which data can move over the bus and whether that rate is adequate for your needs. The ISA bus moves one or two bytes (8 or 16 bits) at a clock rate of 8MHz with typically three clock pulses per data transfer, for an aggregate bandwith of 2.67 megabytes per second (MB/S). Any one ISA device is unlikely to "see" more than 0.5 MB/S. The PCI bus can move four bytes (32 bits) of data at a time at a maximum rate of 132 MB/S and has a standard defined for moving eight bytes (64 bits) of data at the same clock speed that effectively doubles the bandwidth, but this is rarely implemented. Most PCs use a special bus to transfer program instructions and data between main memory and the CPU separate from the bus used for peripheral devices, and many use a PCI bus for high speed graphics accelerators. Most PCs also have disk controllers on the same board as the CPU, so again the main bus is not loaded with hard disk, or floppy disk traffic, though backup devices on PCs, such as tape drives, usually use the ISA bus. All this means that the ISA bus should have sufficient bandwidth to handle most telescope control applications, but you should look carefully at the loading of your bus as part of the process of designing your system.

For amateurs and others working within a tight budget, the PC clone is the best choice of computer hardware. The computer itself is inexpensive, it offers acceptable performance, real-time operating systems are available for PCs, and there is a wide selection of interface cards available from several different vendors, some of which are featured in Chapter 8 and Appendices B and D. Expansion chassis are inexpensive and readily available, in case you have more interface cards than you can fit in your computer. Most card vendors supply device drivers and a library of BASIC or C routines to perform data transfers with the cards, minimizing the amount of software you have to write. The lower cost of this approach tends to mean that PC hardware and their operating systems are less reliable than other approaches, but they are becoming more reliable every year.

Those who need a higher level of reliability, and who can afford to pay for it, should consider the VMEbus. Invented by Motorola in 1979, the Versa Module Eurocard (VME) bus was made an open standard in 1981 by Motorola, Mostek, and Signetics. Over one hundred manufacturers make boards for this international standard bus, which is used in several professional telescope control systems, including the twin 8-meter Gemini telescopes, the Keck 10-meter telescopes, the GONG solar telescopes, and several telescopes on Kitt Peak. More information on the VMEbus can be obtained from the VFEA International Trade Association in Scottsdale, Arizona. If you want to use a UNIX workstation or other computer to serve as the operator interface, then use a VME "crate" (cabinet) to house the interface cards. BIT 3 Computer Corporation in Minneapolis, Minnesota makes VMEbus interface cards for about 16 different computers, including the PC (PCI, ISA, and EISA buses), DEC Q-Bus, Multibus, Apple NuBus, and several workstations from DEC, Sun, HP, IBM, and Silicon Graphics.

Although custom bus adapters can be found to mate an ISA bus or

VMEbus to most computers, from the standpoint of available operating systems, device drivers, and other software, you should focus on using either a PC compatible computer or a VME crate to house your control system interface cards, and even to serve as the main control system computer.

11.3 Selecting the Operating System

As rapidly as computer hardware changes, software changes even faster. We present the information below knowing full well it was obsolete before it was published. We suggest you use it as a guide to the *process* of selecting an operating system, not to the operating systems available at the time you read this.

Although most modern computers have the right hardware design for real-time systems, this is not the case for most popular operating systems. Let's take a look at the process of sending one byte of data to a serial port, to see how a typical operating system works.

When you read or write data to or from a device, you include in the right place in your program a command (in the language you are using to develop your system) to ask the operating system for help in moving data. This can be a `READ` or `WRITE` in FORTRAN, or a `putc` or `gets` in C. To send the mythical byte to a serial port, you put the byte into a storage location and use the address of that location in your `WRITE` or `putc` command. After writing your program, you run it through a compiler that translates your computer language file into instructions recognized by your computer's processor. At the point where you want output in your program, the compiler places a call to the operating system into the code. When you run your program and this output request is executed, the operating system passes the address of the location containing the byte to the device driver, and activates the device driver to perform the output.

Control now jumps to the memory location holding the device driver. The driver takes the byte, initiates the write operation by moving the byte to (depending on the design of the processor) either a port on the processor or an address on the bus, then it sets a flag to permit the device to interrupt the processor when it has received the byte. The device driver then returns control to the part of the operating system that schedules tasks (programs) to run. This *scheduler* then activates another task that is not waiting for input or output, and starts running it.

When the device receives the byte, it sets an interrupt line. This interrupt causes the processor to temporarily stop executing the second task and to manipulate the bus to ask the device to send its vector address. The device passes the vector address to the processor down the bus, then the processor starts executing the interrupt service routine (ISR) at that address. The ISR checks to see if there is another byte in the data buffer to send. If there is, it starts this sequence over again. If not, it returns to the operating system with an indication that the write operation is complete and successful, whereupon the scheduler part of the operating system continues running the task that issued the write, starting with the first instruction after the write.

We are going through this complicated process to show the differences between real-time and non-real-time operating systems. *True* real-time systems guarantee the response time to the interrupt to be some (small) number of milliseconds or even microseconds. Other systems can safely be used for real time applications if they (1) allow control over interrupt dispatch priority and process execution priority, and (2) loading tests show they respond to interrupts within an acceptable range of times. These systems we will call *soft* real-time systems. Both types of real time systems control interrupt response behavior using *preemptive scheduling*, that is, they permit an interrupt to suspend execution of the current program and they immediately start executing the ISR, then they schedule the task with the highest priority to run after the interrupt has been processed. This means they are *multiprogrammed*, that is, they can schedule multiple programs to run, one at a time, in between each other's reads and writes. This is the way that several operating systems work, including RT-11, RSX-11, and VMS from Digital Equipment Corporation. In particular, VMS is a very robust and capable operating system that, unfortunately, was fading from the marketplace in the mid-1990s.

In the PC world, Microsoft's Disk Operating System (MS-DOS) was the most frequently-used operating system in the mid-1990s. MS-DOS is not a multi-programmed real-time system, in that it cannot run more than one program with a guaranteed response time to an interrupt. Also, since it can run only one program at a time, and if that program is not written properly, it is possible that MS-DOS would *never* respond to an interrupt! That is extremely dangerous to humans and equipment, so we do not recommend using MS-DOS for your telescope control system unless you are a true expert in using this operating system, in which case the response time for an interrupt is more predictable than that for other Microsoft operating syetms. The versions of Microsoft Windows (e.g., Version 3.1) and Windows for Workgroups used in the mid-1990s are not true operating systems. They are single programs that run under MS-DOS and make it look like you can run multiple programs at the same time. You open a window for a word processor and run it, then open another window and run a spreadsheet, and you can click back and forth between these windows so it looks as though you are running multiple programs at the same time. But you are not. All this is an illusion created by the Windows program, which is the single program that runs under MS-DOS, because MS-DOS can run only one program at a time.

Programs running under Windows 3.1 must cooperate by checking their message queue (a part of the Windows 3.1 environment) frequently. When an application, such as a word processor, checks its message queue, it asks Windows for messages. Windows uses this as an opportunity to check to see if another application needs to run, and if so, runs it. This is analogous to the scheduling that occurs when a read or write interrupt is processed in a true preemptively scheduled operating system. The difference is that under Windows the application must cooperate with other application programs by asking to read its message queue. In a real-time operating system, the application is preempted when input or output is begun—something the executing program has no control over. Under Windows 3.1, an uncooperative process can hog the CPU, preventing the handling of interrupts from limit switches from ever taking place. Under a real-time operating system this is impossible, since no process can prevent itself from being preempted.

MS-DOS is also unreliable in that a program that runs amuck can cause the operating system to crash, leaving motors running and limit switch closures undetected.

Microsoft realized in the early 1980s that MS-DOS was very limited, and that applications it and others were developing would work faster and better under a multiprogrammed operating system with better interrupt performance, reliability, and security. To meet this need, it began a partnership with IBM to develop OS/2. When the design and development were fairly far along, Microsoft and IBM parted company. IBM continued to develop OS/2, while Microsoft named its version Windows NT (for "New Technology"). Both operating systems are soft real-time systems, with all the features needed to develop safe and reliable real-time applications. In the mid-1990s, OS/2 was being marketed primarily for servers, and Microsoft had better development tools, including Visual C++, for developing NT applications.

Given a choice between the simplicity and lower cost of MS-DOS and Windows 3.1 or the complexity of Windows NT, both users and software developers in the late 1980s and early 1990s chose MS-DOS, in part because NT was so complex that early releases had lots of bugs, tended to "bomb" (cease execution abruptly), and required all available memory, no matter how much you had. These problems were fixed in later releases.

Since MS-DOS was the only PC operating system available for several years, most third-party software (word processors, spreadsheets, CAD packages, etc.) worked only on MS-DOS. NT requires a different Applications Programming Interface (API) to the operating system, and most MS-DOS device drivers will not work with NT. This basic incompatibility forced most third-party software companies to charge more for NT versions of their software or to not offer NT versions at all, making it very expensive or impossible altogether to upgrade from MS-DOS to NT, so most users chose to stay with MS-DOS.

As applications continued to become more complex and more interwoven with each other, the need for NT only increased, in the view of Microsoft. In an effort to reduce the number of operating systems it supports and to move toward the kind of operating system needed for the next generation of PC applications, Microsoft released in 1995 a new version of Windows called Windows 95 to replace the old Windows/MS-DOS combination. To an applications developer, Windows 95 looks more like NT than the old Windows. It has the multiprogramming, preemptive scheduling, and real-time features of NT in a single-user system but it uses MS-DOS device drivers, and all MS-DOS applications work on Windows 95 except those that do not conform to MS-DOS application design guidelines. Unlike the old Windows, Windows 95 does not use MS-DOS directly—it is a real operating system in its own right, though it has a kind of multiprogramming MS-DOS built into it.

Windows 95 is designed to replace MS-DOS and Windows on desktop and notebook computers. It is highly compatible with both MS-DOS and Windows software from both Microsoft and third parties. It offers an improved user interface, and it has new features to help notebook users (e.g., battery energy monitoring) and to permit dynamic reconfiguration of peripheral devices, chiefly through the PCMCIA interface (using what is called "Plug and Play"). When it introduced Windows 95, Microsoft wanted current MS-DOS and Windows users to buy Windows 95 so that eventually they would not have to support MS-DOS and both new and old Windows.

Telescope control system developers should welcome this change, because it gives them the operating system they need to develop telescope control systems to run on inexpensive hardware, namely the PC. Any application written to run under Windows NT using the Win 32 Application Programming Interface (API) will run under Windows 95. When choosing a PC real-time operating system, the choice has boiled down to Windows 95 or Windows NT Workstation from Microsoft. Since they are similar in many ways, the choice is a matter of personal preference, though there are some differences.

Windows NT can be purchased in either "workstation" (single-user) or "server" (multi-user) configurations. Its advantages are:

1. Designed for high-performance scientific, engineering, and business applications;

2. Supports symmetric multiprocessing (multiple processors in the same computer, e.g., one for telescope control and one for instrument data readout, reduction, and display);

3. Runs on several different processors (Intel IBM-compatibles, DEC Alpha, MIPS, and IBM/Apple PowerPC);

4. Puts each MS-DOS/Windows 16-bit program in a separate protected memory area, so if one program bombs it does not affect any other 16-bit or 32-bit program, or the operating system itself; and

5. Superior file system protection and overall security.

Windows 95 is available in only a single-user version with the following advantages:

1. New user interface that is easier and faster to learn and use;

2. Runs on computers with small amounts of memory (4MB);

3. Complete compatibility with Windows and MS-DOS (including most device drivers);

4. Special features for notebook computers; and

5. Plug and Play dynamic peripheral device reconfiguration.

Both operating systems provide extended file names over the MS-DOS/Windows "8+3" naming scheme of 1–8 characters for the file name and 0–3 characters for the extension. A key disadvantage of Windows 95 is that while all 32-bit (Win 32 API) programs have their own protected memory areas (as in NT), that all 16-bit MS-DOS/Windows programs run under Windows 95 in a single protected area instead of separate protected areas for each 16-bit program. This means that, under Windows 95, if a 16-bit application bombs it could take all the other 16-bit applications with it. This is nearly impossible under Windows NT.

Microsoft announced plans while this book was in preparation to incorporate most new features of Windows 95 into Windows NT including the user interface and Plug and Play. After third party software developers convert their programs to the Win 32 API (for Windows 95 compatibility) and the old 16-bit computers are in landfills, it will be relatively easy (and a lot less expensive) for Microsoft to drop Windows 95 altogether to sell, support, and enhance only Windows NT.

So which operating system should you buy? Trueblood recommends Windows NT unless one of the following is true:

1. Your computer does not have at least 16 MB of memory.

2. Your computer processor is not an Intel 80386, 80486, Pentium, or one of the other processors supported by Windows NT.

3. You have less than 200 MB of disk space.

4. You have a notebook computer and Windows NT does not yet have full notebook support.

If Windows NT is recommended, how quickly can it respond to an interrupt? The time to respond to an interrupt includes

1. *hardware interrupt latency*, which is the time required to stop processing the current instruction, locate the interrupt vector, and to jump to that address;

2. *interrupt dispatching*, which is the time required by the operating system to start the interrupt service routine (ISR); and

3. ISR execution time to service the interrupt.

Catlin (1995) used a Hewlett-Packard XU 5/90 personal computer with one 90 Mhz Pentium CPU, 256 kB of cache, 16 MB of memory, and 540 MB of disk space to measure these latencies with the following results:

Hardware Interrupt Latency	1.8	–	2.9 μs
Interrupt Dispatching	4.6	–	10.5 μs
Interrupt Service Routine	10.3	–	16.7 μs
Total	16.7	–	30.1 μs.

Catlin attributed the wide range of values to the effects of virtual memory and, in particular, the cache manager. Even if it took 33.3 μs to process an interrupt, a 90 Mhz Pentium could process 30,000 interrupts per second under Windows NT. Tuning dispatch and execution priorities would safely guarantee response to a limit switch in well under 1 ms, quite acceptable performance.

Those who are interested should note that older real-time operating systems (e.g., RT-11, RSX, and VMS) offer multiprogramming (running several programs concurrently) by scheduling each separate process (program), blocking it while it waits for reads and writes, and letting it run when its input or output (I/O) completes. Modern operating systems such as Windows 95 and Windows NT can also schedule processes (programs), but this can take a lot of overhead.

To get around this operating system overhead, both NT and 95 offer *multithreaded* scheduling. Instead of writing each asynchronous part of your control system as a separate process, you make it a *thread*, or asynchronously scheduled and executed element, of the same process. This makes scheduling more efficient and reduces operating system overhead, giving you back some CPU cycles to use in your application.

One final word for UNIX fans. Although UNIX is multiprogrammed, uses a form of preemptive scheduling, and has good response to interrupts, it is not a true or soft real-time operating system because it has no maximum response time to an interrupt. It also is too unreliable to be safe. UNIX is used successfully in many real-time applications, but our advice is not to use it in a telescope control system. If you are in love with UNIX, use VxWorks by Wind River Systems instead. It looks and feels like UNIX, and you develop software for it on a UNIX system, but VxWorks is a *soft* real-time system that is safe to use on a telescope control system. VxWorks is also available for the PC, using the PC both as a "host" development computer under Windows, and as a "target" control system processor. It also works on a number of other computers, including Motorola 680x0 (a popular VMEbus processor), SPARC, MIPS R3000 and R4000, and AMD 29000 architectures.

11.4 Selecting the Development Environment

Software development can be a major cost component of any telescope control system budget. Procurement of good hardware and software tools to aid software development can yield significant software development cost savings. Selection of a processor should entail evaluation of the software development environment that is available with each different manufacturer's processor, as well as the ability of a particular processor to handle the computational load. The greatest system development cost savings are realized if components of the software development environment can also serve as necessary components of the final control system. The individual elements of the software development environment are discussed below.

Usually, the software development environment must be purchased as part of, or in addition to, the operational software environment (operating system). Software is often the most tedious, time-consuming, and expensive part of the whole project. A little money spent on a good software development environment repays itself very quickly, and many times over.

The minimum hardware and software configuration for software development should support the following requirements:

1. Rapid entry and correction of source code,

2. Rapid production of legible hard copy of source code, link maps, and file listings,

3. Rapid compiling, assembling, and linking of programs,

4. At least 300 million bytes (MB) of mass storage beyond what is needed for the operating system and software development tools,

5. Ability to copy ("back up") source code in case of medium or peripheral device failure,

6. Adequate memory for executing editing, copying, compiling, and other utility programs, as well as for running the control system,

7. Integration of the operational hardware configuration with the software development system, to allow software testing and integration with the final hardware configuration.

PCs are sold equipped with high resolution screens, keyboards, and good editors for entering code. Very sophisticated text editors and document processors are available for a fraction of the cost of the PC that make entering and editing source code easy.

There are several small, fast, high quality bubble-jet printers available for about \$100–\$400 capable of speeds of 200 characters per second or more with line widths of 80 or 132 columns, and with a graphics capability. Laser printers are beginning to be reasonably priced. You should always keep a paper copy of all the software you write, no matter how good your disk and backup are.

Although early PCs came with one or two floppy disks and no hard drives, today 300 MB hard drives are very inexpensive, and even larger drives up to 1 GB are quite affordable. These disks are the most reliable direct access high speed mass storage available today. Their reliability stems from the fact that they are sealed against contaminants that would otherwise shorten the life of the disks. You can never have too much disk storage. Operating systems, development tools, astronomical catalogs, and CCD cameras all require large amounts of storage.

As hard disk drives become larger in capacity and less expensive, the problem of disk backup gets worse. Although disk drives are now very reliable, occasionally problems do occur, and you do not want to be the victim of a disk crash without having a backup. You might want to consider buying one of the many backup devices available today, such as the Iomega JAZ, a quarter-inch tape cartridge (QIC) drive, or even the more expensive Digital Audio Tape (DAT) and Exabyte 8mm tape drives. A reliable backup drive is absolutely essential to making your project a success.

Modern windows operating systems require a large amount of memory, at least 16 MB and preferably 32 MB or even more. Newer processors can address 256 MB of memory or more, but many of the processors in general use have a lower maximum limit. Memory prices have fallen dramatically in the last few years, so it is more economical to purchase a large amount of memory than it is to lower programmer productivity by causing the programmer to run out of available memory at crucial moments.

You should develop the software on the same machine that will be used to control the telescope, due to the large number of tests with the telescope control interfaces that you will need to do. You will need a machine with adequate cabinet space, bus slots, and power supply reserve for peripherals, so that both software development peripherals and telescope control peripherals can be installed, powered up, and functioning simultaneously. It should take no more than five minutes, and preferably even less time, to make a simple change to the source code and generate and load a new program version.

The most critical components of the development environment are the compiler and associated programmer productivity tools, since these have the greatest influence on the software development time and cost. In most systems built today, software is the most expensive component. Booch (1983) describes the current situation as "the software crisis." He cites a study indicating that for systems including both hardware and software, the proportion of costs was divided in 1965 in the proportions 85% : 15% between hardware and software respectively, but in 1970 the proportions were 35% : 65%, and by 1985 they would be 10% : 90%. The Department of Defense spent over $3 billion on software alone in the early 1970s, and Booch quotes a study predicting it would grow to $32 billion by 1990.

Booch presents the results of another study showing that of all software dollars spent by the Department of Defense in 1973, data processing (written primarily in COBOL) used 19% of the total, scientific software (written primarily in FORTRAN) used 5%, with the largest segment being embedded computer systems, at 56%. Other indirect software costs used the remaining 20%. Embedded computer systems are those systems using computers to do something other than compute and display (or print) numbers. That is, the *primary* purpose of the system is not data processing. A telescope control system is an example of an embedded computer system since the real purpose of the system is to actuate motors to point a telescope, even though some numbers are computed and displayed on the operator's screen to achieve this.

The symptoms of the software crisis, as enumerated by Booch (1983, p. 6), are as follows:

- **Responsiveness:** Computer-based systems often do not meet user needs.
- **Reliability:** Software often fails.
- **Cost:** Software costs are seldom predictable and are often perceived as excessive.
- **Modifiability:** Software maintenance is complex, costly, and error prone.
- **Timeliness:** Software is often late and frequently delivered with less-than-promised capability.
- **Transportability:** Software from one system is seldom used in another, even when similar functions are required.
- **Efficiency:** Software development efforts do not make optimal use of the resources involved (processing time and memory space).

In an attempt to solve this problem, software leaders collected successful software development methods into a set of practices called structured programming (see, for example, Dijkstra (1976) and Tausworthe (1977)). Also, a new programming environment, centered around the Ada programming language, was adopted by the Department of Defense and a few other organizations as the only acceptable environment in which to build embedded systems. The emphasis of published papers in the field was on software, because as mentioned above, software was and still is the most expensive component of a system, especially of large systems (see Appendix A).

The main goals of a software development environment are as follows:

- Generate efficient code that executes quickly in a real-time control environment,
- Minimize the initial software development costs,
- Minimize the total system life cycle costs.

In the past, the main choice to be made in a software development environment was the programming language. Several programming languages are discussed below, but as we shall see, selection of a programming language is no longer the major concern when using modern operating systems.

Although assembler language can be used to generate more efficient code than high level languages, programmer productivity is much higher using high level languages, so initial software development costs are much lower when a high level language is used. Since it is easier to understand the purpose of any given section of code written in a high level language, it is easier to find errors in the code, and to make enhancements to the code after the system is operational. Also, after the original programmer is no longer available for software maintenance, there is a better chance of finding programmers conversant in a well-known high level language than in the assembler language of any given processor. It takes a new programmer less time to learn the telescope control software if it is written in a high level language than if it is written in assembler language. All this means code written in a high level language has a lower life cycle cost than code written in assembler language.

Processors are now fast enough and compilers are now smart enough that the very slight speed advantages in well-written assembler language programs are more than outweighed by the extra cost in programmer time and skill level required to write them. As a consequence, hardly anyone writes applications code in assembler language, though some operating systems and device drivers are still written in assembler. Modern processors and operating systems are so complex that writing software in assembler language is simply too expensive and time-consuming.

The relative advantages and disadvantages of some of the more popular high level languages are summarized in the paragraphs below. There is no one language that is clearly the best for implementing a telescope control system, and experts disagree over the usefulness of any particular language for a given application. You must select the language that is best suited to your software development and operational environments.

The Ada programming language was invented specifically for developing embedded computer systems, and to address the problem of the software crisis mentioned earlier. Ada employs many modern software engineering language constructs designed to force a programmer to develop correct, readable, and maintainable code. It was designed for real-time applications, and uses features of the language itself to control inter-task synchronization and communication, rather than relying on a programmer's knowledge of a particular operating system to perform these operations. This allows the code to be transported to different machines without modification.

To enforce this transportability, the Department of Defense has trademarked the name "Ada," and only the DoD certifies particular compilers to be Ada compilers. Therefore, there is only one standard version of the language. Ada compilers tend to be expensive, and it is difficult and frustrating for programmers used to less discipline enforced by the compiler to use Ada. Modern PC operating systems with windows and menus typically do not have development tools for Ada.

The BASIC language is probably the most widely-known high level computer language. Its most popular form is an interpreter, which translates each line of code into machine instructions as the program executes. Although interpreters are useful for developing and testing software quickly (since there are no compile and link phases when a program is changed), the execution speed is 2–50 times slower than a compiled version of the same language. Early versions of BASIC limited variable names to two characters, which hindered any attempt to make the code readable and maintainable. Modern disk-based versions usually do not suffer from this drawback.

In addition, most BASIC versions are not structured. This means that certain fundamental program structures that make programs easier to understand, test, and maintain, such as `IF-THEN-ELSE`, `DO-WHILE`, and `DO-CASE`, are not available in these versions, so the programmer tends to fill a routine with `GO TO` statements that deter others from understanding the program. Many early versions of BASIC also lacked the ability to divide a program into subroutines, and some do not use double precision (8-byte) or even single precision (4-byte) real variables. Although BASIC is available in compilers as well as interpreters, the compilers often do not optimize the machine code for execution speed.

However, there are some versions of BASIC compilers that are quite sophisticated, and that provide all the tools needed to program a real-time control system. Since so many different versions of BASIC are available, if BASIC is considered as the programming language of choice, the version to be purchased should be investigated thoroughly before the decision to purchase it is made. One of the best BASIC development systems is Visual Basic from Microsoft. It makes programming a Windows application relatively straightforward for the typical programmer, who does not know (or want to know) how to deal with window objects, handle mouse clicks, or deal with Windows library primitives. Many vendors of PC interface boards sell Dynamic Load Libraries (DLLs) that serve as device drivers and a programmer interface from Visual Basic to their boards. This makes the process of integrating a vendor's board into a Windows environment relatively easy.

FORTRAN is one of the oldest programming languages in use, so there is a large base of (aging) programmers available to do the software development and maintenance. It is available as a compiler, often with execution speed optimization. Variable names are up to six characters long, with some newer versions allowing 31-character names. Many newer versions also permit structured `IF`, `DO`, and `CASE` blocks that lower both development and life cycle software costs. Most real-time operating systems support both FORTRAN and a large library of FORTRAN-callable routines to perform real-time software functions. FORTRAN encourages the use of subroutines, and many operating systems that support FORTRAN allow overlaying of subroutines to minimize the use of memory.

Pascal is a relatively new language invented as a structured language, that is, it contains many of the features of Ada that tend to enforce good programming practices. In fact, Pascal was one of the models used to develop Ada. It is available as a compiler, and is in wide use, primarily in universities to teach good programming practices. Although it was not intended for real-time control applications, many real-time operating systems support Pascal. One advantage that Pascal has over other languages is that both the programming language and the machine code bit patterns that the compiler generates have been standardized. As a result, there are several "p-code microengines" available that are designed to execute this standard machine code very rapidly. This reduces the execution overhead time that is usually incurred when using a high level language. Pascal lacks many of the real-time and embedded computer system support features of Ada, which often makes developing a control system in Pascal very difficult.

FORTH was invented as a real-time control language at the National Radio Astronomy Observatory, and one of its first applications was telescope control. It enjoyed some popularity among early microcomputer owners because of its efficient use of memory, and the fact that it is self-extensible, that is, the programmer can define new commands in the language. Its main disadvantage is that FORTH uses reverse Polish notation, so that it is not very easy to read and understand a FORTH program. Also, it is not as popular as any of the preceding three languages, so there are fewer programmers familiar with the language. The language structures do not enforce good programming practices, which tends to exacerbate, rather than ameliorate, the software crisis.

C was invented by the developers of the UNIX operating system at AT&T Bell Laboratories, and it is the language in which UNIX is written. It is rapidly becoming the most popular programming language in general use, in part because it lets you do what you want without getting in the way. That is, it is almost the exact opposite of Ada, in terms of enforcing good programming practice. C was later enhanced to add additional features for complex data structures and to support the notion of *classes*, with the resulting product named C++ (where "++" is the C expression for "increment by 1"). Developers of applications for Windows, Windows 95, and Windows NT can use Microsoft Visual C++, a programming environment that handles the problems of displaying modern GUI icons and of receiving inputs from the mouse and keyboard, allowing the programmer to focus on the control algorithms and other unique features of the control system.

One aspect of modern operating systems, especially OS/2 and NT, is their use of *Object Oriented* (O-O, pronounced "oh-oh") programming. An entire college course is often devoted to this subject, but we will explain enough about it to convince you to use development tools such as Visual C++ to develop modern GUI operator interfaces for your control systems. The concepts of Objects were developed to address Booch's transportability problem, that is, programmers keep writing new code that has to be tested and debugged to perform the same old functions, even though similar (fully debugged) software already exists.

A class is one or more pieces of software that performs a well-defined set of operations ("methods") on a well-defined data structure or set of data. When coded, a class is generic. When your program is executed, it uses the class code to create objects, which are "instances" of the class. The class can be extended by defining a new class that consists of the methods and data of the previous class plus new methods and data that you add. By following the rules for classes, software can be reused more easily than before.

If this were all there is to Objects, then the whole concept would have died a long time ago. But the nature of modern computing, especially GUIs, makes O-O programming very attractive. Before modern GUIs were invented, programs were written to be executed in a fairly well-defined order. Control could always jump ("branch") based on conditions at the time of execution, but things usually went in a reasonably prescribed order predefined by the programmer. With modern GUIs, the order in which major parts of the program are executed is determined by the user's mouse clicks on menu items. Although processors still execute programmed instructions in the order in which they occur in memory, the order of execution of large sections of a program no longer matters. What matters is displaying radio buttons and menus, and handling mouse clicks and keyboard strokes properly, all the time maintaining the proper interactive dialog with the user by providing feedback for every user input.

Furthermore, today's users want to take a piece of data (a character string, a piece of clip art, or part of a spreadsheet) from a window executing a program from one software vendor and insert it into a window running a program from a different vendor. How do we make that all work?

The answer in widespread use today is O-O programming. By writing a class according to well-defined rules, the programmer can define within that class code how the various choices open to an operator are displayed (e.g., radio buttons, menu, or dialog box) and what happens when the operator selects each of the possible choices. Classes defined by one software vendor can be linked to those defined by another and reused, making possible the exchange of data (again, according to well-defined rules) between applications written by different vendors. In modern operating systems, the programmer does not directly control the display of these choices. In both OS/2 and Windows NT, this is done by a *presentation manager*, a part of the operating system that interacts with the application software using classes to define the choices and inform the program what selection the user made. Learning to write these classes is somewhat difficult and time-consuming, especially for programmers trained to think in the older sequential execution paradigm.

Realizing this, Microsoft developed Visual Basic and Visual C++ for Windows, Windows 95, and Windows NT to make it possible for ordinary programmers to develop applications for their operating systems. Programming in Visual Basic is about the easiest thing imaginable—you use the mouse to define your pages, then write some Basic code to flesh out what happens if a particular option is selected. To keep it simple, Visual Basic does not offer access to all Windows 95 or NT services, such as multithreading. Visual C++ is more complicated to learn and use than Visual Basic, but it vastly simplifies the process of programming GUIs for Microsoft's multiprogramming operating systems. It also allows multithreading and access to all the features of Windows 95 and NT. You won't realize just how much time Visual C++ saves you until you plunge into the depths of these operating systems and try to write just the simplest application imaginable—a single menu with only one possible choice.

The main point here is that you don't buy only a high level language compiler. The choice is no longer FORTRAN versus Basic versus C. You must choose a development tool that works with the operating system you plan to use, and that allows you to develop software that incorporates GUIs, device drivers, and other aspects of your system seamlessly. The language you use to develop the code is almost a secondary concern, since learning a high level language is a lot easier than learning presentation manager classes.

In this part of the book, we have described the various parts of telescope control systems, and how they work as parts. In the next part, we give examples of how to fit the parts together to make complete, working systems.

Part IV

Examples of Telescope Control Systems

Chapter 12

The Phoenix IV Telescope Control System

12.1 History of the Phoenix IV Telescope

With considerable help and assistance from Andrew Tomer, we describe in this chapter an example of a computerized telescope control system that builds on an interesting equatorial mount to speed acquisition of objects and enhances the overall observing experience. The system uses an approach that can readily be adapted to PCs equipped with a standard ISA bus.

In the mid-1970s, Tomer (Cox and Sinnott, 1977) built a trailer-mounted 12-inch Cassegrain telescope that used an analog servo motor drive without position feedback. He had a few difficulties when he connected his Apple II computer to this telescope, but the project eventually succeeded (Tomer and Bernstein, 1983). The result, however, was cumbersome, with complex interface hardware and even more complex interrupt handling software written in assembler language. This meant that the software was not easy to modify to serve a particular observing program (Tomer, 1984). A subsequent project (Trueblood and Genet, 1985, pp. 127–137) using the Apple II was successful in simplifying the computer interface and in replacing the assembly language software with interpretive BASIC, using `PEEK` and `POKE` commands to control the motor interface boards.

Still, the Apple computer control system proved to have minor but nagging drawbacks after Tomer accumulated extensive field experience with the system. These were:

1. Poor transportability of the Apple II computer equipment, including too many boxes, cables, and other items requiring individual packing for safety,

2. Apple II power supplies proved to be very sensitive to normal generator voltage fluctuations, resulting in “crashes,”

3. The third-party stepper controller cards for the Apple II bus had limited acceleration rate resolution and starting speed, and

4. Most frustrating, the stepper controller cards had no de-acceleration ramp, resulting in gear train strain, and inertial "bounce" at the end of each axis move.

These were addressed over the next several years, beginning with the acquisition of an integrated, portable color computer of unusally rugged characteristics. In the mid-1980s, when the Macintosh replaced the Apple II, Tomer decided to investigate the PC clone market to find a more reliable field computer. The only PC-like machine designed for field portable use (a must in a trailer-mounted telescope) was the Texas Instruments Color Pro. This machine uses an Intel 8080 CPU and a bus very similar to the ISA bus, except for one pin (which is explained below). It does not run MS-DOS, but something TI wrote that is very close. The TI BASIC interpreter is very close to PC BASIC, so readers familiar with the PC should have no trouble adapting the hardware and software described here to their PCs. Although the graphics commands to the screen are different, this is not unusual for PC clones. Furthermore, the TI BASIC does not have explicit `PEEK` and `POKE` commands. Instead, TI BASIC uses `INP` and `OUT`. Otherwise, everything else is recognizable to PC BASIC programmers.

Like the second Apple II system, the PC system uses only interpretive BASIC, and relies on stepper motors to run open-loop without angle feedback encoders. It uses a hand paddle and hard-wired logic to send pulses to the steppers in manual mode. A "local/remote" switch determines whether the hand paddle electronics or the computer sends pulses to the steppers. A high precision worm gear is used on each axis, with mechanical differential gear boxes driving the worms. This approach permits a low-cost system to be built of reasonable accuracy, capable of bringing most objects into the field of a low-power eyepiece.

While based on what today is undeniably an obsolete microprocessor, the TI Color Pro's technical antiquity is handily offset by robust packaging, unheard of in modern transportable computers (except in units costing almost $10,000), as well as remarkable insensitivity to generator power operations. The interior is entirely of aluminum monocoque construction, well shielded against dust on the exterior. Best of all, no particular packing care is needed; fold up the keyboard and load with other telescope equipment. This system is several years old and is no longer made, so Tomer replaced the TI system with a PC notebook computer running Windows 95. Both the TI system and the newer system (under development as of this writing) are described below.

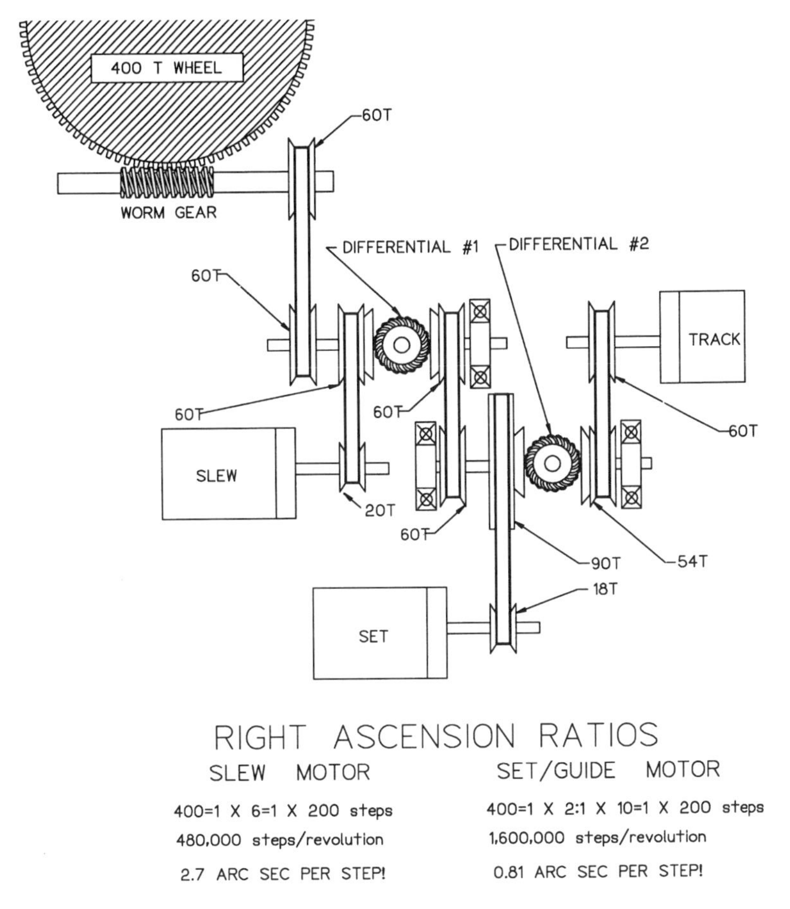

Fig. 12-1 *"Phoenix IV" Right Ascension drive.*

12.2 Drive Train

The strategy pursued by Tomer was to use one motor to "stop the sky" in RA, then make all other motions with respect to, and within, this moving reference frame. A single synchronous AC motor driven by a conventional drive corrector performs this function. Two steppers are used on each axis: one for fast (slew) motion, and the other for fine (set and guide) motion. These motors are connected to a differential gear using two different gear ratios. Thus three motors are used in RA (with two differentials), and two in Dec (with a single differential). The RA drive train is shown in Figure 12-1, while the Dec drive train is shown in Figure 12-2. The differentials serve as the mechanical equivalents of computers, adding the speeds of the two

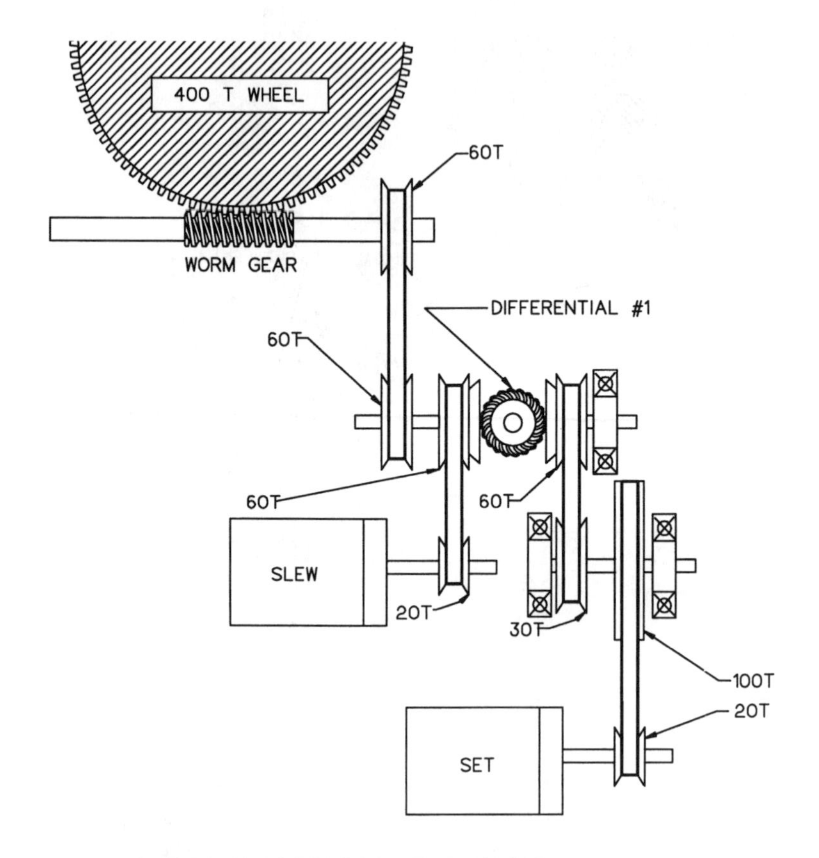

Fig. 12-2 *"Phoenix IV" Declination drive.*

motors on their input shafts and placing the sum (or difference, depending on shaft rotation direction) on their output shafts.

In RA, the AC "track" motor (that "stops the sky") and the fine motion "set" stepper both drive the first differential, the output of which serves as an input to the second differential. The RA "slew" stepper serves as the other input to the second differential, the output of which drives the RA worm. All differentials give an automatic 2:1 reduction. The differentials

Fig. 12-3 *Tomer "Phoenix IV" trailer-mounted telescope.*

have spiral bevel gear construction, not commonly found, and of unusually low internal backlash. They add the two motor speeds that serve as inputs, so the "track" speed always appears at the worm gear, with the other motor speeds added or subtracted according to which direction they are turning. The Dec gearbox is similar, except there is no "track" motor, so it has only one differential. A DC motor is used to move the secondary mirror for focussing.

Figure 12-3 is an overview of the trailer and telescope, while Figure 12-4 is a closeup of the RA bearing, polar axis, and counterweight assembly. The telescope has RA, Dec, and azimuth axes for easy polar alignment in remote locations. Trailer jack stands permit levelling the azimuth bearing. The telescope is stowed in a box in the trailer for towing, and is raised up using a system of pulleys to tilt the polar axis to the observer's latitude.

Figure 12-5 shows the fast (slew) and slow (set and guide) rate Dec steppers connected by toothed belts to the differential. The belts are used for gear reduction, and to provide smoothing of the stepper pulses (full steps are used, instead of half steps) and damping of mechanical resonances that could stall the motors. A similar arrangement is used in RA, with the addition of the AC synchronous motor to provide the sidereal rate.

Fig. 12-4 *Close-up of the RA axis and counterweight assembly.*

Fig. 12-5 *Declination drive train.*

The RA differential drives a worm and 400-tooth worm gear, with the same arrangement used in Dec. The gear reduction to the worm from the 30 oz-in steppers was initially 2:1 for the slew motor. With this small ratio, the motors tended to stall. The reduction ratio was finally increased to 6:1, which proved adequate. This added gear reduction raised the no-ramp (control paddle) step rate from a dead stop from about 300 full steps per second up to over 1000 full steps per second. When the computer ramps the motors, the top speed of 1500 steps per second yields a slew rate of $1\overset{\circ}{.}125$ per second in both axes. The fine stepper on each axis has an additional reduction over that of the slew stepper, yielding a step size of $0\overset{\prime\prime}{.}9$ in Dec, and $0\overset{\prime\prime}{.}81$ in RA. Long frame stepper motors, having twice the output torque of the original slew axis stepper motors, were installed (along with bi-polar circuitry, mentioned below) in later versions of the system to increase motor torque even further.

12.2.1 Electronics

The large box in Figure 12-6 holds the logic and high voltage power supply, stepper controller and paddle interface logic cards, plus the inverter/drive corrector for the AC synchronous motor. The small box in the same figure is no longer used. A closeup of the electronics box is shown in Figure 12-7. The cards are arranged to allow the free flow of cooling air to circulate inside the box.

The approach to the electronics is to provide two autonomous systems—one manual and one computerized. A gated series of stepper pulses is generated by four stepper motor rate generator integrated circuits manufactured by Sil-Walker (similar IC indexers are made by Anaheim Automation and other manufacturers). These ICs are located on the interface board. Switches on the hand paddle select which rate generator output is connected to the hand paddle pulse output line.

The TI bus interface card generates pulses at a rate determined by numbers loaded by software into 24-bit (3-byte) registers on these cards, one byte at a time. Bus switch ICs on the control box interface card select either the hand paddle pulse train or the TI computer pulse train, depending on the position of the hand-operated "local/remote" toggle switch. Since the hand paddle is not connected to the computer, no computer interface or software are needed to read the hand paddle inputs. A significant disadvantage, though rare in usage, is that the hand paddle can be inadvertently left in the "local" mode at that exact time that the control computer has commanded a move to another location. The Sil-Walker PPMC-101B has an input mode (MON) that could be utilized to prevent this occurrence.

Fig. 12-6 *"Phoenix IV" electronics.*

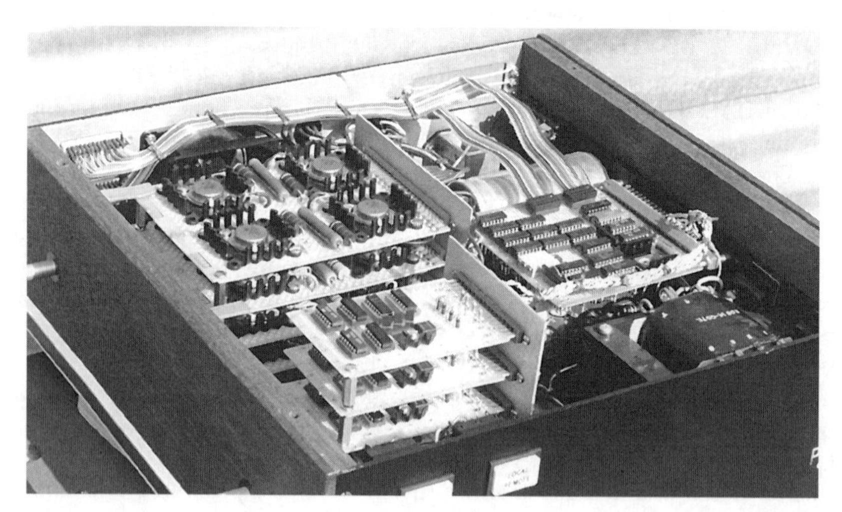

Fig. 12-7 *Close-up of the stepper control box showing the power supply and stepper driver boards.*

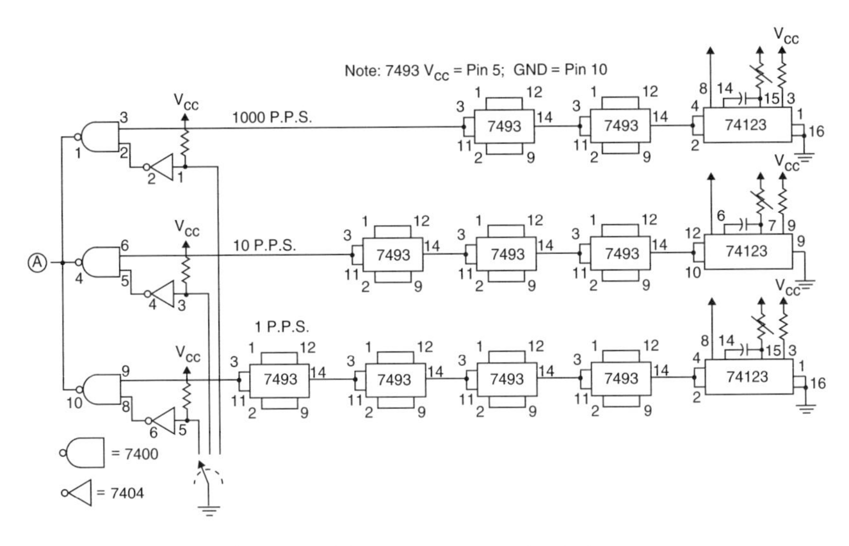

Fig. 12-8 *Tomer rate generator/selector.*

Figure 12-8 shows the hand paddle rate generator schematic diagram. Three separate 74LS123 one-shots are used to generate the slew, set, and guide rates. Since each has its own 20k ohm potentiometer to adjust the oscillator frequency over a 10:1 range, all three rates can be adjusted independently. The nominal frequency of each oscillator is 100 kHz. 74LS93s are used to divide down the 100 kHz to 1000 Hz, 100 Hz, and 1 Hz. A switch on the hand paddle selects one of these rates to be sent to the stepper controller cards. Since the selected rates can be sent to either the slew or the fine stepper, each axis has six possible speeds. This has proved to be quite useful, since it provides an appropriate set or guide speed for most eyepiece magnifications or telescope accessories. The other hand paddle switch (LOCAL/REMOTE) directs the pulses to the 74LS157 bus switches on the computer/paddle interface card, which is shown in Figure 12-9. Tomer acknowledges the help of Joe Bell in the design of this board.

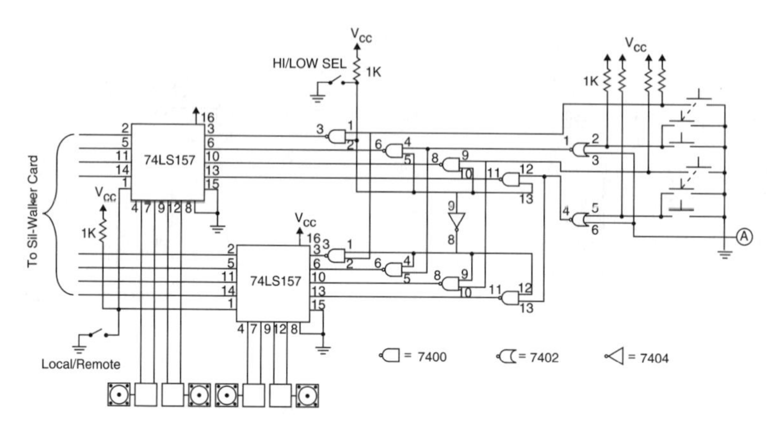

Fig. 12-9 *Tomer computer/paddle interface.*

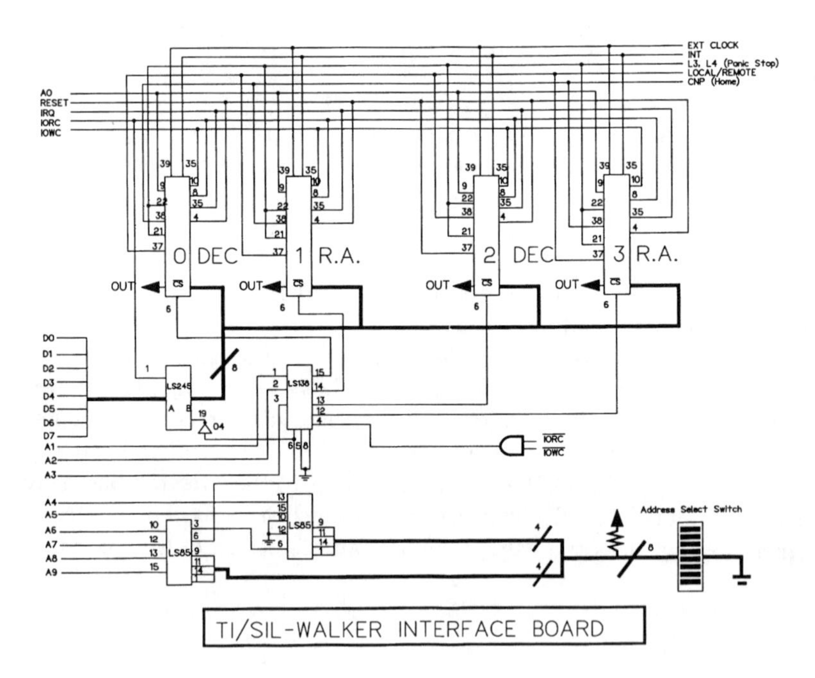

Fig. 12-10 *TI/Sil-Walker interface board.*

SIL-WALKER 101B STEPPER CONTROLLER I.C.

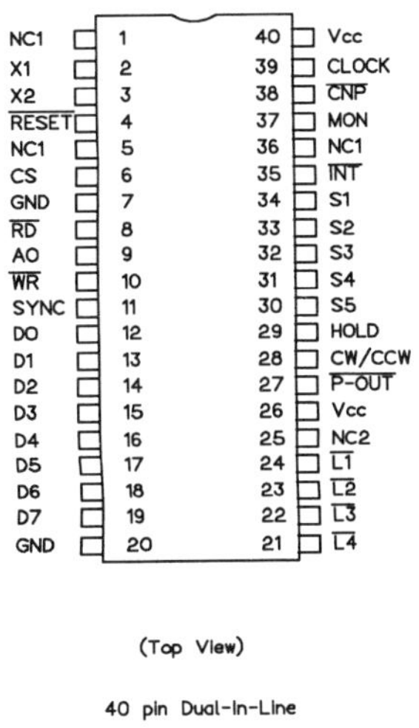

	Pin	Pin	
NC1	1	40	Vcc
X1	2	39	CLOCK
X2	3	38	$\overline{CNP}$
$\overline{RESET}$	4	37	MON
NC1	5	36	NC1
CS	6	35	$\overline{INT}$
GND	7	34	S1
$\overline{RD}$	8	33	S2
AO	9	32	S3
$\overline{WR}$	10	31	S4
SYNC	11	30	S5
DO	12	29	HOLD
D1	13	28	CW/$\overline{CCW}$
D2	14	27	$\overline{P\text{-}OUT}$
D3	15	26	Vcc
D4	16	25	NC2
D5	17	24	$\overline{L1}$
D6	18	23	$\overline{L2}$
D7	19	22	$\overline{L3}$
GND	20	21	$\overline{L4}$

(Top View)

40 pin Dual-In-Line

Signal	Pin#	I/O	Description
X1 , X2	2 , 3	I	X-tal
$\overline{RESET}$	4	I	RESET input
CS	6	I	Chip Select
RD	8	I	READ strobe
AO	9	I	Address 0
WR	10	I	WRITE strobe
SYNC	11	O	Timing output
DO - D7	12 - 19	I/O	Data Bus 8-bit
$\overline{L4}$	21	I	Reverse Limit high speed input
$\overline{L3}$	22	I	Forward " " "
$\overline{L2}$	23	I	Reverse Limit input
$\overline{L1}$	24	I	Forward Limit input
$\overline{P\text{-}OUT}$	27	O	Pulse output
CCW/$\overline{CW}$	28	O	Forward / Reverse status '0' = forward '1' = reverse
HOLD	29	O	Motor HOLD output
S5	30	O	Motor 5th phase output
S4	31	O	" 4th " "
S3	32	O	" 3rd " "
S2	33	O	" 2nd " "
S1	34	O	" 1st " "
$\overline{INT}$	35	O	Interupt signal
MON	37	I	External control '0' = Motor ON '1' = Motor OFF
$\overline{CNP}$	38	I	Base point signal input
CLOCK	39	I	External clock input
Vcc	26,40	I	+5V DC
GND	7,20	I	OV
NC1	1,5,36	I	Pull up to Vcc with 3.3K-ohm or OPEN
NC2	25	O	OPEN

Fig. 12-11 *Sil-Walker 101B stepper controller pinout.*

Figure 12-10 shows the TI bus interface card schematic diagram, while Figure 12-11 shows the pinout of the Sil-Walker 101B stepper controller chip. For those wishing to put the circuit in Figure 12-10 onto an ISA board, use the following table to connect board signals to ISA bus pins:

Board Signal	ISA Bus Pin
A0 through A9	A31 (SA0) through A22 (SA9)
D0 through D7	A9 (SD0) through A2 (SD7)
IOWC	B13 (IOW)
IORC	B14 (IOR)
RESET	B2 (Reset)
IRQ	B25 (IRQ3)

The external clock of 100 kHz can be generated by one of the 74123s in Figure 12-8, or by another means. The INT (interrupt) pin is used to kill a programmed move before it is complete. The L3 and L4 inputs to the Sil-Walker chips are for limit switches, and are used in Tomer's circuit as a panic stop by telling the chip that a hard limit has been reached. The local/remote switch from the computer/paddle interface board is brought over to the Sil-Walker board and sent to the Sil-Walker motor control pin of the four stepper control chips. The CNP or "carriage null point" pin is used to tell the chip when an axis is at its "home" position, if one is defined.

The 74LS157 bus switches shown in Figure 12-9 receive step rate pulses and direction bits from either the hand paddle or the Sil-Walker controller boards inside the TI computer and route them to the four driver/translator cards, one for each motor. The stepper driver cards were patterned originally after those used several years ago by DFM Engineering (Melsheimer, 1983). The parts count of the DFM circuit was reduced by using 2N6284 power Darlingtons and a direction bit was added to each board.

This portion of the control box circuitry was upgraded to bi-polar chopper circuitry. Four Superior Electric SLO-SYN 230-T modules provide 30 volts and 2 amperes of current to the stepper motors when supplied by a 30V, 10A linear power supply. Stepping rates on each axis and available torque have dramatically increased, completely eliminating any propensity to stall under strong wind load or imbalance conditions. Indeed, caution has to be exercised to avoid running the telescope into obstructions, winding up cables, or exceeding horizon limits as torque is now available in quantities to perform physical harm to machines or people.

12.2.2 Software

Like the previous Apple II system, the TI Color Pro computer uses the Basic programming language to operate the stepper controllers located on the interface card. This software is a derivative of the Apple II-based software. While it carries over the direction control algorithims from its predecessor, it adds extensive use of color, graphics, sound and, most important, incorporates an added database of approximately 10,000 objects. The main database is based extensively on the material in *Burnham's Celestial Handbook*, Volumes I, II, and III and was hand transcribed over a period of four months since, at the time this system was built, astronomical databases were almost the exclusive province of professional institutions and were distributed almost exclusively on 9 track tape.

Additional smaller databases are selectable by the telescope control program as well. They include Messier lists for general observing, polar alignment aids (paired star groups), current planetary tables and other information materials.

The various cases handled by the RA and Dec routines are shown in Figures 12-12 and 12-13, while parts of the routines themselves are listed in Figures 12-14 and 12-15. Not included in these listings are corrections

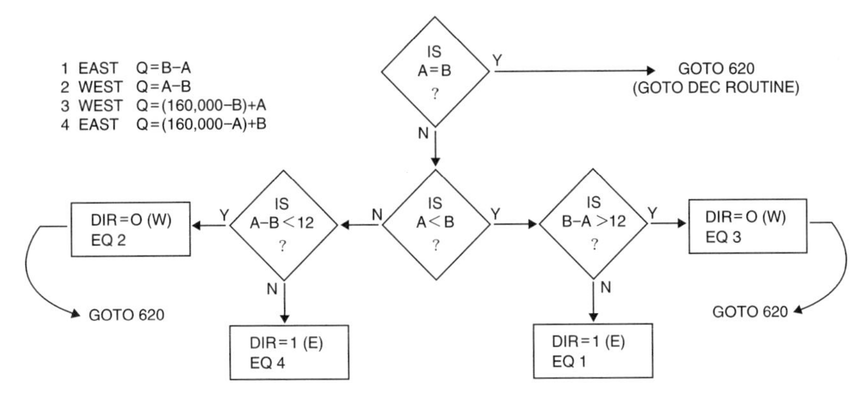

Fig. 12-12 *RA control algorithm.*

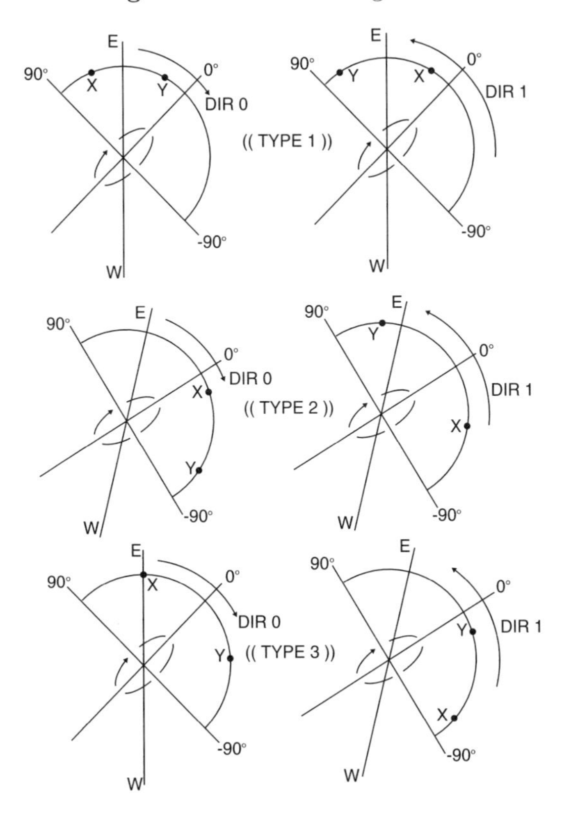

Fig. 12-13 *Declination Control Cases.*

```
4496 ' ==================================================================
4498 '
4500 '                          RIGHT ASCENSION CASE
4502 '
4505 MNUM = 0
4512 '
4520 NEWRA = (INT(RH * 20000) + INT(RM * (20000/60)) + INT(RS * (20000/3600)))
4522 IF OLDRA = NEWRA THEN 4785 : '                        * FIX FOR = **
4530 '
4550 IF (OLDRA < NEWRA) THEN  4592
4560 IF ((OLDRA - NEWRA) < 240000!) THEN  4585
4570 DIR = 0 : MCNT = (480000! - OLDRA) + NEWRA : ''        equation 4
4580 GOTO 4640
4585 DIR = 1 : MCNT = OLDRA - NEWRA : ''                    equation 2
4590 GOTO 4640
4592 IF ((NEWRA - OLDRA) > 240000!) THEN  4605
4595 DIR = 0 : MCNT = NEWRA - OLDRA : ''                    equation 1
4600 GOTO 4640
4605 DIR = 1 : MCNT = (480000! - NEWRA) + OLDRA : ''        equation 3
4607 '
4608 ' ..............................................................
4610 '
4612 '                     MOTOR LOAD AREA
4614 '
4640 CMD  = &H303 : ARG = &H302 : STATUS = &H303
4642 '
4645 IF (PIV = 1) THEN  MCNT = INT(((MCNT*3)+10)/20) : GOTO 4700 : ' BRNCH!
4650 MCNT = MCNT - 1
```

Fig. 12-14 *RA Control Routine.*

for refraction or precession, which are ignored in Tomer's code for now.

The control paddle inputs are handled entirely in hardware, which has three advantages. The first is that no software is needed to handle control paddle inputs, so as soon as the hardware was built, the telescope could be used. Software was then added incrementally without forcing the telescope to be out of service for extended periods of time. The second advantage is that the control paddle is a true manual override. Software that has run amuck, or some failed computer hardware, will not leave the telescope unusable. Finally, any residual pointing error can be tweaked out manually without the need for special software to tell the computer to leave the current RA and Dec alone, despite the fact the telescope is not now pointing where the computer left it after a slew.

The result is that the control software is extremely simple. The telescope is pointed at an object and its coordinates are entered. From that point on, the computer "knows" where the telescope is pointed when it is in control. As shown in Figure 12-12, if a new position is selected from a constellation or buffer page, the new RA is compared with the current RA. If they are the same, the telescope has the correct RA, and the software branches to the Dec control routine. If not, the new and current RA are compared to find

```
4000 ' ================================================================
4010 '       TALK TO SCOPE, MOVE IN CORDINATE GRID, HARDWARE CODE
4012 '
4014 '
4015 ''  OLDDEC, OLDRA and DK have been previously obtained from "INIT" ...
4022 ''                       - or -
4023 ''  OLDDEC, OLDRA and DK have been preserved from last move "new" position
4027 ''
4030 GOSUB 10000 : ''  parse string into D, M, S & DL
4032 NEWDEC = (INT(DD * 1200) + INT(DM * (1200/60)) + INT(DS * (1200/3600)))
4033 IF (OLDDEC = NEWDEC) THEN 4520 : ''      ** FIX FOR = **
4035 '
4040 MNUM = 1 : ''          we'll address DECLINATION first
4050 '
4055 IF ((DK+DL) = 2) THEN  4114 : ''                 type 1 move
4060 IF ((DK+DL) = 1) THEN  4120 : ''                 type 2 move
4090 IF ((DK+DL) = 0) THEN  4200 : ''                 type 3 move
4095 ''
4098 ' ..........................................................
4100 '
4114 IF (OLDDEC > NEWDEC) THEN DIR =1 :MCNT = OLDDEC-NEWDEC:GOTO 4275:' south
4115 '
4116 DIR = 0 : MCNT = NEWDEC - OLDDEC : ''      1                     north
4117  GOTO 4275
4118 ' ..........................................................
4119 '
4120 IF (DK = 1) THEN  DIR = 1 : MCNT = OLDDEC + NEWDEC : GOTO 4275 : ' south
4130 '
4140 DIR = 0 : MCNT = OLDDEC + NEWDEC : ''      2                     north
4150  GOTO 4275
4155 ' ..........................................................
4170 '
4200 IF (OLDDEC < NEWDEC) THEN DIR = 1: MCNT = NEWDEC-OLDDEC : GOTO 4275 :'south
4210 '
4250 DIR = 0 : MCNT = OLDDEC - NEWDEC : ''      3                     north
4260 '
4265 ' ..........................................................
4268 '
4270 '                    MOTOR LOAD AREA
4274 '
4275 CMD = &H301 : ARG = &H300 : STATUS = &H301
4276 '
4277 IF (PIV = 1) THEN   MCNT = INT((MCNT + 3) / 6) : GOTO 4300 : ' BRANCH !
4278 MCNT = MCNT - 1
4280 OUT CMD, (&H1A OR &H0) : '        RESET CHIP 0 TO *PIV* DEFAULTS
4281 OUT ARG, &HFF : '                 START SPEED
4282 OUT ARG, &H20 : '                 HI    SPEED
4283 OUT ARG, &H90 : '                 RAMP  LSB
4284 OUT ARG, &H3 : '                  RAMP  MSB
4290 GOSUB 5730
4294 GOTO 4505 : ' branching
4295 '
4296 '
4300 MCNT = MCNT - 1
4315 OUT CMD, (&H1A OR &H0) : '        RESET CHIP 0 TO *TRACKMOUNT* DEFAULTS
4320 OUT ARG, &HFF : '                 START SPEED
4325 OUT ARG, &H14 : '                 HI    SPEED
4330 OUT ARG, &H90 : '                 RAMP  LSB
4335 OUT ARG, &H6 : '                  RAMP  MSB
4340 GOSUB 5730
4490 '
```

Fig. 12-15 *Dec Control Routine.*

which direction to move and by how much. The number of steps to move is computed and then executed.

The various cases that arise in controlling the telescope in Dec are shown in Figure 12-13. In this figure, X is the current Dec and Y is the new Dec. Type 1 moves are those confined to the quadrant between the north pole and the celestial equator, Type 2 moves are those that cross the celestial equator, while Type 3 moves are confined to the quadrant south of the celestial equator. As shown in Figure 12-15, these cases are separated and dealt with individually in Tomer's software. Note the simplicity of the software that loads slew moves (in number of steps) to the motor indexers (lines 4610–4990 in RA and lines 4030–4340 in Dec, with a common routine in lines 5730–5900; see Appendix H). This is a result of using the Sil-Walker intelligent stepper indexer chips to generate the pulses to the driver cards.

Referring to the software description in Appendix H, when the system is first started, the screen is cleared, and the 5-tone alien theme from the movie *Close Encounters of the Third Kind* is played (lines 17000–17080). Note that the "sound" command has two parameters: the pitch of the tone in Hertz, and the length of the tone in units of 1/16th of a second. Most of the software, encompassing lines 1–3700, is involved with generating displays and accepting operator keyboard inputs, and has nothing to do with controlling the motors. Part of the program initialization is to set up the motors with maximum velocities and accelerations.

The first page to be displayed is the main display page, shown in Figure 12-16. It permits the operator to find an object by its location, and contains a list of 64 constellations, abbreviated to three characters. This list serves as an index into the database of over 10,000 objects. A cursor highlights the current constellation by changing its color from orange (used to preserve dark adaption) to green. The cursor is moved using cursor arrow keys on the keyboard, and a particular constellation is selected using the `Enter (Return)` key. Besides listing the constellations, there are nine spare (undefined) entries designated ..1, ..2, through ..9, and seven object types. These types permit finding an object by the type of object it is rather than where it is located, and includes bright nebulae (nbb), Quasi Stellar Objects (qso), globular clusters (glb), Messier catalog (msr), star pairs for polar alignment (pol), and planetary nebulae (plt).

Suppose you choose the first entry on the page, AND (Andromeda). The Andromeda page is then displayed, as shown in Figure 12-17. This page contains entries from the general catalog that are within the IAU-approved borders of Andromeda, ordered by Right Ascension. Each catalog line contains the NGC, IC, or other name for the object, RA and Dec, the type of object (one of the types listed at the bottom of the first page), visual magnitude, and a comment. Note that NGC names are used, if possible, in

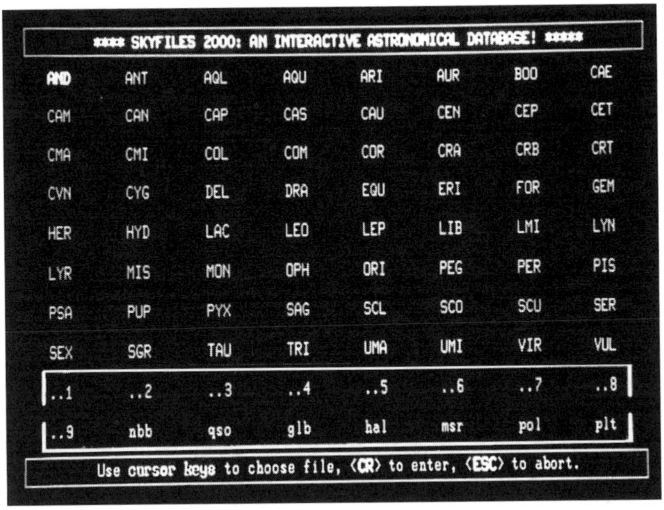

```
**** SKYFILES 2000: AN INTERACTIVE ASTRONOMICAL DATABASE! *****
AND   ANT   AQL   AQU   ARI   AUR   BOO   CAE
CAM   CAN   CAP   CAS   CAU   CEN   CEP   CET
CMA   CMI   COL   COM   COR   CRA   CRB   CRT
CVN   CYG   DEL   DRA   EQU   ERI   FOR   GEM
HER   HYD   LAC   LEO   LEP   LIB   LMI   LYN
LYR   MIS   MON   OPH   ORI   PEG   PER   PIS
PSA   PUP   PYX   SAG   SCL   SCO   SCU   SER
SEX   SGR   TAU   TRI   UMA   UMI   VIR   VUL
..1   ..2   ..3   ..4   ..5   ..6   ..7   ..8
..9   nbb   qso   glb   hal   msr   pol   plt
Use cursor keys to choose file, <CR> to enter, <ESC> to abort.
```

Fig. 12-16 *Main display page.*

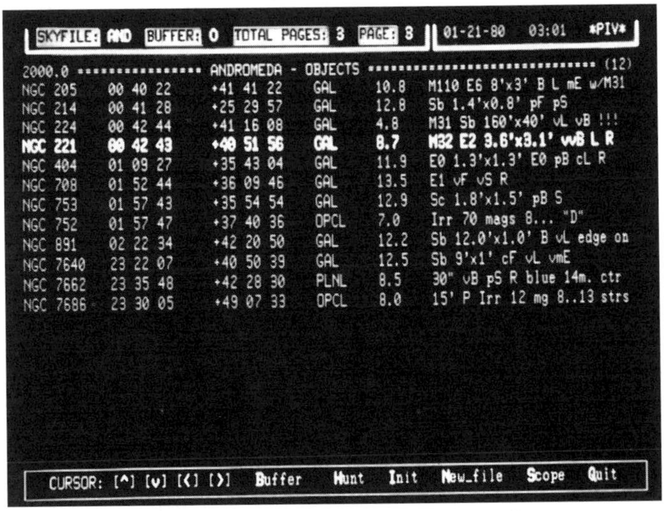

```
SKYFILE: AND  BUFFER: 0  TOTAL PAGES: 3  PAGE: 8  01-21-80  03:01  *PIV*
2000.0 **************** ANDROMEDA - OBJECTS ****************************** (12)
NGC 205    00 40 22   +41 41 22   GAL    10.8   M110 E6 8'x3' B L mE w/M31
NGC 214    00 41 28   +25 29 57   GAL    12.8   Sb 1.4'x0.8' pF pS
NGC 224    00 42 44   +41 16 08   GAL    4.8    M31 Sb 160'x40' vL vB !!!
NGC 221    00 42 43   +40 51 56   GAL    8.7    M32 E2 3.6'x3.1' vvB L R
NGC 404    01 09 27   +35 43 04   GAL    11.9   E0 1.3'x1.3' E0 pB cL R
NGC 708    01 52 44   +36 09 46   GAL    13.5   E1 vF vS R
NGC 753    01 57 43   +35 54 54   GAL    12.9   Sc 1.8'x1.5' pB S
NGC 752    01 57 47   +37 40 36   OPCL   7.0    Irr 70 mags 8... "D"
NGC 891    02 22 34   +42 20 50   GAL    12.2   Sb 12.0'x1.0' B vL edge on
NGC 7640   23 22 07   +40 50 39   GAL    12.5   Sb 9'x1' cF vL vmE
NGC 7662   23 35 48   +42 28 30   PLNL   8.5    30" vB pS R blue 14m. ctr
NGC 7686   23 30 05   +49 07 33   OPCL   8.0    15' P Irr 12 mg 8..13 strs
CURSOR: [^] [v] [<] [>]   Buffer   Hunt   Init   New_file   Scope   Quit
```

Fig. 12-17 *Catalog object list page for the constellation Andromeda.*

the first column while Messier or other names appear in the comment. The cursor works the same way as on the first page, but instead of selecting an entry using the `Enter` key, you hit `S` to command the telescope to slew to that object.

At first, a 670 Hz tone is sounded for $3/16$ second (line 1106), followed almost immediately by a 440 Hz tone for a half second. This warns anyone standing nearby that the telescope is about to move. Lines 4280–4290 load the start and maximum speeds and acceleration rates into the Sil-Walker chip controlling the Dec slew motor, then lines 5730–5900 load the direction bit and the total number of steps to move and start the Dec motor. Lines 4640–4665 and 5730–5900 do the same for the RA motor and start it running. Both RA and Dec start off moving together, accelerating almost in unison to the maximum speed set up earlier. One of the axes has a move shorter than the other, so its motor ramps down sooner than the other. The loop at lines 12020–12030 returns only after the move is complete. The software then sounds a 1320 Hz tone for one second when the move is complete. Notice that the warning tones use different pitches, different numbers of pitches, and different pitch durations to signal the start and end of a move.

Hitting the `B` key adds the currently selected row from the constellation page to the buffer page, then displays the buffer page (see below). The `H` key has not been implemented yet, but was intended to start a spiral search from the current position that is ended when a button on the hand paddle is pressed. To initialize the current telescope position at the beginning of the evening (and any time thereafter), you move to an object and center it in the eyepiece using the hand paddle, find the object on a constellation page, then hit the `I` button to tell the software the telescope is now pointing at that object. The `N` button returns you to the main menu so you can select a new constellation or object type file. As explained above, the `S` button slews to the highlighted object, and `Q` drops you back into the Basic interpreter for software development and maintenance.

If you hit the `B` key, the buffer page shown in Figure 12-18 appears, and the row selected on the previous page is added to the buffer automatically. This buffer is a list of objects you can create at any time. Normally, you might want to build this list during the day for that evening's observing. What is displayed is a buffer that can be stored in a file and recalled. To add an entry to the buffer after the page is already displayed, hit the `I` key. Using the cursor keys, you can select a row and remove it from the buffer using the `D` key. The `E` key empties (clears) the buffer of all entries. The `K` key allows you to enter a row using the keyboard. The `M` key moves the telescope to the currently selected object, just as the `S` key does in the previous page. The `R` key reads a file into the buffer, and the `W` key writes the buffer to a file. The `S` key returns you to the main menu, and the `X` key exchanges two contiguous rows on the page, permitting you to sort the list (for example, to build a Messier Marathon list ordered by setting time).

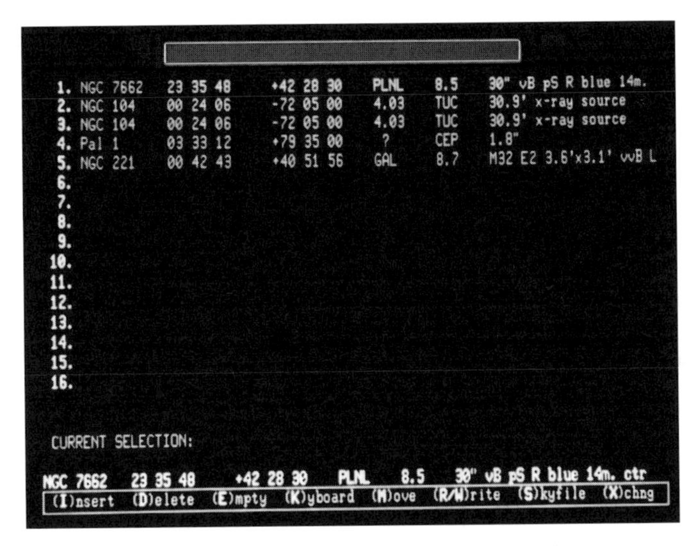

Fig. 12-18 *Observing list buffer page.*

That's the complete system. It is simple to understand, easy to use, and effective in moving the telescope to the desired object. There is no copyright on the software—publishing it here is intended to put it into the public domain. No guarantee is made or implied that it will do what you want as well as you want, and it needs modification to work on a PC. We hope it helps illustrate how to put mechanics, electronics, and software together to form a working system.

12.2.3 Development History

Since the original DC servo motor drive worked with the Apple II computer, the development strategy was to build a small test bed mount to determine if the new system would perform better than the existing one. Four stepper motors were purchased from a local surplus electronics store, the electronics were designed and built, and the system was assembled. The test bed system worked well, and the simplified control software was developed rapidly. After tests demonstrated the soundness of the approach, Tomer decided to implement the system on his 12-inch telescope.

The first problem that he encountered was that the 30 oz-in steppers had adequate torque for the test bed, but when the slew steppers were connected directly to the inputs of the differentials driving the worms on the 12-inch telescope, the steppers tended to stall. Although each differential provides a 2:1 reduction between the slew stepper and the worm, this was not adequate. An additional 2:1 reduction was installed, providing an overall reduction of 4:1. This improved performance, but the motors still tended to stall on occasion. The additional 2:1 reduction was changed to 3:1, for an overall reduction of 6:1. This proved to be entirely adequate, but reduced the slew rate from over 3° per second to just over 1° per second.

Another problem linked to the first was the mesh of the worm with the worm gear. Although the gears are cut to high accuracy (10″ or better), a tight mesh increases the amount of friction in the drive. Tomer found that by permitting as much as 1′ of "slop" in the mesh, the friction could be reduced to tolerable levels. This means the total pointing error is at least 1′, but this is well within Tomer's system performance expectations.

A third problem that was encountered was heat dissipation from the stepper driver/translator cards. With all four steppers powered up at all times, the inside of the stepper electronics box got quite warm. The 12V, 10A power supply runs at 8A, which generates additional heat. Tomer tried powering only the slew motors, but found they were capable of backdriving the fine steppers through the differentials when the latter were not producing holding torque. The solution to the heat problem was to add more ventilation slots to one end of the case.

The final problems were minor electrical faults. Arcing was occuring between the hand paddle case and the telescope trailer. The solution was to relocate some wiring and add proper grounding to all metal parts. Additional small filter capacitors were added to the electronics boards. Large 500μF capacitors were added to the 12V supply line on each board to prevent the voltage from sagging during motor steps. Transient clippers (metal oxide varistors) were placed across the coil of the AC synchronous sidereal rate motor and several switch contacts to prevent the motors from stepping erroneously when switches were closed. Finally, the hand paddle switches were debounced to prevent rapid and erroneous direction reversals when using the guide speeds.

All projects encounter problems, and this one was no exception. All the problems were solved, and Tomer considers this project to be a successful demonstration of how proper hardware design can make the software development manageable.

Tomer's system demonstrates how stepper motors can be used in conjunction with smart interface cards and a personal computer to produce a useful telescope control system. Such open-loop systems are simple and inexpensive, and can be used in applications that do not require extremely high accuracy, or when final centering is performed by the observer. This system consistently gives absolute pointing accuracy on the order of a few arc minutes or better.

12.3 A Platform-Independent Approach

The telescope control system built for Tomer's mobile 12.5-inch Cassegrain telescope became outdated in the mid-1990s. Electrical and software construction techniques, imminently serviceable in the early 1980s, were revisited with an eye towards using more modern technology. The TI Color Pro control system, while still serviceable, evidenced a number of drawbacks and posed future maintenance problems. In particular, the motor drive electronics and overall system architecture relied too heavily on a particular obsolete computer. Tomer searched for a solution that would work with almost any computer available now or in the future, and that would minimize the amount of software development to perform future computer platform upgrades.

The controlling computer for the earlier system, the Texas Instruments Corporation Portable PC, manufactured in 1983-84, had reached the margin of late 90's utility. As a portable, its 29-lb. weight stretched the definition of portability (and the strength of Tomer's arm). The TI PC's bulk is also a detriment on telescope outings where camping equipment and telescope gear and accessories compete for space in vehicles. Since the TI's computing operations, other than word processing, were limited to custom written DOS programs, its potential for future telescope control applications would be severely limited.

Worse, there remained several annoying characteristics of the old custom telescope control system . The hand paddle circuitry section was not "debounced" properly, despite earlier efforts. Pressing switches for direction control would frequently result in the telescope's moving in the wrong direction. This was not a big problem in coarse positioning, but was a definite annoyance when guiding a photograph.

Another problem with the basic circuitry of the telescope controller was the inherently limited performance of the simple resistive-inductive (R/L) drivers for the stepper motors. These homemade, low cost translators began running out of torque at the modest step rate of 1000 pulses per second. This results in rather long and tedious times to slew the telescope reliably over large distances in the sky.

Consequently, polar alignment was a lengthy and unappetizing process. Obtaining maximum pointing precision entails repeatedly adjusting in azimuth by registering on meridinal stars at the equator and then moving to circumpolar stars. The error is reduced by iterating azimuthal adjustments until the stars remain perfectly centered in medium to high power eyepieces. This process could easily consume up to 45 minutes early in an observing session, particularly if the initial polar alignment in azimuth is badly off.

Complicating the slewing performance of the telescope were the sizes of the motors themselves. The original rated torques for both the slew and set/guide motors were 30 ounce-inches, barely adequate for the mechanical characteristics of the telescope. Slewing could be interrupted by either wind gusts on the tube or an out of balance condition, though on those rare occasions where the telescope was directed into an obstruction, the low powered motors and translators were an asset.

Two additional problems remained: wire wrapped logic circuitry in the control box was beginning to evidence atmospheric corrosion and dirt contamination. It seemed long term reliable operation would become increasingly unlikely. And lastly, the "smart" motor control microprocessors in the TI computer possessed only limited intelligence, being limited to trapezoidal motor operations only. There was simply no expansion possibility to this system.

The above limitations led to the construction of a follow-on control system for the telescope.

This control system is centered on the adoption of a platform independent methodology. Whereas the prior control system was based on a classical computer expansion bus scheme, this new system would be centered on serial port communications (RS-232 serial protocol). This would allow virtually *any* computer to perform the controlling function, and insure that migration to future computing platforms would confine problems to software issues only.

While any PC could perform the control function, the best choice for a portable telescope is one of the rapidly proliferating Windows based notebook computers. The prices of these computers have dropped dramatically with their marketplace success, and while their inherent ruggedness is somewhat suspect, they nonetheless are a "commodity item," and are not a critical link in the envisioned motion control scheme.

12.3.1 The Nyden Motion Controller

Industrial based motion control products fall into a number of broad based categories, but can be loosely grouped into two:

1. expansion bus based control arrangements, and

2. distributed network control arrangements.

Vendors of RS-232 (network) based motion control products were surveyed over a six-month period, and the final selection was the MAC300 motion controller from Nyden Corporation of San Jose, CA. This 7″ x 5½″ x 2″ product is shown installed in Figure 12-19.

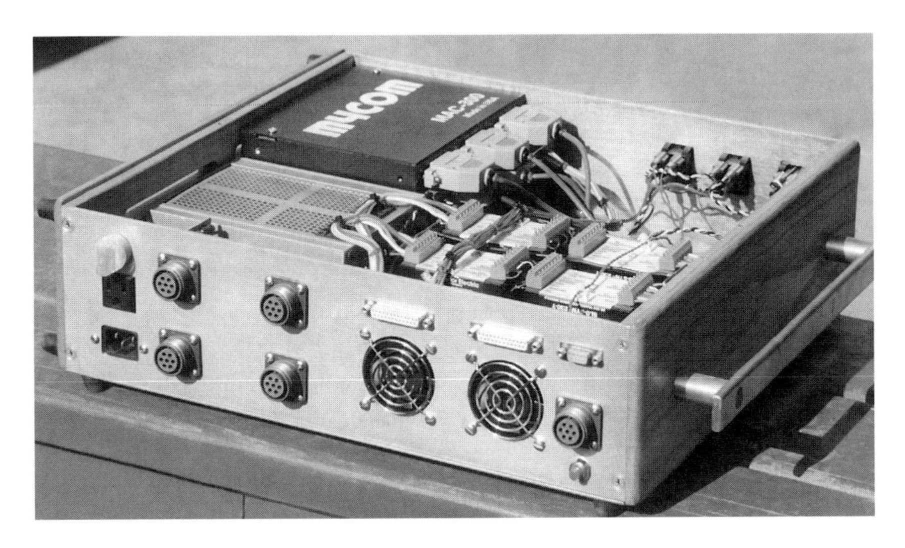

Fig. 12-19

This stepper/servo motion control module met all requirements for functionality, cost (in this case very important) and expansion capabilities. Among its features are the following:

1. Up to 10 programs can be placed in non-volatile storage and invoked with simple commands or switch settings. These programs can be retrieved, replaced or modified easily via the serial port. Variations in programs could produce easily modified motor rates and accelerations for slewing, setting, and guiding, as well as perform I/O controlling functions.

2. The MAC-300 comes standard with 8 input lines and 8 output lines as well as four 8-bit Analog to Digital conversion ports. These single bit I/O lines are ideally suited for handpaddle operations such as rate or motor range select, and the A/D sections can be used to implement a variable stepping joystick function easily. This would relieve one of the laborious construction of electronic hardware that could ramp steppers up and down for manual positioning tasks. (Higher ultimate run speeds can be attained when in manual positioning mode when motors can be ramped up to speed.)

3. Optical encoder feedback of position is a low cost option with the MAC-300 device, and can be retrofitted at any time. (Installation of four integrated circuits is required to permit encoder operation.) This

feature allows a motor, for example, that has stalled short of its destination, to move at a reduced rate of speed (with higher torque) and complete it's ultimate positioning, giving true "closed-loop" control.

4. Very compact packaging and low power consumption are further highlights of the device. Overall dimensions are a very modest 2″ high x 7″ wide x 5 1/2″ deep and result in easy packaging within a larger housing. All CMOS construction results in minimal power consumption, approximately 1/2A at 5V.

5. Unique translator control signals from the MAC-300 permit either "shutting down" current to the stepper motor windings or "reducing power"; i.e., partially reducing current to the motor windings. Unfortunately, they are not logic compatible signals; rather they are 24V signals and necessitate using relays or voltage "down converters" to effect voltage translation to logic levels employed by the translator modules. (In defense of this 24V scheme, it is intended for use in electrically noisy environments, such as factory floor operations, and the selection is sensible for the primary market of the MAC-300.)

6. Inner and outer "limit switches" are standard inputs, as well as a "home" input. The home function can be utilized to "park" the telescope accurately prior to stowage into the trailer cradle, and this capability will probably be implemented at a future date.

7. Additional MAC-300s may be daisy-chained together, since each unit is configured with an "address."

8. Other useful features of the MAC-300: emergency stop, program start/stop, axis select, store teach/data point, error signal output, busy signal, reset, etc. Emergency stop is implemented on the front of the control box, and reset is implemented on the lower left corner at the rear of the box. As mentioned earlier, there are 16 control lines available. While some are dedicated to the hand paddle function, the remainder are brought out for future usage on a DB-25 connector on the rear of the control box.

12.3.2 Bi-Polar Chopper Stepper Translators

The stepping motor translators employed were located surplus, factory new, and at attractive prices. They are from the Superior Electric Corporation in Bristol, CT and are of modern high efficiency MOS-FET bi-polar chopper construction. They are contained in a compact, easily mounted package. Figure 12-20 shows the 230-T translator module.

Fig. 12-20

The 230-T uses resistive current sensing and provides for full and half step operation. Inputs are optically isolated, with a choice of utilizing internal or external opto power supplies. The unit also features reduced current and all windings off capability. (These particular translators do *not* provide output short circuit protection, unlike some manufacturer's models, and care must be taken to be sure outputs are correctly connected to the motor windings. It is OK to mix up leads to the windings, the motor will vibrate and not move, but it is definitely *not* OK to short the translator leads together.)

At this time, the reduced current and "windings off" features are not implemented, and the translators are left in the default half step mode. The 230-T expects logic level signals to these control inputs.

Each translator can deliver 2 Amperes at 28V to the motor, provided it is mounted on a properly sized heat sink. Within the new control box, all four translators are grouped on a large finned heat sink; external air is directed onto it by two fans. This air exits the side of the control box after passing over and through the additional power supplies and electronics in the enclosure.

The manufacturer recommends isolation of each translator's power from other translators in multiple installations. The diodes and capacitors used to perform this isolation can be seen in the schematic of the PIV con-

Fig. 12-21

trol system and in Figure 12-21 as a small PCB immediately in front of the translator modules. Additional 250mF 50VDC capacitors are provided across the ground and power inputs of the modules, per the maunufacturer's recommendations, and are also seen in the schematic.

12.3.3 Analog Joystick

The hand paddle is experimental because the MAC-300 presents a limitation. There are four easily utilized A/D sections, whose input activity can be delivered to either the telescope high range motors or the low range motors. This is fine for the "analog joystick" solution. One the other hand, the MAC-300 provides little support for the "digital joystick;" operations are solely limited to one selected axis at a time. (This constrains the rate at which cross-sky slewing can be accomplished, since one axis needs to have, in essence, fully completed its motion before beginning the other axis travel.)

Tomer elected to use the A/D implementation. A miniature, solid state joystick was purchased from CTI Electronics in Stratford, CT after another survey of the available devices was accomplished. Figure 12-22 shows this diminutive, solid-state device. The Model F1050-N5 outputs a 0-5VDC signal on each axis, which is directly compatible with the MAC-300 analog-

Fig. 12-22

to-digital inputs. The unit is physically quite small, sealed with a rubber bellows, and, because of its CMOS construction, operates on minimal power.

The center "dead zone" of the joystick outputs is definable by the MAC-300 in any region desired. In this application, defining the central 10% of mechanical throw as "neutral" seems to work fairly well, though more experience with the arrangement is needed to fully gauge its practicality.

The hand paddle housing the joystick contains a "rate select" switch, i.e., a simple multi-pole rotary switch. Each switch position selects a particular program stored in the MAC-300 memory. Each of these programs selects the appropriate motor set (high range/low range) and rate of speed (slew/set/guide). Also wired on the joystick is a "local" switch, whose operation takes the controller into the "teach mode." Two switches, wired in parallel, accommodate right or left handed operators.

Figure 12-23 does not depict the entire control box wiring. Omitted for clarity is the usual AC wiring to power supplies, fusing, and auxiliary circuitry needed to power the telescope "tracking" motor, i.e., a 12VDC/110VAC 60 HZ stable inverter circuit. Nevertheless, Figure 12-23 illustrates the simplicity of the controller, and the ease with which it may be implemented. Indeed, there is virtually no additional IC wiring work, only "point-to-point" connections of various modules.

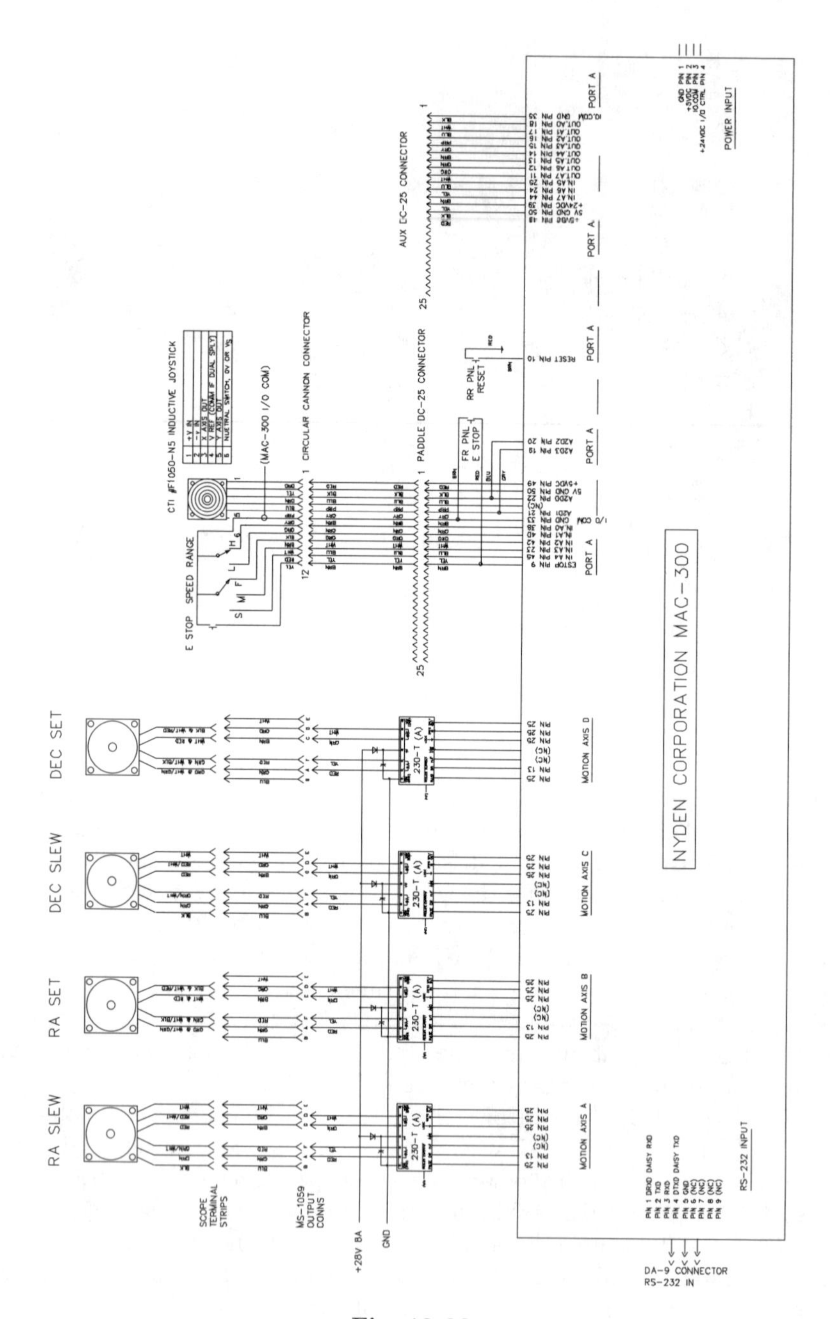

Fig. 12-23

12.3.4 Software

The native programming language of the MAC-300 is a BASIC like syntax, very similar to most common implementations of the language, and immediately usable by those conversant with BASIC languages. It becomes evident after downloading a few new programs to the MAC-300 that there is an on-board interpreter accepting the ASCII text inside the module.

Supplied with the MAC-300 is a general purpose demonstration and testing program called the MAC Commander. MAC Commander allowed immediate testing of the updated telescope motors and the new control system. Any of the four motors may be selected for testing. Accelerations and speeds may be varied easily; very brief testing showed that with the larger motors and more powerful and efficient translators, the telescope was moving seven times faster. Reliability of the moves was vastly better, with no stalling except at extreme imbalance conditions.

Without getting into the extensive capabilities of the MAC-300 language implementation (branching, logical operators, flow control, program selection, etc.), there is one feature that is particularly useful for the future telescope control program. The various position registers within the MAC-300 may be accessed directly. This means that very, very few characters need to be transmitted via the relatively slow serial lines to effect a telescope move, and the interpreter resident in the MAC-300 is bypassed completely. The controller simply accepts the new "destination" data directly and begins an immediate move. This functions fundamentally no differently than the earlier PPMC-101B motor control microprocessors inside the TI Portable control system.

Development of software for the MAC-300 is underway as of this writing using Visual Basic. A portion of the TI software was devoted to display of information, i.e., targets of choice, constellation selection, etc. Implementing these features is a simple process in the GUI environment of Visual Basic. Consequently, in this new system, smaller code segments will replace major portions of the code in Appendix H. It is anticipated that the TI code portions that do the actual telescope positioning and travel direction decision making tasks can, for the most part, be directly moved into this new environment since there are no hardware changes affecting the software.

Microsoft Corporation releases updated versions of Visual Basic periodically. Standard software versions of VB (excepting the Enterprise and Professional Development) do not support usage of the serial COM ports. (Apparently the synchronization of multi-tasking programs and COM port utilization is far more than a trivial set of operations in the "behind the scenes" Windows environment.) Fortunately, this limited availability of the COM ports is changing in the standard VB version release, and should not prove to be a fundamental obstacle for future coding.

Particularly interesting, and yet to be explored, are the DDE (Dynamic Data Exchange) capabilities of the Visual Basic language. There is a growing number of commercial and freeware/shareware general purpose planetarium programs with ancillary telescope control modules. Some of these provide output hooks via the DDE method, and may well prove to be excellent solutions for general purpose telescope control. John Walker's program *The Visual Planet* is an example of this approach. Indeed, a simple spreadsheet with object data can become a straightforward "front-end" controlling program. And then, there are the innumerable database programs, including Microsoft Access.

The earliest mechanical components of this mobile telescope were started in 1974. Despite it's advancing age, with these recent electronic modifications it still compares well with contemporary amateur observing systems. The anticipated software changes implemented now and in the future will expand its capabilities now and make the entire system easier and less expensive to upgrade in the future. Clearly, there is a great convergence of amateur astronomy, electronics, robotics and software for those interested in any or all of these areas.

In the next chapter, we look at a more complex system with higher accuracy requirements.

Chapter 13

The WMO Telescope Control System

In this chapter the design process for a trailer-mounted portable telescope is described. The purpose of this somewhat long chapter is to demonstrate by example the process of defining requirements for a control system, selecting the components to build the system, and putting them together to turn a collection of parts into a working system. Trueblood established a private non-profit observatory named The Irvin Marvin Winer Memorial Mobile Observatory, Inc., or Winer Mobile Observatory (WMO) for short, after a man whom he knew only a relatively short time but who left a permanent impression. When Irv died prematurely in middle age in 1982, Trueblood felt a lasting tribute was in order.

Despite the title of this chapter, the design presented here is not exactly the one used at the WMO. The WMO system is constantly evolving as we learn about field operations, and the purpose of this chapter is to illustrate a design process. But the requirements, the design philosophy, and many of the tradeoffs presented here are reflected in the WMO system. Although the design presented here may be of interest to those contemplating an instrument of similar capabilities, the important issue is not the design details or particular component selections, but the thinking that went into the design and component selections. Given the delay between our writing words for a book and your reading these words, the chapters and appendices on components are already outdated, but the methods used for systems design and integration will last a lifetime.

For a control system as straightforward as Tomer's, a formal design approach is not necessary. Although Tomer did spend considerable time designing his system, he spent more time building it. The drive train and the software are simple both in concept and execution. The pointing accuracy, though quite useful, is moderate when compared to that found at professional observatories.

The telescope described here, on the other hand, is considerably more complex, due to its stringent performance requirements. The task of developing a complex system can quickly become disorganized if it is approached in a haphazard manner. The price of this disorganization is wasted time and money. Small projects can afford small schedule slips and cost overruns. A project of this size quickly generates unacceptable delays and cost overruns if not managed properly.

The approach to designing and building the control system described here is the one presented in Chapter 3. We start with an observing program and develop a system concept. We turn the concept into a set of requirements used to design the control system. The resulting design is then analyzed to ensure that it meets the requirements.

13.1 The WMO Observing Program

The emphasis of the research programs at the WMO is on solar system astronomy—primarily grazing lunar occultations, minor planet occultations, and, possibly in the future, occultations by comets. As lunar laser ranging becomes more routine, the scientific importance of observing grazing lunar occultations diminishes, although these observations are still useful for detecting previously unmapped features on the lunar limb, and for detecting double stars. Photoelectric observations of total lunar occultations are still useful for determining stellar diameters and binary separations, when diffraction patterns can be obtained. Although both lunar and minor planet programs are being pursued at the WMO, the minor planet occultation program receives greater emphasis.

Both programs require that observations be made at a narrowly defined place and time. The typical grazing lunar occultation observation must be made inside an area that is a few hundred miles long by about a mile or two wide. Similarly, minor planet occultations must be observed inside an area roughly several hundred miles long by about 50–200 miles wide. In most cases, there is not a fixed observatory of even modest size in the path, and on those few occasions when there are, telescope time usually cannot be granted to do these observations, or weather interferes. Timing is critical, so that one cannot wait for the weather to clear to make the observations. All this points to the need for a transportable telescope system.

Minor planet occultations are predicted by Edwin Goffin in Belgium, by Larry Wasserman and Bob Millis at Lowell Observatory, and by David W. Dunham. Jim Hart of Pickering Anomalies offers for sale a software package called Asteroid PRO for the PC so you can do your own searches. The main searches for events are made using the AGK3 and SAO catalogs, so stars down to 9th or 10th magnitude are included. Lowell Observatory conducts searches down to 13th to 15th magnitude. When searches are made for visual observers in the United States, events are sought in which the star is no fainter than 10th magnitude, the minor planet is fainter than the star, the predicted drop in magnitude during the event is at least 0.8, and the minor planet has a diameter of at least 100 km.

There are some seven or eight such events predicted to occur on land near populated areas each year using current prediction techniques. Of these, only two to four are actually observed, because of bad weather, last minute shifts in the predicted paths away from populated areas, or a host of other reasons. For photoelectric observations, events are sought in which both the star and the minor planet may be as faint as 15th magnitude, either one may be the brighter object, and the predicted drop in magnitude may be as little as 5%. There are some two to four dozen such events predicted each year, but at the present time, the lack of good astrometry data limits the number of these for which observations are actually attempted.

Although both professionals and amateurs make minor planet occultation observations with telescopes under 12 inches in aperture, one of the more popular observing configurations used by professionals at Lowell Observatory and the University of Maryland is a C-14 and a high speed photometer capable of 1 millisecond time resolution (see Schnurr and A'Hearn, 1983). Although this makes a very convenient and portable system capable of obtaining publishable results on a wide range of events, it is capable of observing only about half of those events predicted each year. Of these, only about half are observable with the full 1 ms time resolution.

The problem is collecting enough photons for high time resolution photometry. A rough rule of thumb is the 6-6-6 rule: a 6-inch telescope receives 10^6 photons per second from a 6th magnitude star. A 12th magnitude star would yield about $1/250$ as much light, or 4,000 photons per second. With a 1 ms integration period, one would observe only four photons per integration period. This is indistinguishable from the combination of the dark count of even a good PMT, and statistical fluctuations in photon arrival times. The C-14 would receive 5.4 times as many photons, or 22 per millisecond. If the event produced a two magnitude drop, this would be detectable, but such a large delta magnitude is not typical of most of those events predicted. The integration period could be increased to observe these other events, but the loss in time resolution translates directly into an equal loss in accuracy of measuring the minor planet diameter, which is one of the key goals of a minor planet occultation observing program. Such information, in conjunction with measured albedos, is useful in determining the composition of the minor planets, and both size and composition information are useful in discriminating among the half dozen or so serious contenders for a theory of the formation of the solar system.

Therefore, by using an aperture considerably larger than 14 inches, one can observe a larger number of events each year. If a 30-inch aperture were used, there would be 25 times as many photons as with a 6-inch aperture, or about 100 photons per millisecond. This would be adequate to observe a 0.5 magnitude change in light from a 12th magnitude star, and would be enough to detect a similar change in light from a 15th magnitude star with $^1/_{15}$th of a second time resolution. Therefore, practically the entire range of predicted events would be available to a portable telescope of such a large aperture. The occultation is not observed to happen between two 1 ms integration periods. Instead, a standard algorithm is used to fit a curve to the data, and to pick an event epoch time based on standard criteria. However, the greater the time resolution, the better this algorithm works.

The minor planet occultation data are gathered using a single-channel high speed photometer capable of 1 ms time resolution. When this photometer was first built, we installed photon counting electronics and a Thorn EMI Gencom 9892B low dark current PMT with a quantum efficiency of 8-12% across the bands of interest. Better quantum efficiency can now be obtained, in the range of 40-50%, using an EG&G SPCM-200 avalanche photodiode operated in Geiger mode. With virtually no noise, requiring supply voltages of +12 and ±5 volts, and with count rates up to 1.8 million per second, these detectors are ideally suited to high speed photometry. As is typical of photometers used in this application, the diaphragm will be only 15″ in diameter, and equipped with a Fabry lens to reduce photon count fluctuations due to short term tracking errors (A'Hearn, 1984).

13.2 System Operational Environment

The WMO environment during normal observing conditions is usually very dark, with temperatures ranging from 0–90° Fahrenheit and humidity from 0% to the dew point. The observer should not be required to move around a great deal to use the computer, since he may stumble over or into a piece of equipment and injure himself. The equipment handled by the observer (such as a hand paddle) should be able to tolerate the range of temperatures and humidity the observer himself may encounter. We observe at remote locations, so wind and rugged terrain are more important factors than at a fixed observatory. Adequate power for the telescope, computer, other electronic equipment, and heating (or air conditioning) must be provided.

The system is designed to be set up quickly, so that useful work can be done as soon as possible after applying power to the telescope drive and the computer. The time required for the observer to enter a command and for the computer to execute it should be no longer than the time required for the observer to perform the same function manually. This is not necessarily a requirement for most computerized systems, such as the Fairborn Observatory APT (see Part V), where the goal of unattended operation permits, and design tradeoffs may even force some tasks to take longer under computer control than when performed manually. However, in the WMO system, setup time is an extremely precious resource, so the sole justification for automating any function in this system is to save time. Physical and electronic setup, and alignment of the telescope for accurate tracking should be as rapid as possible, to allow last minute changes of observing site to avoid clouds and to permit travel mishaps, such as flat tires or getting lost, to be absorbed into the schedule without missing the event.

The observer using a small telescope typically is alone, so features of the system used for centering objects in the field of view should be available to a user at the eyepiece of the telescope, as well as at the computer console. This is to permit alignment of the telescope with the celestial coordinates by sighting stars of known position.

The amount of direct interaction with the computer, for example, through a CRT screen with keyboard and mouse, should be minimized, and should employ a GUI understandable to astronomers. At least half of the telescope time is set aside for guest observers, who typically are not computer specialists.

13.3 System Performance Requirements

System performance requirements should be set keeping in mind not only the observing program and the operational environment, but also the user budget. This requires a good knowledge of the available technology and, specifically, a knowledge of how useful the technology is in the operational environment.

The functional requirements and available budget for any project should be the driving forces behind all tradeoffs and design decisions that are made. The requirements are established to implement the observing program, but the first version of the functional requirements is written without benefit of a top-level design. As the project proceeds in its development, the functional requirements should be reviewed periodically and updated, if necessary. The following performance requirements for the WMO telescope are listed as an example of how to begin the project.

13.3.1 Portability

The telescope will be used primarily for observations that require the telescope to be portable. The vehicle used to transport the telescope should be inexpensive, readily accessible by the observatory staff, and, preferably, be capable of being used for activities other than merely transporting the telescope. In addition, a source of electric power must be provided for operation in remote locations where local power is not conveniently available.

13.3.2 Setup Time

The telescope and control system should require no more than one hour for one person working alone to perform all setup and initialization functions, measured from the moment the telescope arrives at the observing site until the telescope is tracking the target star with the accuracy specified below. These setup and initialization functions include activating the remote power source, locating the telescope on solid ground, levelling the trailer and raising the tires off the ground, physical setup of the telescope tube or truss and drive, aligning the telescope for tracking, readying the photometer, initializing the computer control system, and acquiring the target star. Further reductions in setup time would add greater flexibility to adjust to last minute site changes and travel problems.

13.3.3 Optics

The telescope optics should be of sufficient aperture as to be capable of recording occultations of magnitude 12 stars by minor planets one-half magnitude fainter. This is to make available more events than are currently attempted by existing portable systems.

13.3.4 Telescope Pointing Accuracy

The telescope pointing accuracy should be at least $30''$ if fewer than four star sightings were made after arriving at the observing site, and better than $15''$ if at least one dozen star sightings were made after arriving at the observing site. This requirement should be met in all areas of the sky within 30° of the event in RA and 10° in Dec. This requirement reflects the fact that the telescope location changes frequently, and that unforeseen last minute events (flat tires, traffic tickets, clouds, last-minute astrometry indicating a path shift) may limit the time available for star sightings after arriving at the site. It also reflects the fact that if the target star is very faint, it must still be the correct star, and it can be located almost anywhere in the sky, including very near the horizon. The reason for relaxing the requirement from 15″ to 30″ is that if the telescope arrives at the observing site with too little time remaining before the event to achieve 15″ pointing accuracy, observing conditions may permit the use of a larger photometer diaphragm.

13.3.5 Telescope Pointing Time

The slew rate should be as fast as is practical, with a goal of no more than one minute from any part of the sky to any other part of the sky, and no more than 30 seconds if a slew of less than 90° is made. This includes both the time spent at the maximum slew rate and the required ramp times. These figures are somewhat arbitrary, and might be modified as field experience is accumulated, but are based on predicted slew rates of the microstepped drive described below. Rapid slewing is required to sight enough stars during the setup period to align the telescope for accurate pointing and tracking.

13.3.6 Long Term Tracking Accuracy

After reaching the commanded position, the telescope should track the desired object with a position error accumulation rate of less than 10″ per hour if fewer than four star sightings were made after arriving at the observing site, and 5″ per hour if at least one dozen star sightings were made after arriving at the observing site. Often one can find the star soon after setting up the telescope, but after fussing for an hour with a photometer that is indignant about being asked to function normally in 10°F weather in the middle of a corn field, the star has given up and wandered out of the field of view (the photometer diaphragm is 15″ in diameter).

If all goes well, it could be as long as two or three hours from the time the target is acquired until the main event, which lasts from a few seconds up to about 30 seconds. Observations are often made for 5–20 minutes before and after the event to search for secondary occultations caused by possible satellites of minor planets. Ideally, the observer should have the luxury of centering the object in the diaphragm, then feeling free to devote his attentions to monitoring the data, time receiver, generator voltage, and other field equipment without being required to check the eyepiece every few minutes to determine whether the object is still in the diaphragm. This means that accurate tracking is at least as important as accurate pointing.

13.3.7 Short Term Tracking Accuracy

The drive and control system should move the telescope smoothly enough to track the target star without fluctuations greater than 2″. Although telescopes at choice observing sites often enjoy $0''.3$ seeing, and astronomers want short term tracking on most telescopes to be this accurate for photography, our hypothetical telescope will rarely enjoy such excellent seeing conditions, and will not be used for photography. It will be subject to wind loading not usually found inside a dome, and we sometimes try to observe under wind conditions that would force most backyard and mountaintop astronomers to call it a night. A Fabry lens in the photometer will prevent any $2''$ short term errors from affecting the photon counts significantly.

13.3.8 Data Input and Control Device

The keyboard or other data input and control device, and other exposed equipment should be able to withstand temperatures ranging from -10 to $+90°$ Fahrenheit, humidity ranging from 0% to the dew point, and windblown dust or sand. This device will be lit with a red lamp or otherwise be easy to use in the dark without impairing the observer's dark adaptation, and will be conveniently positioned within arm's reach of the eyepiece. Manual slew commands will be reviewed by the control system software before they are executed to prevent damage to the telescope and drive, and to give optimum pointing and setting performance.

13.3.9 Computer Environment

The control system equipment (the computer) should meet the temperature and humidity requirements described in the previous requirement, or it should be placed in a protective enclosure that can keep the equipment within its operating ranges throughout the specified ranges of temperature and humidity. Disk drives and other equipment sensitive to foreign particles should be protected against wind-driven dust and sand.

13.3.10 Commands

Command inputs should be in the form of menu selections or plain language commands whose meanings are known unambiguously throughout the general astronomical research community.

13.3.11 Extraneous Light Control

The optical assembly will be constructed to minimize the amount of stray light that reaches the focal plane. This is particularly important in a portable telescope, since the typical occultation observing site is right beside a country road with headlights from passing cars shining directly on the telescope tube during the most critical part of the observation. However, errors due to wind loading are more threatening to the success of the observation than loss of photometric accuracy due to stray light.

As the system design evolves, these requirements will be updated and additional requirements will be added. This is a normal part of the iterative design and build process.

13.4 Overall System Design and Evolution

After the performance requirements have been formulated, major design decisions need to be made to arrive at a top-level system design. These decisions include the following for the telescope we are designing in this chapter:

- Method of transport
- Telescope mount type
- Optical system
- Drive design approach
- Control system approach.

Each of these topics is addressed below.

13.4.1 Method of Transport

There are three basic telescope transport options:

1. Make the telescope tear down into pieces that can be packed into a van or small truck.
2. Mount the telescope permanently inside a van or truck, and have a roll-off roof section or side panels that fold down to reveal the telescope.
3. Mount the telescope permanently on a trailer and tow the trailer to the observing site.

The stringent requirement for rapid setup eliminates the first option. It takes two people almost half an hour to set up a C-14, a relatively lightweight and compact 14-inch telescope. To meet the optics requirement, an aperture in the range of 24–30 inches is required. If the telescope had to be set up and torn down at each site, the time required to do this would severely limit the flexibility of the system to respond to changes in the weather or delays in transporting the telescope to the observing site. Of the remaining two options, a truck with a 30-inch telescope mounted permanently inside it would need to be rather large and expensive. Jack screws to lift the vehicle off its tires to reduce vibrations would have to be of the type used on backhoes and firetrucks (big and expensive). Room needed for the control computer and the instruments would further increase the size and expense of the truck needed. Such a vehicle would not fit in a typical home garage, and most residential covenants prohibit parking large commercial vehicles on the street. If the truck is stored in a commercial garage, it would add to both the operating expenses and the time required to get the telescope on the road. All this rules out mounting the telescope inside a truck, which means the telescope will be mounted on a trailer.

13.4.2 Telescope Mount Type

The fact that a telescope trailer must be towed to the observing site plays a larger role in the selection of the telescope mount than you might imagine. While a student at Wesleyan University in 1971–72, Tomer built a trailer-mounted telescope. It was similar to a "Porter's Folly" equatorial mount, and consisted of a polar axis disk (three manhole covers welded together and machined true) that separated two structural "cones," as shown in Figs. 13-1 and 13-2. The apex of the cone below ("south" of) the disk fit into a thrust bearing at the front of the trailer. Automobile wheel bearings mounted inside assemblies attached to the two triangular plates bore the weight of the telescope at two points on the disk. Therefore, three points (the two disk bearings and the thrust bearing) defined the right ascension axis. The second cone was located above ("north" of) the polar disk. Its apex was the declination bearing housing. Andy used a 12-inch Cassegrain on this mount, but it was sturdy enough to hold a 24-inch telescope. The entire assembly weighed about one metric ton.

We towed this trailer (nicknamed "Phoenix III" because it contained parts from two previous telescopes) to Nova Scotia behind a rental truck for the July 10, 1972 solar eclipse. Although it had two axles (four tires) to bear the weight, it bounced and slid its way there and back. It did not corner or maneuver well at highway speeds. During this trip, we suffered a blown tire, and later we were forced to stop to weld first a fender and then an axle spring back onto the trailer. After these experiences, we concluded that

1. the trailer weighed far too much,
2. the center of gravity was too high (which caused the cornering and

Fig. 13-1 *"Phoenix III" telescope trailer—front view.*

Fig. 13-2 *"Phoenix III" telescope trailer—side view.*

highway maneuvering problems), and

3. polar alignment was too difficult, as it required considerable effort on the part of two strong people for several minutes to swing all this mass to within several degrees of the pole—finer alignment proved impossible.

Tomer solved these problems in his next version (Phoenix IV) by making the mount and telescope tube fold down into a low-slung box that formed the basis of his trailer. Setup consists of lifting the telescope out of its box using a gearhead motor and a steel cable wrapped around a series of pulleys. It takes about 15 minutes for the motor to lift the mount up until the polar axis angle equals the latitude. The whole equatorial mount is then rotated on an azimuth bearing to complete polar alignment. The new telescope is shown set up and aligned on the pole in Figure 13-3, and nearly ready for towing in Figure 13-4. The counterweight arm shown in Figure 13-4 is removed, and a canvas cover is fastened in place when towing the trailer. Tomer's solution embodied in Phoenix IV works well for his 12-inch Cassegrain, but there are problems when a 30-inch or larger telescope is used. The geometry involved, and state and federal limits on trailer width, are the constraining factors. Furthermore, the motor and pulley system required to lift the larger telescope and its mount would consume a great deal of power and would not meet the setup time requirement.

Referring to Figure 13-2 again, if the south polar cone were removed from the mount and the disk flipped forward to a horizontal position, the declination bearing housing could be lowered almost two feet. This would remove a fair amount of weight from the mount, and lower the center of gravity considerably. The result would be an altitude-azimuth (alt-az) telescope mount.

Experience in towing both of Tomer's telescope trailers emphasizes the difference in towing characteristics that result from the lighter weight and lower center of gravity of the Phoenix IV trailer. The ease of towing, and the elimination of the need for accurate polar alignment in a short period of time both point toward an alt-az mount as the best way of meeting the portability and rapid setup requirements.

Although an alt-az mount solves many of the problems discussed so far, it introduces new ones. The most difficult are (1) two axes must be driven at varying rates, instead of driving only one axis at a relatively constant rate, and (2) the image field of an alt-az mount rotates at the rate of change of the parallactic angle. These topics were discussed in Chapter 6, in which correction equations are given. A third problem for which there is no correction is that celestial objects cannot be tracked through a cone centered on

Fig. 13-3 *"Phoenix IV" telescope trailer—polar aligned.*

Fig. 13-4 *"Phoenix IV" telescope trailer—stowed for towing.*

the local zenith. A modern PC and the use of the right kind of motors and shaft encoders should solve the first problem, within the performance requirements specified above. Since long-exposure photography is not one of the programs of interest, field rotation is not a problem as long as the target star is in the center of the field. (This could be solved, though, using a third motor and rotation axis, and by using the equation developed in Chapter 6.) Experience on previous expeditions indicates that occultation observations are rarely made within 5° of the zenith. Also note that 5° represents about 1% of the total visible hemisphere.

The optical assembly consists of an open truss separating the primary and secondary mirror cages, to reduce wind loading, a leading source of pointing and tracking errors. Black canvas baffles can be snapped into place on the truss if the wind is not a factor, and if heavy road traffic near the observing site makes stray light a problem. A prototype trailer was built using a 12-inch optics assembly inside Phoenix III's aluminum tube.

The optical assembly is supported by a symmetrical fork. This minimizes the moment of inertia about the azimuth axis, keeping the natural frequency of the mount high. It also eliminates the heaviest counterweights, otherwise useless masses that must be towed and maneuvered on the highway as well as contributing to increased rotational inertia of the mount. Each fork arm consists of three lengths of steel tubing welded in a tripod configuration.

Despite the fact that the choice of an alt-az mount complicates the control system software, it results in a compact and lightweight assembly that is easily transported to a remote observing site, and is easily and quickly aligned to the equatorial coordinate system.

13.4.3 Optical System

The primary requirement affecting the optical system design is portability. This means the optical assembly should be as short as possible, so the primary-secondary mirror separation must be minimized. To do this the primary should be at least $f/3$, if not faster. However, the photometer is designed for an $f/11$ beam. A Newtonian system has three disadvantages. First, it places the 10-pound photometer out at the secondary mirror assembly. This increases the moment of inertia, which decreases the natural frequency of the truss assembly, and greatly increases the overall mass of the system needed to support the bulkier truss.

Second, a Newtonian system cannot simultaneously meet the requirement for a fast primary to reduce the length of the telescope, and the requirement for an $f/11$ beam for the photometer, which is the standard Lowell Observatory high speed photometer designed to be used with C-14s.

Finally, a Newtonian optical system places the eyepiece in awkward positions for sighting stars during the alignment process. There is rarely good footing at remote locations to place a tall ladder for viewing through large Newtonian optics. If a high power refractor were attached to the main truss for sighting bright stars, it would take too long to realign it with the main scope after a trip over rough roads. Differential flexure, temperature effects, and other problems would also introduce significant errors into the encoder calibration constants.

To obtain easy visual use for telescope alignment, Cassegrain optics are used. A Richey-Chrétien design is not necessary because a wide, flat field is not required, and it is undesirable because of the additional expense involved in figuring the optics. The 12-inch prototype system used a Dall-Kirkham Cassegrain, which yielded excellent images on-axis and terrible ones just a few arc minutes off-axis. Since the research targets are all point sources, this is not a problem. Ordinarily, a Cassegrain optical system requires about 18 to 24 inches additional space behind the primary mirror for attaching the photometer or other data collection equipment. The problem with this approach is that it makes the fork arms longer to provide clearance for the photometer when observing near the zenith. This increases both the total weight and the moment of inertia of the mount.

To avoid this problem, a flat third mirror 45° to the optical axis is used in a Nasmyth configuration to direct the beam through the altitude bearing to a focus near an equipment mounting plate attached to the fork arm. The flat is mounted on a rotating platform so that an eyepiece assembly can be mounted on one fork arm, and the photometer can be mounted on the other fork arm. This allows the telescope to be aligned to the equatorial coordinate system using the eyepiece and a hand paddle to sight stars without the need for changing equipment assemblies. It also permits using a photometer with no built-in eyepiece or flip mirror, though the WMO photometer has both.

In sum, the decision to use large aperture folded Cassegrain optics is a result of the requirements for both portability and the ability to observe the events that are the focus of the observing program.

13.4.4 Drive Design Approach

The basic options for driving a telescope were reviewed in Chapter 4. These are the worm gear, band, chain, and friction drives. As mentioned in that chapter, a worm gear drive requires special consideration from the control system to counteract periodic errors and wind loading. The chain drive suffers from excessive periodic error, and both the chain and band drives do not have the stiffness required to keep the natural frequency of the drive high. The disk and roller friction drive was chosen because it does not have these drawbacks, and offers the kind of performance needed in an alt-az mount, with its wide variation of drive rates.

13.4.5 Control System Approach

A classical closed-loop servo was chosen as the control system approach for the following reasons:

- The observing program does not always include photometry, so a photometer is not always mounted at the Nasmyth focus for acquiring and centering targets. Furthermore, it is difficult to offset from bright stars to acquire faint targets, particularly when working under severe time constraints and field conditions. All this requires high intrinsic pointing accuracy.

- Faint targets must be acquired in dense fields of stars of similar brightness. The Fairborn Observatory APTs can tolerate moderate pointing errors because there are relatively fewer stars of the brightness of the target stars used in the APT research program. It is as yet an open issue how faint stars can be and still be accurately acquired using APT techniques. High accuracy position feedback can be provided only by shaft encoders or intrinsically accurate motors.

- Targets must be tracked accurately during an hour or more with an alt-az mount. Modern PCs and real-time operating systems are available that make developing a closed-loop servo straightforward, and Trueblood has over 30 years' experience in programming, so insoluble software problems are not anticipated.

The control computer is capable of computing all the correction equations discussed in Chapters 5 and 6 rapidly enough to ensure accurate pointing and smooth tracking. A great deal of software is needed, not only to handle these basic servo functions, but also to aid setup and alignment (measuring the encoder zero offsets and azimuth tilt angles) and to reduce the time these activities require. If actual field experience indicates it is needed, an extended Kalman filter can be added in the future to improve the pointing and tracking by adjusting the calibration constants in real time in response to manually entered pointing corrections.

Fig. 13-5 *The WMO Van and Prototype Telescope Trailer.*

13.4.6 Top Level System Design

In response to the requirements, we have developed a basic system concept in which a single-axle trailer holds an alt-az mounted telescope of approximately 30-inch aperture. The trailer is towed by a van containing the control computer and related equipment, a gasoline powered generator, power conditioning equipment, and extra heating and cooling capability to protect all this equipment and the observer from the elements.

The trailer is built around a 45-inch diameter bearing that serves as the azimuth axis plane. Most states permit trailers to be a maximum of 8 feet wide. A handy tip to those contemplating similar-scale projects is to build a 1:4 scale balsa wood model before construction begins, to ensure that geometrically related designs are correct. The WMO developed a test bed system using parts from Phoenix III, as shown in Figure 13-5. This was used to verify the basic system concept, and to gain field experience using the telescope in the WMO observing program. During this phase of such a project, the operational procedures (especially those of setting up at a remote site) should be refined. The final step is to procure and install the larger optics. A larger telescope would require re-tuning the drive and control system, but by the time this phase is reached, experience at these tasks will have already been gained with the smaller telescope.

As shown in Figure 13-6, the computer rack, generator, power conditioning and monitoring equipment, and other equipment are mounted on Aeroflex shock mounts to reduce damage to sensitive electronics. The generator is mounted at the rear of the van. It is quite noisy, so we recommend

Fig. 13-6 *Inside the WMO Van.*

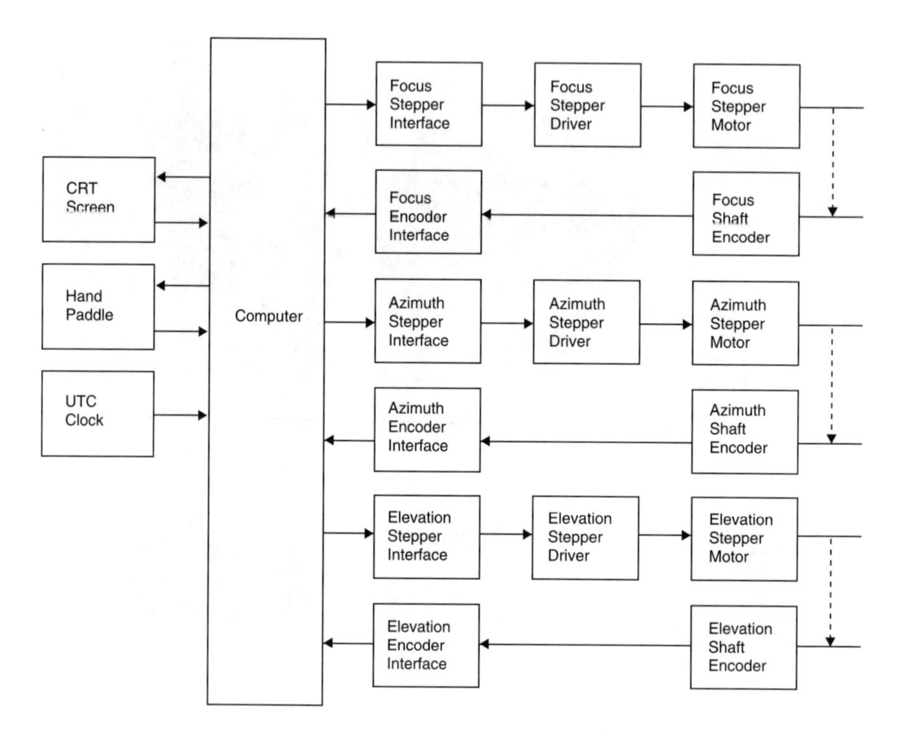

Fig. 13-7 *System block diagram.*

installing a large plywood panel with fiberglass sound baffling to keep the noise down to a tolerable level, or using a fuel cell instead (fuel cells are quite expensive). The generator is fitted with a duct to vent exhaust gasses outside the van.

Figure 13-7 shows the system hardware configuration. The same computer is used to control all telescope functions. The compiter uses time from the UTC clock to compute local sidereal time, which is used to compute hour angle from right ascension. In addition to the components of the operational system shown in Figure 13-7, thought must be given to those items needed to build and maintain the system. Since some custom digital design and fabrication is included in the design, an oscilloscope, digital test probe, VOM, wire wrap tool or a printed circuit board etching kit, and other such equipment are needed. Tools for software development are also needed and can be part of the computer system procurement, including a software developer's kit and documentation, and a printer for generating compiler listings.

This completes the discussion of the top level design. Before a detailed design can be set forth, fundamental questions must be answered. Included are the issues of how to calibrate the encoders, processor performance, and selection of the control computer. These issues are addressed below.

13.5 Position Encoder Calibration

Before the telescope can be used with closed-loop servo control, the shaft angle encoders must be calibrated. This consists of developing an algorithm for converting the raw encoder readings into the true position of the telescope optical axis. Figure 13-8 depicts the calibration curve that results from this process. One method of calibration is to sight a large number of stars and compute the coefficients of a simple polynomial power series, which is then used as the model to transform raw encoder readings into equatorial coordinates. An example is

$$\text{position} = \sum_{i=0,n} A_i E^i$$

where E is the raw encoder reading and the coefficients A_i are determined by recording the encoder readings for n stars, then performing a least-squares fit. This method has several problems, as described below.

1. n can be quite large to obtain a good fit, especially if the function has several high frequency components, so it can take several hours or even several nights to sight enough stars to obtain an accurate calibration.

2. There is no control over the excursions of the algorithm from the true functional relationship (i.e., the peak-to-peak variations may be large).

3. This method will not work for portable telescopes, which must determine new zero offsets and polar axis or azimuth axis tilts for each new observing location, without necessarily re-determining those errors intrinsic to the telescope itself (for lack of time).

4. There is no way to monitor how telescope characteristics change over time. This monitoring can reveal design or construction weaknesses, or a need for maintenance or repair before the problem causes a loss of observing time. This is particularly important with portable telescopes that are subjected to the shocks and vibration of transport over rough roads.

Objections 1 and 2 can be met in part by using Chebyshev polynomials instead of a simple power series. According to Arfken (1970, p. 629), such

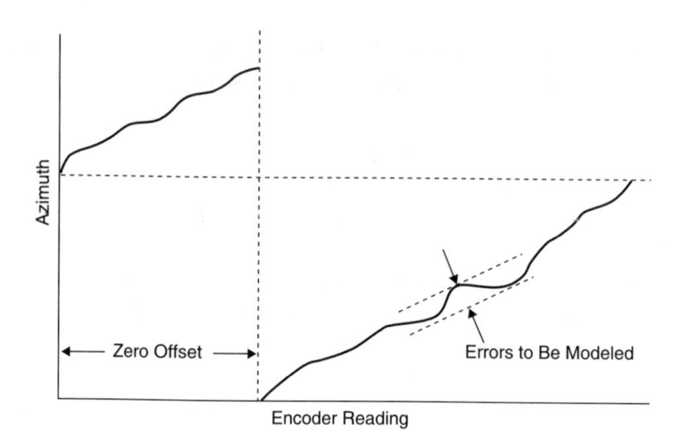

Fig. 13-8 *Encoder calibration curve.*

an approach limits the maximum error (but does not minimize the RMS error), and prevents the errors from bunching up at the ends by distributing them equally throughout the range of the function. It also gives the desired accuracy in the fewest number of terms. Spherical harmonics would also provide a more efficient fit than a simple power series.

However, objections 3 and 4 still stand. Because it is quite useful to know how telescope characteristics change with time, the method of power series or Chebyshev series will not be discussed further.

An approach that addresses objections 3 and 4 is to model the errors. The equations for modeling sources of systematic error were presented Chapters 5 and 6. These equations use coordinate variables, such as (h,δ) or (A,Z), and constants embodying characteristics of the telescope itself. The model should include all sources of error between the encoder and the optical axis. Errors between the motor and the encoder need not be modeled, since such errors do not affect the functional relationship between optical axis position and raw encoder reading. The model constants are shown in Table 13-1.

The constants are determined by sighting stars and inferring the constants, or by modeling (e.g., secondary mirror truss flexure, using a finite element analysis program). Direct measurements of these constants are usually more accurate than modeling. Constants related to location and orientation cannot be measured or modeled ahead of time for portable telescopes. During observations of regular program objects, these pre-determined constants are used in the equations given above to perform the conversion from raw encoder units to optical axis position. The calibration problem is reduced to the problem of obtaining optimal estimates of the error model constants from measurements.

Table 13-1

Equatorial Mount	Alt-Az Mount
1a. Arc seconds per encoder count (2 axes)	1b. Same (2 or 3 axes)
2a. Zero offset (2 axes)	2b. Same (2 axes)
3a. M_{el} (polar axis alignment)	3b. Angle a (azimuth tilt)
4a. M_{az} (polar axis alignment)	4b. Angle b (azimuth tilt)
5a. p (non-perpendicular axes)	5b. Same
6a. $C_{\rm ns}, C_{\rm ew}$ (collimation)	6b. Same
7a. $F(90°)$ (tube flexure—2 axes)	7b. Same (2 axes)
8a. Mount flexure	8b. Ignored—constant
9a. τ' (servo lag—2 axes)	9b. Same (2 axes)

The easiest method of measuring these constants is to construct observations that tend to isolate one error source from all others. A few examples of procedures for doing this were discussed in Chapter 6. For example, if you want to vary Dec and eliminate RA, observe a series of stars along the meridian, then plot observed Dec versus predicted Dec (corrected for topocentric place and telescope errors known at that time). Look at the plot and try to figure out what is causing the deviation from linearity by figuring out what kind of function (e.g., $\sin\delta$) when added to the observed Dec would make the plot linear. It may take two or three iterations of these observations to get the model constants to converge. After the model constants have been measured, their measured values can be averaged to eliminate much of the random error. This simple procedure is the recommended approach for most applications.

However, for those requiring even greater accuracy who are familiar with the mathematics involved, the error constant values can be refined further using modern digital filter optimal estimation techniques. A large number of stars (e.g., 200) can be sighted and the data used in statistical procedures to obtain very good estimates of the error constants. Not only does this yield improved accuracy when computing telescope pointing corrections, but it also permits calibration observations to be made all over the sky, not at particular places determined by a procedure to measure one error constant. The observation equations in the digital filter relate the observables (RA and Dec) to the estimated parameters (the error constants) in a way that permits separating the various error effects, which are not intrinsically orthogonal. Such methods require that the observation equations be overdetermined, that is, there are many more star sightings made than there are constants to be estimated.

The familiar technique of using weighted least squares is not the most computationally efficient, since the covariance matrix used for weighting the observations must be inverted, and matrix inversion cannot be performed rapidly. The Kalman filter technique also yields optimal estimates of the error constants, but without inverting a matrix. The extended Kalman filter can be used during normal observing sessions, rather than being confined to special encoder calibration sessions. Observer inputs to correct telescope pointing errors during tracking can also be used by an extended Kalman filter to improve the values of error constants in near real-time, so that tracking accuracy can be improved while observations are being made. This, too, is of considerable help when using a portable telescope, since the location and alignment of the telescope (and other inputs to error constants) can change considerably from one night to the next. For those with advanced skills in mathematics, these digital filtering techniques are discussed in greater detail in Appendix I.

No matter how hard you try to model all errors everywhere in the sky, there will be some unmodeled errors left over. These can be put into a pointing map which is then interpolated (e.g., using a cubic spline) for points on the sky in between the points in the map.

13.6 Computational Requirements

One of the criteria for selecting a control computer is the speed of its processor. Only when you know the rate at which calculations must be performed can you understand the computational requirements. In a digital servo, the rate of performing calculations is determined in part by the rate at which errors accumulate during the interval between the calculations. Since all the equations of interest in Chapters 5 and 6 depend on hour angle, elevation, or some other parameter that changes with time, they can be used to assess error growth rates.

We stated previously that based on several existing telescope control systems, computers based on the Pentium and PowerPC chips should have more than enough speed for most telescope control applications. However, for those with unusually stringent requirements or with slower computers, we present the exercise below as a guide to how to assess the ability of your computer to handle the task of processing your control algorithms.

Each error correction can be assigned an error budget. The simplest way of doing this is to give each error source an equal weight, and then determine the error budget for each calculation based on the total number of corrections to be made. There were eight significant calculations discussed in Chapter 5 in the reduction from mean to topocentric place (assuming UT1 is part of the encoder zero offsets), and nine calculations in Chapter 6 for mechanical and alignment corrections. If it is assumed these 17 random errors accumulate in quadrature (are not related to each other), then the root sum square total error budget should equal the short term tracking error maximum of 2″. This means that

$$\sqrt{17e^2} = 2''$$

for equally weighted errors, which gives individual error budgets of roughly e = 0.″5. In the following sections, the time to perform each error correction calculation once and the number of times per second it must be performed to stay within the error budget are found. The calculation times are given in terms of the computational units developed in Chapter 5. This system of computational units assigns 1 unit for additions and subtractions, 2 for multiplications, 3 for divisions, and 22 units for trigonometric and other complex functions. The results are then summarized at the end of the section. Note that times computed this way may vary for your processor, due to different relative times for arithmetic. For example, the Intel Pentium processor typically takes one clock cycle for additions, subtractions, and multiplications and up to 39 clock cycles for divisions. Fortunately, most of the equations in Chapters 5 and 6 have at most one or two divisions.

13.6.1 Topocentric Place Correction

13.6.1.1 General Precession

In Chapter 5, we estimated precession requires 272 computational units. To determine the allowable time interval between precession calculations, you look at the rate at which errors grow. The time dependence of precession enters into the evaluation of ζ_o, z, and Θ in (5.3) through T and t. These quantities are in units of 100 years, and produce changes on the order of 50″ per century, so that during the course of observing the same object over an evening, precession will change insignificantly. Therefore, precession needs to be computed only once an evening per object, when the object's catalog coordinates are entered.

13.6.1.2 Nutation

Nutation requires 524 units to perform the necessary calculations. The periodic term dominating nutation is Ω, through the $\sin\Omega$ term of $\Delta\psi$ and the $\cos\Omega$ term of $\Delta\epsilon$. The periodic term of Ω that changes most rapidly is $(5r + 482890.''539)T$ where T is in Julian centuries. During a single evening of observing, T will change by roughly 0.00001, so Ω will change by roughly 0.°018. This term goes through 360° once in approximately 19 years. Depending on the value of Ω in that 19 year cycle, a change of Ω by 0.°018 could affect the $\sin\Omega$ or $\cos\Omega$ by at most 0.0003 ($\sin 0.^\circ018 - \sin 0$). This factor times 17.″2289 is only 0.″005, so there is no need to compute nutation more than once per evening per object.

13.6.1.3 Aberration

The following table summarizes the operations involved in computing annual aberration and applying it to (α,δ) using Equations (5.16):

Table 13-2

Item	Equation	Trig	+	−	x	/
Annual Aberration	(5.16)	18	17	9	35	4
E-Terms	(5.17)	3	4	0	9	0
Totals		21	21	9	44	4

It is assumed that to compute secant, cosecant, and cotangent functions, one computes the corresponding cosine, sine, or tangent function, then inverts the result by dividing 1 by the result; that l' was computed previously when computing nutation; that seven iterations were used to solve Kepler's equation for E; and that a square root takes as long as a trigonometric function to compute. With these assumptions, annual aberration requires 592 computation units.

Since the value of annual aberration depends on the star's (α,δ) and the position of the Earth in its orbit about the Sun, it will change very little during a single evening of observing, so it can be computed once when the star's coordinates are entered into the control system.

The number of computation units required to compute diurnal aberration using Equations (5.18) with $r = 1$ is 128 units. It is assumed that L_c, the observer's latitude, is entered by hand, or read from a disk file. However, diurnal aberration corrections need not be made, since the coefficients in Equations (5.18) are less than $0''.5$ and the other multipliers in these equations are trigonometric functions with values less than or equal to 1.

13.6.1.4 Parallax

Using the method of Equations (5.19)–(5.20), the number of computation units to compute stellar parallax is 245 units, counting a square root to have the same computational complexity as a trigonometric function, and assuming L_t, e, ν, ϵ, etc., were computed earlier.

When computing annual parallax by this method, and assuming that C, D, and ϵ were computed previously when computing annual aberration using Equations (5.16), the number of computation units required in using Equations (5.21) and (5.23) to compute the additional terms for stellar parallax is 106 units. It is assumed that π is given directly in the star catalog.

If annual aberration is computed using Equations (5.16) and annual parallax is computed using the method of Equations (5.19)–(5.21), the total number of computation units, allowing for overlapping computations in the two methods, is 834 computation units. When both annual parallax and annual aberration are computed using (5.23), and C, D, and e from (5.16), the total is 698 computation units, saving 136 units.

Equations (5.19)–(5.21) can be used to estimate the amount of time that is necessary for errors to grow to $0''.5$. From Equations (5.19), the time-dependent quantities are the heliocentric coordinates (X,Y,Z). In Equations (5.16), L_t and e depend on T (in units of Julian centuries). During the course of an evening, the T term in e changes by a factor of about $1/(36525 \times 3)$, or about 0.00001. The change in e, to first order, is roughly $0''.0004$, and the change in $\sin e$ or $\cos e$ cannot be greater than that. During the course of an evening, the T term in Lm changes by the same factor, but because this is multiplied by some $36000°$, Lm changes by about 0.3. This is offset by the corresponding term in l', so L_t does not change by more than $0''.5$. Therefore, stellar parallax need be computed only once for each different object viewed.

Assuming $\sec\delta$, $\cos\delta$, and $\sin\delta$ were computed previously, the number of computation units required to compute the planetary parallax using Equations (5.24) is 109 units.

From Equations (5.24), the only time-dependent quantity is h, the hour angle. Thus the corrections to (α,δ) will change by at most $\pi(d(\sin h)/dt)$, where π is about $9''$ per A.U. in distance. During one hour, h changes by $15°$, which means that in going from $h = 0°$ to $h = 15°$, α,δ will change by about $2''$. Thus to keep the error below $0''.5$, geocentric parallax should be computed every $1/4$ hour, or every 15 minutes, for one A.U. of distance, less frequently for greater distances.

13.6.1.5 Refraction

To evaluate the number of calculation units that must be performed to compute refraction, Equations (5.26) are used for computing $\Delta\alpha$ and $\Delta\delta$, and (5.28) are used for computing R. Assuming five iterations of (5.26) are required for convergence, the result is a total of 584 computation units, assuming the trig functions of the latitude are computed at the start of the night.

In Chapter 6, it was argued that refraction should be recomputed every time the pressure or temperature changes by a detectable amount. In addition, refraction should be recomputed whenever the zenith distance changes by an amount capable of changing the corresponding refraction by a tolerable error. Even though the tolerable error for other systematic error sources is only $0.''5$, considering the accuracy of the refraction calculation, the tolerable error can be $1.''0$. This relaxation of the refraction error budget will not seriously affect pointing performance, since most of the other error sources can be held to $0.''1$ or less, so the total RSS error is still within the total budget. Tracking performance is not affected, since the changes in the errors in computing R are considerably less than $0.''5$ over long periods of time.

The zenith distance has the greatest effect upon the angle of refraction at the horizon ($Z = 90.6°$ for a ground-based telescope). Rather than evaluating first and second derivatives of R with respect to Z to find the maximum value of dR/dZ, since we know dR/dZ is a maximum at the horizon, a simple evaluation of Equation (5.28) at $Z = 90°$ and $Z = 89°$ is used instead. Using $P = 30$ and $T = 50$, $R(Z = 90°) = 2118.''2$ and $R(Z = 89°) = 1481.''9$, so R changes roughly at the rate of 636 arc seconds per degree change in the zenith angle. Thus to keep the error in R less than one arc second, R must be computed whenever the zenith distance changes by $1/636$ of a degree, or every 5.7 arc seconds.

The rate of change of zenith distance with time at the horizon can be taken to be no more than 15 arc seconds per second, so refraction must be recomputed roughly three times per second. This is an extremely conservative estimate, since dR/dZ does not change rapidly until the telescope is pointing very close to the horizon. Photometrists care little about refraction effects, since they do their observing at zenith distances less than 60°, but those who observe occultations occasionally find themselves observing in difficult conditions, including zenith distances in excess of 89°.

13.6.1.6 Orbital Motion

The typical method of applying orbital motion corrections is to multiply a constant by the Julian date and add the result. This requires a total of 6 computational units for both RA and Dec. If the stars are close enough to require the calculation to be performed more frequently than once per evening, they probably cannot be resolved by the telescope.

13.6.1.7 Proper Motion

Proper motion corrections are usually applied in a manner similar to orbital motion corrections. That is, they require 6 computational units, and need be computed only once per evening per object.

13.6.2 Mechanical Corrections

13.6.2.1 Conversion of the Encoder Reading

The binary encoder reading (an n-bit integer) must be multiplied by a scale factor to obtain degrees, seconds of arc, or some other physical unit. The scale factor is determined by the gear ratios between the encoder and the telescope axis, and the number of counts per revolution of the encoder. This conversion requires two computation units, and must be performed each time the encoders are read.

13.6.2.2 Zero Offset

The addition of a constant to the converted encoder reading represents one computation unit. This calculation must be repeated each time the encoders are read.

13.6.2.3 Polar Axis Alignment

Evaluation of Equation (6.1a) requires the use of three trigonometric functions, one subtraction, three multiplications, and one addition (to apply the correction Δh to h), for a total of 74 computation units. Similarly, evaluation of (6.1b) requires the use of two trigonometric functions (already evaluated in (6.1a)), two multiplications, and two additions (one to apply the correction $\Delta\delta$ to δ), for a total of 6 computation units. Again, this calculation should be repeated each time the encoders are read.

13.6.2.4 Azimuth Axis Alignment

The calculation to find the azimuth axis tilt angles a and b requires 227 computation units. This calculation must be repeated as often as is necessary to keep the error in $Z - Z'$ less than a reasonable amount. From (6.7c), since a is small, the greatest dependence of Z is on Z' for values of Z down to about $Z = 80°$. In this range, dZ'/dh can have a wide range of values, depending on the declination of the star. For example, for $Z' = 20°$, $a = 1°$, $b = 0°$, and $A = 90°$, $\cos Z = a \sin Z' + \cos Z' = 0.94566$, so $Z = 18^{\circ}\!.97$ and $Z - Z' = -1^{\circ}\!.0252$. For $A = 0$ and $b = 0$, $\cos Z = \cos Z'$, so $Z - Z'$ changes by about one degree in the time it takes to go from $A = 90°$ to $A = 0°$, or six hours. This is a change of one arc second in $Z - Z'$ every six seconds of time. To cover all cases and keep the error below $0''\!.5$, the calculation should be repeated once every three seconds.

13.6.2.5 Equatorial to Alt-Az Conversion

Equations (6.8)–(6.10) are evaluated in alt-az control systems to convert the apparent position of a star from equatorial to alt-az coordinates. To obtain zenith distance Z, Equation (6.8a) is evaluated using previously-computed values of $\sin\phi$, $\cos\phi$, $\sin\delta$, and $\cos\delta$. Therefore, two trigonometric functions ($\cos h$ and $\arccos Z$), three multiplications, and one addition are performed, for a total of 51 computation units. Equation (6.8c) is evaluated to resolve the quadrant ambiguity, at a cost of 70 units. Similarly, Equation (6.8b) is evaluated using one trigonometric function ($\sin Z$), one subtraction, two multiplications, and one division, for a total of 30 units. These calculations must be performed at the rate used to read the encoders, so that the error (apparent star position minus current true telescope position) can be used to generate a motor command.

To compute the azimuth drive rate, Equation (6.9b) is used. It contains two trigonometric functions not evaluated previously ($\arccos Z$ and $\sin Z$), one division (to obtain $\cot Z$), three multiplies, one addition, and one subtraction, for a total of 55 computation units. Equation (6.9a) is used to compute the altitude (or zenith distance) drive rate. This equation uses two trigonometric functions ($\arccos A$ and $\sin A$), and two multiplies, for a total of 48 computation units.

To track a celestial object, Equations (6.8) and (6.9) are computed frequently enough so that the drive rate integrated over the time between drive rate computations and updates does not exceed the error budget. Figure 6-10 can be used to find where dZ/dt is changing most rapidly. An example of a high rate of change occurs when $\phi = 40°$, $\delta = 35°$, and $A = 0°$. At hour angle $= 359°$, $Z = 5\overset{\circ}{.}062$ and $dZ/dt = -1.862''/\mathrm{S}$, while at hour angle $= 0°$, $Z = 5°$, and $dZ/dt = 0$. The time it takes for the hour angle to change by one degree is 240 seconds. If dZ/dt is set to $-1.862''/\mathrm{S}$ at $Z = 5\overset{\circ}{.}062$, 240 seconds later, $Z = 4\overset{\circ}{.}938$ instead of $5°$, a difference of $223\overset{''}{.}68$. This error accumulates in 240 seconds, or at the rate of roughly $1''$ per second. Thus to keep the error at or below $0\overset{''}{.}5$, (6.9a) should be recomputed twice each second.

Similarly, Figure 6-11 is used to find where dA/dt is changing most rapidly. When $\phi = 40°$, $\delta = 35°$, and $h = 356°$, dA/dt is changing very rapidly. At hour angle 356°, $A = 326\overset{\circ}{.}359$ and $dA/dt = 101.890''/\mathrm{S}$. If dA/dt is set to $101.890\ ''/\mathrm{S}$ at $A = 326\overset{\circ}{.}359$, 240 seconds later, $A = 333\overset{\circ}{.}152$ instead of $333\overset{\circ}{.}620$ at hour angle 357°, a difference of $1684\overset{''}{.}8$. This error accumulates at an average rate of about $7''$ per second. To keep the error at or below $0\overset{''}{.}5$, (6.9b) should be recomputed 14 times per second. This rough estimate compares favorably with the 20 times per second rate used successfully by a LAGEOS satellite laser ranging system at Goddard Space

Flight Center to measure continental drift (Mansfield, 1984). This system works equally well when tracking stars or the LAGEOS satellite.

A given error in azimuth or zenith distance does not translate to the same error in the respective equatorial coordinates. Depending on the latitude where the telescope is located, a larger error in azimuth or zenith distance can be tolerated without affecting the resulting pointing accuracy of the telescope in equatorial coordinates. This would reduce the rate at which the coordinate conversion equations need to be computed. Furthermore, the rates quoted are first order (linear) approximations. If the error in azimuth accumulates at the rate of $7''$ per second of time worst case, computing the equations 14 times per second would most certainly keep the errors under $0\rlap{.}''5$, but the errors will not actually grow at this rate. These two factors reduce the control of an alt-az mount to a tractable problem for modern computers.

The final drive rate is for field rotation correction. Although this feature is not currently planned to be a feature of the hypothetical telescope in our design problem, it is included for completeness. Equation (6.10) contains one trigonometric function not evaluated previously ($\sin h$), three multiplies, and one addition, for a total of 29 computation units. Equation (6.11) contains two trigonometric functions not evaluated previously ($\arccos \eta$ and $\sin \eta$), six multiplies, one divide, one addition, and two subtractions, for a total of 62 computation units. Together, (6.10) and (6.11) require 91 units. Figure 6-13 shows that when $\phi = 40°$, $\delta = 35°$, and $h = 356°$, $d\eta/dt$ is changing rapidly. At hour angle 356°, $\eta = 36\rlap{.}^\circ326$ and $d\eta/dt = -111.940''/\mathrm{S}$. If $d\eta/dt$ is set to this value at $\eta = 36\rlap{.}^\circ326$, 240 seconds later, $\eta = 28\rlap{.}^\circ863$ instead of $28\rlap{.}^\circ395$, a difference of $1684\rlap{.}''8$. This error accumulates at an average rate of about $7''$ per second. To keep the error in η at or below $10''$, $d\eta/dt$ should be recomputed roughly once per second.

13.6.2.6 Non-Perpendicular Axis Alignment

For the equatorial mount, Equation (6.16a) requires only one multiplication and one addition (to apply the correction) or 3 units to compute, since $\sin \delta$ can be evaluated before pointing and tracking begin. The same is true of (6.16b), since Δh can be computed in advance. This means the correction need be computed only once, since δ does not change during tracking.

For alt-az mounts, Equation (6.17a) uses the $\cos Z$ found in (6.8a), so only 3 units are required for correcting both A and Z. The correction to A changes at the worst case rate of $p \sin Z(dZ/dt)$, which can be as high as $15p''$/second at the horizon. To keep the error less than $0\rlap{.}''5$, the error should be computed $30p$ times per second. However, this worst case rate occurs where $\cos Z$ (in the equation) is approaching zero, which reduces the size of the correction (and its error growth rate) considerably. Therefore, the correction can be computed roughly 2 or 3 times per second. The same is true of the correction to Z.

13.6.2.7 Collimation Errors

To compute collimation error corrections for an equatorial mount, C_{ns} is used directly as the correction to declination, and is applied using one addition, or 1 computation unit. The correction to hour angle is computed using Equation (6.19). This requires one multiplication and one division (to compute $\sec\delta$ from $\cos\delta$, which was obtained in evaluating Equation (6.8a)) to compute the correction, then one addition to apply the correction, or 6 units.

The alt-az mount correction uses C_{ns} as the correction to Z, and Equation (6.20) to compute the correction to A. This requires one division to obtain Z ($\sin Z$ was obtained previously when Equation (6.8a) was evaluated), then a multiplication to complete the evaluation of ΔA, and an addition to apply the correction, for a total of 6 units.

From (6.19) and (6.20), collimation errors are either constants, or depend on trigonometric functions of δ and Z. In an equatorial mount, δ does not change with time when tracking the same object, so collimation errors can be computed once when slewing to a new object, then added in each time the encoder is read. In an alt-az mount, the correction to A depends on dZ/dt. As in the case of non-perpendicular axes, the correction can be computed 2 or 3 times per second to obtain adequate accuracy.

13.6.2.8 Tube Flexure

For an alt-az mount, Equation (6.21) requires one multiplication and one subtraction, or 3 units, to evaluate, since $\sin Z$ was found previously in evaluating (6.9b). However, for the equatorial mount, the alt-az conversions must be performed, since Equation (6.21) requires knowing Z. Equations (6.8) were found earlier to require 151 units. To evaluate (6.10), $\cos h$, $\sin h$, and $\sin\phi$ were found previously, and $\cos A$ was found in (6.8b). Thus two trigonometric functions are required to find $\sin A$, then 3 multiplies and 1 addition are needed to evaluate (6.10), for a total of 51 units. Equation (6.21) requires two trigonometric functions to find $\sin Z$ from $\cos Z$, then one multiply and one subtraction to evaluate and apply the correction, for a total of 47 units. Finally, Equations (6.22) require two trigonometric functions to find $\sin\eta$ from $\cos\eta$, two multiplies to evaluate the equations, and two additions to apply the corrections, for a total of 50 units. The grand total for Equations (6.8), (6.10), (6.21), and (6.22) is $151+51+47+50 = 299$ units.

From Equation (6.21), tube flexure depends on $\sin Z$, hence it changes at a rate of $\cos Z(dZ/dt)$. $\cos Z$ is minimized (in absolute value) where dZ/dt is maximized, so again, a calculation rate of 2 or 3 times per second ought to suffice.

13.6.2.9 Mount Flexure

This correction is not needed in alt-az mounts. For equatorial mounts, Equation (6.23) requires only two multiplications and one subtraction, or 5 units, since $\sin h$ was evaluated previously for Equation (6.1a) and $\sec \delta$ can be evaluated at the time the new coordinates are entered. To determine the calculation rate, note that (6.23) varies with $\sin h$. Taking $\sec \delta$ to be 1, at the meridian $\Delta h = 0$, and at $h = +1°$, $\Delta h = -18.9(0.017) = -0''.33$. For the example of the AAT (using the $-18''.9$ figure), mount flexure can be recomputed every 364 seconds, or every 6 minutes, to obtain $0''.5$ accuracy. The calculation rate is different for each telescope, since it depends on the actual tube flexure.

13.6.2.10 Servo Lag Error

Using Equation (6.24), the servo lag error requires one subtraction to compute I, a multiplication to compute the lag, and one addition to apply the correction, for a total of 4 units. This error must be computed each time the encoders are read and a new motor command is computed.

13.6.3 Processor Loading Calculations

The corrections described in Chapters 5 and 6 can be computed in either open-loop or closed-loop servos to improve pointing and long term tracking accuracy. To be accurate, these calculations must be synchronized with telescope motion to provide accurate current telescope pointing information as inputs to these equations. To point a telescope at an object and track it, the processor must perform the following functions:

1. Compute the desired telescope position
2. Determine the true telescope position
3. Compute the position error
4. Find a set of motor commands that both minimizes the position error and minimizes its rate of growth in the future.

Table 13-3

Computation	Comp. Units	Comp. Freq.	Comp. Units per Second
1. Annual aberration	698	once	-
2. Stellar parallax (part of Step 1)			
3. Precession	272	once	-
4. Nutation	524	once	-
5. Orbital motion	6	once	-
6. Proper motion	6	once	-
7. Diurnal aberration	128	once	-
8. Planetary parallax	109	1/(15 minutes)	< 1
9. Refraction	584	3/second	1752
		Total	1752

The first function is performed in the reduction from mean to apparent place. Table 13-3 summarizes the loading information developed earlier.

The calculations to be performed once per object per evening total 1634 computation units, which can be executed in about 78μS on a 60 MHz Pentium (all Pentium times assume three clock cycles, or 48 nS, per computation unit). This time is insignificant, since these calculations are performed immediately after the mean place coordinates of the next object to be observed are entered, and are complete before the astronomer can turn around to check to see if the telescope is moving to the new coordinates.

The following tables summarize the process of converting encoder readings into axis position. In a typical digital servo, the encoders are read, the readings are converted to appropriate coordinates, the result is compared to the desired position, and new commands are sent to the motors, all in sequence. This loop is executed at periodic intervals, so the encoder calibration equations and motor command generation software are executed at the same frequency.

Gearing and bearing errors were ignored in determining the CPU loading, as they usually are small enough to ignore until after the major error sources have been modeled. Table 13-4 gives the processor loading data for an equatorial mount. Although this is not relevant to the hypothetical telescope example, most readers will be interested in the results. The computation frequency of 3 iterations per second is seen as the lowest rate consistant with the apparent place and mechanical correction computation frequencies. The total processor utilization for error modeling with an equatorial mount is $(1173 + 60) \times 0.048\mu S = 59\mu$S per second, an insignificant amount. This can be reduced even further if tube flexure is small enough to be ignored.

The comparable figures for an alt-az mount are given in Table 13-5.

Table 13-4

Equatorial Mount—Polar Axis

Computation	Comp. Units	Comp. Freq.	Comp. Units per Second
1. Convert encoder reading to degrees	2	3/s	6
2. Zero offset	1	3/s	3
3. Polar axis misalignment	74	3/s	222
4. Non-perpendicular axes	3	3/s	9
5. Collimation	6	3/s	18
6. Tube flexure	296	3/s	888
7. Mount flexure	5	3/s	15
8. Servo lag	4	3/s	12
		Total	1173

Equatorial Mount—Declination Axis

Computation	Comp. Units	Comp. Freq.	Comp. Units per Second
1. Convert encoder reading to degrees	2	3/s	6
2. Zero offset		3/s	3
3. Polar axis misalignment	6	3/s	18
4. Non-perpendicular axes	3	3/s	9
5. Collimation	1	3/s	3
6. Tube flexure	3	3/s	9
7. Servo lag	4	3/s	12
		Total	60

Table 13-5

Altitude-Azimuth Mount—Azimuth Axis

Computation	Comp. Units	Comp. Freq.	Comp. Units per Second
1. Convert encoder reading to degrees	2	14/s	28
2. Zero offset	1	14/s	14
3. Azimuth tilt	227	1/3s	76
4. Convert to alt-az	151	14/s	2114
5. Drive rate	55	14/s	770
6. Non-perpendicular axes	3	14/s	42
7. Collimation	6	14/s	84
8. Servo Lag	4	14/s	56
		Total	3184

Altitude-Azimuth Mount—Altitude Axis

Computation	Comp. Units	Comp. Freq.	Comp. Units per Second
1. Convert encoder reading to degrees	2	2/s	4
2. Zero offset	1	2/s	2
3. Convert to alt-az (already done above)			
4. Drive rate	48	2/s	96
5. Non-perpendicular axes	3	2/s	6
6. Collimation	1	2/s	2
7. Tube flexure	3	2/s	6
8. Servo Lag	4	2/s	8
		Total	124

Table 13-6
Altitude-Azimuth Mount—Field Rotation Axis

Computation	Comp. Units	Comp. Freq.	Comp. Units per Second
1. Convert encoder reading to degrees	2	1/s	2
2. Zero offset	1	1/s	1
3. Field rotation	29	1/s	29
4. Rotation rate	62	1/s	62
5. Servo Lag	4	1/s	4
		Total	98

The total processor utilization for error modeling with an alt-az mount is $(3184 + 124 + 98) \times 0.048\mu S = 163\mu$S per second, or 0.02% of the CPU.

After the control software has found the desired and true telescope positions, the position error E is found by subtracting the desired position from the true position. Next, the current drive rate is corrected by the error as follows:

$$S = K_m\left(R + \frac{E}{FP}\right)$$

where S = motor speed command (steps per second, or volts)
K_m = steps per arc second (stepper motor), or volts per arc second (servo motor)
R = the computed axis drive rate (arc seconds/second)
E = position error (arc seconds)
P = time to correct a position error (seconds)
F = frequency of commanding motor speed updates (Hz).

If the computed motor speed is significantly different from the current motor speed resulting from the previous command update, the motor must be ramped up or down. In addition, if the computed speed is greater than the maximum motor speed, slewing will have to be performed, or, in the case of an alt-az mount near the zenith, tracking of the object must cease until it appears on the other side of the cone defining the area of the sky through which the mount cannot track.

The total computational load to compute the position error and resulting motor command should require about 50 computation units. When this is repeated 3 times per second on each of the two axes on an equatorial mount, it requires an insignificant fraction of a 60 MHz Pentium processor. When computed 14 times per second on the azimuth axis, 2 times per second on the altitude axis, and once per second on the field rotation axis, it consumes $41\mu S$ per second, still an insignificant amount of a 60 MHz Pentium.

One approach to synchronizing telescope motion and error calculations is to use timer interrupts to repeat motor commanding at well defined intervals. In such a system, the interrupts of concern to the programmer are clock interrupts indicating that it is time to perform a given set of calculations. Encoders can be read and motors commanded without interrupts, since the control registers for these devices typically sit right on the bus.

A few years ago, an engineer calculating CPU loading would estimate the number of interrupts per second and the time to process an interrupt. With modern processors, it is nearly impossible to estimate all the system interrupts, and it is unnecessary, because most of the CPU will be used for operating system and GUI overhead, not input/output. One can safely assume that 20–30% of the CPU will be used for non-applications computing including the presentation manager, device drivers, and other operating system functions, leaving about 20% of the CPU for the application to keep the CPU loading under 50%. Given the estimates above, this appears to be more than enough. If you want to use an Intel 80386/25 or slower CPU, you should do all the above calculations carefully. If your estimate comes in higher than 20% of the CPU for the control algorithms, you have the following options:

1. Drop some of the calculations
2. Reduce the rates at which the calculations are performed
3. Code the applications software in assembler language
4. Offload the processor of some of its assigned functions.

The first two options are the ones taken by Boyd and Genet with the APT control system, which computes only precession in an environment in which only one pulse train is being generated at a time, and the number of interrupts per second is very low. In our design example, with its tighter requirements on pointing and tracking accuracy, a faster processor is used which is capable of handling all the computations and interrupts at the required rates. Coding the software in assembler language is not recommended, since the value of the additional time wasted in debugging assembler language code is many times more the value of smart controller cards (the fourth option) or a faster processor, and the resulting assembler language program is many times more difficult (and costly) to maintain and enhance over the life of the system. Considering the relatively low cost of computers, use the older computer for word processing and use your calculation to justify to your spouse buying a new computer.

Another computer resource to be considered, besides that of the processor, is the bus. Since all buses have a maximum rate at which data can be moved across them, there is a maximum bus bandwidth, in bytes per second, for each bus. Since the amount of data (encoder readings, motor commands, etc.) moving over the computer bus is small compared to the number of instructions moving over the bus each second, and the processor manufacturer designs the computer bus to accomodate the instruction rate of the processor plus a reasonable amount of data, the bus loading percentage will likely be equal to or less than the processor loading. This may not be the case when large amounts of instrument data move across the bus, for example, when doing multi-channel high speed photometry. Therefore, bus utilization should be considered when high data rate instruments are added to the computer bus.

13.7 Selection of the Development and Control System Environments

Based on the considerations in Chapter 11, including the availability of an adequately fast processor, a real-time operating system, and commercial boards for motor control, data acquisition, and digital parallel input/output lines, as well as overall cost, there is no good alternative to the Pentium PC clone running Microsoft Windows 95, except possibly to run Windows NT. Windows 95 is less expensive than Windows NT and is compatible with more 16-bit software than NT. Microsoft Visual C++ was chosen as the development environment, as it is the only tool available for both Windows 95 and Windows NT. The main reasons for these selections are as follows:

1. Since the interface between the control program and the operating system is the same for both Windows 95 and Windows NT (the Win32 Applications Program Interface, or API), and the device drivers are the same, the less expensive of the two choices (95) was selected.

2. From the processor loading calculations above, it is obvious that a 60 MHz Pentium has more than enough speed to control the telescope. A 90 MHz Pentium would be able to handle control of an instrument, the observatory (temperature and humidity sensors, and other items not included in the loading calculations), and background data processing to get a quick look at the data while it is being taken, as well as controlling the telescope.

3. The PC, operating system, and development environment can be purchased relatively inexpensively, for about $3,000 not counting the interface cards. Other alternatives, including the Macintosh, UNIX workstations, or VMEbus crates, would cost considerably more and offer no real advantages.

4. The Visual C++ development tool works well, comes fully documented, and is capable of developing all the required software. Trueblood has extensive recent experience programming in C, eliminating the need to learn a new language.

5. As processing needs grow, the processor can be upgraded quickly, easily, and inexpensively.

6. Appendix D contains a list of manufacturers making PC-compatible interface cards for a wide variety of applications. From the length and content of this list, it appears there will be no trouble connecting strange devices (shaft encoders, stepper motor controllers, astronomical instruments, etc.) to the computer.

The key to computer selection is to think out every feature of both hardware and software that is needed to make the system work, and then buy a computer that does not lack any of these features. Many have made the mistake of waiting until the computer was delivered to discover that the hardware or software lacked some feature that was assumed to be included. For example, any operating system billed as being "real-time" does not necessarily have all the features that are needed to make a particular system work. Most computer vendors will allow a prospective buyer to peruse the manuals before making a large investment in hardware and software.

13.8 System Development and Evolution

To help ensure the success of this project, the system will be built in a methodical fashion. Figure 13-9 shows the basic functions of the system, and how the system development will proceed. The system consists of

1. the observer/computer interface (hand paddle, keyboard and mouse, and CRT screen),

2. the error and command processing and servo implementation, and

3. the computer/telescope interface.

In this system, the observer/computer interface is developed first. In modern GUI-based systems, the GUI is the heart of the application, so

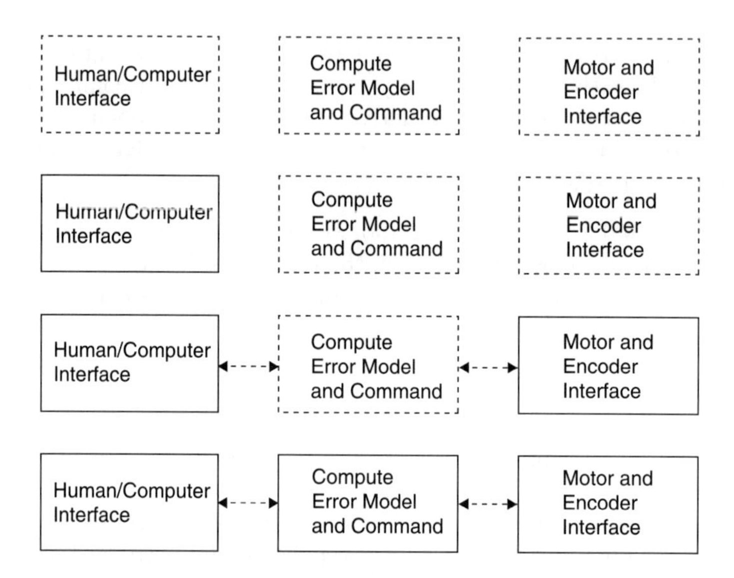

Fig. 13-9 *System development strategy.*

it is needed first. By developing the display pages first, a method will then exist for displaying various intermediate results to help test the rest of the software in the system as it is developed. This is very important. When developing software for a real-time system, often one must see what is going on inside a program. If a screen output statement is placed in the program to display intermediate results, the relatively long time it takes to perform this extra output may throw off the timing of the real-time software. By first writing the "background" or lower-priority task (thread) that is used in the operational system to display data, data from a higher priority task under development can be sent to the lower priority task for display without disturbing the critical timing of the high priority task.

The motor and encoder interfaces are added next. First, a simple command generator is developed that commands the motors to run at a constant speed. The next step is to get the motor to respond to keyboard and mouse inputs. The final step of this phase is to get the motors to respond to hand paddle inputs. This includes the motor ramping for slewing, and the alt-az drive rates. If at a later point you decide encoders are needed, the encoder interface hardware and software can be tested first by slewing the telescope at a constant (slow) rate, and watching the altitude or azimuth change on the CRT screen. Once all this is finished, about one-third of the total lines of code, two-thirds of the subsystems shown in Figure 13-7, and three-quarters of the total software development effort will have been completed.

The last phase is to put in the reduction to topocentric place, encoder calibration, and error calculations. At this stage, a simple power series calibration can be used. After some field experience is gained, the error models discussed in Chapter 6 are implemented, and only if field experience dictates will the extended Kalman filter be used. You will have several months of tweaking and tuning to do, even though the software development for telescope control are essentially complete.

When you are ready to model telescope errors, you should make observations that reveal the errors you want to model, as explained for polar alignment of an equatorial mount in Chapter 6. Make your observations, plot the errors in the telescope coordinates (e.g., altitude) as a function of major terms in the relevant equations (e.g., $\sin Z$ for tube flexure), and take the slope of the curve as the coefficient you need (e.g., $F_u(90°)$ for tube flexure) for the model equation.

After completing your model, you will be left with unmodeled errors. Make observations all over the sky, and fit the errors to a grid pattern of, e.g., every 5° or every degree. This is your pointing map. Use it by interpolating between grid points to where your mount thinks it is to compute the corrections to where it actually is. Only after making a well-designed and well-machined mount, transforming to topocentric position, modeling the errors, and making a pointing map should you consider using a Kalman filter.

This approach is an example of the "build a little, then test a little" philosophy, since each new capability is tested before the next one is even coded. If all the code were developed at once, there would be thousands of lines of code to peruse to find the bug(s) that prevents proper operation. It's far easier to inspect 100 lines of code for a bug than 10,000 lines, so anything you can do to limit where a bug can be hiding will speed up development. This way, one builds on software that works, which then isolates the location of the next irksome bug.

13.9 Detailed Servo Design

Now that the requirements have been defined and we have a top-level design, the detailed hardware design can begin. This involves identifying each functional building block of the system, finding the available options to perform that function, and then performing a trade-off study to select the best option. After all components have been specified, the resulting system design should then be compared to the original functional requirements to ensure that it will give the desired level of performance.

In the following sections, each major function of the telescope is listed.

Where it is instructive, we identify options and explain the reasons for choosing each option. Once again, what follows is a didactic example, and may not represent what is actually done at the WMO.

The easiest method of implementing pointing and tracking functions on an equatorial mount is to "stop the sky" in hardware. This simplifies the software by allowing it to maneuver in equatorial coordinates. This is the approach taken by Boyd and Tomer.

A similar approach is taken in the design example telescope, but the alt-az drive rates (shown in Figs. 6-10 and 6-11) do not permit the sky to be stopped in hardware very easily. Instead, the approach is to use software to generate the alt-az drive rates needed to stop the sky in one routine, then add corrections to these rates generated by the error correction routines. The resulting drive rate is then sent to the motor. The frequency with which this is done is the effective bandwidth of the servo, with the control software serving as the servo filter by determining how frequently these calculations are performed and commands sent to the motors.

Many telescope control systems, especially for equatorial telescopes, distinguish between slewing and tracking. This is because tracking on an equatorial telescope consists of moving the mount only in RA at a relatively slow and constant rate, while slewing consists of moving both axes over a wide range of speeds. On an alt-az mount, it is difficult to distinguish between slewing and tracking, because tracking involves moving both axes over a wide range of speeds, as does slewing. In our design, slewing is considered to be a special case of tracking, and uses the same software for both purposes.

For the tracking/slewing servo, there are two factors determining how often the motor speed must be updated: (1) the natural frequency of the telescope, and (2) the rate at which drive rate errors accumulate an error in position. The goal is to make the motor speed update rate very much different from the natural frequency of the telescope. This is to avoid subjecting the mechanical structure to speed changes (accelerations, which produce forces on the mount) at a rate that causes it to oscillate continuously. Since telescope natural frequencies tend to be in the range of 1–5 Hz, speed changes should occur either slowly (once every 10 seconds or more), or rapidly (10 or more times per second).

In the previous chapter, it was determined that the alt-az drive rate errors accumulate faster than errors due to either astronomical or mechanical corrections. These error accumulation rates require that the azimuth drive rate be computed 14 times per second, and the altitude drive rate be computed twice per second. To correct for drive rate errors at other than the telescope's natural frequency, speed changes must be made at a rate higher than this frequency. Since the altitude drive natural frequency is likely to be near 2 Hz, the altitude motor speed update rate should be higher. It is easiest just to make it the same as the azimuth rate. Experimenting with the servo execution rate is part of the tuning process that is a necessary component of any telescope control project.

When new RA and Dec coordinates are entered, the topocentric place is computed. These coordinates are next converted to alt-az coordinates and passed to the track/slew servo. Instead of performing the ramping in the control computer, accelerations and maximum velocities are sent to the motor indexer board, then a move command is given with the number of steps to move. This number is computed to first order using the azimuth and altitude that correspond to the new position's equatorial coordinates at the moment the motor speed function is computed. A second order correction is then added to account for the changes in alt-az coordinates that occur during the time required to execute the slew. When the slew is complete, the servo goes back to reading the encoders (if present), using them to handle the final lockup on the target.

At appropriate intervals, the planetary parallax (if needed) and refraction corrections are recomputed, and the topocentric RA and Dec are updated. The encoders are also read each time through the servo software, and corrections are applied to compute the current telescope pointing angles. Each time this is done, the topocentric equatorial coordinates of the target are converted to alt-az coordinates. The difference between the actual and desired positions is then found, and converted to a step rate change designed to reduce the error to zero by the next execution loop of the servo software. Finally, the drive rate for that axis is computed, and the error correction rate is added to it. The resulting rate is sent to the motor controller card. An enhancement to keep in mind for the system tuning stage is to add a term that depends on the rate of error growth (as measured by two successive encoder readings). This added complexity would make the tracking a bit smoother.

Initially, use fixed rates for slew and set commands entered on the hand paddle. The software will give appropriate ramp parameters to the motor controller card when executing these commands. The rates used can be adjusted in accordance with field experience.

This approach to the pointing and tracking functions represents a reasonable balance between hardware and software complexity, while promising to meet or exceed the performance requirements. It uses the capabilities in the motor indexer cards that interface the motors to the computer.

13.10 Drive Train Design

13.10.1 Drive Train Mechanical Design

The top level design discussed earlier employs a friction disk drive. Figure 13-10 shows the azimuth bearing and its drive. The size and location of the outer flanges of the bearing, and the need to have part of the telescope mount overhang the top flange of the bearing preclude driving the bearing by its flange. Instead, an additional steel annulus is bolted to the top flange and driven by the friction drive roller.

The drive roller is a 2-inch diameter steel cylinder hardened so that a force of several hundred pounds can be used to bring the roller into contact with the annulus and drive the mount without slipping. The drive roller is supported by roller bearings press-fitted into steel plates that form the drive assembly housing to which the motor is bolted. The encoder is attached to the center of the bearing annulus. The drive assembly is brought into contact with the bearing using an eccentric cam pushing on the assembly from behind. If the outer diameter of the annulus is 50 inches, the effective "gear" reduction ratio from the roller to the azimuth axis is 50/2 or 25:1. This last stage in the drive will reduce the errors occuring in earlier stages by this effective reduction ratio without introducing significant new errors.

The drive for the altitude axis is similar to that for the azimuth axis. In the prototype, a 30-inch diameter steel flywheel ring found in a junk yard was accurately machined and attached to the main telescope tube assembly. This same size is useful for the large scope, giving a final altitude axis gear reduction of 15:1 with a 2-inch drive roller. A drive assembly using only the rollers attached to the motor is pressed directly against the disk.

The drives are not loaded when the trailer is being transported, since movement of the roller against the driven disk when the trailer hits a bad bump will tend to score the roller or disk. Scoring introduces periodic errors that are difficult to model, especially if the scoring occurs on the roller.

13.10.2 Motor and Motor Gearing Selection

To simplify the drive design, the computer interface for the motors, and the control software, only one motor per axis is used.

The following three motor configuration options were considered:

1. A 200-step per revolution Superior Electric Slo-Syn motor or equivalent, coupled to the two-inch drive shaft through a 359-tooth high accuracy Byers worm and worm gear.

2. A 655,360-step per revolution Compumotor Dynaserv directly coupled to the two-inch drive roller.

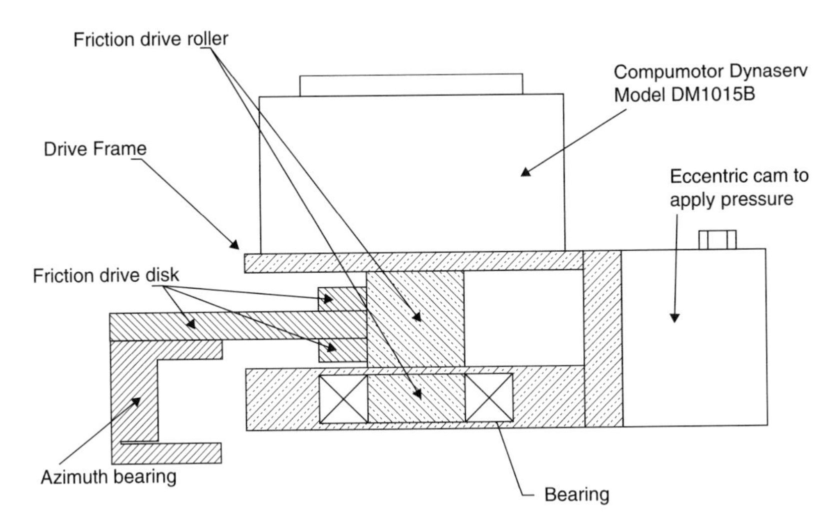

Fig. 13-10 *Azimuth bearing and its drive.*

3. A several thousand step (2,000 to 12,000) per revolution microstepped motor (available from a variety of manufacturers), with a planetary gearbox coupling the motor to the drive roller.

Option 1, the stepper driven at full (or preferably half) steps through a worm, has a major disadvantage. Such motors have large drops in torque starting around 1000 steps per second when driven with conventional stepper drivers, and require a very sophisticated driver to operate above 5000 steps per second. The 359-tooth worm gear is used to make each motor half step represent $0\rlap{.}''1$ rotation of the azimuth axis. This is done to obtain smooth tracking. At 5000 half steps per second, the maximum speed is $500''$ per second, or about 0.14 degree per second. This would be a slow slew speed, so separate slew motors would be needed, complicating the drive considerably. Even if full steps are used (which would make the motion jerky) the slew rate would only be 0.28 degree per second.

Option 2, the Compumotor directly driving the shaft, has the wide range of speeds needed to drive an altazimuth mounted telescope. The azimuth shaft-to-bearing reduction of 25 gives $655,360 \times 25 = 16,384,000$ steps per revolution, or $0\rlap{.}''079$ per step, while the altitude reduction of 15 gives 9,830,400 steps per revolution, or $0\rlap{.}''13$ per step.

To insure enough torque to move a 30-inch telescope, the motor would have to have a torque rating of at least 1000 oz-inches. The smallest Dynaserv DM series motor has a torque rating of 11 ft-lbs (2112 oz-inches), more than enough. The motor is capable of step rates up to 1,600,000 per second with a reduction in torque of about half. This step rate would permit a theoretical maximum slew rate of 35° per second with ramping—truly impressive performance. With its built-in optical encoder, the motor has a repeatability of 2″ and an accuracy of 25″, both of which are divided by the friction drive stage reduction ratio of 25, for an accuracy of 1″. The major drawback to the motor is the cost, nearly $5,000 per axis for the motor and its microstepping controller.

Option 3 is the microstepped motor and drive that can be programmed to have a variable number of steps per revolution, typically in the range of 200–25,000. This motor would be used with an accurate gear reduction to reduce the motor shaft angle error of 12′ to a few arc seconds at the driven axis. The first design step is to choose the effect of one motor step on the azimuth axis of the telescope. Step sizes used successfully by observatories in high accuracy drives vary from 0″.125 to 0″.05. To try to keep the step rates higher than most of the anticipated mechanical resonances in the drive and mount, each step will be 0″.05.

The next step is to choose the reduction gear ratio between the motor and the shaft. With 1,296,000″ per circle, if each step is 0″.05, there are 25,920,000 steps per circle. With a reduction of 25:1 in the final friction drive, this means the motor/gear combination produces 1,036,800 steps per revolution of the two-inch drive shaft. Using the Compumotor Plus motor/microstepper drive as an example, the maximum of 25,600 steps per revolution makes further gear reduction necessary. In this case, a gear ratio of 1,036,800 / 25,600, or 40.5:1 produces the proper step size. One might expect to find a standard model with a 40:1 reduction, which is close enough. This ratio divides down the motor shaft error of 12′ by a factor of 40 * 25 (= 1000) to 0″.72, an acceptable figure. By changing the number of steps per revolution (easily done over a serial interface), the motor can turn up to 50 revolutions per second, but not with 0″.05 steps. Using an Oregon Microsystems controller with a maximum step rate of 1,044,000 steps per second, this combination gives 14.5° per second, a very respectable slew rate.

The problem with Option 3 is that the gearbox introduces errors into the drivetrain, as well as complexity. A typical "high accuracy" gearbox would have an error at its output shaft of 4′, which, when divided by the altitude friction stage reduction of 15:1 would be over 16″. It would be difficult to model the errors of the planetary gearbox and compute them out.

Considering that Option 3 costs about $2,500 for the motor and controller, plus another $1,000 for a high accuracy planetary gearbox capable of transmitting the motor's torque, and the large residual errors that would have to be measured and modeled, Option 2 was selected for the design exercise telescope.

13.10.3 Motor Controller Computer Interface Selection

The Compumotor Dynaserv controllers have two input lines. One carries a series of pulses that determines the motor's speed (the motor steps once for each pulse received). The other controls the motor's direction. This is what stepper motor indexer cards provide on each axis of their outputs. The purpose of the motor indexer is to provide the pulse rates needed to drive the motor over the range of rates needed for accurate tracking.

The long term tracking accuracy requirement is $5''$ per hour, which represents 1 part in 10,800. The short term tracking accuracy requirement is a $2''$ circle of confusion, which means the pointing angle in equatorial coordinates should not change more than $2''$ between two successive motor speed commands. For the azimuth axis, there are 14 motor speed commands generated each second, and for the altitude axis, two each second.

Any motor interface card can offer only a finite number of motor speeds. If that number is too small, the difference between two "adjacent" speeds may be so coarse that in the half second between commands to the altitude motor, the speed selected in the last command cycle produces an error of more than $2''$ before the next command is sent. To prevent this from happening, motor speed resolution (the number of different motor speeds available) should be greater than or equal to the long term tracking resolution requirement (10,800). If each step in altitude is $0''.13$, then the difference between any two motor speeds in the entire speed range cannot be more than $2 \times \frac{2''}{0''.13} = 30.7$ steps per second. Most indexers can meet this requirement at low speeds, but few can meet it at high speeds.

From the commercial boards for the PC ISA bus, the options considered for the pulse rate generator were as follows:

1. The Compumotor AT6400 4-axis indexer
2. The Anaheim Automation CLCI2004 4-axis indexer
3. The Oregon Micro Systems PC58 8-axis indexer
4. A custom indexer board.

The Compumotor board sells for \$2,320, making it the highest cost per axis of control. A single Compumotor board can control azimuth, altitude, and focus motors with a spare for a future upgrade for a field derotation motor, but one or more additional boards would be needed to control filter and aperture motors on the photometer. The AT6400 has a maximum speed of 1,600,000 steps per second, and comes with software for

MS-DOS, Windows, and Windows NT (and therefore also Windows 95). Using a combination of a scale factor (in the range 1–999,999) and a velocity command (in the range 0.00000–1600000 with a total of 7 digits, where the decimal point counts as a digit) one can obtain the speed resolution needed for telescope control over the entire speed range of the indexer. The AT6400 has quadrature inputs for four optical encoders that feed up/down counters that can be used for things other than encoders. There are also 24 input bits, 24 output bits, 4 analog inputs, and 4 interrupt-driving inputs.

The Anaheim Automation board has a maximum step rate of 2,500,000 steps per second, the highest of the group. The 4-axis board sells for $1,000 making it the second most expensive per axis (but less than half the cost of the Compumotor board). To control both the telescope and the photometer would require the purchase of a second board. The board comes with QuickBASIC, C++ subroutines and function libraries, and a Windows DLL, but there is no support for Windows 95 (this is likely to change as Windows 95 gains acceptance). The CLCI2000 series boards have inputs for encoders including a mark pulse input for resetting the on-board counter, and inputs for soft, home, and hard limit switches in both directions for each axis. There are also 10 input bits and 10 output bits.

The speed in steps per second from the board is computed using the following equation:

$$\text{speed (steps/S)} = \frac{300 \times \text{velocity}}{\text{scale factor}}$$

where the velocity is an integer in the range 1–16,383 and the scale factor is an integer in the range 2–16,383. The division is done to full precision, and the speed is then obtained by dividing down an oscillator running at 9.8304 MHz.

The Oregon Micro Systems PC58 has a step rate of 1,044,000 steps per second. This 8-axis board has 22 digital input/output lines and sells for $1,695, giving it the lowest cost per axis of the commercial boards considered. It can be obtained configured for 4 motor ports and 4 encoder ports instead of 8 motor ports, and other configurations are also available. Software supplied with the board includes sample programs, but there is no Windows 95 device driver (again, this is likely to change). The board is controlled by transferring ASCII commands, including a "jog" command (JG) with an integer number of steps per second up to the maximum, and a "jog fractional velocities" command (JF) that permits a total of 10 digits plus a decimal point. To meet the speed resolution requirement of 1 part in 10,800, for speeds up to 10,800 steps per second one would use the JF command, and for higher speeds, the JG command. Thus the Oregon Micro Systems board is capable of meeting all WMO performance requirements, but the lack of a Windows 95 VxD driver rules it out for now.

The final option is to build a custom board (with all the problems mentioned in previous chapters with designing and building custom boards). Let's go through a hypothetical custom board design case. Our board generates pulses by dividing the output of a crystal oscillator by a 32-bit number loaded into a register on your computer bus. To avoid using exotic circuitry to deal with high oscillator frequencies, the oscillator runs at 20 MHz. Using a 32-bit divide-by-n counter, pulse rates from 20 MHz down to 0.005 pulses per second can be generated. Provisions can be made for controlling when the pulses are sent to the motor controller, for sending individual pulses under direct software control, and for a direction bit to tell the controller to turn the motor clockwise or counter-clockwise.

Assuming step sizes of $0''.05$, the pulse rates range from one million arc seconds per second (278° per second) down to $0''.00025$ per second. Most microstepped motors have a practical limit of about 500,000 steps per second, which translates to 25,000″ per second (about 7° per second). A maximum slew of 180° will take about 26 seconds, plus ramp up and ramp down time. This is very good slewing performance.

At a drive rate of 15″ per second, the pulse rate is 300 steps per second (comfortably above any expected telescope natural frequencies), requiring a divisor of 66667, which gives $14''.999925$ per second. The next divisor, 66666, gives $15''.00015$ per second. The difference is $0''.000225$ per second, or one part in 66,666, which more than meets the one part in 10,800 criterion. At slew speeds, a divisor of 40 gives 500,000 steps per second. This gives speed change increments of roughly 10,000 steps per second for each successive integer divisor, which should be adequate resolution to ramp the motor by incrementing or decrementing the 32-bit divisor by one several times over a period of a few seconds. Note that this type of circuit always has a speed resolution of n, where n is the divisor used to divide the crystal oscillator output down to the desired speed.

This simple circuit solves the speed resolution problem at the low end of the speed range, but for tracking near the zenith, the speed resolution degrades as the speed increases until tracking becomes erratic near the zenith. At the Dynaserv motor's maximum step rate of 1,600,000 steps/S, the speed resolution is only one part in 12.5. This can be increased to 31.25 using a 50 MHz oscillator, but much above this frequency you would have to use very high speed integrated circuits to do the divide-by-n, which would significantly add to the cost and heat dissipation. At only 500,000 steps per second, the speed changes by 10,000 steps per second between adjacent divisors, far exceeding (in the wrong direction) the short term step rate resolution of 30.7 steps per second. A custom board must be far more complex than this simple model to compete with the commercial units in meeting the requirements.

Even though it is more expensive than the other boards, the design uses the Compumotor AT6400 because it provides the needed step rates, speed resolution, and Windows 95 device driver. We clearly wanted to avoid the problems of designing our own board, and believe the extra expense is justified in getting the system running sooner.

13.10.4 Position Encoder Selection

Given the choice of motor, it may be possible to achieve the required pointing and tracking accuracies without an encoder. The Dynaserv motor has an optical encoder built into it, and the motor's specified error will be divided by the friction drive reduction ratios of 15 for altitude and 25 for azimuth. Given the expense of these motors and the added expense of encoders, it might be worthwhile to build the telescope without encoders initially and see if it works well enough open-loop. One should not be too surprised if it doesn't work, because wind loading or bits of dirt on the drive disks or rollers could cause the roller to slip, effectively disconnecting the motor's encoder from the driven axis. To proceed with the discussion, we will assume you discover that encoders are needed.

Of all the encoder types discussed in Chapter 9, the ones that have the resolution and accuracy (about 5″) needed for this application are the optical encoder, the Canon laser rotary encoder, the Sony Magnescale rotary encoder, and the Inductosyn. Assuming direct connection to the axes, the Inductosyn, Sony, and Canon encoders are the least expensive, costing about $2,500–$3,000 each for 1″ resolution and accuracy. Note that they are all incremental, not absolute encoders. Many designers favor absolute encoders over incremental. This is because every time the power is turned off, incremental encoders catch "instant amnesia." When the electronics are powered up every night, to use incremental encoders you must have a Sony Magneswitch or similar device to mark a home position, or sight at least one star to recover the zero offset calibration constant for the incremental encoders.

Another way to avoid this problem is to buy an incremental encoder with a zero track that gives a pulse at a predefined zero point of the encoder disk. If the telescope is stored in a "home" position at the end of an observing session, then when the telescope is moved from the "home" position at the start of the next observing session, the zero pulse can be used by the computer to zero out the raw position count.

If an error occurs in an incremental encoder, it cannot be detected or corrected. For example, if a positive noise pulse occurs on the line from the encoder, the raw position count is too high. If a genuine signal pulse is mixed on the line with a negative noise pulse, the counter may not respond to the good pulse. Either way, a random error is introduced into the estimated position, and no way exists to know the error occurred, much less to correct it. A succession of errors of the same type tend to accumulate until the error is large enough to affect the pointing accuracy of the telescope.

An absolute encoder, on the other hand, never allows errors to accumulate. If a bad reading is received by the computer for some reason, the next time the absolute encoder is read, the error is corrected. However, absolute encoders have the following disadvantages:

1. They are considerably more expensive (up to 10–20 times) when purchased new than incremental encoders.
2. They require a more elaborate (and typically more expensive) computer interface.
3. There are many more signal lines to be run from the encoder on the telescope to the interface card in the computer.
4. Although surplus absolute encoders are relatively easy to find, they often have one or more of their bits burned out, or some other electrical malfunction, which must be found and repaired.

The BEI H-25 incremental encoder has up to 50,800 counts per revolution (almost 16 bits) with interpolation, and costs about $450. Although this is more expensive than the typical surplus encoder, the price is low enough to justify purchasing new encoders that will almost certainly work the first time, that have only a few signal lines to the computer interface card, and for which PC interface cards are available. These encoders can be ordered with the zero track pulse, but for our design example telescope, Sony Magneswitches will be used instead. The H-25 can be ordered with open collector outputs and pullup resistors to aid noise immunity. This option, when used with heavy guage (16–18 awg) wires, should afford enough noise immunity to bring noise-induced errors down to an acceptable level with proper grounding and shielding.

Although the BEI encoder does not have enough resolution to be connected directly to the telescope axes, one unit can be "geared" to each axis through a friction drive using a 1-inch diameter roller. That would make each count equal to $0''.85$ in altitude, and $0''.51$ in azimuth. Periodic errors would be introduced by variances in machining the disks or rollers, but these can be mapped and computed out in the encoder calibration curve. Since the disk is driving the encoder (instead of the motor driving the disk), it is less likely to slip as a result of dirt or wind loading because the encoder offers no resistance to the backdriving force. However, if you need an encoder because the motor driver roller slips, you should keep the disk very clean to help prevent the encoder roller from slipping.

13.10.5 Position Encoder Computer Interface Selection

Of all the output options offered for the BEI H-25 encoder, there are two that would be useful to our example telescope. The first is interpolation logic, using the standard quadrature outputs on two lines. This produces 12,700 pulses per revolution on each quadrature output line, with the two outputs offset in phase by 90°. The second option, pulse steering logic, takes the output of the interpolation logic and puts out a string of 50,800 pulses per revolution on two lines. One line has pulses on it if the encoder shaft is rotated CW, and the other line has pulses on it if the encoder is rotated CCW. This is a $60 option on each encoder, but permits the use of a standard counter interface (as long as the counter can count both up and down). In our design, the Compumotor board accepts quadrature inputs and performs the 4x pulse multiplication, so only the interpolation option is needed.

13.11 Computer System Hardware

Figures 13-11 through 13-16 show the major hardware subsystems down to the board or major component level. Figure 13-11 shows the major subsystems of our example computer system. The PC chassis contains a power supply, the motherboard with the CPU, memory, cache, and controllers for the Winchester and floppy disks, and serial and parallel ports, and a few ISA bus (PC/AT) slots. Each subsystem in Figure 13-11 consists of one or more interface cards in the ISA card cage located in the van connected by one or more cables to hardware located at the telescope or in the van. Each subsystem is described below in detail.

13.11.1 Control Computer

The control computer components are shown in Figure 13-12. The motherboard in this design is a 90 MHz Intel Pentium board. The board supports 128 MB of main memory, but Windows 95 runs well in only 16 MB. The system can have a disk as small as 300 MB, but you should plan to upgrade it to 1 GB fairly soon. The Exabyte drive stores 2.3 GB on a single 8mm data cartridge.

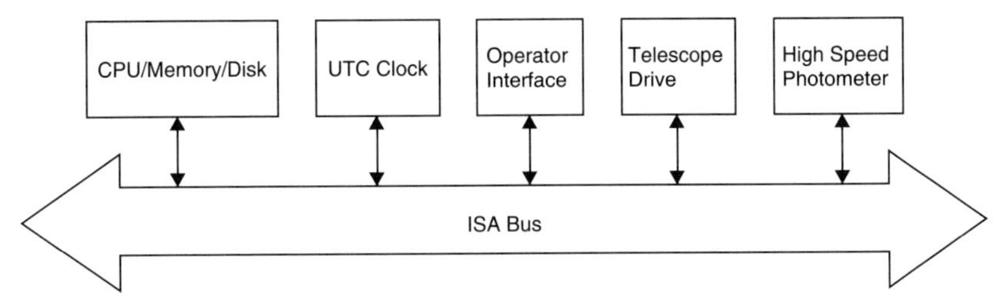

Fig. 13-11 *Control computer functional block diagram.*

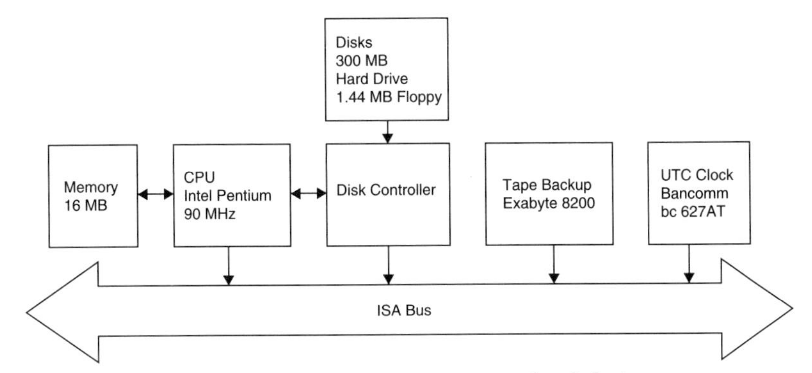

Fig. 13-12 *Control computer top-level design.*

Usually, typical commercial grade PCs do not have enough Industry Standard Architecture (ISA) slots for all the interface cards in a telescope control system, so you buy a separate ISA chassis with a power supply, several ISA slots, and a cable with a card that plugs into your computer. One of the items plugged into the ISA chassis is the Bancomm bc627AT Global Positioning System (GPS) receiver to give UTC time with an error less than 2μS. This is used to time-tag data and to compute local mean sidereal time, which then is used to compute local hour angle from RA. The correction from local mean sidereal time to local apparent sidereal time is found by sighting two stars at the observing site, to distinguish between dUT1 and the encoder zero offset or by download of UT1 predictions from the United States Naval Observatory WEB page. Although both position and time information can be obtained by sighting stars, it is easier and faster to know the time and position independently, so that there are fewer unknowns to be found, which means fewer stars need be sighted before commencing operations. To obtain hour angle to an accuracy of $0\farcs5$, both longitude and Coordinated Universal Time (UTC) must be known to within $0\farcs25$. Since there are 15 arc seconds per time second, this means that

UTC must be accurate to within $0\overset{s}{.}017$. The Bancomm unit exceeds this requirement by several orders of magnitude. The required level of time accuracy is not obtainable using short-wave WWV broadcasts, because the path of the signal from the transmitter to the observing site cannot be determined and constantly changes.

GPS signals can be received almost anywhere in the world, which is a real advantage for a portable telescope. The 1 MHz output frequency provided by the Bancomm board is divided down to 1 kHz and fed to the custom photometer interface board to synchronize integration periods within a few microseconds.

Obtaining accurate position information is more difficult than obtaining accurate time. A minute of longitude is roughly equivalent to a statute mile at U.S. latitudes, so a requirement of $0\overset{\prime}{.}25$ maximum error is roughly equivalent to 25 feet. An accuracy of 3 feet can be achieved by observing both channels of several Transit satellites over a period of several days, and reducing the data two weeks later when the exact satellite ephemerides are made known for the periods of observation. An accuracy of a few centimeters can be obtained using special surveying differential GPS receivers that, again, require post-observation data reduction, and that cost tens of thousands of dollars. Neither approach is much help to a portable telescope that needs an accurate position fix within minutes of arrival at the observing site. By observing one channel of a Transit satellite over the roughly five-minute period the satellite is in view, an accuracy of about 200 meters ($7''$ of latitude) can be obtained. However, one may have to wait up to 45 minutes for a Transit satellite to come into view.

GPS satellites are in much higher orbits than the Transit system, providing three or more satellites in view all the time from most places on the Earth, so a position fix accurate to 50 feet (on the military encrypted channel) or 300 feet (on the public channel) can be obtained in minutes. This is not as accurate as we would like, but the error introduced is about $3''$, which will not affect many observations, and which can be computed out if enough calibration stars are observed.

Owing to the errors in GPS receivers, USGS maps will be used to obtain accurate positions after an observation is made. Roughly 95% of all features shown on these maps are accurate to within 40 feet of the North American datum, though experience has shown that buildings are depicted less accurately than road intersections and rivers (Dunham, 1984). It would be prohibitively expensive, in terms of both cost of the maps and space in the van to store them, to keep a complete set of USGS maps in the van to determine the van's position accurately for pointing the portable telescope. Electronic storage of USGS maps on CDROM would be ideal, and such products have appeared on the market but still are somewhat expensive for

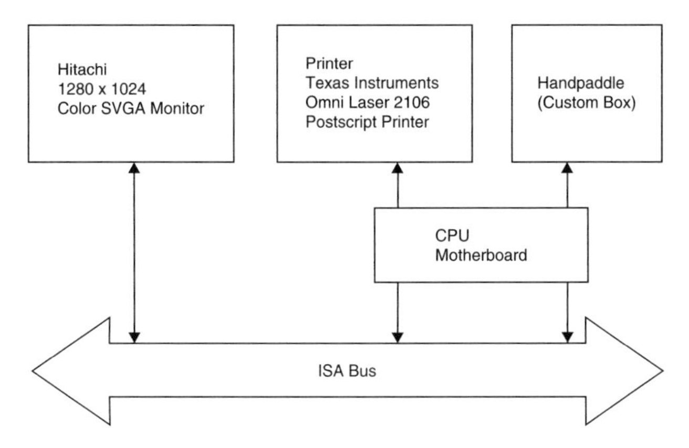

Fig. 13-13 *Operator interface block diagram.*

a complete set of 7'5 maps. For more information on using GPS for remote observing, consult Trueblood (1993) and Trueblood (1995).

13.11.2 Operator Interface

The various devices that form the operator interface are shown in Figure 13-13. The CRT is a 19-inch rackmount Hitachi color monitor with 1280 by 1024 resolution with 0.28 mm pixels. An SVGA video board with that resolution plugs into a PCI port on the motherboard to generate all the displays using a graphics accelerator. The keyboard and a trackball that emulates a mouse plug into their usual serial ports on the back of the system box.

There is no printer in the van. For software development and printing results at the home base, there is a Texas Instruments OmniLaser 2106 PostScript laser printer that connects to the parallel printer port.

The standard COM2 serial port is connected to the hand paddle used at the telescope for finding and centering bright stars for calibrating the encoders. The hand paddle consists of a box with pushbuttons for direction (N, S, E, and W) and speed (slew, set, and guide). The box contains circuitry that generates ASCII characters when individual buttons are pressed, and converts these characters from parallel to serial format for transmission to a serial port.

13.11.3 Telescope Drive

Figure 13-14 shows the telescope drive components. The Compumotor AT6400 indexer card plugs into the ISA bus and connects directly to all

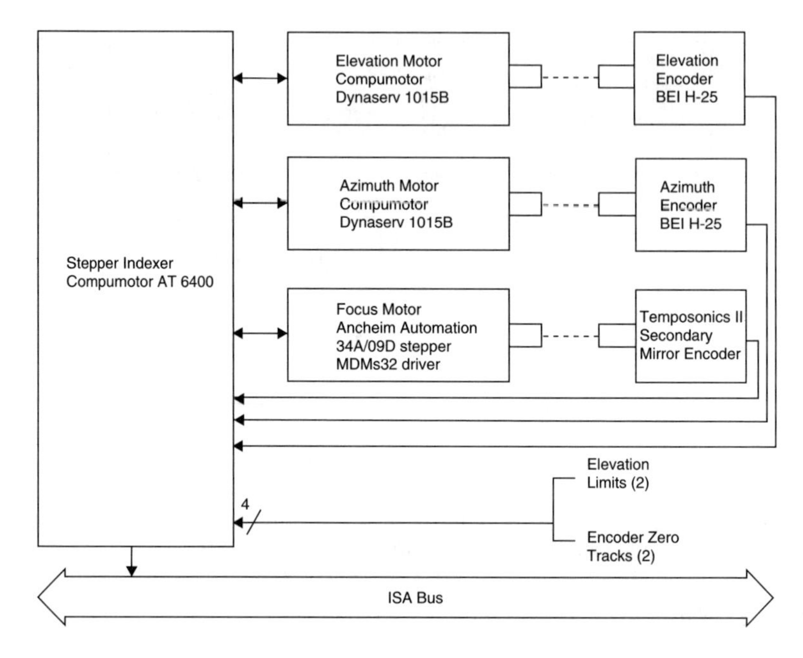

Fig. 13-14 *Telescope drive block diagram.*

motor controllers using 18 awg shielded twisted pair wire to control interference between lines to different motors running near each other. This grade wire is also used to connect the encoders and limit switches to this board.

The BEI H-25 optical incremental shaft encoder is used on the azimuth and elevation axes. The encoders use the optional interpolation and pullup resistors. The focus position is encoded using a Temposonics II linear position transducer. Limit switches prevent overtravel, as well as limiting travel in the altitude axis to the 90° between the horizon and the zenith. A Sony Magneswitch defines the zero point or home position of focus on the secondary mirror.

13.11.4 High Speed Photometer

Figure 13-15 shows the high speed photometer (HSP) block diagram. The photometer head was designed by Bob Millis, with additional engineering and assembly by Ralph Nye, both of Lowell Observatory. Attached to the photometer head is a stainless steel tube containing the photomultiplier tube (PMT), resistor voltage divider, and Fabry lens. Attached to the outside of the PMT housing are a high voltage power supply, preamplifier,

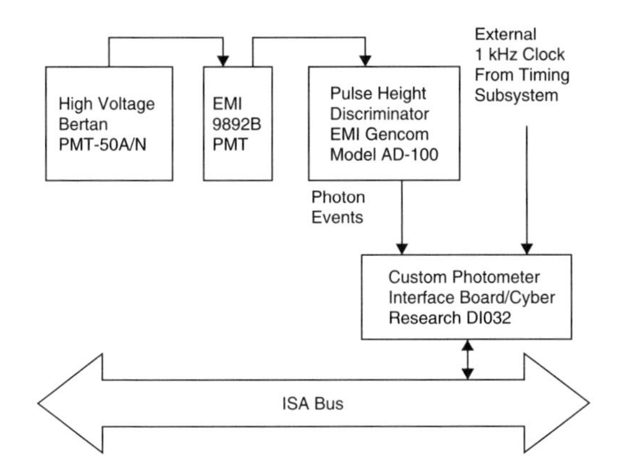

Fig. 13-15 *High speed photometer block diagram.*

and pulse height discriminator electronics. The HSP head contains manually operated filter and aperture wheels and a pellicle for post-aperture viewing during the event to ensure the telescope is tracking properly. The PMT is an EMI 9892B bi-alkali end-on tube with very low dark current at room temperatures. This tube is suitable for photon counting without cooling. The Bertan high voltage power supply provides the required voltage regulation.

The instrument requirements for data handling and time synchronization for observing minor planet occultations are much more stringent than for UBV photometry of variable stars. One system designed to fulfill modern research requirements in this field is that of Schnurr and A'Hearn (1983). Based on an Apple computer and custom hardware, this system has 0.001 second photometer integration periods, and 0.001 second time resolution and synchronization to WWV signals. With 1,000 integration periods (data points) per second, if each data point were to generate an interrupt, as it would under ordinary programmed data transfers, the computer could quickly become overburdened simply handling interrupts.

The solution is to use a "smart" peripheral card that controls event counter resetting, time synchronization, and data transfers without requiring software execution for every data point. The functional diagram of this card is shown in Figure 13-16, and is the only custom board in the system. It was designed to work with a DEC DR11W Direct Memory Access (DMA) card for the old MicroVAX II control system developed at the WMO. Fortunately, CyberResearch offers the DIO 32F dual DMA board with two 16-bit ports to perform the same function in the PC that the

DR11W performed in the MicroVAX II. The external inputs to the card are the photon event pulses from the photometer's pulse height discriminator, and a 1 MHz pulse source from the Bancomm board that is divided down to 1 kHz on the custom integration board.

The card is initialized by custom device driver software on request from the telescope control program. The device driver loads the integration period in milliseconds (up to 65,535), the number of data points to transfer, and the address in memory to which to transfer them. When the operator signals the control program to start taking data, it commands the device driver to start the data collection process, which, in turn, sets a bit in the control and status register to initiate data transfers.

Each time a pulse from the 1 kHz clock is received, the integration timing counter is incremented. When this counter equals the integration period (in milliseconds) that was sent to the card during initialization, it directs the counter switch to route all further photon event pulses to the second photon event counter. The contents of the stopped photon event counter are transferred through the bus switch and data register on the ISA bus to the address in memory previously specified. This is done under the control of the DIO 32F board. If the integration period is specified as 1 ms, then this happens on every pulse from the 1 kHz clock. The memory buffer address is then automatically incremented by 2 bytes by the DMA logic, the ISA bus register holding the number of data points to transfer is decremented by 1, and the second event counter counts photon events until the next integration period has ended. When this occurs, the first counter is reset and all further photon events are routed to it, and the contents of the second counter are transferred to the (previously incremented) memory address.

This process continues until the ISA bus register holding the number of data points to transfer is zero. At this point, the card generates an interrupt. Up to this point, the entire process has been handled by hardware, without one instruction in software being executed for this activity. This has allowed the CPU to perform other functions in between the DMA data transfers. If the number of two-byte data points between interrupts is set to 250, and the sampling rate is once per millisecond, the computer has to deal with an interrupt only four times per second. When this interrupt is received, the interrupt handling software reads the GPS clock, writes the clock and photometer data to disk, and reinitializes the buffer address and transfer count on the photometer card.

Two memory buffers are used, so that new data can be transferred to memory by the photometer DMA card while the previous set of data is being written to the Winchester disk. The DMA card is set up to release the bus after every data transfer, so it does not monopolize the bus between trans-

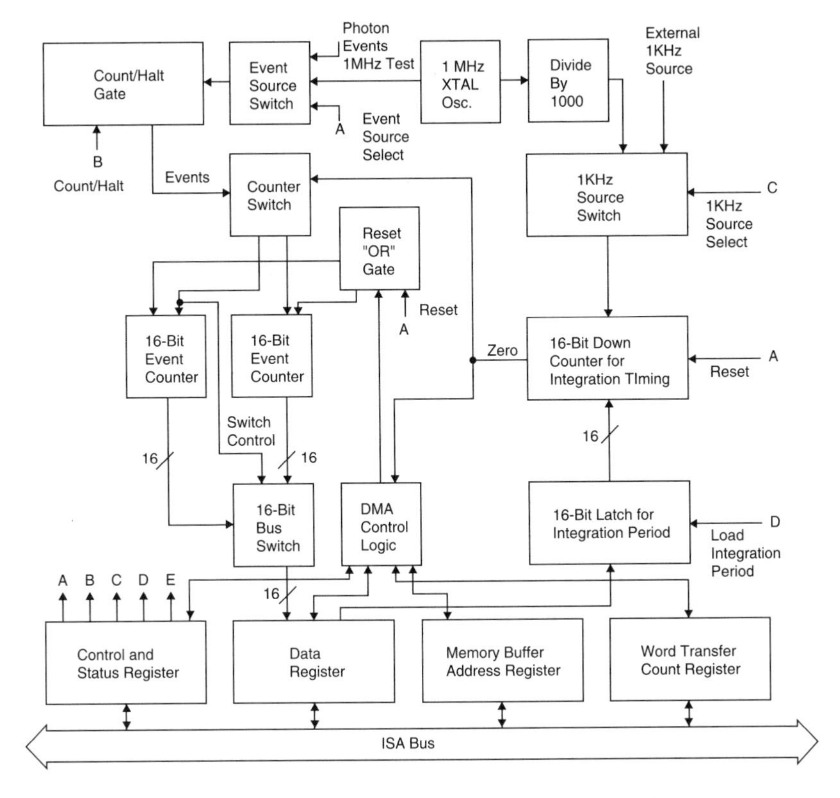

Fig. 13-16 *Custom photometer integration card block diagram.*

fers. The ISA bus is organized so that bus cycles are stolen from the CPU without its being aware of the transfer, so no special software is needed to perform the data transfers, only to handle the interrupt when the entire data transfer of 250 16-bit words is complete.

This is yet another example of the benefits to be obtained from using a smart peripheral card and a standardized bus. A smart card allows a computer to handle very high data rates, and the use of the ISA bus means that a card containing all of the ISA bus DMA logic is available off the shelf. Only the relatively simple logic shown in the upper half of Figure 13-16 needs to be designed, built, and tested by the user to implement all these rather complex functions. The hard part, the manipulation of the bus control lines to steal bus cycles from the CPU, has already been designed, built, and tested by CyberResearch. For more information on the WMO HSP interface, consult Trueblood (1987).

13.12 Software Design

13.12.1 Operational Considerations

As discussed earlier in Chapter 11, the telescope control program consists of a GUI organized as a series of pages, with menus, boxes, and buttons to issue commands and display status. The "home" or main page is designed to be used most of the time, providing the status information and commanding capabilities needed for most operations. This page includes a system status area for major telescope and weather sensor data, an object area describing the target object, the catalog it came from, and its position, and a third area for date, time, sidereal time, and related information. Additional pages display motor step rates, limit switch positions, raw encoder readings, error constants, values for each apparent place correction, a list of stars to sight for encoder calibration, and other special items of interest.

To help alert the operator to error conditions, an area that looks the same on every page and is located in the same place on every page is used for displaying status alarms.

This approach provides a system that is easy to use by those whose orientation is the use of telescopes, not the programming of computers. In addition, using different pages for different types of data allows the system to grow and evolve without incurring the high costs of replacing and enhancing dedicated hardware. Some older control systems in large observatories look like 1960's-style NASA control rooms, full of dials, meters, 7-segment displays, and special pushbuttons in a hardwired control panel. Such systems are not easily modified to meet changing needs. In contrast, the only hardwired pushbuttons in this system are those on the hand paddle and a large "kill" button that physically disconnects power from the telescope. The rest are "soft" buttons on the computer screen that you "push" with a mouse and cursor, not your finger, that you can reprogram in minutes.

The servo loop portion of the control program is a list of calculations performed in sequence, culminating in a command sent to a motor. The calculations are repeated periodically, so that they are synchronized with each other. The complete control system, however, is a set of several asynchronous threads of the single control program occurring with no temporal synchronization with each other. Often, these independent threads require some sort of communication among themselves. It is the job of the operating system to provide this communication. The functions occurring in a telescope control system are as follows:

1. Reception and interpretation of operator inputs, and generation and display of responses

2. Reading the UTC clock and computing sidereal time
3. Fourteen times per second, reading the azimuth encoder, computing the actual azimuth position, computing the desired azimuth position, computing the motor command, and sending the command to the azimuth motor
4. Twice per second, reading the altitude encoder, computing the actual altitude position, computing the desired altitude position, computing the motor command, and sending the command to the altitude motor
5. Updating the page on the monitor
6. Optionally, running an extended Kalman filter to update the error constant estimates
7. Reading the condition of limit switches, detecting alarm conditions, and displaying alarm messages in the alarms area of the current page
8. Receiving guiding inputs from the hand paddle, mouse, or the keyboard, correcting the actual position calculation, and moving the telescope in response to the operator's commands
9. Receiving focussing commands from the hand paddle or the video terminal keyboard, and sending commands to the secondary mirror motor
10. Commanding and receiving status from the photometer
11. Displaying data from the weather sensors.

When designing the software, one should remember that Items 2–4 should be given highest priority; if execution of this software is delayed while awaiting completion of a lower priority item, tracking accuracy will suffer.

The items listed above occur while tracking a celestial object. Additional functions performed when one is not tracking are slewing, calibrating the encoders, performing a trend analysis of the error constants, analyzing instrument data, searching a star catalog data base for objects to observe, and preparing the manuscript of a journal article. The software should be designed in a manner to accomodate all of these functions separately, yet allow one to switch from one to another quickly and easily.

13.12.2 Top-Level Software Design

Selection of the Windows 95 operating system determines the overall software design. The control software is a single process consisting of a set of threads, or independently scheduled and executed "programs" that collectively form a single program. Each of the software functions listed above is a separate thread. Communication between the threads occurs through a common data area in which each thread knows the addresses of the data in this area. The interthread communication mechanisms provided by Windows 95 are also used to synchronize threads.

When tracking or slewing is turned on, the track/slew thread is activated. It executes the control algorithms, issues commands to the motors, then reschedules itself before exiting, asking the operating system to execute the track/slew thread again in $^1/_{14}$ second. This thread has the highest priority of all the threads in the process when it is executing. The other threads that execute periodically operate pretty much the same way: the display thread starts them when the operator turns them on, they do their processing, then they reschedule themselves and exit. The rescheduling involves telling the operating system how much time must elapse before the thread runs again.

Threads that talk to devices do not display data or error messages. Instead, they pass status information or data from sensors to the thread that generates the page with that data on it. This passing of data occurs asynchronously by coding the data sending thread to put the data in predetermined places in a global common area. The display thread reads data from this common area and displays it on the screen. The display thread is always running at a lower priority than the thread that feeds it the information to display. This guarantees there will be information to display, even in a heavily loaded system.

When a timer times out, or input is received from the keyboard or hand paddle, the operating system activates the appropriate thread as a completion routine, so named because it executes when input or output completes. One completion routine is dedicated to receiving data from the thread that reads environmental sensors and placing it in common storage for the video update and alarm threads to use.

The soft limit switches are simple single-pole units, but the hard limit switches are double-pole switches, so that step pulses are prevented by hard-wired logic from reaching the motor controller, and the hard limit condition is reported to the computer. The logic circuits allow commands to reach the motor that tell it to back off the limit by moving in the opposite direction.

The design approach described above solves the problem of assigning priorities to different asynchronous processes. A more detailed software design is not presented here, because it requires our readers to have an extensive knowledge of the Windows 95 operating system.

Each manufacturer of commercial ISA boards usually includes not only device drivers, but sample programs. These sample programs will be used to make diagnostic programs for each interface card, to diagnose problems in the hardware. They are also useful in learning how to communicate with the commercial cards.

13.13 Assessment of System Performance Requirements

The major system performance requirements presented earlier are reviewed below to assess the ability of the system just described to meet these requirements.

13.13.1 Portability

The telescope is mounted on a trailer for portability. Preparing the telescope for transport consists of hitching the trailer to the tow bar of the van, securing a cover over the telescope, and driving away. The mount and primary and secondary mirror supports are not torn down for transport, and the optics remain in place on the trailer at all times, even during transport. By using an alt-az mount, the center of gravity and total mass of the trailer are both kept low enough to make a 30-inch aperture telescope towable by a car or small van. A generator located inside the van provides power in remote locations.

13.13.2 Setup Time

This requirement is probably the most difficult to meet. We continue to work on reducing setup and teardown time, to make it possible to make last-minute location changes to avoid clouds.

13.13.3 Optics

The first phase of the telescope system used a 12-inch aperture Cassegrain telescope to test the basic system concept. Once the system is working well, a set of 30-inch folded Cassegrain (Nasmyth mount) optics can be installed. Good optics of approximately this size will meet the optical requirement.

13.13.4 Telescope Pointing Accuracy

The main pointing error source is atmospheric refraction, which can be computed to around $4'' - 6''$ accuracy at best, depending on the temperature and pressure sensors used. The total error should be less than $\pm 15''$ if enough stars are sighted during the encoder calibration process, and possibly less than $\pm 10''$ over most of the sky. The atmospheric pressure measurement becomes a factor only at very low altitudes. The following table lists the error sources and their expected absolute accuracies:

Table 13-7

Apparent Place Corrections	**Error**
1. Annual aberration	$0''.1$
2. Stellar parallax	$0''.1$
3. Precession	$0''.1$
4. Nutation	$0''.1$
5. Orbital motion	$1''$
6. Proper motion	$1''$
7. Diurnal aberration	$0''.1$
8. Planetary parallax	$0''.5$
9. Refraction	$2''$ typical
	$6''$ worst case

Mechanical Corrections	**Error**
1. Zero offset	$3''$
2. Azimuth tilt	$3''$
3. Drive rate	$0''.5$
4. Non-perpendicular axes	$2''$
5. Collimation	$3''$
6. Tube flexure	$3''$
7. Servo lag	$2''$
8. Hour angle	$0''.5$
(based on time and location accuracy; the result is used in other calculations)	

The RSS error value obtained when all these errors are combined is $6''.7$ when the refraction error is held to $1''$ or better, and $6''.99$ when the refraction error is $6''$. It's reasonable to assume that mechanical errors that do not vary with location, such as tube flexure, could be well understood after a few months' observing, while errors that change night to night with changes in the trailer location, such as azimuth tilt, will be larger. The accuracy of computing the topocentric place corrections is predictable, but the accuracy of measuring and estimating the encoder calibration constants cannot be known until you build the telescope and use it. Many of these errors can be as large as $10''$ and still meet the overall pointing accuracy requirement, assuming the errors combine randomly (in quadrature) with other errors.

13.13.5 Telescope Pointing Time

The slew rates with the proposed drive system are about 35° per second in azimuth and 58° per second in altitude, exceeding the slew rate requirement. At that rate, it will require only about 5 seconds to slew 180° at the maximum slew rate in azimuth, and about 3 seconds in altitude. Ramping should add about 20% when ramping from a nominal tracking rate (300 steps per second) to slew rate (1,600,000 steps per second) and back down. Note that as higher slew rates are used, the added ramp time grows while the total slew time falls, so the percentage of total time spent ramping increases. The slew rates calculated here are dangerous to humans and equipment, so the slew rate will be limited in software to about 10°/S.

13.13.6 Long Term Tracking Accuracy

The error modeling proposed above is expected to be accurate enough to meet the long term tracking accuracy requirement.

13.13.7 Short Term Tracking Accuracy

The motor speed resolution, encoder resolution, servo loop repetition rate, and the expected natural frequency of the telescope all combine to lead to the expectation that the short term tracking accuracy requirement will be met. Table 13-8 lists the error sources and their expected short term accuracies. The RSS error value obtained when all these errors are combined is $2''.46$ when the refraction error is held to $1''$ or better, and $3''.32$ when the refraction error is $6''$. Again, it was assumed that mechanical errors that do not vary with location will be more stable than those that do. To obtain smooth tracking, these error sources need not be known absolutely to these values. Instead, the errors in computing the corrections must not change over the course of a few seconds by these values.

13.13.8 Data Input and Control Device

The hand paddle is built with simple pushbuttons that meet the temperature and humidity requirements, and is housed in a sealed enclosure to keep out dust and dirt.

13.13.9 Computer Environment

The van computer is kept within its environmental operating limits by using heating and air conditioning inside the van.

Table 13-8

Apparent Place Corrections	**Error**
1. Annual aberration	0.″1
2. Stellar parallax	0.″1
3. Precession	0.″1
4. Nutation	0.″1
5. Orbital motion	0.″5
6. Proper motion	0.″5
7. Diurnal aberration	0.″1
8. Planetary parallax	0.″5
9. Refraction	1″ typical
	6″ worst case

Mechanical Corrections	**Error**
1. Zero offset	1″
2. Azimuth tilt	1″
3. Drive rate	0.″5
4. Non-perpendicular axes	0.″5
5. Collimation	1″
6. Tube flexure	0.″5
7. Servo lag	0.″5
8. Hour angle	0.″5
(used in other calculations)	

13.13.10 Commands

Commands are issued to the GUI thread using the mouse and keyboard.

13.13.11 Extraneous Light Control

The optics of the prototype are inside a well baffled tube, while the larger telescope uses a truss to minimize wind loading, with canvas curtains to keep out extraneous light.

All of the requirements listed earlier are met, or are likely to be met, by the proposed system. The next step in this exercise would be to proceed with building the system. As progress is made, the requirements and design would be modified as problems are encountered and experience is gained with the system.

Chapter 14

Professional and Commercial Telescope Control Systems

This chapter briefly highlights a few of the control systems at professional observatories and commercial products for both professional and amateur observatories. We aren't suggesting that an amateur try to use on a 10-inch telescope the same control system used on the Keck 10-meter telescope. But there are features of these professional and commercial systems that illustrate the points we made in earlier chapters, and that would improve your own control system.

14.1 The Keck 10-m Telescope Control System

Everyone building a telescope control system should read Patrick Wallace's paper *Pointing and Tracking Algorithms for the Keck 10-Meter Telescope* (1988). In it, Wallace describes his ideas for the first Keck telescope, which at the time the paper was published was still under construction, so the paper may not represent what is in use on the Keck I telescope today.

In his paper, Wallace begins by describing the requirements for the control system, primarily pointing and tracking accuracies. He set the goals for the Keck control system at $1''$ RMS at night, and $2''$ RMS during the day for elevations above $20°$ (to avoid the problem of computing refraction accurately at large zenith distances). He set the tracking goal at $0\overset{''}{.}05$ in 10 seconds, $0\overset{''}{.}10$ in 10 minutes (day or night), and $0\overset{''}{.}5$ (night) or $1''$ (day) in 1 hour. Considering the large size of the Keck telescopes, these are very stringent performance requirements. Note that the Keck telescopes use alt-az mounts, so the alt-az corrections need to be incorporated into the servo.

Next, Wallace develops a design approach similar to the one discussed in the previous chapter, using software executing at frequent, regular intervals to implement a control servo, in this case at a rate of 20 Hz. The servo works in mount coordinates, so motor commands are computed with corrections for non-perpendicular axes and other effects.

The following calculations are performed in the servo: MEAN $[\alpha, \delta]$ to J2000 FK5 MEAN $[\alpha, \delta]$ by handling pre-1984 E-terms; then correcting for annual aberration, light deflection, precession, and nutation to APPARENT $[\alpha, \delta]$; compensating for Earth rotation to APPARENT $[-h, \delta]$; computing diurnal aberration to find TOPOCENTRIC $[-h, \delta]$; performing alt-az coordinate transformation to TOPOCENTRIC [A,E] (Wallace uses Elevation instead of Zenith distance); correcting for refraction to OBSERVED [A,E]; and compensating for azimuth axis tilt, tube flexure, Az/El non-perpendicularity, and the position of the instrument rotator pointing axis to obtain the MOUNT [A,E] that is then fed to the error determination software.

Starting from the "opposite direction," the encoders are read and the RAW ENCODER [A,E] is corrected for gear and centering errors, encoder index errors, and encoder errors to obtain the CORRECTED ENCODER [A,E] that is then fed to the error determination software. The latter compares the MOUNT [A,E] to the CORRECTED ENCODER [A,E], computes the pointing error, and sends it to the software that computes the motor speed commands.

Wallace gives the matrices for all corrections and goes into some detail as to how to treat collimation errors, instrument rotator errors, refraction, and other corrections, and discusses which calculations are performed once every 5 minutes, once every 5 seconds, and 20 times per second.

To implement this control approach efficiently using only 7% of a DEC MicroVAX II, Wallace defines the concept of slowly changing *osculating transformation matrices* (OTMs) for representing most telescope pointing transformations mentioned above. For every target he defines three probe vectors at the vertices of an equilateral triangle roughly 0.3° on a side. These probe vectors are multiplied by the OTMs to perform the transformation, and the OTMs are updated at two different rates to make the computation more efficient. This slight complication is not needed on modern PCs, as noted earlier.

Wallace manages to condense much of this book into his 16-page paper. If the conference proceedings book containing the paper is out of print and you cannot obtain a copy of it elsewhere, write to Trueblood through Willmann-Bell and enclose a stamped, self-addressed 9 by 12-inch envelope with sufficient postage for 16 sheets of paper.

14.2 ACE PC-Based Control System

A newcomer to the telescope control scene in the mid-1990s, Astronomical Consultants & Equipment Inc. (ACE) offers complete customized telescope control systems using PC-clone computers. The company president, Dr. Peter Mack, is an astronomer turned engineer, programmer, and design consultant. Dr. Mack worked on various systems on the Michigan-Dartmouth-MIT telescope on Kitt Peak and was instrumental in improving the seeing at that telescope in the early 1990s. He developed a PC-based telescope control system for the SARA Robotic Observatory on Kitt Peak that permits local or remote real-time control and fully autonomous queue-scheduled observing for photoelectric photometry and CCD imaging. If you prefer to purchase an off-the-shelf telescope control system instead of developing your own, it would be difficult to find a better solution.

ACE's initial offering is based on a dual 90 MHz Pentium system and an ISA expansion chassis that permits a total of 10 interface cards. ACE's first system used MS-DOS and Windows 3.1, but Dr. Mack now offers Windows NT for the reasons mentioned in Chapter 11. A copy of their main control page is shown in Figure 14-1.

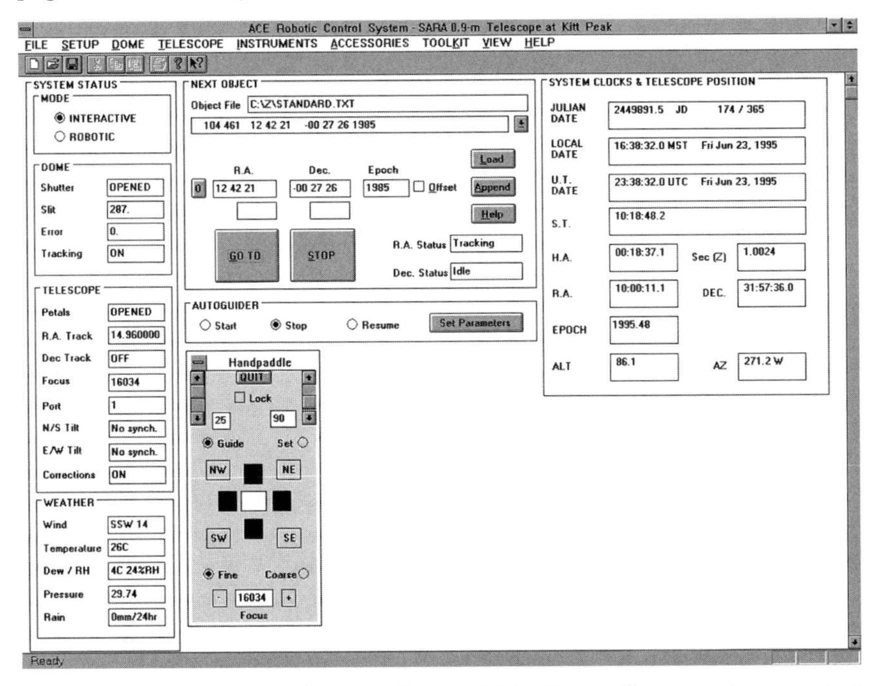

Fig. 14-1 *ACE Robotic Control System Main Page. Courtesy Astronomical Consultants & Equipment Inc.*

The main page has major command and display areas defined for System Status, Next Object, Autoguider, and current date, time, and position information. The system can be operated interactively by a human operator using the mouse and keyboard to issue commands to the system through its GUI, or it can be placed in full robotic mode to execute a pre-programmed observing queue completely autonomously. The Dome, Telescope, and Weather sub-areas in the System Status area are self-explanatory, meeting the requirement for an intuitive operator interface.

The Next Object area permits the operator to select an object catalog file by moving the mouse to the down arrow located to the right of the Object File dialog line and clicking the left mouse button. This brings up a menu of object files known to the system. You select a file name from this file by clicking on the down arrow to the right of the Object File line. This brings up the menu of objects in the selected object file. You choose an object by running the mouse down the menu and clicking on the object you want. Additional objects can be added to the current object file using the Append button on the right side of the Next Object window. The coordinates of the selected object are displayed in the R.A., Dec., and Epoch boxes along with proper motions in the line below the main RA and Dec line. If after reviewing the object file name, object name, and coordinates you want to slew the telescope to that object, you click on the Go To button. You can also enter the RA and Dec directly by clicking on the appropriate box and typing the numbers on the keyboard. To stop the telescope at any time, whether tracking or slewing, you click on the Stop button. In keeping with the GUI rule to offer multiple ways to issue a command, you can activate these buttons using the G and S keyboard keys, respectively.

The ACE system has several other features. Context-sensitive on-line help is always available using the F1 function key. This means the software "knows" where you are in using the system, and offers a help message keyed to the state of the system and what you did recently. The system also offers both hardware and software hand paddles. The former is essential if the control computer is located in a warm room but you are working at the telescope performing maintenance or observing with an eyepiece, while the latter is convenient for manual guiding or adjusting the focus during an interactive observing session. The soft paddle stays on the screen until you tell it to go away, always ready for your command. While the paddle is on the screen, you may issue commands in other page areas without having to change window focus between the paddle and the rest of the screen. Guide and Set speeds can be set with a scroll bar to make it easy to control the telescope at different plate scales or image magnifications.

Many control systems are easy to operate, but difficult to set up and administer. Often they require a text editor to initialize or change configuration files containing information about the system, such as the number of steps per arc second, acceleration and maximum velocity rates, and other system parameters. The ACE control system offers a pop-up window that permits you to configure the system quickly and easily, and to change a parameter just as easily. A dialog box requiring the engineering password prevents over-eager guest observers from changing critical parameters in the middle of a run. A similar dialog box is used to set up the focus parameters and to define up to 4 pre-set focus positions for different instruments, and up to 8 focus positions per instrument for different filters or operating modes.

At the start of every night when the ACE control system is first run, a dialog box appears with areas for the Telescope, Dome, and Focus, and boxes to check to command the system to load track rates, turn on sidereal tracking, open the mirror cover, open the dome shutter, and other initialization activities. Once you have selected which activities to perform for initialization, you click on the INITIALIZE button to get things going. A similar dialog box is used to shut down the system at the end of the evening.

Other dialog boxes pop up for defining the telescope tracking rates, choosing which epoch to display, initializing the dome encoder and defining the dome home position, defining the RA/Dec location of the dome flat field screen, and setting up an autoguider. All settings are stored on disk, to speed up initialization at the start of the evening and recovery after a power failure. ACE is quite willing to customize their system to a customer's specific needs, e.g., roll-off roof instead of a dome or alt-az mount instead of equatorial. See Appendix E for ACE's address and telephone number.

14.3 AB Engineering

The description of the AB Engineering control system given below was excerpted from *Computer Controlled Drive System Design For Astronomical Telescopes* by Alan D. Bell, P.E., AB Engineering, Inc, Fort Wayne, Indiana (manuscript circulated privately) and edited with permission of the author.

AB Engineering, Inc., located in Fort Wayne, Indiana, designs and manufactures complete telescopes and individual components for research, education, and advanced amateur observatories. The firm customizes each telescope's computer controlled telescope drive system to meet the customer's specific needs. Taking a complete system view, the company integrates the telescope's optics, support structure, and drive to develop a system that satisfies the telescope user's requirements within the budget.

AB Engineering utilizes a systems engineering approach on both new and retrofit installation projects. The process begins by translating the requirements of the telescope user into detailed design requirements. The detailed requirements are divided and allocated to each specified engineering discipline–optical, mechanical, electrical and software. Within each engineering discipline the requirements are further partitioned between subassemblies. During the design process, engineering analyses and tests are used to validate individual design attributes. Optical ray trace analysis, structural finite element analysis, and software development tools are a few of the engineering tools employed during the design effort to model the hardware ensuring that the final design satisfies customer requirements. Throughout the systems engineering process, designs are reviewed for manufacturability and to maximize use of commercial off-the-shelf components to develop cost effective solutions.

AB Engineering began manufacturing telescope components (star diagonals), subassemblies (large focusers) and small telescope mounts in 1986. In 1990, business focused on the development of a small aperture (16-24 inch), computer automated telescope suitable for research by small colleges and astronomical organizations. The thrust of the design effort was development of a moderate cost computer controlled telescope drive system. The resulting drive system is used on AB Engineering's new telescope mounts and can be installed on existing telescopes being retrofitted for automated operation. The design approach was to utilize commercial off-the-shelf hardware and integrate the individual mechanical, electrical and software components into a complete system. In 1993, a prototype computer control drive system was completed. In 1995, AB Engineering introduced to the astronomical marketplace a computer automated, German equatorial telescope mount capable of supporting a variety of tube assemblies with optical apertures ranging between 16 to 22 inches. Also in 1995, AB Engineering began retrofitting Butler University's (Indianapolis, Indiana) 38-inch Fecker built telescope with computer control drives for automated operation.

It is important to understand how the optics, opto-mechanical support structure, mechanical drive, and electronics influence the design of the telescope control system. In most observing applications a CCD camera is used. Since the quantum efficiency and dynamic range of CCD cameras is one to two orders of magnitude greater than that of photographic film, CCD devices "see" every drive perturbation at the focal plane, including those lasting only a small fraction of the error period. Furthermore, all but the largest CCD arrays are very much smaller than photographic plates or 35mm film, so pointing must be more accurate than with larger format detectors. Thus, the drive system developed for telescopes using these small, sensitive devices must be very accurate and responsive. Table 14-1 highlights the principal system performance factors that one must consider while developing a computer controlled drive system.

Table 14-1
Drive System Performance Factors

Optics	Mount	Drive Motor	Electronics	Other
Focal Length Aperture	Stiffness Inertia Bearing Torque Orthogonality Balance	Dynamic Range Torque Speed Resolution Drive Frequency Drive Assembly	Resolution Stability Precision Accuracy Architecture	Polar Alignment Atmosphere User Interface Cost

The drive system must deliver accurate tracking, accommodate the telescope mount's inertia during rapid slew, and facilitate seamless automated operation. If the telescope control system is built around a computer, it becomes possible (and desirable) for the control system to correct for telescope pointing errors due to telescope structure deflection and misalignment (to the sky and between mechanical and optical axis). It is also easier to make the system responsive to the telescope operator, both in function and operation (user interface).

AB Engineering partitions drive system engineering considerations into four principal categories: mechanical components, electrical hardware, software development, and cost.

The design of a drive system begins with the design of the mechanical drive and an understanding of the telescope mechanics to which it is being installed. The first activity of the Butler University's 38-inch telescope retrofit installation was to determine the polar and declination axes' bearing torques and establish the mass moment of inertia of the fork equatorial mount. The telescope, originally built in 1954, incorporates a 38-inch diameter 360:1 polar worm gear set. Mechanical inspection and gear analysis demonstrated that the large gear could be used as-is. This eliminated the costly complete disassembly of the telescope to remove the gear. Since worm gear sets are not capable of being backdriven, gear tooth analysis was conducted to confirm that the gear teeth bending stresses that are developed when the telescope suddenly stops during a 5°/sec slew (worst case scenario condition) are below the allowable gear material yield stress values. Lastly, the gear set was measured for periodic error and compared to a focal plane stability measurement, which determined that it was possible to use the existing polar gear, but that a new worm was needed. A replacement alloy steel, hardened and ground precision worm was installed, then candidate stepper and servo motors were installed on the telescope and evaluated for performance versus cost. Due to the cost constraints of the Butler retrofit installation, stepper motors were selected.

The declination axis tangent arm assembly was then removed and replaced with a 26-inch diameter worm gear assembly. Both the polar and declination drives utilize stepper motors with a helical gear speed reducer driving the worm. Helical gears are utilized to minimize backlash and provide continuous gear teeth engagement not available with standard spur gears, which helps to reduce error magnitudes normally associated with spur gears. All of the gear elements are supported by preloaded ball bearings mounted in a cast, ductile iron housing. Integral with the drive assembly is an adjustment stage to ensure the worm and worm gear are properly engaged. As specified by the worm gear set design, clearance is provided to allow low temperature grease to lubricate the gear set. To eliminate the backlash associated with a drive direction change, a constant torque (via preload weights) is applied to the worm gear shaft.

In the partitioning of the tracking and pointing errors of the complete drive system, the mechanical subassembly (motor and gears) is given a minimal portion of the error budget. The design philosophy is to build the drive system with minimal errors, subsequently minimizing the task of compensating with control software. On the Butler telescope, with a 360:1 worm gear set ratio, the periodic error of the worm (1 revolution/4 minutes during sidereal tracking) is controlled at or below the resolution of the telescope's optics. The result is that atmospheric seeing becomes the limiting parameter to focal plane image stability. Minimizing the drive error leaves the control system with only the task of compensating for measurable, deterministic misalignments (mount, optical, and polar). The emphasis in design and cost is placed on robust mechanical components which, once installed, will not have to be replaced as new control electronics and software are made available. It is more cost-effective to eliminate mechanical errors by good mechanical design than to use complex control algorithms in software to compensate for poor mechanical design.

The engineering challenge in developing a computer control drive system is not in the design of the mechanical drives. The mechanical hardware is straightforward and performance is readily measured at the subassembly level. It is the design and implementation of the electronics and control software that is the challenge, especially when the technology and cost are always changing. The starting point for the control system design is to realize that two types of processing occur within the telescope drive system to make the telescope operate correctly and responsively: *hard-real-time* and *soft-real-time.*

Hard-real-time performance, characterized by predictable, rapid response to external inputs, is absolutely required by the lower-level control system managing the individual axis drive motors, reading position encoders and other analog and discrete position sensors, performing motion profiling (acceleration and deceleration curves), and coordinating the dome motion. When the (properly aligned, equatorial) telescope is tracking, the control system processing is minimal as it periodically monitors the status of limit sensors. Checking the status of sensors can be performed in intervals of whole seconds; however, while accelerating or decelerating the telescope, significant activity occurs in each axis. Ideally both polar and declination axes are coordinated so that simultaneous motion occurs, reducing seek time. Once slew speed is attained, the processing again becomes minimal other than determining when to initiate deceleration. However, sensor monitoring becomes more of an issue since the axes are moving three orders of magnitude faster than during tracking. Thus, processing that was performed in whole seconds during tracking must be performed in milliseconds during slew to obtain equal safety margins. As the hard-real-time processor performance decreases, drive system performance must be reduced to maintain safety margins in motion control.

Soft-real-time performance is desirable, although not absolutely required, in the user interface processing. What the telescope operator perceives as drive system responsiveness is a function of the soft-real-time software. User responsiveness is considered a subjective issue by most designers, one that is typically an afterthought; however, the telescope operator quickly decides whether the user interface is adequately responsive. It is undesirable for the operator to perceive the control system as being sluggish.

Another fundamental design point is that the mechanical drive configuration imposes several difficult requirements on the electrical drive system. Specifically, using a single motor on each axis for seamless tracking and slew operation requires the motor and its control system to have very high velocity resolution at tracking rates while providing adequate bandwidth for slewing. This range of motor speeds can span three orders of magnitude. In addition, tracking requirements impose precision, accuracy, resolution, operational smoothness, and stability issues on the electrical drive system.

During the initial development of AB Engineering's drive system in the late 1980s, the controller design utilized a single custom controller to manage all drive and sensor subsystems. A simple character-based user interface was provided to the telescope operator. Because of the availability of then-current technology, this solution could not leverage the use of commercial off-the-shelf hardware or software to satisfy the drive system requirements and was not built.

As computers became more reasonable in cost and performance, and operating systems with real-time features and good performance became available at reasonable costs, the drive system design evolved from the custom, all-in-one controller to a system based on a standard computer

(IBM-compatible PC). The design employed simple, "dumb" I/O cards to interface to the drives and sensors, and used a real-time operating system to allocate the processor to the motion control and user interface tasks. This solution combined custom and commercial off-the-shelf hardware to satisfy the real-time system performance requirements. However, performance limitations existed that compromised critical low-level control system functions, subsequently limiting the entire drive system performance. In addition, the initial procurement cost of the real-time operating system was high, which when combined with the software maintenance fees, significantly inflated the total system cost. While this system would work, the higher cost and lower performance were a poor design solution.

The AB Engineering drive system employs a dedicated real-time controller for motion control, combined with a commercial off-the-shelf operating system providing standardization and economy of scale. Over the drive system development period, Microsoft Windows has evolved to become the *de facto* standard for hosting user interface software. Of the three main versions of Microsoft Windows available when the system was developed, Windows NT was selected. Windows running on DOS (e.g., Windows 3.x) was not chosen since it requires programs running simultaneously to cooperate—a single poorly-written program can easily consume an entire PC. Also, it supports only 16-bit applications. This means using Windows 3.x imposes several programming constraints for the development of application software. Although Windows 95 is a 32-bit operating system, it was not chosen since it still contains portions of 16-bit code and at the time the system was being developed, it was much newer and less stable than Windows NT. Windows NT provides a true 32-bit environment with preemptive tasking, kernel threads, some degree of soft-real-time performance, full graphical windowing user interface, and the ability to support other applications in addition to telescope software. Although Windows NT is more expensive than the other operating systems considered, the higher initial investment was justified when spread over several systems.

The telescope control system is divided into two subassemblies: a telescope control computer, and a second computer operating the high-level telescope user interface. The telescope control computer controls the drive system and orchestrates the commands to, and receives status from, the telescope. The telescope control computer concentrates only on the motion control of the telescope and dome. It is a standalone system and simple commands may be issued through it to move the telescope. The second computer runs the high-level telescope control graphical user interface. Rather than develop yet another user interface, AB Engineering uses *The Sky* from Software Bisque. *The Sky* provides a complete high-level telescope control environment and incorporates algorithms for determining positions of planets, object rise and set times, database observing list management and atmospheric related pointing compensation. This program also incorporates TPOINT, an algorithm developed by Patrick Wallace to compensate for pointing errors due to telescope flexure and alignment errors; it also operates under Windows NT.

The communication link between the telescope control computer and the user interface computer is over a standard RS-232 serial interface. The telescope control computer also receives inputs from the handpaddle and autotracker. As mentioned previously, the telescope may be operated in a rudimentary mode through just the telescope control computer. Secondary telescope operator input devices may be added as required.

AB Engineering's computer control drive system operates on commercial off-the-shelf hardware and will support commercial off-the-shelf user interface software to provide responsive telescope operation. The company evolves its products as technology advances, with the goal of producing a standalone controller mounted in a dedicated telescope control computer that manages all low-level motion control functions on a processor independent of the user interface computer processor. For most of AB Engineering's market the hardware is relatively affordable, and it satisfies the system performance requirements and is a one-time cost. Each specific telescope installation requires modifying only the motion controller software to tailor the generic motion controller. The telescope control user interface software provides basic functions required to perform simple telescope control for both the installation and tuning phases as well as simplified control without the use of the high-level control software. Finally, it also accepts commands from external higher-level control computers and performs any required processing prior to communicating to the motion controller.

The system built by AB Engineering is one of many selections offered by commercial firms. Their partitioning of telescope control and user interface functions makes it unique among systems known to the authors, and also gives the customer greater flexibility in integrating the AB Engineering system into an existing control system.

14.4 COMSOFT PC-TCS

ACE and AB Engineering could not predict what other applications their customers would run on their systems while their control systems were running, and they wanted a reliable and safe operating system that would be available and in use for several years upon which they could build their businesses. They are also counting on the operating system to provide real-time, multi-programming features. ACE uses relatively expensive motor indexer cards to interface stepper motors to their PCs. Both chose Windows

NT for their platform upon which to build systems that can cost tens of thousands of dollars.

COMSOFT took a different approach to make a system typically costing in the range $600–$3,000. The Personal Computer based Telescope Control System (PC-TCS) uses MS-DOS without Windows and a counter-timer board costing about $300 installed in a very inexpensive computer. The minimum system is a 33 MHz 80386-based PC with a math co-processor, but COMSOFT recommends an 80486. With the advent of the Intel Pentium and Pentium Pro processors, 386 and 486 computers can be found in surplus stores for a few hundred dollars.

Much of the CPU's power is spent controlling the counter-timer board. By insisting that no other program runs in parallel with PC-TCS on the same computer, COMSOFT has better control over MS-DOS interrupt servicing performance and other operating system behavior that would otherwise rule it out as the operating system of choice.

PC-TCS was written by David A. Harvey starting in 1988 under a contract with a custom telescope and instrument builder. Dave, who works at the University of Arizona's Steward Observatory building telescope control systems for large professional telescopes (such as Steward's 90-inch telescope on Kitt Peak), established COMSOFT to market an affordable system that could be used on large professional telescopes, yet many of his approximately 30 installations are at amateur, college, and small university observatories. One of his systems is installed on the Yerkes Observatory 41-inch reflector (not the 40-inch refractor).

If you wish to buy only the software and to provide your own computer, motors, and wiring, the stepper motor version costs $600 while the servomotor version sells for $800. The stepper version uses a commercial counter-timer board (Computerboards CIO-CTR10/H50, $279) while the servo version uses a custom dual-axis servo card available from COMSOFT. You can purchase a complete system with software, computer, and motors for under $3,000, and you can do the wiring yourself or have COMSOFT perform the entire installation. The system cost depends on how much custom work COMSOFT performs at your site. You can run the stepper version open-loop or with encoders of just about any resolution.

COMSOFT designed PC-TCS to be a Type 2 servo, in which there is a fixed acceleration error (no velocity or position errors). The counter-timer board controls one or two motors using a single interrupt vector that is user-selectable. The card interrupts at a rate of 100 Hz for the time base. Every fifth interrupt (20 Hz), PC-TCS executes a position loop with a modified PID algorithm for tracking and a square root algorithm for long slews.

The MS-DOS text-based menus provide access to dozens of functions. The system includes the Messier, FK5, Yale Bright Star, SAO, PPM,

GCVS, Oke-Stone spectroscopic standard star, and Revised New General Catalog catalogs, and a utility permits you to generate your own catalogs. PC-TCS also computes the positions of the Sun, Moon, comets, minor planets (asteroids), and all eight major planets. It displays over 50 control parameters including current RA and Dec, hour angle, altitude and azimuth, parallactic angle, airmass, local sidereal time, UTC, Julian date, calendar date, focus position, dome position, instrument rotator angle, catalog object information (equatorial position, name, and magnitude), and time until rise or set (for current, next, and reference positions).

PC-TCS will automatically slew and track to any moving target (planet, Sun, Moon, comet, minor planet, or artificial Earth satellite) if you give it orbital elements. It automatically computes Dec and RA rates for "hands-off" tracking of non-stellar objects. All pointing angles are corrected for proper motion, precession, nutation, annual aberration, atmospheric refraction, and repeatable telescope flexure and mount misalignments. Tracking rates are corrected for differential refraction, flexure, RA gear periodic error, and Dec gear backlash. You can command the system to display mean positions for any epoch with precession, aberration, refraction, and flexure backed out.

Currently, PC-TCS handles only equatorial mounts, but it does so gracefully, with user-definable manual or automatic German flips. You can define limits in hour angle, Dec, horizon, and airmass (sec Z).

The paddle interface is the same as that used by SBIG, so SBIG autoguiders can be used easily. Up to eight serial ports can be connected to other computers, including by modem to computers around the world, and PC-TCS accepts commands from *The Sky* and other computerized atlases using that communications protocol.

There are really too many features to mention in a short space. PC-TCS has many features found only on professional telescopes, including spectroscopic drift scans for extended objects, raster and spiral search patterns, and wobbling (nodding) between two beam positions for infrared observations. More information on this very capable system can be found in Appendix E.

14.5 Soft-Tec Systems

The Sky Probe 1000 telescope positioning system by Soft-Tec Systems is designed to provide computer aided positioning for existing equipment, either equatorial or alt-az mounted, at a price comparable to the COMSOFT system. Using an IBM PC compatible computer provided by the customer, the Sky Probe 1000 moves the telescope to the desired object or coordinates, correcting for refraction. The system provides menus similar to COMSOFT's.

Objects may be identified by NGC or IC number, Messier Catalog number, or by RA and Dec Epoch 2000.0 coordinates. Planet positions are calculated for the time they are selected and are accurate to approximately 1′ (Pluto 15′). The Sky Probe 2000 takes a similar approach to COMSOFT, running on IBM PC clone computers using only MS-DOS (no Windows 3.1, 95, or NT), and requiring that no other programs run at the same time.

With appropriate gearing, the software can slew the telescope at rates ranging from 20° to 60° per minute. Besides the gearing, the speed of the computer also determines the maximum slew rate, since unlike other control systems discussed here (including COMSOFT's) the computer generates each individual motor step.

A typical package from Soft-Tec consists of software and an object database, two stepper translators mounted in a separate enclosure, hand paddle, two stepper motors, and a manual. The hardware carries a one-year warranty and the package includes one year of free software upgrades. The hand paddle plugs into a COM serial port, and the motor translator plugs into the parallel printer port using a supplied cable. Most Soft-Tec customers perform their own installation using the complete instructions provided (and list of vendors to obtain gears, belts, and other items), but Soft-Tec will install the motors on a telescope on a custom basis.

The software executes a servo loop once every 27 mS, in which it calculates the next position (including refraction and axis misalignments) and the number of motor steps needed to get to that position. Assembler language routines output the motor phase signals (turn a stepper motor phase line on or off) through the parallel printer port, so an 8-bit port can control two 4-phase motors. The parallel lines go to the stepper translators that amplify the digital signals into currents for the motor windings.

To do all this processing, Soft-Tec recommends using a 386 33 MHz machine or faster, with at least 640 kB of memory, two 720 kB floppy drives or a hard drive, at least one serial communications port, and at least one parallel printer port. A math coprocessor is not required, but the software uses it if present.

Once positioned at an object, the system tracks at sidereal rate. Since both the RA and Dec axes are controlled by the computer, if Dec motion is required to maintain proper tracking (as is the case with alt-az mounts), the Dec motor will be stepped. The system also provides variable tracking rate. This allows easy setting of solar, lunar, and planetary tracking rates.

The Sky Probe 1000 includes a hand control to move the field of view around the sky. The hand control has four different rates of motion: FAST, MEDIUM, SLOW, and STEP. The fast speed is equal to the speed at which the system moves between objects. The medium, slow, and step speeds are

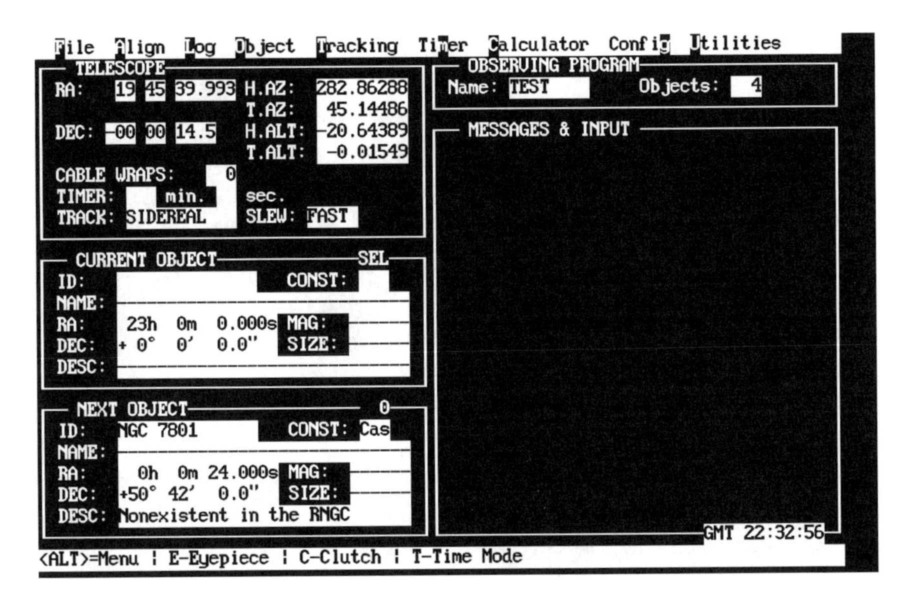

Fig. 14-2 *Sky Probe 1000 main menu.*

used for fine tuning the position of an object in the field of view. To help you know which slew speed you have chosen, the system emits a different tone each time the speed is changed. This eliminates the need to keep checking the computer screen to determine the slew speed.

The hand paddle also includes NEXT and PREV buttons used to command the system to slew to the next or previous objects in a user-defined observing program. Used together, these two buttons capture alignment stars and touch up alignment coordinates. The HOLD button allows you to interrupt motion and tracking from the hand control temporarily. Another important feature of the hand control is its ability to reverse the motion direction for either the RA or Dec button pairs. This feature aids in guiding by allowing the observer to orient the hand control to the field of view and to conform with personal preferences.

The Sky Probe 1000 software is completely menu driven, with the main menu shown in Figure 14-2. The system includes a user's manual with a comprehensive tutorial. The manual also provides many helpful hints on getting the most out of the system. An installation facility allows one to complete the software installation in minutes. Most commonly used commands are available using a single function key, minimizing the amount of typing required. This makes the system easier to use under the night sky. The night vision mode uses only red and black on the monitor to preserve

```
*** NGC OBJECT LIST - Epoch 2000.0 ***      Available: 7047   Objects:  0
NGC   Type hh mm    dd   mm   Con Size    Mag  Mes  Comments
==========================================================================
I 5371       0  0.0  32  49.0 And
  7801 -     0  0.0  50  42.0 Cas
I 5374       0  1.0   4  30.0 Psc
  7804       0  1.0   7  45.0 Psc
I 1527       0  2.0   4   7.0 Psc
I 5378       0  2.0  16  37.0 Peg
I 5379       0  2.0  16  35.0 Peg
I 5380       0  2.0 -66  11.0 Tuc
  7812 Gx    0  2.0 -34  15.0 Scl
I 5381 Gx    0  3.0  15  58.0 Peg  1.5
  7814 Gx    0  3.0  16   9.0 Peg  6.3  10.5
  7815 -     0  3.0  20  41.0 Peg
I 5382       0  3.0 -65  11.0 Tuc
  7822 Nb    0  3.0  68  37.0 Cep  60.0
I 5383       0  3.0  16   0.0 Peg

¦SPACEBAR - Select¦ <HOME>File START   ¦ <END>File END   ¦ <F3>Block SELECT
¦<F6>Search Window¦ Ctrl<F6>Search Again¦ Ctrl<F7>Nxt SLCTD
↑↓ PgUP/PgDN for More ...
<F10> to Exit/Save - ESC to Quit ...
```

Fig. 14-3 *Sky Probe 1000 observing program build screen.*

night vision, and a feature protects you from accidentally returning to color mode.

Polar alignment is not required—the system will track even when used with an alt-az mount. To align the system, the observer places the telescope in one of six user-selectable alignment start positions, then points the telescope at any two of the 198 available alignment stars using the system hand control. As each star is centered in the eyepiece, the observer presses the appropriate key on the hand control or computer keyboard. The telescope is then ready for a night of observing.

The system also provides for touching up the alignment as might be required. Simply choose an alignment star from the Align Star list (or any other object that has a well defined center) and tell the system to go to it. Once there, use the hand control to center the star or object in the eyepiece. Press the appropriate key on the hand control or keyboard, then resume executing the observing schedule.

If the telescope is not going to be disturbed or relocated between observing sessions, the alignment data may be saved. The next time the Sky Probe software is started, the last alignment values are restored, without requiring another two star alignment.

A feature of the Sky Probe 1000 system is its ability to provide planned viewing sessions. Before going to the observing site, you can build an

"Observing Program" consisting of all the objects to be observed during the evening. Each program can contain up to 250 objects. The Observing Program build screen is shown in Figure 14-3.

To build an Observing Program, you enter the name and a brief description of the program you are building, then the system makes available the complete NGC 2000.0 database. You can page up and down through the database and with a press of the spacebar, select any of the 13,226 objects. You can search for an object based on the contents of any of the fields presented with each object. Additionally, before entering the program build function, you can limit the objects displayed for selection and addition to the Observing Program by magnitude, angular size, object type, NGC or IC only, and constellation.

The system also provides a horizon limit (planisphere search) capability. By entering the expected time of observation and local latitude, the system will limit the database presentation to those declinations that are above the horizon. You can add non-database objects to any Observing Program. Once you select all the NGC/IC objects you desire and exit the build mode, the system will ask if you wish to add objects. By answering **yes**, one can enter non-database objects up to the program limit of 250 objects. You must enter the RA and Dec (epoch 2000.0) for the object and can optionally enter any other data and description items you desire. As with the database items selected previously, these items are sorted in order by RA then saved to disk.

When the Observing Program is activated, it moves the telescope to the next object when the observer presses the NEXT button on the hand paddle or the appropriate keyboard key. New programs can be selected and loaded at any time, even in the middle of another program, without interrupting observations. You may skip forward or backward in the active Observing Program list without slewing to each object. This means you can start observing with any object in the program. When you reach the end of the program, it automatically starts over at the first object. You can also select objects or slew to new coordinates that are not part of the Observing Program without losing your place.

Observing Programs can be edited (objects added or removed), reviewed (listed on the computer screen), and printed. Observing Programs can also be used over and over again until you delete them from the system. When the Sky Probe system is started, it remembers and activates the last used Observing Program.

An important function for both professionals and amateurs is maintaining records of observations. As each observing session begins, the Sky Probe system automatically starts a log of objects viewed. Every object or new set of coordinates that you seek provides an entry in the Session Log.

Each entry contains an ID number, object name, constellation, magnitude, angular size, RA and Dec coordinates, and an object description. This information comes from the database when an object is selected. Or, for an object specified by its coordinates, the observer may enter all or part of the information.

The session log also stores the time in minutes the system spends at the object, an observational quality rating, the primary eyepiece used, primary filter, and up to 280 characters in 4 lines of notes. The system automatically stores a list of up to 7 additional objects that are in the field with the primary object sought. The user may create as many Session Logs as desired. One can also produce a Master Log listing all objects contained in all Session Logs present in the current program directory. The Master Log lists object IDs and names, the Session Log file name, the time spent at that entry, and a total time for each object across all logs. Sky Probe also provides an object Composite Log. This log contains all entries from all Session Logs for a particular object. The Composite Log can be edited and printed just like any other Session Log.

Another system feature is the identify function. If you have an object centered in the eyepiece and you are not sure what it is, pressing a key on the keyboard displays its NCG or IC number if the object at the current coordinates is in the NGC/IC catalog. If there are others close by, a list of up to 10 of the closest is displayed. This function also allows changes in the magnitude and size of the objects included in the search, and the diameter of the search pattern.

The Sky Probe system tracks at sidereal rate in normal operation. You may also enter a variable mode that allows setting lunar, solar, and planetary rates. You can return to sidereal mode at the touch of a button and the system will remember the variable rate that you have set. Other tracking modes include manual and SBIG ST-4 guiding modes. When in these modes, the system locks out all keyboard and hand control functions that might interfere with a good guide. You can re-enter the normal operating mode at any time. Sky Probe also provides meridian swap for German equatorial mounts.

Other features of the Sky Probe include an activity timer, an "astronomical calculator" (to perform time, angle, and other conversions), four time display modes, and the ability to drive the motors from a battery.

Both COMSOFT and Soft-Tec offer systems running on inexpensive PCs under MS-DOS for similar prices. They have many features in common, but differ in other areas. If you are considering one of these products, you should look at both and determine which one is best suited to your application.

14.6 EPICS

The Experimental Physics and Industrial Control System (EPICS) is a product of a collaboration among major U.S. national high energy physics labs with Los Alamos National Laboratory (LANL) taking the lead in monitoring enhancements, tracking bugs, and making distributions of the software. The product offers essentially all of the executable software to implement a complex and very capable control system. The design is comprehensive and adaptable enough to build a wide range of control systems, for applications as diverse as particle accelerators, telescopes, and industry. EPICS is distributed free of charge over the Internet by LANL; all you do is sign an agreement not to distribute the source code outside your organization, then *ftp* the source code to your site. There are two firms that market commercial versions: Tate Integrated Systems sells it as TIS-4000 and Kinetic Systems sells a version under the name Reality. Baltimore Gas and Electric uses TIS-4000 to control its liquified natural gas plant, compression station, and storage tank farm in a populated area very near downtown Baltimore, so they consider the product to be sufficiently mature and reliable for this potentially dangerous application.

All the software to read sensors, implement control algorithms, display sensor readings in a wide variety of ways (dials or meters, vertical or horizontal thermometers, plain numbers, and many other ways using a variety of colors to represent nominal performance, warnings, and error conditions), accept command inputs, and control motors and other actuators is already developed, tested, and in use at dozens of locations. For many applications, the EPICS system developer is not a software engineer—often no software in the traditional sense needs to be developed. Instead, many EPICS developers are engineers with a background in control theory or process control who know little or nothing about high level language compilers.

Development of an EPICS system consists of populating a database with information about your system—what type of sensor is at each address, the range of voltages you expect from it, the values of nominal, warning, and error signals, how the sensor data are to be displayed, the various error computation algorithms, actions to be taken when certain errors are detected or commands entered, and control parameters to motors. The developer then runs a process that builds EPICS data structures from the database and merges them with the standard EPICS code to produce an executable program. If EPICS device drivers are not available for the particular motor controller cards or sensor modules you plan to use, you can either use a regular programmer to write the necessary software, or change your hardware design to one of the dozens of cards for which EPICS drivers already exist. You might also need to produce some software to implement specialized control algorithms, but aside from these two examples, your conventional software development is minimized by EPICS and the resulting software has far fewer bugs than any completely new system you are likely to write in-house.

EPICS is written as a distributed system to run on different kinds of hardware. The operator interface part of EPICS that displays sensor icons and accepts commands runs on a Sun workstation, while the sensor and motor drivers and control algorithms run on a Motorola 680x0 in a VME crate under the VxWorks real-time operating system. The Sun workstation and VME crate communicate by Ethernet.

Botlo et al. (1993) report on the performance of EPICS. The results are too complex to report in a few numbers, and anyone considering using EPICS should review this paper to help determine if EPICS is right for their application. The paper indicates that EPICS can handle several tens of thousands of process variables in a major system at a rate typical of telescope control servos.

EPICS is a complex and capable system that requires a large investment of time to learn. However, once that investment is made, it is much easier to develop new systems and to maintain existing systems than writing custom software. The resulting systems are more reliable, since most of the software is already debugged. Astronomy programs using EPICS include the Gemini 8-meter telescopes, Kitt Peak National Observatory, Keck II, the WFFOS spectrograph at the William Herschel Telescope, and UKIRT. Over 20 high energy facilities and detectors use EPICS, as does the LIGO gravity wave telescope.

14.7 Gemini 8-meter Telescopes Control System

The Gemini 8-m Telescopes Project is an international partnership among the United States, United Kingdom, Canada, Chile, Brazil, and Argentina to build two alt-az mounted 8-m monolithic-mirror telescopes, one to be located on Mauna Kea, Hawaii and the other on Cerro Pachon, Chile. The telescopes were designed to deliver exceptional images to take advantage of these two outstanding sites, and to have very low emissivity at near infrared wavelengths. The northern telescope has mirror coatings and adaptive optics optimized for the infrared.

The control system for the Gemini 8-meter telescopes is based on EPICS. It is divided into an Observatory Control System (OCS), Telescope Control System (TCS), Interlock System, and Instrument Control System (ICS). Data from instruments is stored in a separate Data Handling System. All systems talk to each other over Ethernet or FDDI except the Interlock System, which communicates with all other systems using its own

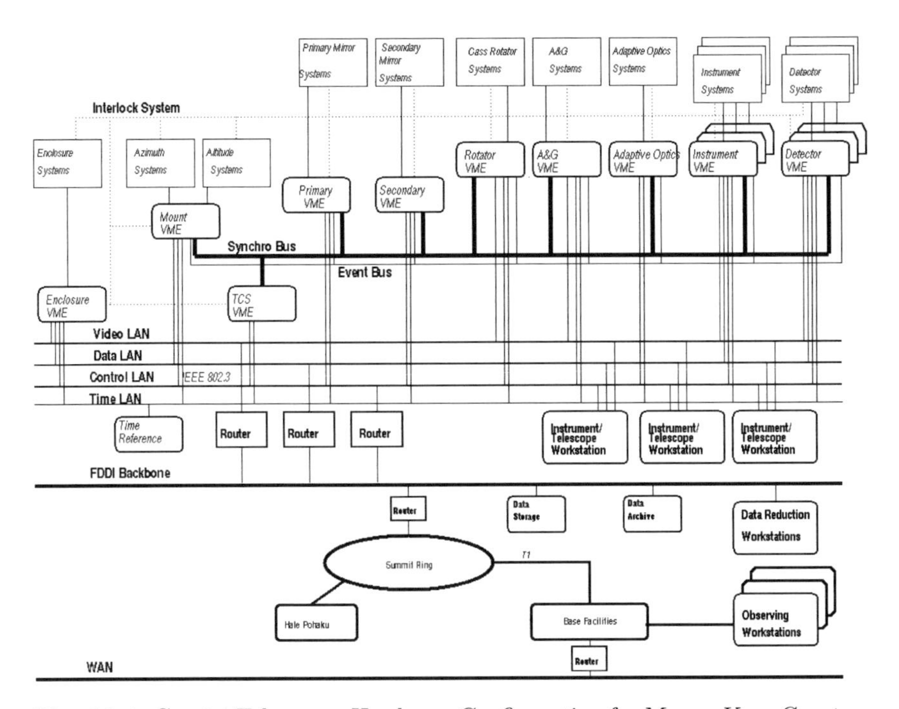

Fig. 14-4 *Gemini Telescopes Hardware Configuration for Mauna Kea. Courtesy International Gemini Project Office.*

bus. The OCS is used to monitor the weather, move the dome, open and close the dome shutters, and control other parts of the telescope enclosure. The TCS is used to monitor and control the telescope itself, command the telescope to point to an object, and similar functions. The ICS is used to move filter wheels, start and end integrations, and control the various operating modes of an instrument. The Interlock System is a special high-speed hardware system with ties to the other systems to prevent damage to humans and equipment.

The figures shown below for the Gemini control systems were taken from a document that reflected an early design for the operator pages. They are shown here with the understanding that they are now most likely quite different in appearance. A control system hardware configuration concept is shown in Figure 14-4. All Gemini control systems use Sun workstations running the EPICS CA Client under Sun's version of UNIX, and Motorola 680x0 single board computers in VME crates running the EPICS CA Server under Wind River Systems' VxWorks real-time operating system. Software development is minimized by using motor, encoder, and sensor interface cards in the VME crate for which EPICS drivers already exist. The basic

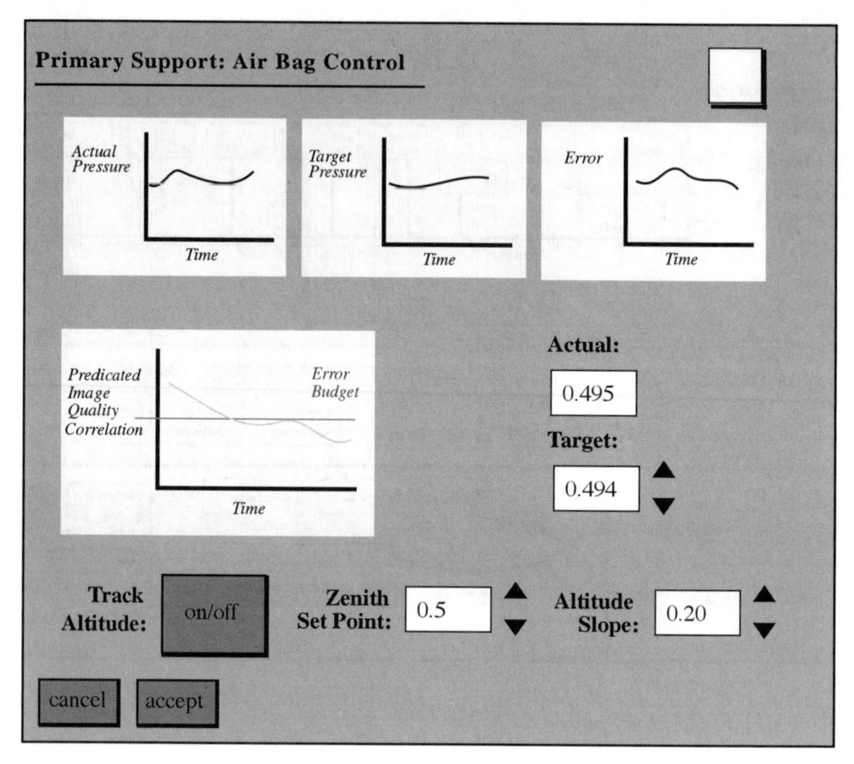

Fig. 14-5 *An Example OCS Command Page. Courtesy International Gemini Project Office.*

control servo is software with an overall concept similar to that in Wallace's paper.

Figure 14-5 shows an example of an OCS page, in this case for adjusting the pressure in the air bag that bears much of the weight (over 30 tons) of the primary mirror. At the top of the page are software strip chart areas plotting parameters of interest, with another such area in the middle. Note that the programmer that lays out the page defines the size of these areas—they cannot be resized within the window, the way the entire window can be. Text boxes give status from sensors or permit operator commanding. The latter have up and down arrows that the operator can click on to change the commanded value.

Figure 14-6 shows the main TCS page that illustrates many of the good design principles discussed in Chapter 10. Pull-down menus at the top of the page provide functions not normally used during observing. These include opening the telescope at the start of the night and closing it down at the end, and configuring various functions. This keeps the page clean

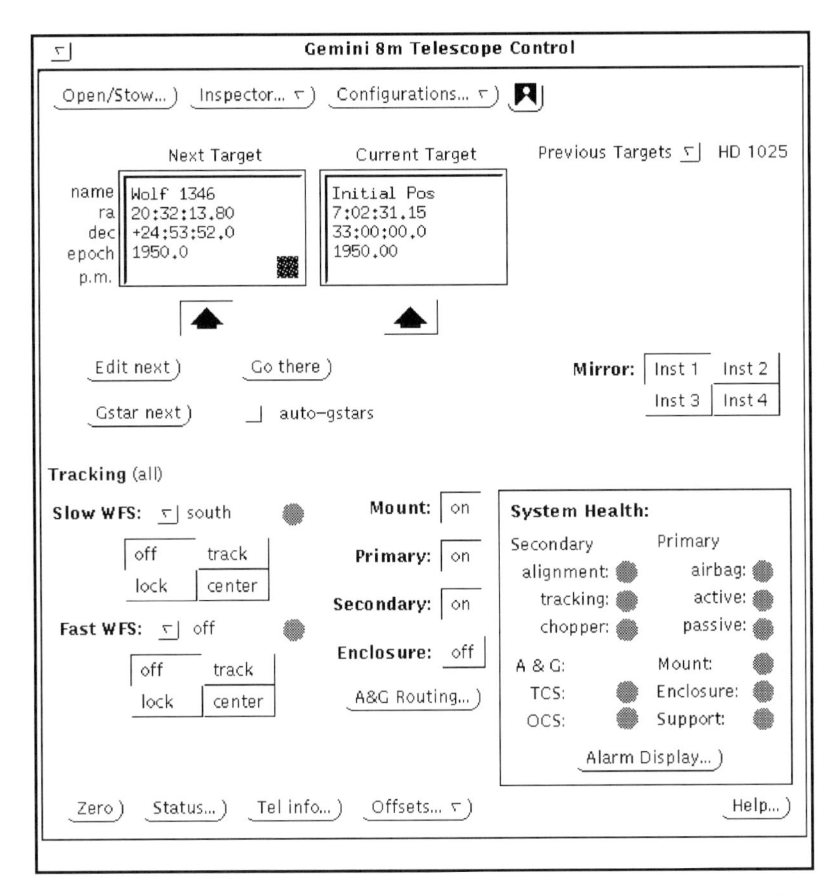

Fig. 14-6 *An Example TCS Command Page. Courtesy International Gemini Project Office.*

and uncluttered, making it easier for the operator to use, yet it is a simple matter to call up these functions, if needed. Below these menus are two position areas, the left one for the next object and the right one for the current object. Clicking on the arrow next to "Previous Targets" displays a history of where the telescope has been earlier that night, a useful feature for reminding an observer suffering from fatigue and altitude of progress through the object list.

Other buttons and areas are defined for selecting guide stars, moving the science fold mirror to direct the beam to another instrument, and controlling tracking and adaptive optics wavefront sensors. An area in the lower right gives a quick overview of telescope status without using a lot of

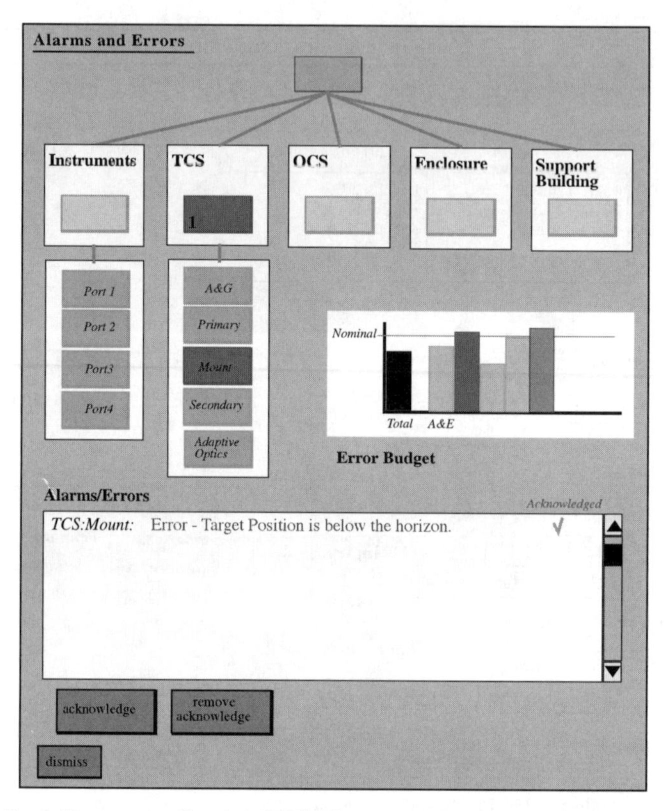

Fig. 14-7 *A Prototype Gemini OCS Alarm and Error Monitoring Page. Courtesy International Gemini Project Office.*

space. The circles next to each status item could be coded as traffic lights: green for OK, yellow for a warning condition, and red for an error condition. When the operator notices a warning or error condition, a quick click on the "Alarm Display" button brings up the page with details on the condition that caused the traffic light to go yellow or red. Buttons along the bottom call up other useful functions, such as resetting the encoder zero offsets to read the correct coordinates, and providing pull-down menus of status page options, a list of individual telescope subsystem status and command pages, and a list of various types of offsets computed by the pointing and tracking servo. Context-sensitive help is available by clicking on the button at the lower right of the page. This page is organized to help the operator keep track of important status items, and to make the correct decision and implement it quickly under altitude conditions that turn intelligent people into morons.

A prototype OCS alarm and error monitoring page is shown in Figure 14-7. The concept here of interest is the idea of grouping the control system into major subsystems, and displaying the count of errors in each subsystem. When the operator clicks on the subsystem icon, a menu showing components of the subsystem pops up. In this case, the parts of the TCS shown are acquisition and guiding, primary mirror cell, the telescope mount, the secondary mirror assembly, and the adaptive optics unit. The mount icon is colored differently to indicate that it is the source of the error, with the error message in the Alarms/Errors scrolling area indicating the nature of the mount error. This use of graphics helps the operator to identify the source of a problem, and to make the correct decision on a course of action.

Also note the Acknowledged column in the alarm messages area. If a major disaster were to befall one of the Gemini telescopes, it would be useful to reconstruct what happened after the fact to find out what information the operator had and did not have. Once the sequence of events is reconstructed accurately, usually it is straightforward to determine how to prevent the same thing from happening again. To know what information the operator had, and to help the operator develop the habit of checking alarm messages, it is useful to require the operator to acknowledge alarm and error messages. On the page in Figure 14-7, the operator does this by clicking on the Acknowledge button in the lower left of the page. A check mark then appears in the Acknowledged column to indicate that the operator has acknowledged the message. Since operators could rush through clicking the button before reading the message, there is also a Remove Acknowledge button to permit the operator to correct a mistake.

This covers the major subsystems of the Gemini control system. Another Gemini system not part of the control system is the Data Handling System, of which the quick look subsystem is a major component. Although this book is about telescope control, not data handling, this brings up an important point—telescope control systems do not work in a vacuum. Almost any telescope worth the trouble to computerize will, at some point, have an electronic instrument attached to it. This instrument must be controlled and the data from it managed properly, so the telescope control system must be integrated with instrument control systems and instrument data handling systems.

In the case of the Gemini telescopes, this integration is quite tight and occurs at several levels. For example, an on-instrument wavefront sensor that in some cases may be located inside the instrument cover must communicate a large amount of data very rapidly to the acquisition and guiding subsystem and the secondary mirror tip/tilt subsystem to keep the

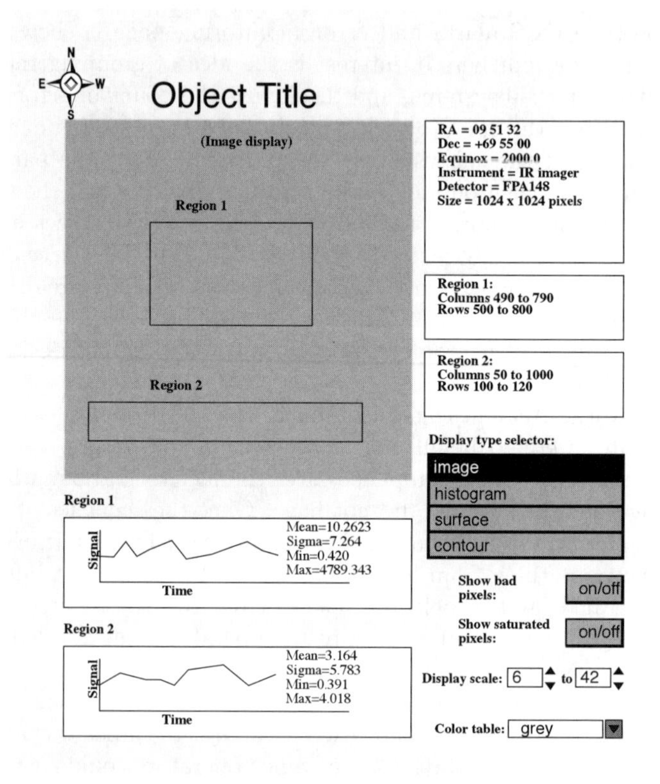

Fig. 14-8 *An Example Quick Look Page. Courtesy International Gemini Project Office.*

telescope pointed and tracking accurately, and to make best use of the observing conditions. For other telescopes, integration may be no more than sharing a disk file.

An example of a quick look page is shown in Figure 14-8. Data from an instrument is displayed in the area marked "Image display." The black rectangle marked "Region 1" indicates the aspect ratio of images, while "Region 2" is for spectra. Both might be used if a spectrograph has a post-slit viewer as well as the spectrum detector. The observer might put the spectrum in Region 1, then select some rows (e.g., containing an order of an Echelle spectrum) to be displayed in Region 2, then the observer calls up the post-viewer image in Region 1. This page offers minimal image processing capabilities and display types to help the observer determine whether the data are good, without enticing the observer into playing with the data all night instead of observing.

Another major Gemini software package that must be integrated with the control system is the observation planning tool. Gemini has several different ways of observing, including queue scheduling, remote observing, and service observing (with and without eavesdropping), as well as the traditional approach. Proposals are placed into different queues, depending on the instrument to be used, the seeing required, and other factors. As the seeing changes night-to-night, different queues are executed. The observations in these queues are preplanned, and are performed by an astronomer who may not be familiar with the observer's field of research. They are planned using an observation planning tool and placed into a file that must be readable by the TCS. This is similar to the ATIS file read by automatic photoelectric telescopes (see Part V).

The Gemini control system and related systems represent the latest concepts for controlling large, modern astronomical observing facilities used by a broadly based international community. As such, it incorporates many of the features of good control systems discussed in Chapter 10 to help observers and operators from around the world operate what promises to become two very productive telescopes.

14.8 Indiana University Control System

Honeycutt and Turner (1992) describe the control system software for the 16-inch automatic telescope at the Goethe Link Observatory at the University of Indiana. This telescope is used to obtain CCD images that are reduced for stellar photometry, and is controlled by a MicroVAX II computer running VMS. The CCD imager generates about 100 to 200 exposures per night, each of which is 1.3 MB. To avoid having to store all these data, the Indiana system includes data reduction software that automatically detects each star above some minimum magnitude, centroids it, and extracts an instrumental magnitude. This number is retained while the image from which it came is discarded, resulting in an effective data compression of nearly 1000:1. The control system points the telescope within $15''$ of the center of an $8'$ CCD field of view.

The system controls the telescope, observatory, camera, and auxiliary equipment, such as cryogen fill and weather equipment. Weather is monitored by checking for precipitation and for visibility of Polaris. The system selects a star from a list, then performs a number of checks including making sure the star is visible, whether the weather sensors indicate observing is possible, whether liquid nitrogen is needed in the camera dewar, whether dome flats are needed, and the time (whether it is still night). If all tests are passed, then the system moves the telescope, correcting for precession, aberration, refraction, and flexure. All motions are made using intelligent motor indexers that plug into the MicroVAX II Q-Bus. The system adjusts the focus, moves the dome, and moves the right filter into place, then takes the exposure. The image is reduced and stored, then the system checks to see if this is the last exposure for the star. If not, the loop is re-executed starting with the checks. If so, another star is selected and the cycle is repeated. More details are available in the referenced paper.

14.9 Quadrant Systems

Quadrant Systems offers two commercial systems: the Coordinate I for telescopes up to 16 inches in aperture, and Coordinate II for larger telescopes. Both systems are designed to accommodate either portable or fixed site operation. They both use a hand paddle or joystick and LCD display as the basic operator interface. Encoders provide position information on the LCD display. Object coordinates are specified by entering an NGC number or using the RS-232 serial port. When the controller uses the remote interface, it sends out an ASCII user menu for use with a terminal, a computer running a terminal emulator, or with a computer running *The Sky* or *Astrolink* to provide a modern GUI operator interface. In miniature, these systems resemble the EPICS workstation/VME combination in that the control and status processing are performed in a physically separate box from the user interface.

The Coordinate I controller sends half steps to stepper motors, with a maximum slew rate of 10,000 half steps per second. Gearing must be chosen to obtain reasonable tracking and slewing rates. This limitation makes the Coordinate I most suitable to equatorial mounts. Quadrant Systems claims the tracking accuracy is $30''$/hour. The Coordinate II uses a digital frequency synthesizer to generate speeds in increments of 1×10^{-7}, permitting tracking of moving targets as well as sidereal objects, with "no appreciable drift over 10 hours." The Coordinate II permits microstepping in the range of 400 to 51,200 microsteps per full step. Both systems are compatible with SBIG CCD autoguiders, and support selectable gear ratios, speeds, and backlash settings. More information is available from the manufacturer (see Appendix E).

This concludes the section of this book devoted to examples of control systems for conventional telescopes. The next section is devoted to controlling fully automatic telescopes that perform routine photometric observations every clear night without help or guidance from human operators.

Part V

Robotic Telescope Control

Chapter 15

Automatic Photoelectric Telescopes (APTs)

15.1 Robotic Telescopes

A telescope is "robotic" when its operation is fully automated, including all the appropriate instruments and controls, and it is located within a fully automated and normally unattended observatory dome or other enclosure.

In the previous four sections of this book we have assumed that humans were, to some extent, in control of telescopes. True, a computer was assumed to carry out the tasks of determining coordinates, of calculating various astronomical and mechanical corrections for the sources of systematic errors, of commanding appropriate stepper or servo motors to actually move the telescope, and of reading encoders to assure that the telescope moved precisely to the desired position. We even assumed that a computer served as a human interface to the telescope and its instruments, often via friendly graphical displays. But we always assumed that a human would at least check the weather, open up the observatory, turn the computer on, and decide what object should be observed first, when and what object should be observed next ... and finally, would decide when to stop observing (dawn, bad weather, or finished observing program) and turn off the equipment (including the computer) and close up and secure the observatory.

In this final section of our book we will still assume that a computer will determine the coordinates of objects, calculate corrections, command movements, and read encoders. All this has already been well treated in previous sections. It will not be repeated here as there is no difference. We will not, of course, be concerned with human interfaces, at least at the telescope, for there will (usually) be no humans there.

Instead we will now assume that the computer will itself check the weather, open up the observatory, decide which object will be observed first, the next object to be observed..., when the observatory will be shut down, and then actually shut down and secure the observatory. In addition, we will concern ourselves with the sending of observational results (typically via Internet) to users (usually human) thousands of miles away, how a user can check the quality of the data (Murphy is a special friend of robots), and finally, how users can, at a high level, *suggest* changes in what the robotic telescope might observe. These will be the topics in this final section along with, again, the human-computer interface. This time, however, it is only a remote user who is checking the quality of the observations or giving high-level guidance (during the day) to a robot that works while humans sleep.

Robotic telescopes are of interest because they can efficiently and conveniently conduct science that otherwise would be prohibitively expensive, difficult, or boring to do. Massive observational programs, requiring months or even many years of observations, night after night, can readily be handled by robotic telescopes. The quality of observations these telescopes produce can be superior to that of manual observations, due to extensive nightly and fully automated quality control observations.

While individual robotic telescopes cost about the same to purchase as non-robotic telescopes, they often obtain more high quality observations because they are located at the best possible sites, with clear skies and stable seeing, and are extremely rapid and precise in their movements. Furthermore, robotic telescopes never take coffee breaks, stop for midnight snacks, or ask for vacation time. For typical photometric observing programs, robotic telescopes can gather data at a rate several times faster than manual observers. Robotic telescope costs are often reduced via sharing: multiple robotic telescopes can share the same site and enclosure, and multiple institutions can share the same telescope. Additional facilities, equipment, maintenance, and administrative costs can also be shared. Because users do not need to be transported to and housed on remote mountaintops, transportation costs are essentially eliminated, and existing mountaintop observatory sites can be utilized without having to expand visiting astronomer facilities or support. This later point is crucial for future operations at such extremely remote but vital observational sites as the South Pole and a lunar outpost.

Robotic telescopes allow users to sleep at night. Astronomers do not have to skip classes to travel to remote observatories to make observations or worry directly about quality control and standardization observations—these can be automatically made for them every night. This convenience improves the quality of science, as observational programs can have larger samples and be of longer duration. Perhaps most importantly, an astronomer's time can be spent analyzing results rather than gathering data.

15.2 Automatic Photoelectric Telescopes (APTs)

Automatic Photoelectric Telescopes, henceforth APTs, are a subset of robotic telescopes. Because they are the simplest of robotic telescopes, we will consider them first and use them as our primary example of robotic telescopes. In the last chapter of the book, we will then expand our view to other types of robotic telescopes beyond those dedicated to stellar photometry.

In astronomical research, small telescopes are often used for photoelectric photometry. The wide bandwidth of UBV(RI) photometry, the high quantum efficiency of the photodetectors, and the "zero dimensional" nature of the data all tend to make the best use of the meager photons available to smaller telescopes. Most small telescope photometry is of variable stars. Such photometry is generally a highly structured and repetitious task that readily lends itself to automation.

It is convenient to break variable stars into two broad classes: Those with short and long periods. Short period variables are those stars that require observation continuously for many hours or all night long, while long period variables only need to be observed once (one set of readings) per night. Typically, one observes a single short period variable all night long, while one observes many long period variables in a single night and repeats these observations for many nights. As these two situations are quite different, it would be expected that different systems might evolve to meet each situation.

Short period variables, especially eclipsing binary stars with short duration eclipses, have been favorite photometric objects since the beginning of photometry. The first small semi-automated telescope system to observe them was designed and built by Skillman (1981). This system, which is controlled by Apple and KIM microcomputers, has been in operation for many years now. The system required manual startup, shutdown, and "training," and thus was not fully automatic.

However, in this chapter we will not be concerned with systems for observing single short period variables hours on end, but will, instead, concentrate our attention on systems designed for observing many long period variable stars each night.

The small automatic photoelectric telescope (APT) developed by Code and others at the University of Wisconsin in the mid-1960s (McNall et al., 1968) was fully capable of observing long period variable stars, although it was not applied to this task to any extent. Rather, it was used to observe a number of standard stars spread across the sky for purposes of determining nightly extinction coefficients. This 8-inch telescope was controlled by a PDP-8 minicomputer, and its photometer was at prime focus. It was housed in a small roll-off roof building, and its operation was fully automated. This promising start on this type of APT was not developed further until Boyd, quite independently, knowing nothing about the long since discontinued Wisconsin effort, developed a small APT specifically for observing long period variable stars.

This chapter discusses some of the considerations involved in designing a small APT for observing a large number of different stars per night. It is not intended to be either exhaustive in enumerating design alternatives or to give specific recommendations. Rather the intent is to introduce the subject and recount some thoughts, experiences, and ideas picked up from others working in similar areas. The hardware design of an actual system is addressed in Chapter 15, while APT software is discussed in Chapter 16. For additional details, the reader is directed to Boyd and Genet (1983), that describes the first Fairborn Observatory APT during its development; to Boyd, Genet, and Hall (1984a), that announced the first fully automatic operation of the Fairborn Observatory APT, along with equipment details and the first observational results; to Boyd, Genet, and Hall (1984b), that describes operational experience and the second generation system; to Hall and Genet (1988), that describes automated photometry in some detail; and to Genet and Hayes (1989), that gives a comprehensive discussion of robotic observatories. Also see the three volumes on robotic telescopes in the Astronomical Society of the Pacific's Conference Series. These three volumes were edited by Alexei V. Filippenko, by Ronald J. Angione and Diane P. Smith, and by Gregory Henry and Joel Eaton.

15.3 Astronomical Considerations

For an APT to realize its full potential, it must be able to observe enough different variable stars each night to keep it occupied. If the telescope is too small, it could run out of accessible objects and have to spend time waiting between observations or making less useful repeat observations. If it were too large, it would become too expensive for the typical small college or advanced amateur to afford. The actual size needed to keep such a telescope fully occupied is a complex function of sky brightness, pointing accuracy, slewing rates, etc. Since, for simple systems that use the photometer itself for acquisition (instead of a CCD camera), a star is observed many more times during the process of acquisition and centering than it is for the actual measurements, the acquisition observations must be of short duration. Thus the need for a measurable photon level during the acquisition process actually places the lower limit on the acceptable telescope size. In general, an 8-inch aperture telescope can be kept fully occupied, and a 16-inch telescope would have a tremendous selection of stars.

Another important consideration is the accuracy of the observations. Small observatories tend not to be located at ideal mountain-top sites, and short term variability of atmospheric extinction can be a serious problem. The speed at which an automated system can move between and acquire stars can increase the accuracy of differential measurements since there would be less time spent between measurements, minimizing changes in atmospheric extinction. The same thought applies to variations in the sky readings. Generally, the atmospheric scintillation, and not the sensitivity of the photometer, places the lower limit on integration times for an automated system. A larger telescope helps this somewhat, but the difference between an 8-inch and 16-inch instrument might only reduce the required integration time by half.

Finally, for long period variables it is possible to arrange the observing schedule so that most observations are made close to the meridian, thus providing the smallest air mass. While this can improve accuracy, it reduces the year-around coverage of the variables. Coverage is maximized by observing stars in the west at the beginning of the night and in the east at the end of the night. These are but two of many observing strategies that can be used in APTs.

15.4 The Telescope

Photometers have been successfully placed at prime, Newtonian, and Cassegrain foci of the APTs mentioned earlier. All things being equal, photometers are easier to design for higher f-ratios. The more timid observer may still wish to be able to look through an eyepiece to see if the automatic system can actually obtain the correct star and accurately center it, but this feature can add to the complexity of the system, and is of little use once the telescope is operational. A large finder scope is more useful in watching the actual acquisition process on the side, so to speak.

The software development task on a small APT is generally much greater than the hardware development task. Because of this, the best approach to the hardware is one that simplifies the software. There are two simplifications that should be seriously considered. One is to use a well aligned equatorial mount. The other is to provide the stellar rate motion separately in a way that is transparent to the software. The software task then becomes one of acquiring and measuring stars in a sky that does not rotate.

When an astronomer is at the telescope eyepiece, it is possible to take a few liberties with polar alignment, the orthogonality of the Dec axis to the RA axis, and the alignment of the optical axis with the mechanical axis. A computer is not as forgiving! It is wise to make provision for the fine adjustment of the polar axis in both altitude and azimuth, and also for fine adjustment of Dec axis orthogonality and optical axis alignment. Close attention to these points will help keep the acquisition searches small.

An APT is unusual in that it makes a large number of quick, short movements during the course of a night's work. During a full night of observing, one of the Fairborn Observatory APTs makes about 25,000 separate movements—most of these in acquiring and centering stars after a slew. All of this starting and stopping places special demands on the telescope design. It is much easier to move the telescope around quickly if it has a low moment of inertia. A symmetrical fork or yoke design is thus most desirable. Compact and lightweight Schmidt-Cassegrain optical assemblies are well suited to this application.

All of the short, fast movements tend to induce vibrations in the telescope, and observations can not be made until these vibrations die down. The vibrations are minimized and die down most quickly when the natural resonant frequency of the telescope structure is high. A very stiff, relatively lightweight telescope structure tends to have a high natural frequency. Also, the necessity to make many quick, short movements requires drives with a minimum of backlash.

15.5 Mount and Drives

Most large telescopes have separate slewing and tracking motors. The requirements for an APT are different, however, from other telescopes. In a manually operated photographic telescope, for instance, high tracking accuracy for long periods of time is a prime requirement, and rapid slewing rates are desirable for operator convenience. Large telescopes are sometimes equipped with shaft encoders to determine the position of the telescope, but these are unnecessary for APTs. The sudden stopping and starting place special demands on the design of an APT, but the ability to measure the light from a star reduces the requirements for the drives in some respects. Since the APT stays on one star for less than a minute, and since it can update the position of the telescope relative to the sky each time a star is centered, high tracking accuracy is not required. More vibration of the telescope is acceptable in photometry (at least in aperture photometry) than with photography or visual observation, and this allows the size of the steps of the tracking motors to be greater (even as high as $2''$ per step). The step size, in seconds of arc, is not necessarily the same as the angle of movement caused by vibration, which may be greater or smaller than the step size, depending on the mechanics of the mount and telescope.

There are really three different accuracies involved in positioning APTs. First there is the open-loop accuracy needed after making a long move from one side of the sky to the other. Such a long move can be followed immediately by an expanding spiral search for the first "navigation" star. A navigation star can be purposely chosen so that it will not be confused with any nearby stars—thus avoiding the problem of the system's locking onto the wrong star. There are, however, time limits to how big a patch of the sky it is practical to search for the navigation star. The area to be searched (and the time taken to search it) is the square of the absolute pointing error, so the accuracy of the open loop "long move" must be reasonable. Our experience indicates that $12'$ accuracy is sufficient. This could be significantly relaxed if long moves across the sky were rare, or if inefficiency due to long search times were acceptable. To achieve $12'$ accuracy, it is helpful to precess from mean coordinates to the current epoch, but no other corrections are needed. Open-loop accuracies over long moves of better than $12'$ reduce the search time, but not enough to make a noticeable effect on the overall efficiency of the APT system.

The second accuracy of importance is that of a local move from one star to another after the navigation star has been initially acquired. We require this to be $3'$, but have found that it is generally much better—typically about $30''$. When slewing between stars in the same area, the APTs usually put the stars within the diaphragm on the slew, and if not, almost always they are found in the first loop of the square spiral search pattern.

The third accuracy is that of centering a star within the diaphragm. This is a tradeoff between centering accuracy and time taken to center. We initially required an accuracy of about only 20% of the diaphragm diameter—about $6''$ to $12''$. We eventually found, however, that inaccuracies in centering were the main contributor to measurement errors, and our later systems center stars with an accuracy of $1''$ or better.

When using stepping motors, it is desirable to accelerate and decelerate smoothly (ramp) at the ends of each movement. The system built by Boyd initially ran with no ramping, but its efficiency was greatly improved when it was modified to include this feature.

15.6 Control System

The control systems on large conventional telescopes can be expensive because the pointing accuracy without reference to any stars in the sky may have to be only a few arc seconds. This often requires that encoders be mounted on each axis, and that corrections be made for precession, atmospheric refraction, flexure of the telescope, and other minor errors. Because of its ability to search for, acquire, and center stars, an APT can dispense with all of these corrections, except precession. There is no need for encoders—the loop, so to speak, is closed on the stars themselves.

In a conventional telescope control system, the need to calculate the many systematic error corrections repeatedly keeps the computer busy, and can require a very capable microcomputer. However, by dispensing with the need for calculating extensive corrections, an APT can easily make do with a modest computer and, if appropriate steps are taken (pardon the pun), it can also directly control hardware stepper motor drivers. Most programming can be accomplished in a high level language, such as BASIC or Pascal, without the use of interrupts. The use of a high-level language without interrupts considerably reduces the programming effort.

15.7 Background of the APT Project

A discussion of APTs would be incomplete without discussing how Louis Boyd came to develop the first Fairborn Observatory APT, and some specifics of his development program. With an interest in both astronomy and electronics, Louis (an electrical engineer) decided to build a radio telescope. In 1972, while looking for parts for his radio telescope in a surplus store, he ran into a fellow "scrounger," Richard Lines. Richard and his wife Helen had an optical observatory in Mayer, Arizona, and soon Louis was helping them on the electronics for a photoelectric photometer. The photoelectric observations at the Mayer Observatory were always made with Richard and Helen working as a team. Louis watched a number of these observational sessions, and began thinking about how stepper motors and a computer might be used to move a telescope between stars—perhaps even automatically centering them. Helen Lines, after hearing talk about a computerized photoelectric telescope for a while, finally told Louis to "stop talking about one and build one!" This was the beginning of the APT project. The radio telescope was permanently set aside.

The APT project began in earnest in early 1979. First on the agenda was the physical construction of the telescope itself. A modified fork mount was selected for the telescope. The modification consisted of extending the fork beyond the north polar axis bearing to a second cross-piece at the south polar axis bearing. This added stiffness to the fork, but at the expense of limiting the angular motion of the telescope in RA. While this worked well on Boyd's APT, he does not recommend it for other APT telescopes for several reasons. First, a sufficiently rigid fork is possible without the extension. Second, a larger sky coverage would be helpful in some cases. Finally, if 180 degrees, or even better, a full 360 degrees of rotation can be

Fig. 15-1 *The Boyd automatic photoelectric telescope.*

made, the alignment process is simplified. Although it takes more space than a fork, a yoke mount has many of the properties desired in an APT, and is worthy of consideration.

Besides the mount configuration itself, another important decision was the mechanical drive system. Precision gears are quite expensive, and without the complications of preloading, they are subject to backlash. Boyd chose to use a chain and disk drive. The disks (in both RA and Dec) are 32 inches in diameter, $^3/_{16}$-inch thick, and are made of aluminum. They were purchased on the surplus market for $5 each, and were suspected of being part of a Honeywell computer memory system in a previous life. While this drive has worked well on the APT, Boyd points out that mixing different types of materials in the mechanical parts of the telescope, such as aluminum disks and steel chains, leads to differential temperature expansions, and these in turn can necessitate more frequent tension adjustments than might otherwise be the case. Sticking with either steel or aluminum throughout the mount has much to commend it. Louis Boyd and his original APT are shown in Figure 15-1.

The problem of getting an APT to track a single group of stars (variable, comparison, and check) for many hours is much simpler than the problem of getting an APT to observe many groups of stars in an appropriate order, recording and analyzing the data, etc. One needs to know, for instance, where the Sun and the Moon are located, so as to start and terminate an observing run when the sun is safely out of sight, and to avoid trying to observe stars too near the Moon (this can be hard on PMT systems). It was these problems that occupied the last year of the development program. In October, 1983, all the pieces finally came together, and the Boyd APT started observing multiple groups of stars. Genet had been following the development of the system rather closely, supplying occasional ideas, and was kindly invited to observe the system on its first full night of totally automatic operation on October 13, 1983.

The early Fairborn Observatory APTs each handled about 75 groups of stars per night (225 different stars). These observations were made at the rate of about one group per 7 minutes, with 33 ten-second observations being made per group. Thus about 330 of every 420 seconds (7×60), or 67% of the time within a group is spent on the actual measurements, and only 33% of the time was spent on moving between stars, acquiring stars, centering stars, and changing filters. Moving between groups typically took between one-half and two minutes. The efficiency of the systems was eventually improved even further, as discussed in the last chapter. The next two chapters describe the hardware (primarily electronics) and software of the early Boyd APTs.

Chapter 16

Basic APT Control Hardware

16.1 Introduction

The telescope control system described in this chapter is an open-loop type with a single stepper per axis. Since the system is open-loop, there are no shaft angle encoders or other telescope position sensors. A fixed gear ratio is used in each axis, and the necessary speed range results from the use of stepper ramping and special overvoltage (bi-level) stepper drivers. The electronics were designed by Louis Boyd. The schematics, diagrams, and software listings that appear in this book have been furnished courtesy the Fairborn Observatory.

While this design was developed in the early 1980s, we have retained it in the second edition of this book because its simplicity best illustrates the principles of APT control. Later APT control systems became much more complex and are discussed, in a general way, in the last chapter, there not being room for the details of highly complex, specialized systems.

The simple APT control system described in this chapter was implemented by one of us (RG) in 1984, using a single board computer costing only $300. One need not spend a lot of money to develop a computer controlled telescope. The computer used in this example was based on the Motorola 6809 microprocessor. While the 6809 was a good choice in 1983, and the Microware OS9 operating system and Basic09 language were well-suited for complex real-time control tasks such as telescope control, more capable computers, operating systems, and languages are now available.

16.2 Telescope Mount and Drive

When the mechanical portion of this system was designed, the decision was made to use existing piers and optics. The north and south piers

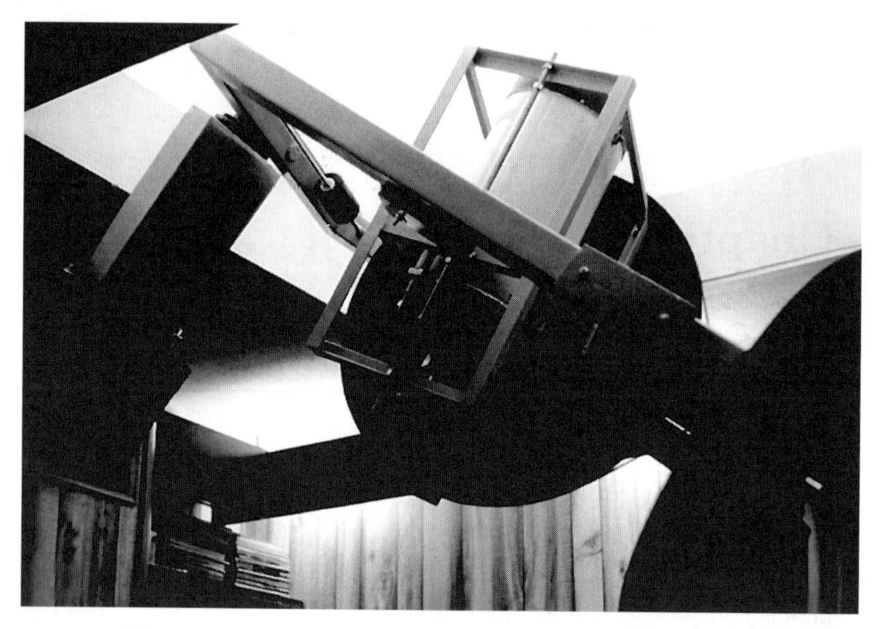

Fig. 16-1 *Overall view of the yoke-mounted telescope.*

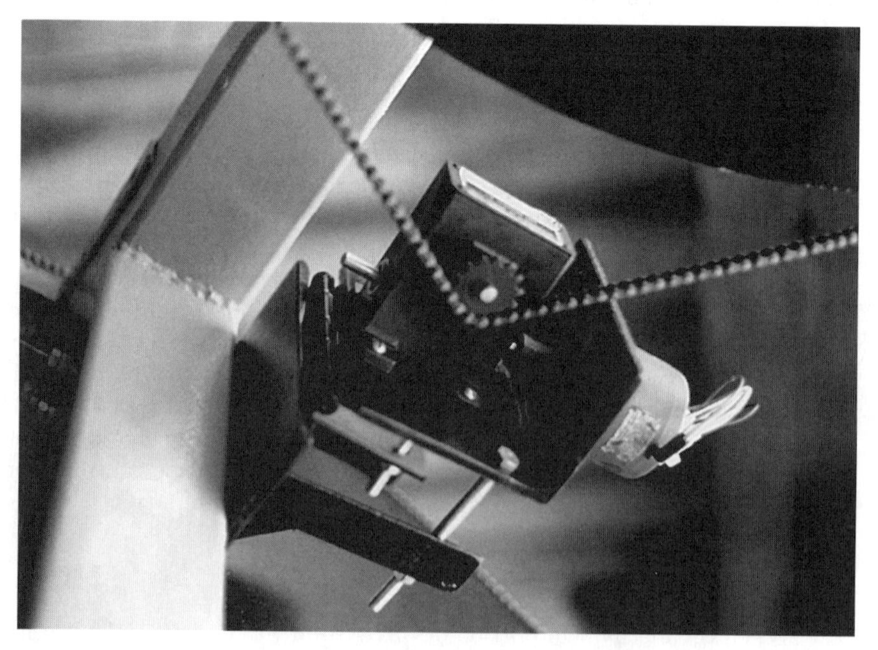

Fig. 16-2 *Declination stepper motor and Berg reduction gear.*

were large concrete structures that were originally used to support the massive 0.4-meter telescope at Fairborn Observatory in its original Ohio location (before it was moved to Arizona). The optics consisted of a 10-inch Schmidt-Cassegrain set donated by K. Kissell. A yoke configuration was chosen to utilize the two piers and maintain a low moment of inertia with a symmetrical mount. An overall view of the mount, drive, and optics is shown in Figure 16-1.

Drives in both axes consisted of large-diameter, thin aluminum disks driven by chains and sprockets. These were, in turn, driven by steppers through a gear reduction unit consisting of a single antibacklash worm gear in a housing that sealed the gears against dust and dirt. The 32-inch diameter aluminum disks ($^1/_8$-inch thick) were driven by $^1/_4$-inch pitch industrial chain from 1-inch diameter sprockets. This gave the final drive stage a 32:1 reduction. The worm gear unit was a 120:1 reduction unit from Winfred Berg. The worm wheel was split in two and spring loaded against itself to eliminate backlash. The Berg unit was driven by Slo-Syn M061-FD02 steppers in the half-step mode (400 steps per revolution). The Berg reduction gear and Slo-Syn stepper motor for Dec are shown in Figure 16-2. The total number of half-steps per revolution for the telescope was $32 \times 120 \times 400$, or 1,536,000 half steps per revolution. As there are $360 \times 60 \times 60$ or 1,296,000 arc seconds per revolution, each half step corresponds to $0\overset{''}{.}84375$.

Chain drives have distinct advantages and disadvantages. The advantages are that they are very easily built without the need for special machinery or careful alingments—ideal for the casual "homebrewer." Also, they have no perceptable backlash and they are reasonably low in cost. For small telescopes, such as that discussed here, the chain is stiff enough that the total system stiffness is not seriously degraded, but this would not be the case for much larger telescopes. The sprocket and chain combination does cause some periodic error, however. While this may not be objectionable in Dec, the RA periodic error might not be appreciated by those engaging in astrophotography, although it is of less concern to photometrists.

16.3 Control System Hardware

16.3.1 System Block Diagram

The overall layout of the control system electronics can be seen in the system block diagram shown in Figure 16-3. The equipment, besides the

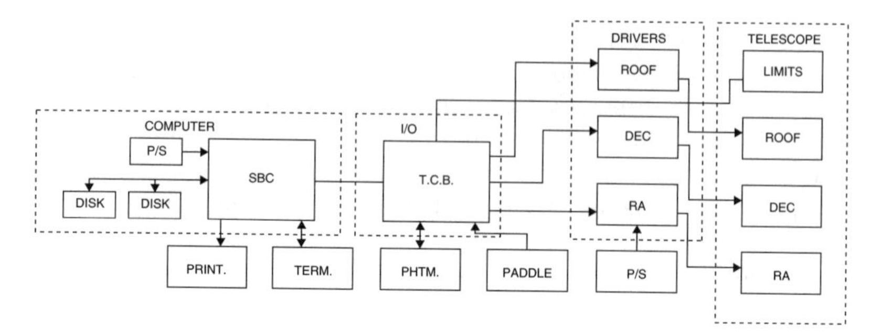

Fig. 16-3 *Control system block diagram.*

steppers and limit switches on the telescope itself, was placed on three aluminum chassis plates, one each for the computer, telescope control board, and the stepper driver board. There were also a number of auxilliary items such as a control paddle, terminal, etc.

16.3.2 PT69 Computer

The PT69 single board computer was purchased from Peripheral Technology as a completely assembled and tested board. The cost was slightly less than than \$300. The board was only 5.5×6.5 inches in size, as shown in Figure 16-4. Included on the board were the 6809E processor, a floppy disk controller, two RS-232 ports with adjustable baud rate, a clock/calendar using the MC146818 chip, 4K bytes of EPROM, and 56K bytes of RAM. Needed to complete the computer was a power supply, two disk drives (5.25-inch floppies), and a chassis or cabinet. This early home-brew computer used at the "Ohio" Fairborn Observatory is shown in Figure 16-5.

The system described in this chapter would work well with many other computers. All that is required are provisions for access to the data lines and sufficient address lines to provide the port select signals.

16.3.3 Telescope Control Board

The telescope control board provides three vital functions, and three optional ones, as can be seen in Figure 16-6. The vital functions are the buffering or latching of the control computer signals, control of the RA stepper, and control of the Dec stepper. The optional functions are control of a third stepper which changes a photometer filter, a counter for use with external instrumentation (such as a photometer), and a clock/calendar.

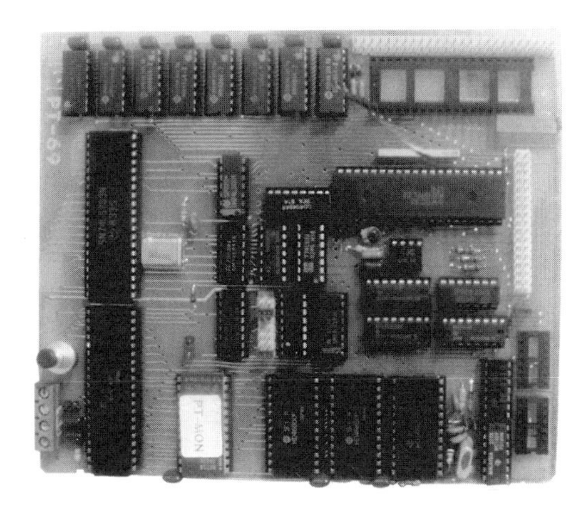

Fig. 16-4 *The PT69 single board computer.*

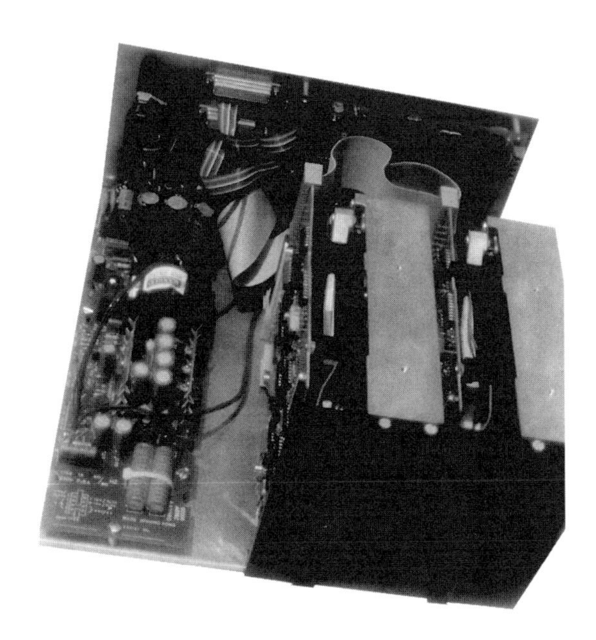

Fig. 16-5 *Complete computer.*

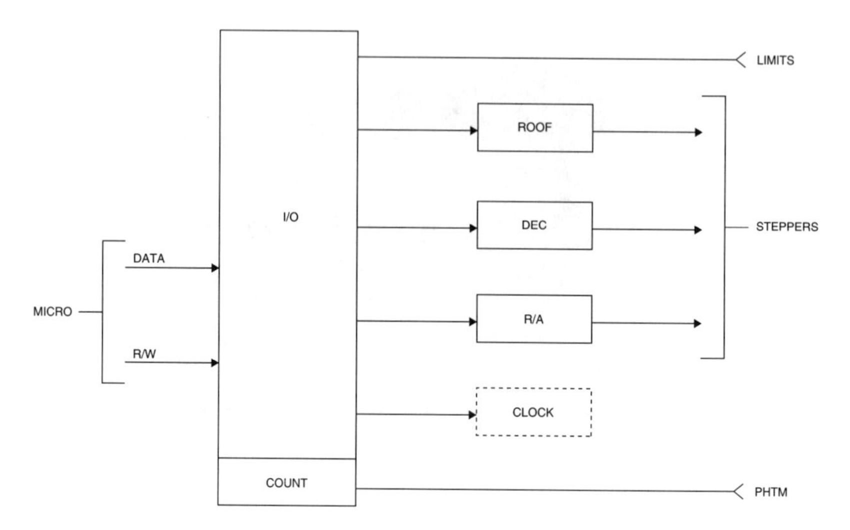

Fig. 16-6 *Telescope control board functions.*

The version of the telescope control board discussed here implements all the vital and optional functions, except for the clock/calendar (not required), as there is already one on the PT69 computer board.

The computer interface portion of the telescope control board is shown in Figure 16-7, along with the optional counter. On the left of the schematic is the 50-pin connector that runs from the PT69 (the connections would be different for other computers). Coming from the computer are the eight data lines, four read port select lines and four write port select lines, +5 volts DC, ground, and an already somewhat divided down clock signal at 308 kHz. This clock signal is just passed on to the RA section, to be discussed later, for the sidereal rate drive.

The top chip in Figure 16-7, IC 10, provides inputs to the three stepper motors used in this system (there are provisions for a fourth stepper if one were desired). The outputs are normally high, and the appropriate outputs are brought momentarily low for the stepper or steppers one wishes to move.

IC 13 is another output chip, and it provides latched outputs to control a variety of functions. Two lines are used to move the photometer filter stepper motor, enable high voltage for the photometer, and switch in a photometer count prescaler. These photometer functions are optional, and these lines could easily be used for other functions. Another function is an "enable" line for the 24-volt overvoltage, which turns on the high drive voltage to the steppers during high-speed operation. This is a necessary, but not sufficient condition for applying the overvoltage. Once the sidereal

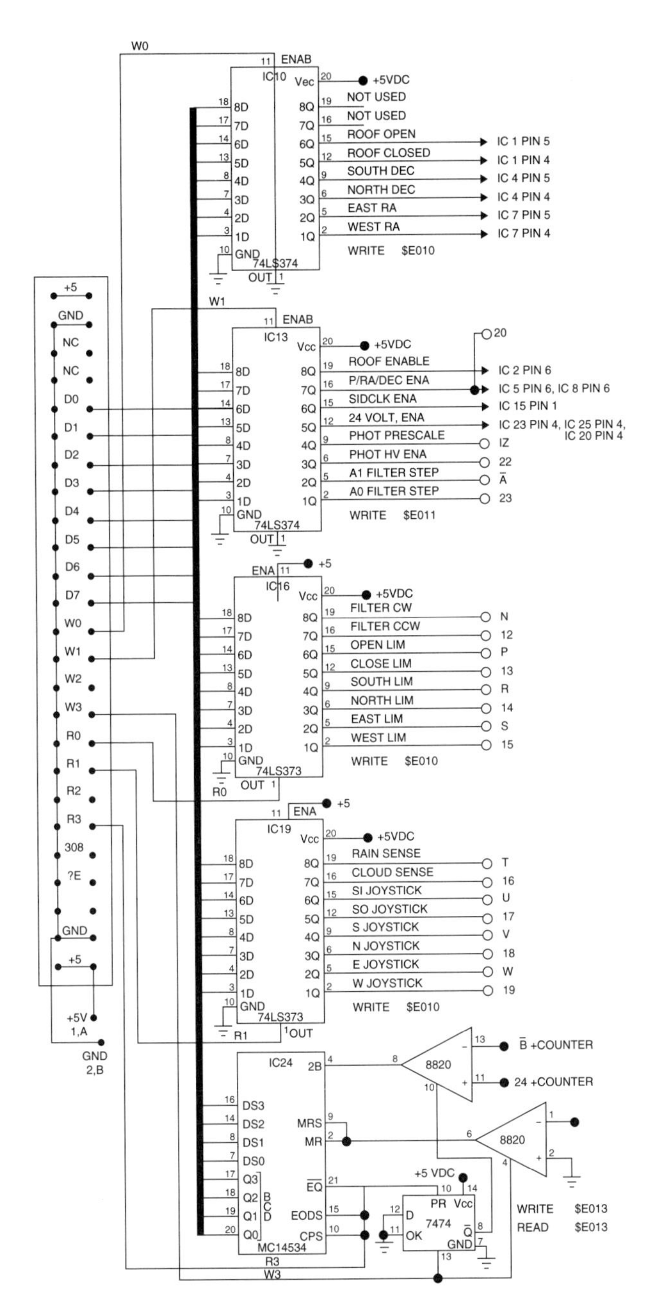

Fig. 16-7 *Telescope control board A—interface and counter.*

clock is enabled (high), the sidereal rate pulses are provided to the RA stepper motor. When the photometer/RA/Dec stepper enable line is high, these steppers are enabled. Finally, there is the roof stepper enable line (separate so it can be disabled unless the telescope is in its home "stowed" position).

IC 16 is an input port that handles all the limit and position sensing switches for the entire system. While a computer controlled telescope can be operated without limit switches, one software problem could cause the telescope to encounter an obstruction or wind up the cables. The "home position" limit switches also define an initial position for the telescope.

IC 19 is another input port that handles two control paddle inputs and two auxilliary inputs. The auxilliary inputs in this system are weather sensors which, when activated, cause automatic system shutdown. These are not needed for a manually operated system, as most astronomers can tell when it is raining on their telescope and know what action to take!

Finally, IC 24, the two halves of the 8820 chip, and one half of the 7474 chip provide a simple counter with differential (balanced line) input. Writing to the 7474 starts the counting from the input, while the reading of the first digit stops the count. After the other digits are read, another count can be taken.

Shown in Figure 16-8 is the roof control portion of the telescope control board. IC 1 is a binary up/down counter. Roof open pulses cause the counter to count up, while roof close pulses cause the counter to count down. The three least significant bits of the counter output are fed to a 3-to-8 line decoder/multiplexor (IC 2), which, in conjunction with the three input 7410 NAND gates (ICs 3, 6, 9, and 12), provides the four-phase pulses required by the roof control stepper motor.

IC 23 is a retriggerable monostable multivibrator used as a one-shot. As long as pulses are continuously received from "roof open" or from "roof closed" and the 24-volt enable is high, then the Q output (pin 8) will be high and the roof high voltage (24-volt slew overvoltage) will be enabled, but if there are no very recent steps or if the 24-volt enable is low, then the overvoltage will not be enabled. The reason for all this is that providing a stepper with a five times overvoltage for high speed operation is just fine, as long as the stepper continues to step along at a reasonable speed. However, if this same overvoltage were applied when the stepper was standing still, it would quickly overheat. While the software should turn off the overvoltage at the instant the stepper stops stepping, IC 23 makes sure that even if this were forgotten, that it would be turned off within a fraction of a second anyway—protecting the hardware from software errors.

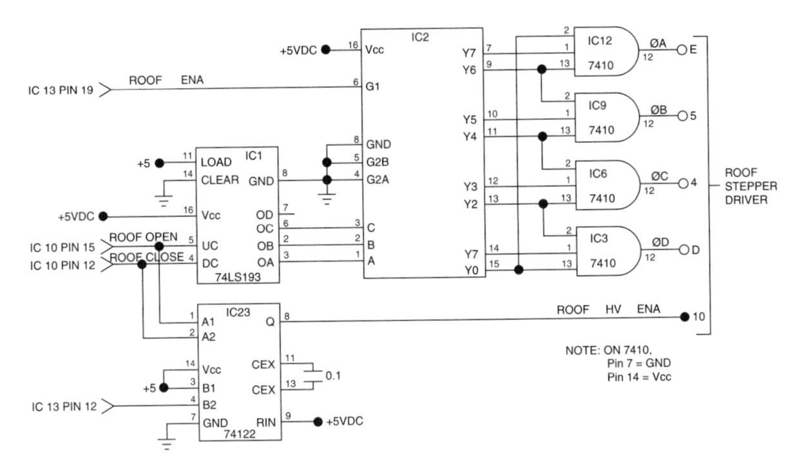

Fig. 16-8 *Telescope control board B—roof control.*

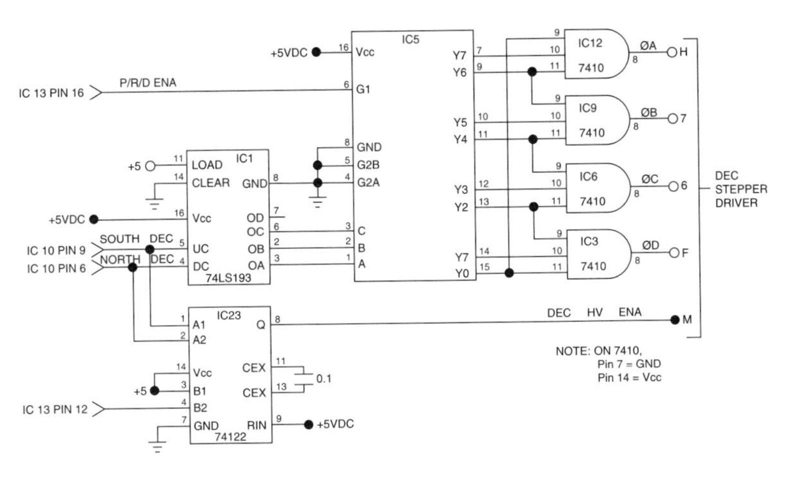

Fig. 16-9 *Telescope control board C—Dec Control.*

The Dec control portion of the telescope control board is shown in Figure 16-9. The schematic is identical to that of the roof control, except that the IC numbers are different, as are the inputs and outputs. As the functions are identical, no additional explanation is needed.

The RA control portion of the telescope control board is shown in Figure 16-10. ICs 7, 8, and 25, and the 7410s perform exactly the same functions as ICs 1, 4, and 7 in the roof and Dec controls. The added chips provide a constant sidereal rate in RA. ICs 15, 17, and 18 form a

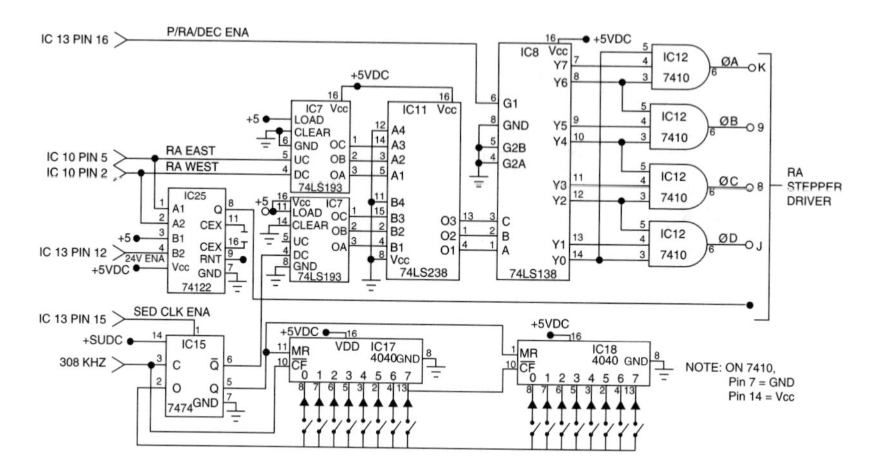

Fig. 16-10 *Telescope control board D—RA control.*

DIP-switch setable binary divider. This allows the 308 kHz constant clock from the PT69 computer to be divided down to the exact rate needed for sidereal tracking. IC 14 is an up/down counter that performs for the sidereal rate signal the same function provided by ICs 1, 4, and 7. IC 11 is a four-bit binary full adder, and it "adds" or combines the sidereal rate signal with any RA motion commanded by the PT69. When RA motion is west, the two are added; when RA motion is east, it is more appropriate to think of them as being subtracted.

In theory, one could provide the sidereal rate pulses and any commanded motions entirely from the computer without this external hardware on the telescope control board, but in practice, this is very cumbersome in software, and it also would add a greater processing burden to the computer.

Figure 16-11 shows the layout of the chips on the telescope control board (component side). The board itself is shown in Figure 16-12 on its chassis. The cable from the computer enters from the left. The cable to the stepper control chassis leaves from the front, and jacks for the control paddle and photometer are provided. Table 16-1 lists the pinout of the telescope control board, while Table 16-2 summarizes the port commands.

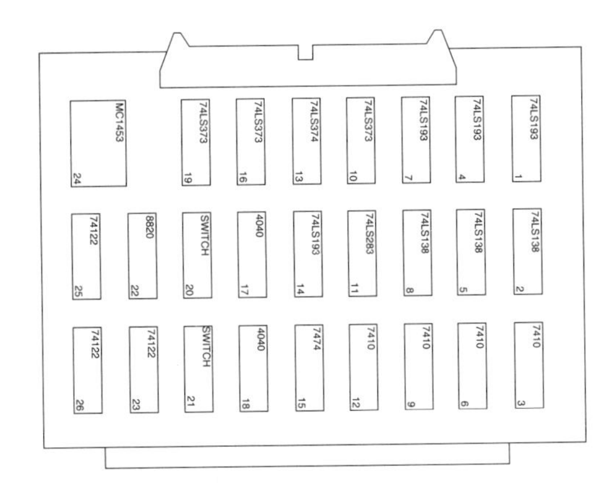

Fig. 16-11 *Telescope control board chip layout (component side).*

Table 16-1
Telescope Control Board Pinout

1	+5 Buss	A	+Buss
2	Ground	B	Ground
3	NC (Common to C)	C	NC (Common to 3)
4	Roof Phase C	D	Roof Phase D
5	Roof Phase B	E	Roof Phase A
6	Dec. Phase C	F	Dec. Phase D
7	Dec. Phase B	H	Dec. Phase A
8	R.A. Phase C	J	R.A. Phase D
9	R.A. Phase B	K	R.A. Phase A
10	Roof HV Pulse	L	NC
11	R.A. HV Pulse	M	Dec. HV Pulse
12	Filter CCV Limit SW	N	Filter CW Limit Switch
13	Roof Close Limit SW	P	Roof Open Limit Switch
14	North Limit Switch	R	South Limit Switch
15	West Limit Switch	S	East Limit Switch
16	Cloud Sensor	T	Rain Sensor
17	Joystick Speed O	U	Joystick Speed 1
18	North Joystick	V	South Joystick
19	West Joystick	W	East Joystick
20	Photo. Stepper Enable	X	NC
21	NC	Y	NC
22	Phot. HV Enable	Z	Phot. Divide by 10
23	Phot. Stepper Addr O	−A	Phot. Stepper Addr 1
24	Counter Input +	−B	Counter Input −
25	NC	−C	NC
26	NC	−D	NC
27	NC	−E	NC
28	NC	−F	NC

Fig. 16-12 *Telescope control board on its aluminum panel chassis.*

Table 16-2
Port Addresses and Functions

Address	Bit	Pin	Read	Bit	Pin	Write	Bit
			Limit Switches			Stepper Pulses	
$E010	0	15	West	1		R.A. West	1
	1	S	East	2		R.A. East	2
	2	14	North	4		Dec. North	4
	3	R	South	8		Dec. South	8
	4	13	Roof Close	16		Roof Close	16
	5	P	Roof Open	32		Roof Open	32
	6	12	Filter CCW	64		Unassigned	
$E011			Misc. Inputs			Misc. Outputs	
	0	19	West Joy Stick	1	23	AO Filter Stepper	1
	1	W	East Joy Stick	2	A	A1 Filter Stepper	2
	2	18	North Joy Stick	4	22	Photometer HV Enable	4
	3	V	South Joy Stick	8	Z	Photometer Prescale	8
	4	17	Joystick Speed	16		24 Volt Enable	16
	5	4	Joystick Speed	32		Sidereal Clock Enable	32
	6	16	Cloud Sensor	64	20	Phot., R.A., Dec. Enable	64
	7	T	Rain Sensor	128		Roof Enable	128

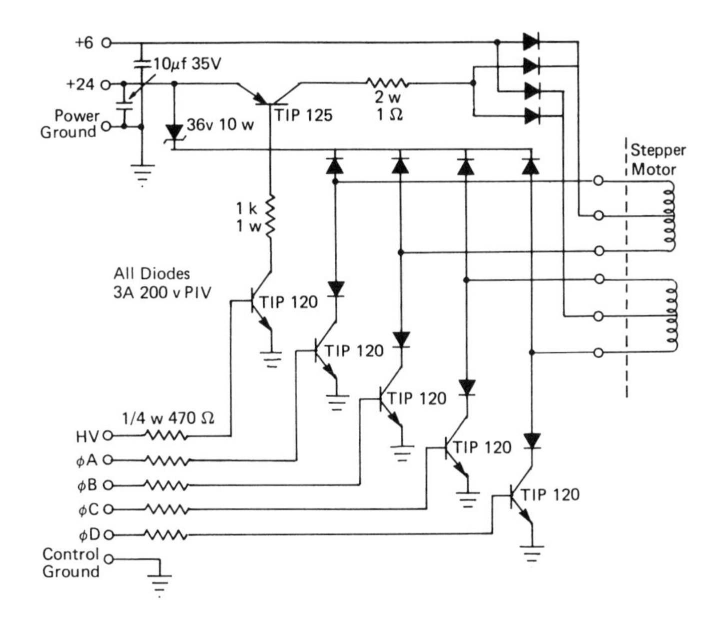

Fig. 16-13 *Stepper driver schematic diagram.*

16.3.4 Stepper Drivers

The schematic for the stepper driver boards is shown in Figure 16-13. The only unusual feature of the circuit is the use of the TIP 120 and TIP 125 to turn on and off the 24-volt overvoltage, and the use of diodes to switch between the 5- and 24-volt supplies.

The parts were wired point-to-point on a perforated circuit board. The driver board is shown in Figure 16-14.

16.3.5 Hand Paddle

Manual operation of the system is made simpler with the control paddle shown in Figure 16-15. The top four buttons determine direction (two can be pressed simultaneously, such as north and east, to move the telescope northeast). With none of the buttons pressed, the telescope moves at its slowest speed. The buttons on the bottom row are used with the direction buttons to obtain medium (set) and high (slew) speed operation. One of the bottom buttons can also be used as a panic "stop everything" button.

As you can see, the mechanical and electronic hardware are relatively straightforward. What makes an APT "automatic" is the software, which is described in the next chapter.

Fig. 16-14 *Assembled stepper driver board.*

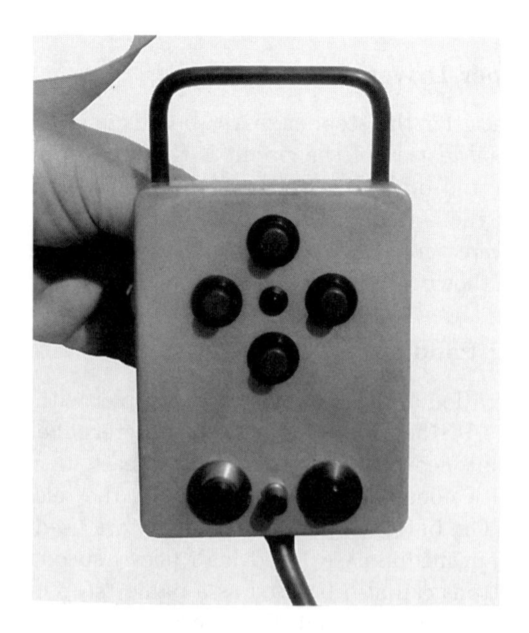

Fig. 16-15 *Control paddle.*

Chapter 17

Basic APT Control Software

17.1 Introduction

In most real-time control systems, the easy part is the hardware. The hard part is the software. In designing a system and making the various trade-offs between hardware and software, no effort should be spared to make things as easy to program as possible. The most dangerous words of all are "Oh, we can take care of that in software." This is particularly true if one is using software to correct a poor mount or a drive with backlash or excessive periodic error. The software to drive an APT could be written in almost any language, and could be made to work on almost any computer, but if consideration is given at the start of the project as to the exact requirements and objectives, backtracking can be prevented. Sometimes a few external chips can eliminate a large portion of the software, saving large amounts of programming time.

In the case of the early Fairborn Observatory Automatic Photoelectric Telescopes (APTs), the software task was greatly eased by the following:

1. Using an equatorial mount
2. Using a single motor per axis without clutches, gear shifts, tangent arms, etc.
3. A tight drive system without backlash
4. Not using shaft angle encoders or any other form of position feedback (except the photometer centered on a star)
5. Supplying the stellar drive rate via hardware, not software
6. Programming in a high level language (with only one exception)

7. Using an "assembly" of independent program modules
8. Having a control program that was "serial," i.e., it has no interrupts.

Without these simplifications, the job would have been much greater, although the system still required considerable work. Even under the best of circumstances the software development will not be an easy task. A decision must be made early in the system design as to how much data reduction will be done in real-time, and exactly how automated the system will be. It is also important to define the file structures for the input and output files and internal data structures before coding begins. The software task for an APT is sufficiently immense that thought needs to be given to breaking it into separate programs. There are undoubtedly numerous ways that this could be accomplished.

17.2 Elementary Software

Perhaps the most elementary software is that needed to move the telescope about the sky in slow motion. This is essentially performing the same function with the computer that one has on fairly simple non-computerized telescopes. In this section, the most elementary control software will be presented. The software is written in a way that individual control functions can be identified and separated from other functions by making each function a module called a *procedure.* The more advanced modules needed to do such functions as slew to coordinates and precess coordinates are presented later in this chapter, along with the modules needed to make the operation of the telescope completely automatic.

`QJOY2` is the main control procedure for manual control of the telescope without the display of coordinates. This procedure is not used very often, as the telescope is normally run in an automatic mode, and even in a manual mode, one would usually prefer to slew to coordinates. However, it has the merits of being very simple.

The `QJOY2` procedure "runs" or "calls" two other procedures—`DIGITAL` and `MOVE`, which in turn calls `RAMP`. This is a high speed assembly language procedure that performs the telescope slewing and all other telescope movements. These procedures are summarized in Table 17-1.

Before considering each procedure, it is appropriate to see what they do overall. After the procedures are loaded, one simply types in the command "RUN `QJOY2`" and picks up the control paddle and operates the telescope. The first thing `QJOY2` does is run `DIGITAL`, and `DIGITAL` returns to `QJOY2` with the parameter PADDLE, which is the value representing the keys pressed on the control paddle. From the combination of keys pressed, `QJOY2` calculates two parameters—`MOVEX` and `MOVEY`—which contain the number

Table 17-1
Software Procedure Functions

Name	Functions
QJOY2	Main control, decodes control paddle, and sets incremental move steps
DIGITAL	Reads control paddle
MOVE	Breaks overall move into two seperate moves and develops command for RAMP procedure
RAMP	Assembler procedure to ramp steppers up and down

of steps and direction to move in an X-Y (RA-Dec) coordinate system. West is `+MOVEX`, while north is `+MOVEY`.

`MOVE` is a very general procedure that can take any `MOVEX` and `MOVEY` input and cause the telescope to make the desired move. The general move case is shown in Figure 17-1. To move from A to C, `MOVE` breaks the move into two segments, the AB and BC segments. In the AB segment, both the RA and Dec steppers operate together, simultaneously taking an equal number of steps. After momentarily stopping at point B, the remaining steps on just one of the steppers (the case shown is RA) is executed. Each one of these two moves is made by `MOVE` running the `RAMP` procedure. Thus generally `MOVE` will call the `RAMP` procedure twice. After the second move has been made, `MOVE` realizes that the move is completed, and returns to `QJOY2`, which runs `DIGITAL` again to see if there is a new command.

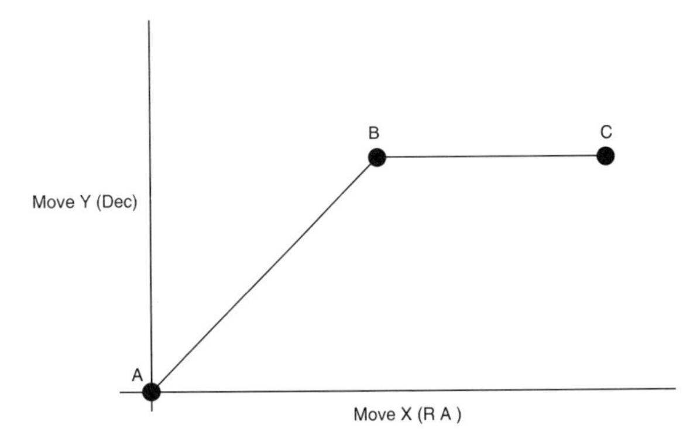

Fig. 17-1 *General* `MOVE` *case.*

If one of the direction buttons on the control paddle is held down—say the north button—then the telescope makes a series of moves to the north. For guide speed, each motion is a very short one separated by a brief pause, and the telescope does not move very fast. For slew speed, the moves are much longer, and the telescope has a chance to get up to full speed and move along some distance before it stops again. This series of discrete movements allows (when other procedures are added) coordinates to be updated and other actions to be taken between each move.

Each procedure will now be considered in more detail.

`QJOY2`: The flow diagram for procedure `QJOY2` is shown in Figure 17-2. As can be seen, the first thing `QJOY2` does is to run `DIGITAL`. It then sets or resets `MOVEX` and `MOVEY` to zero, and `MULT` to its smallest value. `MULT` is the number of steps that will be taken. A check is then made to see if the highest order "STOP" key has been pressed. If it has, all execution is stopped. A check is then made to see if the slew or set buttons have been pressed. If they have, `MULT` is increased from its minimal value. If one of these buttons has not been pressed, then it is assumed that the system is in the guide mode and the `MULT` parameter is left at its smallest value.

A check is then made to see if any of the direction buttons has been pressed. If none has been pressed, then `MOVEX` and `MOVEY` are left equal to zero. If one or two have been pressed, then `MOVEX` and `MOVEY` take on the appropriate sign. The final value of `MOVEX` and `MOVEY` is the product of the signed direction information and the number of steps, `MULT`.

After executing a small delay (to give a very brief pause between each discrete move), procedure `MOVE` is run.

`DIGITAL`: This is a very simple procedure. The control paddle is read by doing a PEEK at location E011 Hex. A button press brings lines low which are normally high. This gives an inverted value for the key, which must be subtracted from 255 to obtain the value corresponding to the key pressed. Since bit 7 is not defined, the resulting number (stored in `PADDLE`) needs to be between 0 and 127 regardless of whether bit 7 is high or low. A flow diagram of a procedure that will implement this is shown in Figure 17-3.

`MOVE`: The flow diagram for `MOVE` is shown in Figure 17-4. The inputs to `MOVE` are the parameters `MOVEX` and `MOVEY`, which are the number

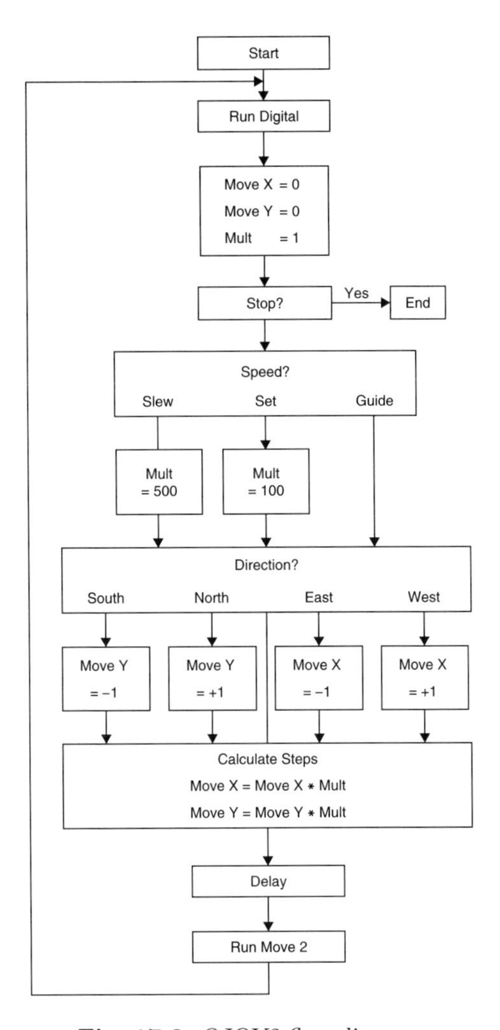

Fig. 17-2 *QJOY2 flow diagram.*

of steps and the direction the telescope is to be moved. In the general case, `MOVE` divides the total move into two separate moves and determines the parameters for calling `RAMP`. These parameters are known as Elements 0 through 5. Elements 0–2 are the number of steps to be taken, specified as a three-byte number. Element 3 is a constant that defines the acceleration to be used in ramping the motor in the assembly language `RAMP` procedure. Element 4 sets the maximum slew speed in `RAMP`. Element 5 is a motor direction command. Once these elements are calculated, `MOVE` runs `RAMP`. When `RAMP` returns to `MOVE`, the move which was just completed is subtracted from what

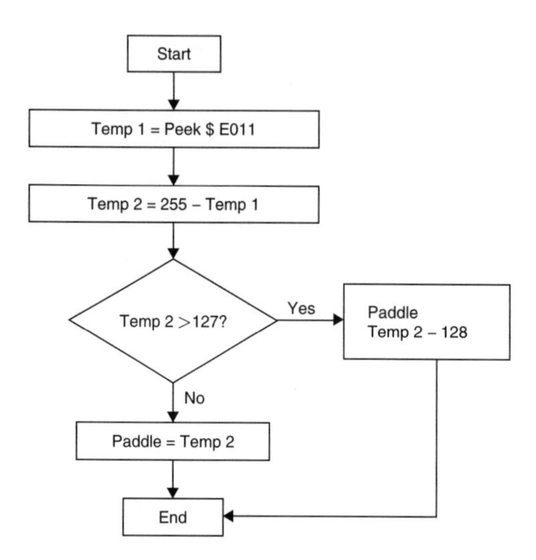

Fig. 17-3 *DIGITAL flow diagram.*

was originally requested. If the move has not been completed, the process is repeated.

`RAMP:` This is the assembly language procedure which moves the chosen steppers in the commanded direction the proper number of steps. It accelerates to maximum slewing speed, and after the proper number of steps have been executed, decelerates to a stop.

`JOY4:` After using the simple `QJOY2` discussed above for some time, various enhancements were made over time. These included a table of stars and other objects, the ability to move the telescope to any coordinates entered, the ability to "tweak" the telescope position without changing coordinates, and a list of objects with preset coordinates. A call to a precession routine was also added, as was a move to the home (storage) position.

Slewing can be handled by making each move in two segments, the first a diagonal move with both motors running for an equal number of steps; the second move with one motor running for the remaining number of steps. This is shown diagrammatically in Figure 17-5.

Searching can be handled by a square spiral pattern, in which the size of each side is incremented by a constant on opposite corners. The constant

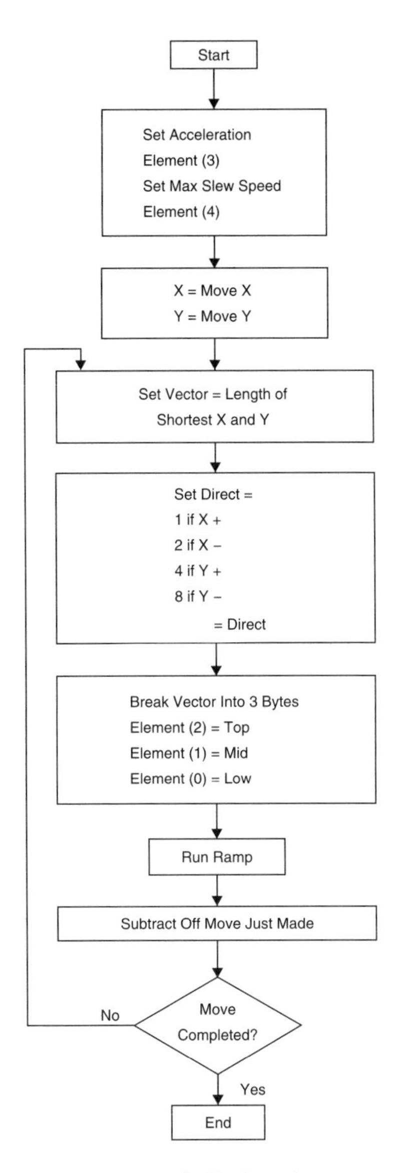

Fig. 17-4 *MOVE flow diagram.*

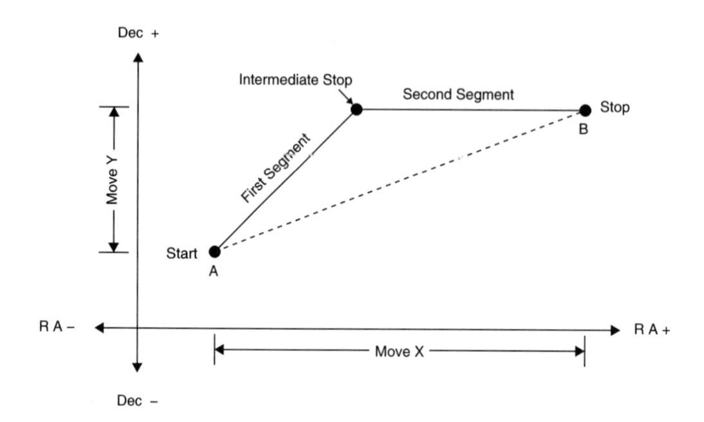

Fig. 17-5 *The automatic photoelectric telescope* `MOVE` *command.*

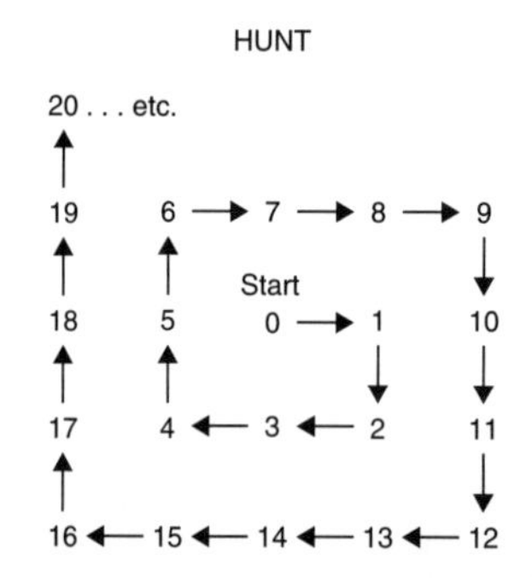

Fig. 17-6 *The automatic photoelectric telescope* `HUNT` *command.*

should produce a motion about half the size of the diaphragm. This insures sufficient overlap on each loop so that a star will not be missed in the presence of vibration or small errors in the drive. The pattern is shown in Figure 17-6. At each numbered position a reading would be taken for about 0.2 seconds to determine if a star is present. In this early Fairborn Observatory system, the star was always found within twelve loops, but provision was made to exit after a reasonable search if the star wasn't found. Clouds cause that!

Centering is no more difficult. The most obvious method is to make a cross-shaped pattern to find the edges of the diaphragm. This method is slow and vibration of the telescope can cause false centering. A much simpler way is to move the telescope to four corners, each just inside the

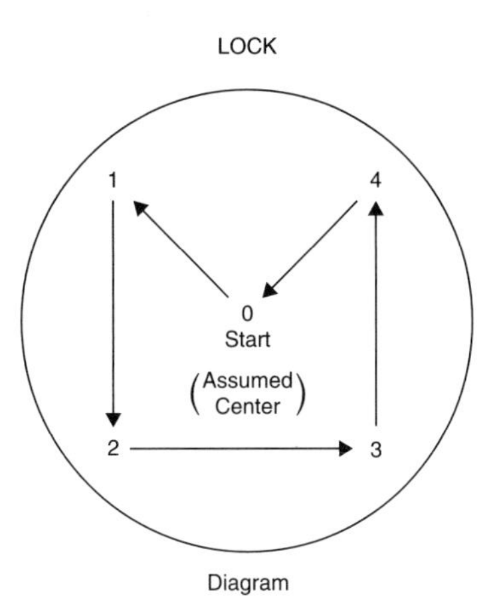

Fig. 17-7 *The automatic photoelectric telescope* `LOCK` *command.*

edge of the diaphragm by a small amount when moving from a properly centered star. This is shown in Figure 17-7. A reading is taken at each position (1–4) and the following logic applied. If all spots indicate a star is present, the star is centered. If one spot is outside the diaphragm, make a small move toward that direction in both axes. If two adjacent spots are outside of the diaphragm (e.g., 1 and 2), then move on one axis a small amount. If three spots are outside the diaphragm, move away from the one which is in the diaphragm. In each of these cases, repeat the process until all four corners show the star inside the diaphragm. If an invalid combination occurs, such as showing spots 2 and 4 out with 1 and 3 in, the process should be repeated. Scintillation can cause this.

When making the scientific measurements on the centered stars, it is suggested that a fixed, short integration time be used on all readings (perhaps 0.1 seconds) and then as many of them be made as required for the actual measurement integration times. A number of short integrations will allow a sum-of-squares computation to be made while making the readings (if desired) for a statistical evaluation of the quality of each reading. If the reading proves to be poor statistically, it may be repeated, reducing the errors caused by scintillation. The other advantage is that the hardware counter need only count the maximum number of pulses from the PMT or V/F converter in the shorter period. This may simplify the hardware.

When searching or centering, it is important to select a decision threshold (star present or not) which will be reliable. That threshold must be greater than the sky readings. It has been found that the most reliable threshold will be slightly less than half the difference between the expected values of the star and sky readings. Scintillation produces greater variation in the value of the star reading than in the sky reading. A threshold must be selected which will not lock onto a fainter nearby star, or miss the desired star entirely. This problem is the major factor limiting the faintest star which may be measured automatically. In practice, a star which gives only four times the sky reading can be easily acquired. Usually, searching and centering should be done with a filter in place so that the threshold may be set as a function of some known magnitude (such as "V" in the UBV system).

There are many ways in which an APT could be programmed. In any program, some of the portions will be hardware dependent, but in terms of functions to be performed, there is likely to be considerable similarity between programs. In discussing APTs, going over each line of code in some specific program would only be useful to someone about to implement the same hardware and software. The approach taken is to discuss the control program in somewhat general terms in this chapter, and then give the specific algorithms used in Appendix F and the software listings in Appendix G. The language used in the control programs was Basic09, which runs under the OS9 operating system.

The supervisory program being used at the Fairborn Observatory was called `MAIN`, and it is a specialized supervisor to the extent that it is intended to do UBV or VRI photometry of medium to long period variable stars. It is not difficult to modify it to do continuous photometry of a short period variable star, as this is a much easier case. Many other modifications for different observing programs are possible. For instance, an asteroid photometry `MAIN` program could calculate the position of each asteroid at the time of observation and automatically select the nearest suitable comparison star. A long period quasar photometry program would be identical to the long period variable star case except that the objects would be fainter and the telescope larger.

`MAIN` will first be summarized, then discussed in some detail. Then directly required procedures and files, such as `HUNT`, which acquires the stars, `LOCK`, which centers the stars, and `MOVE`, which moves the telescope, will be discussed. Finally, auxilliary procedures such as `SET`, which allows the clock to be reset, will be discussed.

17.3 APT Software Functions and Subprograms

17.3.1 `MAIN`

`MAIN` has been divided into ten areas for convenience. Each of these will be mentioned briefly below and then will be discussed in more detail in the following sections.

Build Rise/Set Time Table. An efficient APT stores all the objects to be viewed year around. On any given night, only some of the objects will be observable at some time during the night. Which ones these are can be determined by calculating the rise and set time of all objects for the night in question.

Open Output File. An output file to store the photometric data gathered during the night is opened. About 14 files from two weeks of observing take up about 360K. Later, data on a given star is consolidated with data on the same star from other nights and other weeks.

Initialize Telescope. The telescope and photometer are readied for the night's observations. This includes assuring standard starting positions for the telescope itself, and initializing the filter wheel in the photometer.

Determine Which Group to Observe. By finding which group (variable, check, and comparison stars and sky background) will "set" (leave the photometric observing area of the telescope) first, this group can be observed first. For longest year around coverage, the westernmost groups need to be observed first in the evening, and the easternmost observed last.

Skip If Near to Moon. Photometry too near the Moon can reduce accuracy, and for PMT-based systems, it can damage the PMT.

Move to Group/Star. In this portion of `MAIN`, the telescope is moved to the group in question.

Hunt and Lock. Before the stars can be measured, they must be acquired (in a spiral search pattern), and centered in the diaphragm.

Make Photoelectric Measurements. This is what it is all about. Measurements are made in the U, B, and V passbands (or in the VRI system). Results are stored in a temporary buffer.

Store Measurements. The measurements are placed in the output file. This is only done after all the measurements in a group have been completed, and then it is done all at once. This avoids unnecessary operation of the output disk drive and the recording of incomplete groups.

Evaluate Group Data. When measurements of a group are attempted, there are three possible outcomes: (1) the system will be unable to complete measurements on the group due to clouds or other reasons; (2) the measurements will be completed, but are of unacceptable quality—perhaps suggesting a remeasure; and (3) the measurements are of an acceptable quality and the next group should be attempted. An evaluation of the group data can distinguish between the latter two cases.

Each of these ten portions of the main control program will now be considered in more detail.

17.3.2 Build Rise/Set Time Table

There are two basic approaches to determining what objects an APT should observe during some particular night, and what sequence it should observe them in. One is for some human to choose the objects and their observational order in advance. The other is to let the APT do this itself. The systems at the Fairborn Observatory use the latter approach. From a set of groups for year-round observing, the systems automatically determine which are observable on a given night. The order in which they are observed is determined as the observations are made, and is dependent on the outcome of the observations as they progress.

This approach has the advantage of eliminating the nightly human input that would otherwise be required. If there are no equipment failures, the system will continue to observe the program objects until the output disk or tape is filled, or a new program list is given to the APT. This approach also has the advantage of allowing the system to adjust to the circumstances of the night's viewing as the night progresses. The disadvantages of this approach include complications in the software to implement such "smarts," and a selection process less sophisticated than that of some humans (but very consistent!).

The first task in this routine is to build a table of the rise and set times for all the groups in the total observing program. These are not the rise and set times of the groups at the physical horizons in the east or west, but are the times for a predefined area of the sky above the telescope that is deemed to be physically and photometrically observable. In the case of the Fairborn Observatory APT in Arizona (before it was moved to Mt. Hopkins), this is defined as all of the sky that is within 45° of the zenith *and* that is also within four hours (60°) of the meridian. The 45° "cone" assures that high air masses that could reduce photometric accuracy are avoided, and also avoids letting the walls of the observatory (a converted garden shed) block off any of the light. The hour angle restriction is due to the telescope mount—a modified fork. The fork extension limits telescope motion in RA. In Ohio, the APTs were operated essentially as meridian telescopes (one of the APTs was later moved to Mt. Hopkins). All objects were observed within one hour (RA) of the meridian. This increases photometric accuracy (at the expense of some seasonal coverage) and allows narrow "hatches" in the roofs to be used.

To determine when a group will rise and set in the observing window for that night, an angle is calculated that is the plus or minus angle from the group's RA where the group will rise or set in the observing window. The group's RA is taken as that of the centroid between the variable and comparison stars, which is also usually the sky measurement position. This angle is a function only of the observing window, the Declination of the group, and the latitude of the observatory.

A table is then formed that contains the group name (usually named after the variable star in the group, such as "RS CVN"), the rise and set times, and an indicator of how many times the group has been observed. This indicator is called `STATUS` (for group status), and it is initially set to zero for each new group.

In setting the edges of the observing window, it should be kept in mind that the groups have some size themselves, and that they will move some amount in the time it takes to make the measurements. Typically, 1° or 2° might be added to allow for group size, and seven minutes' time (2° in angle) to allow for the time it takes to measure the group. In the case of some fixed portion of the window (such as the physical fork limit mentioned earlier), it might pay to add a small safety margin—perhaps a couple of degrees. This could avoid hitting any of the limit switches.

17.3.3 Open Output File

Before any actual measurements are made, a name needs to be determined for the output file, the output file opened and tested to see that it is there, and then closed again until the first data are ready. The opening and closing of data files is unique to particular equipment and operating systems, and will not be discussed further. The determination of the name of a file is of interest, however.

In an automatic system, the name of the output file needs to be determined by the system itself. It helps later data consolidation efforts if the file name is unique to the night of the observations. Then if for some reason observations should be interrupted and the system restarted in the same night, a new file name will not be created, but any further data will be added to the previously opened file. As the system contains a clock/calendar that runs at all times (with battery backup), the modified Julian date (MJD) can be used for a unique name each night. A typical file name is "`MJD45641.`"

17.3.4 Initialize Telescope

To operate properly, the telescope pointing angle and the time of day must both be known accurately. As the telescope control system does not include angle encoders of any kind, it must be provided with some other way of determining at least one position each in RA and Dec. If a human operator were allowed to position the telescope initially on a star of known position, then this could be used as the starting point. While easily done, this does not allow fully automatic operation. The approach used in the Fairborn Observatory APTs utilizes the RA and Dec limit switches to place the telescope in a known initial position. The limit "switches" are actually knife edges placed on the RA and Dec drive wheels that pass through LED/photodiode pairs. A Schmitt trigger is used (three in series) to sharpen the switchpoint. How the telescopes are brought precisely to the start position on the southeast limit switches is discussed later in STARTSCOPE.

17.3.5 Determine Which Group to Observe

This portion of MAIN determines what group should be observed next. Once this determination is made, a check is made to see if the group is too near the Moon. If not, then the telescope moves to the selected group.

Before considering the complexities of automatically determining which group to observe next, it is worthwhile to consider why the observing order can't be fixed for all time, or at least computed once and for all at the beginning of the evening. In theory, perhaps the order could be determined once and for all, and the starting point on the list just adjusted for the starting date and time of the system. However, this would assume that the time taken to observe each group of stars is known. In fact, this time is not actually known in advance. A gust of wind could throw the telescope off a bit during a long move, and this could result in a longer spiral search time. Also, if the order of observations were fixed, then if observations on one group were not really satisfactory, they could not be repeated, and the system would have to blindly plunge on to the next group so as not to upset the pre-arranged timetable. Finally, the sky simply is not clear all night every night, and the system can get delayed by poor weather, and this alone makes a fixed schedule far from optimum or even workable. By building "smarts" into the control program, it can respond dynamically to the actual situation as it happens.

Before selecting the next group to observe, a parameter known as (the system) DEPTH is always set to zero. DEPTH is an overall property of the system for any given night, and when it is increased, the system will start repeated measures of groups. The number of times any individual group has been measured is its (group) STATUS. Initially, the system DEPTH and the STATUS of each group are both 0. As measurements are made, the STATUS of many groups will change from 0 to 1. If a small cloud were over some group, the attempt to measure it would not be successful, and its STATUS would remain at zero. If there were a large number of groups in the overall observing program, then on a given night, there might not be enough time to observe all the groups that were observable, so the DEPTH of the system would not go from 0 to 1 in this case. On the other hand, if the number of groups in the observing program were small, the system would run through them all quickly. Rather than do nothing for the rest of the night, the system would start repeat measurements. Two points need to be made.

First, the system assumes that at least one group is observable at all times. If this is not the case, then the system will get in a loop where the DEPTH is continuously increased. With a 90° window (2 × 45°), four evenly spaced objects would meet this criterion, but more typically there would be a minimum of a dozen groups. This point should be kept in mind when system tests are first initiated.

Second, why is the system DEPTH always reset to zero each time the system goes through the process of deciding what group should be observed next? Would it not be more logical for it to progress from 0 to 1 to 2, etc.? If a group is not successfully observed, then its STATUS will remain at zero. Some temporary problem (such as a cloud) may have gone away, and by always resetting the DEPTH to zero, an attempt will be made to observe this group again before other groups are observed twice.

The reason that at a given system DEPTH the group that will set first is observed next is that this will maximize the seasonal coverage of groups. As soon as it is dark enough, the group that will set first is observed first, and the system works eastward. At the end of the night's observing session, some new group, not previously observed, will just make it past the eastern edge of the observing window for the first time. As this group will not have been observed before, its group STATUS will be zero. It will then take precedence over any group that has been observed at least once.

With a moderate number of program groups (50 to 100), the system will start with groups at the western edge of the window and work its way east. This will take most of the night. When it finally arrives at the eastern edge of the window, it will then slew to the west and start working groups in the west a second time. However, any time a new group comes up in the east, it will stop working the groups in the west, run over and work the new group in the east (for the first time), and then return to working other groups for the second time.

There is nothing sacred about the above logic. It works well for the purpose intended, but many refinements in the logic are possible in this application. For other applications, an entirely different logic may have to be used.

Having tentatively selected the next group to observe, some other checks need to be made before the telescope is moved to the group and observations actually made.

17.3.6 Check Moon

Before the telescope actually moves to a group to make observations, a check is made to see that the group is at least an appropriate distance from the Moon. To do this, program `LUNAR` is run. Given any date and time—the modified Julian date including the fractional (time) portion—`LUNAR` returns the current RA and Dec of the Moon. The distance from the Moon in RA and Dec can then be calculated simply by taking the absolute differences between the position of the group and that of the Moon in RA and Dec.

If the group is closer than 10° to the Moon, the `STATUS` of the group is incremented by one. This gives this group another chance later on (the Moon moves about 15° in 24 hours). After incrementing this group's `STATUS`, the `DEPTH` is reset to zero, and a fresh start is made to find the next group to observe.

If the group is not too near the Moon, then—at long last—it is almost time to move to a group. On the way to doing this, the system identifies the group it is about to move to, and also displays the coordinates of the group it is about to move to.

17.3.7 Move to Group

To prepare for the first set of measurements, not only does the telescope have to move (using program `TRAVEL`) to the area of the next group, but any flip mirrors, and filter and diaphragm wheels must all be properly positioned. Also, some rough idea of sky brightness is needed before a search is even started for the first star so that an appropriate brightness threshold can be set. If for some reason the sky is abnormally bright (dangerously so), then shutdown would be appropriate. `TRAVEL` computes the number of steps to make along each axis, and the length of each of the two submoves which comprise the total move. The move is executed, which should bring the telescope close to the navigation star.

The telescope is now in the area of the group, and it has been readied for making measurements. However, as the first star has not yet been centered, the telescope does not know exactly where it is. Also, it does not have a good idea of how bright the sky is, and this information is needed to make the search for the first star. With a very uncertain position, a sky brightness measurement could very well contain a star, certainly fouling everything up. This ambiguous situation is resolved by using procedure `THRESH` to make five background measurements in a "square with center" pattern. The lowest of the five readings is then used as one of the inputs to determining the threshold level for acquiring the first star. Note that this sky measurement is not a data point. This first measurement is only used to set the threshold for the HUNT routine.

Finally, it might be noted that after a long move between groups (which can be clear across the sky), the accuracy of the system on arrival will not be as great as it will be within a group once the first star has been acquired and centered. To account for this difference, the acquisition search (in the HUNT routine) for the first star in a group is widened by allowing the square spiral pattern search to make more "loops," and thus cover a wider patch of the sky. The number of allowable loops (variable LOOPS) before the system decides it can not find the first star of a group is set to 25. Within a group, it is reduced to 5. Also, the first star is the check star, and it is purposely chosen to be in an easily acquired brightness range, and to be free of nearby stars of similar brightness that might confuse the system. Thus the check star is also the "navigation" star.

Now the system is ready to acquire, center, and measure the first star. It will then move on to measure the other stars and the sky, until all the necessary measurements (in all three colors) have been made on the group.

MEASUREMENT SEQUENCE	U	B	V
0. Check Star			
1. Sky			
2. Comp Star			
3. Variable Star			
4. Comp Star			
5. Variable Star			
6. Comp Star			
7. Variable Star			
8. Comp Star			
9. Sky			
10. Check Star			

Fig. 17-8 *The APT measurement sequence.*

17.3.8 Hunt and Lock

The hunt and lock procedure described in this section and the UBV (or VRI) measurements described in the next section are usually performed 11 times in each group. The order of measurements is given in Figure 17-8.

As is customary in photometry, a "balanced" measurement sequence is used, so the check star is measured first and last. The sky is then measured at a position halfway between the variable and comparison stars. If this position is too near a known background star, the "group coordinates" (this halfway position) are shifted slightly to avoid this problem. Traditionally, the sky measurements are made immediately following each star measurement and next to the star, but as this system never works at high air masses (low elevations) or in close proximity to the Moon, and the variable and comparison stars are always within a degree or two of each other, the two sky measurements (1 and 9) at the group center are more than adequate. Also, as is customary, each variable star measurement is "bracketed" by comparison star measurements. Thus for the three variable star measurements made, four comparison star measurements are required.

The RA and Dec of the appropriate star or sky are obtained (in J2000 coordinates), and precessed to current coordinates with procedure `PRECESS`. Procedure `TRAVEL` is then used to move the telescope to the precessed coordinates (always a short move because it is within the group). If the sky is being measured, then the `HUNT` and `LOCK` routines are not needed, and the UBV measurements are made.

Now the system is ready to start the square spiral search for the star. If the star is not found within the allowable number of loops, the `HUNT` program automatically returns the telescope to the center of the spiral search pattern, and the `STATUS` of this group is incremented by one. A system level parameter called `GROUPABORT` is set to TRUE, and another system level parameter called `ABORT` is incremented by one (`ABORT` is initially set to zero at system startup). A check is then made to see if `ABORT` is 2. If it is, then `STOPSCOPE` is called immediately, and the system is shut down. If `ABORT` does not equal 2, it must then equal 1, and the system picks a new group. If this new group, or any subsequent group, aborts, then `ABORT` is 2, and the system will shut down. In short, one abort will not cause a shutdown, but two will.

If the star is found, then the star is centered with procedure `LOCK`. `LOCK` is an iterative procedure that makes only four measurements per iteration. The measurements are made at the four corners of a square that fits within the diagram. At each corner, a determination is made whether or not a star is there, based on whether it exceeds a threshold or not. Thus the result of each corner measurement is a true or false indicator of whether or not a star was in that corner. There are 16 possible combinations of these indicators. Of these 16, there are 14 combinations from which the logic can determine the direction the telescope should take to center the star. Two of the 16 combinations are not logical (with a single star in the field), and call for a remeasurement. One example of a proper combination would be one in which the two measurements on the right of the square showed stars, and the two on the left did not. This would suggest that moving the telescope to the right will move the star closer to the center of the diaphragm. An example of an improper combination is where one pair of diagonal corners contains stars, while the other pair does not. When the star appears in all four corners, it must be reasonably centered, and `LOCK` is deemed successful. If the lock is unsuccessful, then `HUNT` is tried again.

In searching for a star it is necessary to set a threshold such that counts below the threshold will indicate the star is not in the photometer diaphragm, while counts above the threshold indicate it is. The threshold is set by `MAIN` by first measuring the sky background (see the description of `THRESH` later in this chapter), then using an equation to predict the expected count. The threshold is then appropriately set between the sky background and the expected count.

In the second generation software that was used in Ohio, the integration time used in the spiral search pattern is 0.1 seconds during the first search in a group for the check star. This star is always chosen to be reasonably bright, and is sought after a long move from another group when position uncertainty is highest. Once the check star is found, then searches within a group are made with one second integration periods. This allows much fainter comparison and variable stars to be found than would otherwise be the case. As the pointing accuracy of the moves is high within a group, there is little searching, and the longer integration times do not hurt the total system efficiency.

Assuming a successful lock, the system—at last!—is ready to make the actual measurements.

17.3.9 Make Photoelectric Measurements

The measurements are made by sequencing the filters and making timed integrations of 10 seconds each. These counts, along with the time and zenith angle, are placed in an output buffer as each star is measured. For the middle variable star measurement, the heliocentric correction for the group is calculated.

17.3.10 Store Measurements

Once the measurements of a group have been successfully completed, it is necessary to store them on the output file. On the way to locating the end of the output file, the name of each group observed so far is displayed on the monitor. Once the end of file is reached, the output data are transferred from the output buffer to the output disk, and the file is closed. Finally, the system `DEPTH` is reset to zero so that when the system starts to select a new group to observe, those observed the least (or not at all) will be given priority. However, before a group is chosen, an evaluation is made of the quality of the present group's data, and if it is not acceptable, it is measured again.

17.3.11 Evaluate Group Data

There are 11×3, or 33 ten-second integrations in a group. There are 12 different types of measurements, namely U, B, and V measurements for the check, variable, and comparison stars, and the sky. Two types have two measurements in each color (check star and the sky), one type has three measurements in each color (the variable star), and the last type has four measurements per color (comparison star). A statistical test is made of all nine star ratios (three colors on three types of stars). No test is made on the sky ratios, although they are reported on the monitor for the interest of any human who might happen to be watching. If the test is passed, then the `STATUS` of the group is incremented by one and the selection of a new group to observe is initiated. If the test fails, then the group will be reobserved until it passes the test, the group sets in the west, morning arrives, or there is a *system abort.* A system abort condition occurs when not all the stars in a group can be observed. If the test eventually passes or the group sets in the western edge of the observing window, then the system selects the next group to observe.

It may seem futile to keep repeating observations as a group moves towards the west. Experience suggests, however, that any continuing problem not severe enough to cause an abort (i.e., the system is successfully finding all the stars in the group), yet serious enough to produce observations of poor quality could be a transient problem. Observations on the group might as well continue in hopes that it will clear up during the evening. This has happened a number of times.

The evaluation test was used in the system in Arizona. In Ohio, which had "quasi-meridian" telescopes, the evaluation was made, but not acted on. We preferred to keep on observing until all the stars in our "slot" centered on the meridian were observed, then we just repeated the observations a second time if new stars had not popped into the slot from the east.

On a decent night, the system will not abort, nor will it get hung up on a group that fails repeatedly to pass the quality test. Instead, it will make it all the way through the program list of stars observable on that night. Dawn will find it working its way through the list for a second time, picking off the high-priority unobserved groups in the east as soon as they enter the observing window. With dawn, the computed angle is greater than the twilight angle, and the system executes procedure `SHUTDOWN`.

17.4 APT Supporting Procedures and Files

In the previous section, the overall control program (`MAIN`) for the early APTs at the Fairborn Observatory was described. It uses 21 different supporting procedures to carry out its various tasks. Each supporting procedure is discussed briefly next (in alphabetical order).

17.4.1 COEFFICIENTS

`COEFFICIENTS` is a file that contains the assumed "average" extinction coefficients. These are used to set the `THRESHOLD` value in `MAIN`, and in data reduction in `TRANSFORM`.

17.4.2 DIGITAL

`DIGITAL` is a short procedure that reads the manual control paddle. In automatic operation with `MAIN`, the paddle input ports can be checked occasionally for human override requests.

17.4.3 HELIO

`HELIO` provides the correction from geocentric to heliocentric times. It is customary in variable star research to record observational times as if they were viewed from the Sun.

17.4.4 HUNT

`HUNT` executes a square spiral search routine to locate a star. The input parameters to the procedure are `RADIUS`, `LOOPS`, `THRESHOLD`, and `DURATION`. `RADIUS` is the radius of the diaphragm in steps. By making this a variable, different diaphragm sizes can be used. Although for any given group, the current system uses the same size diaphragm for both acquisition and observation, it is conceivable that some systems might use a larger diaphragm for acquisition, and a smaller one for the actual observations. `LOOPS` is the maximum number of square spiral loops that will be taken before the system aborts. Up to 25 loops are used to acquire the first star in the group, then this is reduced to 5 for all subsequent stars in the group. `THRESHOLD` is the number of counts above which it will be concluded that the correct star is in the diaphragm, and below which it will be concluded that there is no star (or a fainter "wrong" star) in the diaphragm. A more complex scheme could be used, such as requiring the star to be within a window defined by both maximum and minimum values, or measuring it in several colors and evaluating the color index, etc. However, at least with smaller telescopes which are, by necessity, observing brighter stars, this simple scheme works very well. DURATION sets the number of 0.1-second integrations used at each stop in the square spiral search pattern. Typically, DURATION is 1 during the first search for the bright navigation star, and a larger value for fainter "within group" stars.

HUNT makes a series of short integrations. After making the first one, it moves one radius to the west and makes another integration. It then moves one radius to the south and makes the third integration. It then goes to the east two radii. Then two radii to the north, completing the first loop. Every second "side" of a loop includes one more move, and the repetitive nature of the moves makes for simple logic.

After each integration, the COUNT from the photometer is compared with the THRESHOLD. If the COUNT is larger than the THRESHOLD, then the telescope stops at this position, and THRESHOLD is set equal to COUNT. If THRESHOLD is equal to or less than COUNT, then the system moves to the next position in the spiral. If the system makes the maximum number of loops, it will stop northeast of the original position (at the end of the last loop), and on deciding an error condition exists, it will return the telescope in a southwest direction to its starting point in the center of the spiral. In this case, the THRESHOLD value will be unchanged from its original value, signalling that the search was not successful.

17.4.5 LOCK

LOCK was described earlier in some detail, and is an iterative procedure used to center a star in the photometer's diaphragm.

17.4.6 LUNAR

To avoid doing photometry too close to the Moon, program MAIN calls LUNAR to check if a group is near the Moon before commencing measurements. The modified Julian date (including the fractional "time" portion) is provided as an input, and the position of the Moon is returned.

There are some very accurate algorithms for determining the position of the Moon in the *Almanac for Computers,* but these are total overkill for this application. Instead, the approach suggested by Burgess (1982) was used as the basis for the algorithm. It is not very accurate, but is usually within a degree, which is close enough for an APT trying to stay at least 10° away from the Moon.

17.4.7 MEAS

MEAS sets the integration time for the photon counting system, and returns the final count. The parameter `TIME` sets the integration time in tenths of a second. If `TIME` = 100, then the integration time is 10.0 seconds. The final count is normalized to counts per tenth of a second, and is returned as `COUNT`. As the specifics are hardware dependent, no further details will be given here.

17.4.8 `MOVE`

`MOVE` receives the input parameters `MOVEX` and `MOVEY`, and moves the telescope these amounts. `MOVE` has no outputs. `MOVEX` and `MOVEY` are signed real numbers. An integer could not be used, as some moves are greater than the limit of 32767 imposed by Basic09 on signed integers. The sign convention is $+X$ is west, and $+Y$ is north.

`MOVE` breaks the total move into one or two separate moves. The one move case is where either `MOVEX` or `MOVEY` equals zero, but not both. In this case, the telescope would make a straight move in either RA (alone) or Dec (alone). The two move case is the general case where motion is required in both RA and Dec. `MOVE` takes the total move (in the general two move case) and breaks it down into two separate moves. The first move requires motion in both RA and Dec of an exactly equal number of steps. The length of this move is set by the shortest of the RA or Dec motions, and with both motors operating at once, this motion is always in one of the four "45°" directions. As full slewing velocity is much higher than the steppers can obtain from a dead start, ramping up and then ramping back down is required. The second part of the total move takes place in either RA or Dec, but not both, and it completes the "unexecuted" steps remaining in the longer of the two-axis moves.

17.4.9 `PRECESS`

`PRECESS` takes year 2000.0 coordinates in radians and precesses them to the current Julian date. The simple algorithm that is used does not account for changes in the precession rate.

17.4.10 `PTLCK`

`PTLCK` reads the clock/calendar chip on the PT69 computer used in the Fairborn Observatory systems. It is hardware dependent.

17.4.11 `RAMP`

`RAMP` is a short assembly language routine that generates the pulses for the stepper electronics. `RAMP` starts at an initial slow step rate, ramps up to a maximum step rate (slew), and then ramps down to a complete stop.

17.4.12 SHOCO

To SHOw COordinates, procedure SHOCO takes RA and Dec as inputs (in radians, epoch J2000.0), and converts them to degrees or hours, minutes, and seconds, and displays them in the format "XX HH MM SS." SHOCO is used by JOY4 and MAIN.

17.4.13 SOLAR

SOLAR, given the modified Julian date (including the fractional time), calculates the RA and Dec of the Sun and returns them as variables RA and DEC. The algorithm is based on material in the U.S. Naval Observatory *Almanac for Computers.*

17.4.14 STARFILE

STARFILE is a file that contains data on all the groups in a given observing program. Data includes the number and name of each group, the identification number (HD) of each star in the group, as well as its coordinates and magnitude. Additional information, such as the sky measurement position, is given.

17.4.15 STARTSCOPE

STARTSCOPE, as the name implies, is used to initialize and start up the telescope. It is a procedure called by the main control program.

First, any diaphragm and filter wheels are placed in their home positions. Next, the telescope is moved exactly to its home position. When it was previously shut down, it was moved to its home position, but rather than trust this (it could have been disturbed), the telescope is backed off the home position 500 steps in both RA and Dec, and then moved back to home. It is then moved off a tiny amount (20 steps) and moved back to home. This latter small movement makes sure that the telescope is travelling very slowly when it hits the limit switches. The limit switches are not physical switches, but are small IR transceivers. A "knife edge" on each main drive disk interrupts the light beam for very precise positioning. With the telescope in its exact home position, all that is left to do is to read the time, and start the telescope at this exact time. The telescope immediately starts moving to the west, and is now in celestial coordinates.

17.4.16 STOPSCOPE

When the MAIN program determines that morning has arrived, or that clouds have moved in and two groups have been missed (ABORT = 2), it calls STOPSCOPE to shut down the telescope. This is done by moving any opaque filter or diaphragm into the light path, and moving any flip mirror to the viewing position (to keep light out of the photometer during the day).

The sidereal rate clock is then turned off (and the telescope stops tracking). A move well beyond the limit switches is commanded, and when the scope runs into the RA and Dec limit switches, it has arrived at the home position. The program then closes the roof and does other shutdown functions as appropriate.

17.4.17 SUNANGLE

SUNANGLE calculates the zenith angle of the Sun. First it runs TIME to get the current time, Julian date, and local mean sidereal time. It then uses the Julian date as an input to SOLAR, which returns the RA and Dec of the Sun. The RA and Dec of the Sun and the local mean sidereal time are then used as inputs to program ZENITH to determine the zenith angle of the Sun. This is used to determine when it will become dark in the evening or light in the morning. This allows automatic startup and shutdown, respectively.

The programming task can be greatly simplified for an APT if commonly needed functions and routines can be run as a commonly available program (or subroutine). Some forethought is required if this commonality is to be maximized.

17.4.18 THRESH

THRESH is used when the system first moves to a new group to determine the background count that is used to acquire the first (check) star. Five measurements in a "square with center" pattern are made, and the lowest count is retained as the result.

17.4.19 TIME

Given the time and date as inputs (using PTCLK), and knowing the local longitude, procedure TIME determines the modified Julian date (MJUL) and the local mean sidereal time (LMST).

17.4.20 TRAVEL

TRAVEL determines the total number of steps in X (RA) and Y (Dec) to go from the current position to a new position. To make the actual move,

TRAVEL calls MOVE, and MOVE splits the total travel into two discrete moves, then executes the moves. The inputs to TRAVEL are the current position and the desired new positions.

In general, a move from point A to point B is made in two distinct submoves. In the first segment of the move, both RA and Dec steppers are stepped together in exact synchronization (driven from the same pulse source). As each motor can be moved independently in a positive or negative direction, there are four different directions the telescope can go (++, +−, −+, and −−). These four directions are at "45°" to the "ordinal" RA and Dec directions. At the end of the first segment of travel, the telescope is ramped down to a complete stop, and then the second segment of travel is initiated. It is also only in one of four directions, but these directions are along the regular RA and Dec directions. In the second segment of travel, only one stepper motor on a single axis is active.

17.4.21 ZENITH

Given the time and the position of any object and that of the observatory, this subprogram calculates the object's zenith angle.

17.5 Auxilliary Procedures

Auxiliary procedures are those not required in the actual operation of the system, but which are needed either to enter program objects, bring data out of the system in raw or reduced form, or, initially, to align and check out the system. In some systems, it may be desirable to keep all the auxilliary programs separate.

17.5.1 BUILDFILE

BUILDFILE provides for manual entry of the objects to be observed onto the master file. After building the initial file, objects can be changed, corrected, or added. BUILDFILE allows changes to be made on just those entries in need of change.

A number of conventions are used. The name of a group is usually the name of the variable star, such as R Leo, V411 Cyg, etc. The name of a star, including any variable star, is its HD catalog number, including the letters "HD" as a prefix. Magnitudes are the V magnitudes, and in the case of variable stars, the magnitudes are the magnitudes at minimum light, not maximum, as is usually given in star catalogs.

Coordinates are entered using epoch J2000.0. The input format is somewhat free-form, as program MANCO is used to digest the inputs and convert them to radians. In general, the system displays epoch J2000.0 coordinates in degrees or hours, and stores epoch J2000.0 coordinates as radians, since most computer trigonometric functions require angles in units of radians. In commanding movements of the telescope, the coordinates are precessed as the last action, assuring that precession is not accidentally done twice. The *Sky Catalog 2000.0* (Hirshfield and Sinnott, 1982) has been found to be particularly useful, as it contains all the stars within range of the telescope in J2000.0 coordinates. Other data provided in this catalog, such as the V magnitudes, is also helpful. Use of epoch J2000 coordinates throughout the system aids an observer who is verifying system performance using the new and popular epoch J2000.0 catalogs and atlases.

17.5.2 `DATREAD`

`DATREAD` displays the contents of raw data files. It adds in the appropriate header information, and formats the data in a convenient manner.

17.5.3 `JOY4`

Manual control of the telescope is needed for making observations, for initial telescope alignment, for letting the neighborhood kids look at the Moon, and for "fuzzballing" (unauthorized human use of an APT during prime dark observing time to look at Messier objects).

A menu provides such options as returning to the home position, preselected stars, and interesting objects; pure manual control; and photometric measurement (manual with computer assistance). Current coordinates are displayed after each motion.

17.5.4 `MANCO`

`MANCO` stands for "MANual COordinate entry." It allows entry of the coordinates of program objects (epoch J2000.0 coordinates) in a reasonably "free-form" format. Procedure `JOY4` can call `MANCO` for coordinate entry during manual operation of the telescope. However, the most frequent use of `MANCO` is by program `BUILDFILE` when adding new objects to the observing program or modifying existing ones.

17.5.5 `SET`

`SET` allows the clock to be set against WWV. The clock should be set (and maintain time) to within about one second of time. Note that a one second time error translates into a 15 arc second pointing error in RA. This error is eliminated as soon as the first star is centered, but an error as large as one minute of time is 15 arc minutes of error in RA, and this would cause the system to miss its first acquisition.

17.5.6 SHOWTIM

SHOWTIM allows the current time to be displayed.

17.5.7 TILDARK

TILDARK has no input or output parameters. It repeatedly runs program SUNANGLE until it finds that the angle of the Sun is greater than the variable TWILIGHT. When this occurs, it runs procedure MAIN. This allows the system to start observations automatically as soon as it gets dark.

17.5.8 TRANSFORM

TRANSFORM takes the raw data and reduces it to differential magnitudes. This was done off line in the early APTs.

This completes the discussion of the early Fairborn Observatory APT software. Details on specific APT algorithms are given in Appendix F, while complete program listings are provided in Appendix G.

Chapter 18

Advanced Robotic Telescope Control

18.1 The Normal Growth of Complexity and Specialization

From the relatively simple APT control system described in the previous chapters in this part, robotic telescope control has rapidly grown in both complexity and specialization. Where, previously, the relatively simple control systems could all be described in a few chapters and appendices, as we did in this book, describing just one of today's complex control systems would now take an entire book in itself. The complexity of the control hardware has increased by an order of magnitude, as has that of the attendant computers. Control programs now run to well over 10,000 lines of code and are still growing. This appears to be a natural progression in almost all fields where computers have replaced or aided human efforts.

Along with the growth in complexity there has been a natural increase in the varieties of robotic telescopes and the specialized astronomical tasks to which they have been applied. Simple aperture photometry, while still an important task for robotic telescopes, has now been joined by fully automated CCD photometry and spectroscopy. Small robotic telescopes have been joined by large ones. The few small trees planted in the early 1980s are rapidly becoming a forest.

While it is no longer possible to cover the details of so many different and highly complex control systems in a single book, we will at least consider, in an overview fashion, the highlights in this closing chapter.

18.2 Automatic Telescope Instruction Set (ATIS)

The basic APT presented earlier selected what "group" of stars, from all possible groups, it was going to observe next, using a fixed and simple selection algorithm. The APT then made its observations in a fixed, unaltering sequence. Users asked for more variety and, eventually, got what they wanted. Rather than implement in a piecemeal fashion a continual stream of requested changes, Boyd and Genet devised an instruction set for automatic telescopes that allowed the using astronomers to themselves "script" what their remote APT would do. They were allowed to choose from a variety of group selection options, and to specify, quite exactly, the sequence, duration, etc., of all the observations made within a group once it was selected.

The original automatic telescope instruction set (ATIS) was specified in full detail in Genet and Hayes (1989). Subsequently, ATIS was expanded to handle CCD photometry, imaging, and other advanced features. See Boyd, et al. (1993).

ATIS was devised so that requests for observations could be sent to the remote telescope via Internet as ASCII text files, and observational results could be similarly received the morning after the observations were made. This allowed users to easily change their observational programs but, more importantly, also allowed them to promptly retrieve their observations.

18.3 Centering and Finding Stars

The simple procedure described in detail earlier in this book for centering stars was adequate for achieving "normal" photometric precision (repeatability) of about 0.01 magnitudes. It was found, however, that photometric precision could be increased by about an order of magnitude if the stars could be centered much more accurately. Two approaches were adopted for doing this.

One approach, suggested by Russell Robb and implemented by Genet, used the same aperture photometry hardware, just adding a new, additional software algorithm. The system roughly centered the star, as before, and then "fine tuned" the centering by observing the star as it was bisected on the four "edges" of the diaphragm (top, bottom, left, and right). The position of the "assumed center" was then iteratively adjusted, using the differences in magnitudes between opposing edges as a guide, until the observed magnitudes of the top and bottom and the left and right edges matched each other. Genet was able to achieve photometric precisions in excess of 0.001 magnitudes using this procedure, and was able to show that the precision achieved was at the theoretical limit imposed by the random and unavoidable atmospheric scintillation and photon arrival noises.

The other approach, implemented on APTs by Louis Boyd, used a CCD camera to center stars. While adding complexity to the hardware, this approach had the advantage that the CCD camera could also be used to find the stars to begin with. This was not only faster than the earlier square spiral search using the aperture photometer, but allowed fainter stars to be found.

18.4 Improved Accuracy and Quality Control

With improved centering, it was realized that APTs had the potential, as yet unrealized, for achieving greater accuracy than was possible for manual photometry. To this end, Genet organized a workshop on Precision Robotic Photometry that was attended by, among others, Andrew Young, Louis Boyd, Wesley Lockwood, Diane Smith, and Donald Epand. During this workshop, the sources of photometric imprecision and inaccuracy were considered, and a program for the improvement of robotic photometry was drawn up (see Young et al., 1991).

This program was implemented, primarily, by Louis Boyd and Gregory Henry. Besides using CCD centering, a special, high precision photometer was designed and built by Boyd that incorporated the features recommended at the workshop, including rigid temperature and humidity control of the entire photometer's optical chain and detector.

The advanced features of ATIS were then used to automatically schedule, each and every night, quality control observations of automatically selected standard and extinction stars. Other quality control observations, such as Fabry "diaphragm scans" were also made nightly. A program was devised to automatically, each morning, separate the quality control observations from the normal "user" observations and reduce the control observations. The reduced observations, automatically plotted confirmed, at a glance, the proper functioning or lack thereof of the entire system. In this way, even the slightest problems were quickly spotted so that corrective actions could be immediately taken. Boyd and Henry worked together as a team (separated by a thousand miles) to perfect these quality control procedures. The result was consistent millimag robotic photometry right at the theoretical limits, photometry considerably better than that achieved by human observers.

18.5 Fully Automated CCD Photometry

Kent Honeycutt at Indiana University developed a fully automated robotic telescope using a CCD camera as the photometric detector (Honeycutt, 1992). While not yet achieving as high a precision as automated aperture photometers, CCD-based robotic systems are able to observe much fainter stars, due not only to their higher quantum efficiency but to their ability to greatly reduce the effective sky background.

Donald Epand devised a set of ATIS commands that allowed remote users to implement CCD photometry in a manner similar to aperture photometry. These commands were included in the ATIS standard referenced earlier (Boyd et al., 1993).

18.6 Fully Automated Stellar Spectroscopy

Kent Honeycutt (yet again!) developed a fully automated spectroscopic telescope. Many of the advantages of automating photometry naturally accrue to spectroscopy. A spectroscope, however, is in some ways more complex than a photometer, and the quality control procedures are somewhat different. The issue of quality control in robotic spectroscopy has been discussed by Hearnshaw (1995).

Spectroscopy, in general, requires more photons than photometry and, as a result, the telescopes require a larger aperture. Thus the era of robotic spectroscopy will, undoubtedly, usher in an era of increasingly large robotic telescopes.

18.7 Networked Robotic Telescopes

David Crawford and Genet have both, for many years, proposed that robotic telescopes should be linked together in cooperative networks. One advantage of such networking is the ability to "pass" a star from one robotic telescope to another as the Earth revolves, thus keeping it under constant observation for long periods of time. Another advantage is the ability to observe stars anywhere in the northern or southern hemispheres, assuming that robotic telescopes in the network are situated in both hemispheres. Earth-based robotic telescopes could be linked to space-based robotic telescopes. Photometric and spectroscopic robotic telescopes could, all quite automatically, make simultaneous observations. There are many possibilities.

Crawford has championed the notion of a network of identical robotic telescopes all under central control. By having identical hardware and software, this approach avoids the difficulties inherent in bringing diverse observations to a common standard. Centralized control also reduces problems of coordination between net users.

Genet had championed the notion of a network of independent, presumably different robotic telescopes, all operating without any central control. The various observatories in the network would be "free agents" to trade observational time about the network to any other observatories that were agreeable. While a common language, such as ATIS, would be required for networking, the telescopes, instruments, and software would, presumably, vary considerably. One advantage of such a network would be that it would not have to be implemented as a large, centrally controlled project. Down with Big Brother! Democracy in action.

Mark Drummond, Cindy Mason, and others at NASA Ames Research Center have worked through some of the problems inherent in such a network of independent "agents" and have simulated such a network in action. The most recent ATIS standard, referenced above, includes a number of features that allow the automated networking of such independent robotic telescopes.

18.8 AI-Based Operations

Mark Drummond and others at NASA Ames Research Center have also developed an advanced, artificial intelligence (AI)-based operations system for robotic telescopes. This system utilizes the most recent version of ATIS, which includes features tailored specifically for such advanced operations.

Their operational system includes automated long-range planning. It also includes nightly schedule optimization features which allow a greater range of observational strategies to be automatically implemented than was the case for the much simpler algorithms in the original ATIS-based control software.

Ann Patterson-Hine and others, also at NASA Ames Research Center, have developed an AI-based remote diagnostic system for robotic telescopes.

18.9 The Future of Robotic Telescope Control

If present trends continue, we should expect that robotic telescope control should become ever more complex and diversified. We should also expect that the aperture of the largest robotic telescopes should keep on increasing. Networking of both the Crawford and Genet varieties could become a reality before the turn of the century.

Another "trend" that may grow is the grouping of many robotic telescopes at superior but remote sites. The Fairborn Observatory has some eight robotic telescopes in simultaneous, unattended nightly operation at a fully robotic observatory in southern Arizona. Amazingly, all eight telescopes and the observatory itself are kept in operation by a single individual, Louis Boyd, and it only takes him part time at that! Manpower-wise this must be the most efficient astronomical operation on the face of our planet!

One of the advantages of robotic telescopes has always been the ability to operate them at remote sites that are astronomically advantageous but humanly difficult. The more difficult the human circumstances, the greater the advantage of robotic over non-robotic telescope control. The future should see this inherent robotic control advantage pushed even further than it has been to date. Three possibilities come to mind.

The first is the placement of robotically-controlled telescopes on even higher and more remote mountaintops. Genet's favorite candidate was suggested as an ideal site for robotic telescopes by the late Harlan Smith. It is a 20,000 foot elevation mountaintop in northern Chile. Amazingly, there is a good road, the highest in the world by far, to within 400 feet of the summit. The road is used to transport sulfur from an outcropping near the summit to a copper mine some hundred miles away. There is, on the mountain, only a minor amount of precipitation about once every ten years, most years there being none at all. There is no visible life. The summit looks and feels like the Moon. The sky is black, not blue. It is, perhaps, the driest, clearest, most astronomically desirable site on Earth. The air is a bit thin for humans, but robots would do quite well.

The second possibility is the South Pole and the even higher and drier plateau nearby. Astronomical telescopes are accumulating at the South Pole in ever increasing numbers. There is, however, a serious problem with this accumulation; the humans needed to control these telescopes. Supporting a human at the South Pole is not only an expensive undertaking, but the fragile polar environment cannot gracefully support much more in the way of wintering-over humans. The robotic solution to this dilemma is obvious. While the Fairborn Observatory solution of having a single person handle *all* the telescopes might be a bit much to expect, we should see a move in this general direction.

The third and final possibility is our Moon. Lacking any appreciable atmosphere, yet providing a firm, essentially infinite-inertia object to which telescopes can be attached, the Moon is, perhaps, the most highly desirable astronomical site in the solar system. The appreciable logistical costs of continually supporting human observers on the Moon would, however, make robotic telescopes a virtual necessity. Consider what might be the ultimate telescope, the ultimate telescope control system.

The ultimate "telescope" would actually be a group of co-located telescopes, networked to each other and to AI-based network control on Earth. Located on the Moon's far side, Earth shine would not be a factor. In one mode of operation, the telescopes would work together as a single, giant optical interferometer, easily imaging Earth-like planets in other stellar systems or resolving the most intricate details of galaxies on the far edge of the universe. In another mode, the telescopes, acting individually, would conduct simultaneous photometric, spectroscopic, and other observations of a diverse variety of objects for a multitude of observational programs.

All this would be automatically controlled by AI-based telescope and network control systems running on computers of as yet undreamed capabilities. Program code might run into the millions of lines. We stand ready, should this observatory begin operation in our lifetime, to describe this ultimate telescope control system in the presumably ultimate edition of our book.

Appendix A

Estimates of Telescope Control System Costs—Some Historical Perspectives

Using modern computers, telescope control systems can be developed relatively inexpensively. Software is now one of the most expensive components of any new system.

An example of a project to retrofit a new system to an existing telescope by a commercial concern is the system installed in 1983 by DFM Engineering, Inc. on the CTIO 1.5-m telescope. Similar systems cost about \$50,000 (Melsheimer, 1983), including the dual-processor computer (about \$10,000), new motors and gearboxes (about \$5,000), terminal and displays (about \$1,000), software (about \$25,000), and installation (about \$5,000).

An amateur project is described by Tomer and Bernstein (1983). Their computer upgrade of a trailer-mounted 12-inch Cassegrain telescope cost a total of about \$1,000 plus the cost of an Apple II+ computer with disk. The simplest system one can build, using a single board computer and two stepper motors, can be assembled for under \$500.

In contrast, professional observatories in the mid-1970s were faced with far higher costs. The following projects were successfully completed using late 1960's and early 1970's minicomputer technology:

(a) \$50,000 plus software (Meeks, 1975, p. 23)

(b) \$120,000 plus software (Linnell and Hill, 1975, p. 60)

The price represents 10% of the cost of the whole observatory.

(c) \$150,000 to \$200,000 plus software (McCord, Paavola, and Snellen, 1975, p. 160)

(d) "well above half a million (dollars)" (Beaumont and Wolfe, 1975, p. 275).

These costs are primarily for hardware, since software was developed by observatory staff and tended to be ignored. In the mid-1990s, professional observatories tend to use UNIX based workstations in their control systems, while commercial firms, such as ACE, often use the PC. By way of comparison, the hardware in the following examples represents the late 1960's and early 1970's technology that was available at the time these systems were built.

(a) Two 6502 microprocessors in a custom circuit (Melsheimer, 1983).

(b) An Apple II+ (Tomer and Bernstein, 1983).

(c) One 6809 on a Peripheral Technology PT-69 single board computer, driving custom circuitry (Fairborn Observatory APT).

(d) LSI-11 with 64 KB memory, driving Compumotor microsteppers at the Lowell Observatory (White, 1984).

(e) An 8-bit Multi-8 computer made by Intertechnique (France) with 12K bytes of memory, ASR 33 teletype, 32-bit I/O board, and 8-line interrupt board (Bourlon and Vin, 1975, p. 63). Today, computers costing $1,000 are many times as fast and capable. In the early 1970s this configuration cost about $25,000.

(f) A 16-bit HP 2100 with 8K of memory (Van der Lans and Lorenson, 1975, p. 253).

(g) A system employing four simple microprocessors to handle the dome, hour angle, declination, and RA and DEC tracking rates (Fridenberg, Westphal, and Kristian, 1975, pp. 218–242). Note that today's computers have far greater processing capability and memory than these early systems, but a large portion of these resources are devoted to graphical user interfaces.

Software development costs can no longer be ignored, since software is now the most costly part of modern control systems. Published estimates of programmer time for designing, coding, debugging, and documenting telescope control software written at professional observatories in the 1970s are given below. These span several years and many different hardware and software approaches. Note that these projects were concerned with high accuracy control of large telescopes, and the software was developed without many of today's software development tools, such as good editors, high level language compilers, and even disk drives in some cases (paper tape was used on many of the smaller minicomputers).

(a) "more than a man-year" (Ingalls, 1975, p. 24).

(b) "approximately 2 man-years" (Taylor, 1975, p. 28).

(c) "worked on it part-time for 3 years written about 50,000 machine language instructions" (Hill, 1975, p. 60).

(e) "100,000 FORTRAN statements ... of which about a quarter are currently active" (Moore, Merillat, Colgate, and Carlson, 1975, p. 95).

(f) "something less than one man year" (Paavola, 1975, p. 160).

(g) "Everyone seems to be taking between one and two man years to program his system, almost regardless of the size of the system ..." (Rather, 1975, p. 390).

(h) 1.5 man-years (Melsheimer, 1983)

Even with modern software development tools and techniques, in 1994 "... it still takes two man-years to develop telescope control software..." including time spent debugging the software, according to Wayne Rosing, former Vice President at Sun Microsystems Computer Corporation, who built a control system for his telescope. To demonstrate how expensive this is at a professional observatory that must pay for software, the following cases are presented:

1. Software consultant: expensive, but fast @$80/hour × 1.0 man-years (2000 hours), Cost = $160,000
2. Full-time professional programmer: also expensive @$60/hour ($30/hour salary + 100% overhead) × 1.5 man-years, Cost = $180,000
3. Full-time astronomy graduate research assistant: slower, not as experienced with the computer or with programming techniques @$20,000/year × 2.0 man-years, Cost = $40,000.

Since the hardware costs for a retrofit control system can be as low as $5,000–$15,000 today, software is a significant fraction of the total system costs, even when the least expensive labor is used. This is confirmed by the proportion of system costs attributed to software in the DFM Engineering, Inc. commercial system.

The cost of labor is unimportant to those of us who program for the fun of it. However, the time required to get these systems to work suggests that simplicity is the best approach, and the less software that is required, the better.

Those who are designing closed-loop control systems might be interested in the history of the basic approach used in systems built in the 1970s. Successful computerized telescope control systems have used a variety of servo techniques. Often the same system will be called different names at different observatories, so that what is a velocity servo to one engineer is a position servo to another. The following types of servos have been implemented successfully:

1. Position only

   Kitt Peak NRAO 36-foot radio telescope (Moore, 1975)

2. Digital position and analog velocity

   University of Hawaii 88-inch telescope (Harwood, 1975)

   ESO 3.6-meter telescope (van der Lans, 1975)

   CFH telescope (Beaumont, 1975)

   Anglo-Australian 3.9-meter telescope (Wallace, 1975 and Bothwell, 1975)

   MMT 4.5-meter synthetic aperture telescope (Stephenson, 1975)
   Kitt Peak National Obervatory—various telescopes (Paffrath, 1975)

3. Digital position (pointing) or digital velocity (tracking)

   INT 98-inch telescope (Beale, 1975)

4. Open-loop, no feedback

   Haystack aperture synthesis interferometer (Burke, 1975).

Appendix B

Manufacturers of Motors and Related Hardware

Company names, addresses, and telephone numbers often change. To obtain more current information and other sources not listed here, consult the Thomas Register, available at many public libraries or from Thomas Publishing Company, One Penn Plaza, New York, NY 10001, (212) 695-0500.

1. Aerotech, Inc.
   101 Zeta Drive
   Pittsburgh, PA 15238-2897
   (412) 963-7470
   Steppers and drivers; servomotors and controllers; PC-based controllers; linear and rotary positioning stages; angle and linear encoders; linear motors.

2. Allied Devices
   2365 Milburn Ave.
   Baldwin, NY 11510
   (516) 223-9100
   Small gears, bearings, and pulleys.

3. American Precision Industries
   4401 Genesee Street
   Buffalo, NY 14225
   (716) 631-9800
   Steppers (65-3,000 oz-in) and drivers, microstep drives and indexers (up to 50,800 steps/rev.); servomotors and controllers; gearheads; angle encoders; brakes and clutches; linear actuators.

4. Anaheim Automation
   910 E. Orangefair Lane
   Anaheim, CA 92801
   (714) 992-6990
   Complete line of stepper motors, drivers (bilevel and microstepped), and indexers.

5. Anorad Corporation
   110 Oser Avenue
   Hauppauge, NY 11788
   (516) 231-1995
   Servomotor controllers, linear servo motors, specialized industrial X-Y and rotary stages.

6. Applied Motion Products
   404 Westbridge Dr.
   Watsonville, CA 95076
   (800) 525-1609
   Gear reducers that mount on NEMA-sized motors.

7. Ball Screws & Actuators Co., Inc.
   3616 Snell Ave.
   San Jose, CA 95136-1305
   (800) 882-8857
   Ball screws, nuts, and other items for moving a secondary mirror to focus a Cassegrain telescope.

8. Bayside Controls
   20-02 Utopia Pkwy.
   Whitestone, NY 11357
   (516) 484-5353
   Gear reducers that mount on NEMA-sized motors.

9. Boston Gear Division
   IMO Industries
   14 Hayward Street
   Quincy, MA 02171
   (800) 343-3353
   Motion control, gears, belts, chain, and motor couplers.

10. CAMAD
    6200 Excelsior Blvd., Suite 104
    Minneapolis, MN 55416
    (612) 926-9401
    Servomotor controllers for PCs with encoder and digital I/O, complete computer systems.

11. Cybernetic Micro Systems
    P.O. Box 3000
    San Gregorio, CA 94074
    (415) 726-3000
    IC intelligent stepper controller chips.

12. Eastern Air Devices, Inc.
    Dover, NH 03820
    (603) 742-3330
    Steppers, synchronous and induction motors, gearmotors and reducers.

13. Epoch Instruments
    2331 American Ave.
    Hayward, CA 94545
    (415) 784-0391
    Telescope mounts, gears, bearings, and pulleys.

14. Galil Motion Control Inc.
    575 Maude Court
    Sunnyvale, CA 94086
    (408) 746-2300
    Programmable servo motion controllers for PC/XT/AT, VMEbus, Multibus, STDbus, RS-232, and standalone. Servomotors with controllers, and design software.

15. Helical Products
    901 W. McCoy Lane
    Santa Maria, CA 93456
    (805) 928-3851
    Helical shaft couplers.

16. Hurst Manufacturing Corp.
    P.O. Box 326
    Princeton, IN 47670
    (812) 385-2564
    Steppers, linear motors, and DC motors.

17. Kollmorgen Inland Motor Division
    501 First Street
    Radford, VA 24141
    (800) 284-6526
    Torque and servo motors; linear amplifiers; linear force motors; alternators/generators; brushless motors.

18. Lovejoy, Inc.
2655 Wisconsin Ave.
Dowers Grove, IL 60515
(312) 852-0500
Shaft couplers.

19. Martin Sprocket and Gear
P.O. Box 888
Arlington, TX 76004
(817) 465-6377
Large gears, bearings, and pulleys.

20. Motion Science, Inc.
1485 Kerley Drive
San Jose, CA 95112
(408) 287-0300
Servo motor system with STD bus or RS-232 interface, and microstepper drivers.

21. New England Affiliated Technologies
620 Essex Street
Lawrence, MA 01841
(800) 227-1066
Stepper and servo motors, linear steppers, X-Y stages.

22. Nordex
50 Newton Rd.
Danbury, Ct 06810
(203) 792-9050
Small gears, bearings, and pulleys.

23. Oregon Micro Systems Inc.
1800 N.W. 169th Place
Suite A400
Beaverton, OR 97006
(503) 644-4999
PC (ISA and EISA), VMEbus, STD Bus, and RS-232/422 motor controllers, DOS and Windows software.

24. Oriental Motor U.S.A. Corp.
2580 West 237th Street
Torrance, CA 90505
(800) 468-3982
Inexpensive stepper motors, torque motors, brushless DC motors.

25. Parker Compumotor Division
    5500 Business Park Drive
    Rohnert Park, CA 94928
    (800) 358-9068
    Microstepped servomotors and steppers (up to 65,536 steps per revolution), direct-drive DC torque motors (up to 1,024,000 steps per revolution), drivers, and indexers.
26. PIC Design
    P.O. Box 1004
    Middlebury, CT 06762
    (800) 243-6125
    Small gears, bearings, and pulleys.
27. Plastock
    Three Oak Rd.
    Fairfield, NJ 07006
    (201) 575-0038
    Plastic gears and pulleys.
28. Precision Industrial Components Corp.
    P.O. Box 1004
    Benson Road
    Middlebury, CT 06762
    (800) 243-6125
    Gears, linear bearings and motion systems, shafting, belts.
29. Reliance Motion Control
    6950 Washington Avenue South
    Eden Prairie, MN 55344
    (800) 328-3983
    Electro-Craft servomotors, drives, and controllers.
30. Rogers Labs
    1939 S. Susan
    Santa Ana, CA 92704
    (714) 751-0442
    Stepper interface cards and complete motor/card kits at reasonable prices.
31. SECS, Inc.
    520 Homestead Ave.
    Mt. Vernon, NY 10550
    (914) 667-5600
    Small gears, bearings, and pulleys.
32. Seitz
    Torrington Industrial Ln.
    Torrington, CT 06790
    (203) 489-0476
    Small gears, bearings, and pulleys.

33. The Singer Company
    Kearfott Division
    1150 McBride Avenue
    Little Falls, NJ 07424
    (201) 785-6000
    Steppers, drivers, and synchronous motors.
34. SKF Specialty Products Co.
    1530 Valley Center Parkway
    Bethlehem, PA 18017
    (610) 861-3729
    Very high accuracy recirculating roller screws and translation stages.
35. Small Parts, Inc.
    6891 N.E. Third Ave.
    Miami, FL 33238
    (305) 751-0856
    Gears, pulleys, screws, and drills.
36. Solidur Plastics
    200 Industrial Dr.
    Delmont, PA 15626
    (412) 468-6868
    Small pulleys.
37. Stock Drive Products
    2101 Jericho Turnpike
    New Hyde Park, NY 11040
    (516) 328-0200
    Small gears, bearings, and pulleys.
38. The Superior Electric Company
    383 Middle Street
    Bristol, CT 06010
    (203) 582-9561
    Steppers, controllers, vibration dampers, synchronous motors.
39. Tech 80 Inc.
    658 Mendelssohn Avenue North
    Minneapolis, MN 55427
    (800) 545-2980
    DSP servo controllers and encoder interfaces for PC, VMEbus, and STD bus interfaces, books on designing motion control systems.
40. Texas Micro Express
    5959 Corporate Drive
    Houston, TX 77036
    (800) 950-9199
    nuLogic servo motors and stepper motors, PC ISA controller cards, rack mount PCs, software.

41. Union Gear and Sprocket
    111 Penn St.
    Quincey, MA 02269
    (617) 479-6800
    Small gears, berings, and pulleys.
42. Warner Electric
    Motion Control Systems Division
    1300 N. State Street
    Marengo, IL 60156
    (815) 568-8001
    Steppers, driver cards, vibration dampers, ball screws.
43. Winfred M. Berg, Inc.
    499 Ocean Avenue
    East Rockaway, NY 11518
    (516) 596-1700
    Gears, belts, motor shaft couplers, and many useful drive components.
44. Whedco Incorporated
    4750 Venture Drive
    Ann Arbor, MI 48108-9559
    (313) 665-5473
    Complete line of stepper and servo motors and controllers.

Appendix C

Manufacturers of Position Sensors

Company names, addresses, and telephone numbers often change. To obtain more current information and other sources not listed here, consult the Thomas Register, available at many public libraries or from Thomas Publishing Company, One Penn Plaza, New York, NY 10001, (212) 695-0500.

1. Analog Devices
   One Technology Way
   P.O. Box 9106
   Norwood, MA 02062
   (617) 329-4700
   Inductosyn interface chips, synchro/digital converters, analog/digital and digital/analog converters, digital switches.

2. Astrosystems, Inc.
   6 Nevada Drive
   Lake Success, NY 11040
   (516) 328-1600
   Synchro/digital converters, multispeed converters for synchros geared in groups.

3. BEI Motion Systems Co.
   P.O. Box 3838
   Little Rock, AR 72203
   (501) 851-4000
   Full line of incremental and absolute optical encoders, up to sub-arc second resolution.

4. Buckminster Corporation
   719 Washington St.
   Newton, MA 02160
   (617) 864-2456
   Encoder to computer interfaces.

5. CAMAD
   6200 Excelsior Blvd., Suite 104
   Minneapolis, MN 55416
   (612) 926-9401
   Encoder to computer interfaces.
6. Canon U.S.A., Inc.
   One Canon Plaza
   Lake Success, NY 11042-1113
   (516) 488-6700
   Very high resolution laser interferometer rotary (18 million pulses per revolution) and linear (0.01 μm, 0.0000004 inch) encoders.
7. Elm Systems Encoders Division
   1101 Brown Street
   Wauconda, IL 60084
   (708) 526-5003
   Inexpensive incremental optical encoders up to 12,700 counts per turn.
8. Farrand Controls
   99 Wall Street
   Valhalla, NY 10595
   (914) 761-2600
   Inductosyns (both rotary and linear).
9. FSI/Fork Standards Inc.
   668 Western Ave.
   Lombard, IL 60148-2097
   (708) 932-9380
   Optical, heavy duty, and high resolution encoders.
10. Hecon Corp.
    15 Meridian Rd.
    Eatontown, NJ
    (800) 524-1669
    Shaft encoders.
11. Heidenhain Corp.
    115 Commerce Drive
    Schaumburg, IL 60173
    (708) 490-1191
    Incremental and absolute shaft angle encoders.
12. HMT Technology Corp.
    1055 Page Ave.
    Fremont, CA 94538
    (510) 490-3100
    Magnetorestrictive encoders.

13. Hohner Shaft Encoder Corp.
    777 Cayuga Street
    Lewiston, NY 14092
    (416) 563-7209
    Incremental and absolute shaft and hollow shaft encoders.
14. Interface Engineering, Inc.
    300 Centre Street
    Holbrook, MA 02343
    (617) 986-2600
    Synchro/digital converters; binary angle/sine, cosine converters; coordinate conversion modules.
15. Itek Optical Systems Division
    Litton Industries
    10 Maguire Road
    Lexington, MA 02173
    (617) 276-2000
    Very high resolution (up to 22 bits or 0.31 arc-seconds) absolute and incremental optical shaft angle encoders.
16. Litchfield Precision Components, Inc.
    Three Precision Drive
    Litchfield, MN 55355
    (612) 693-2891
    Rotary and linear optical encoders.
17. Litton Industries, Inc.
    Encoder Division
    20745 Nordhoff Street
    Chatsworth, CA 91311
    (818) 341-6161
    Full line of incremental and absolute optical encoders.
18. Micro MO Electronics, Inc.
    742 2nd Ave. S.
    St. Petersburg, FL 33701
    (813) 822-2529
    Optical and magnetic encoders.
19. MTS Systems Corporation
    Sensors Division
    P.O. Box 13218
    Research Triangle Park, NC 27709
    (800) 633-7609
    Temposonics II linear magnetostrictive encoders.

20. Onshore Electronics Mfg. Corp.
    141BB Central Ave.
    Farmingdale, NY 11735
    (516) 753-6516 x201
    Shaft encoders.

21. Opti-Cal
    1240 12th Street
    Los Osos, CA 93402
    (805) 528-3601
    Laser linear position encoders.

22. PMI Motion Technologies Division
    Kollmorgen Corp.
    49 Mall Dr.
    Commack, NY 11725-5703
    (516) 864-1000
    Optical encoders to 2540 lines resolution.

23. Renco Encoders, Inc.
    26-28 Coromar Drive
    Goleta, CA 93117
    (805) 968-1525
    Absolute and incremental optical encoders.

24. Ridge Associates Inc.
    270 E. Route 46
    Rockaway, NJ 07866
    (201) 586-2717
    High accuracy resolvers.

25. Servo-Tek Products Company
    1086 Goffle Rd.
    Hawthorne, NJ 07506
    (201) 427-3100
    High resolution encoders.

26. Stock Drive Products
    2101 Jericho Turnpike
    P.O. Box 5416
    New Hyde Park, NY 11042-5416
    (516) 328-3300
    Encoders.

27. Sumtak Encoders
    615 Pierce St.
    Somerset, NJ 08875
    (908) 805-0008
    Optical, hollow shaft, incremental, modular, and rotary encoders.

28. Tech 80 Inc.
    658 Mendelssohn Avenue North
    Minneapolis, MN 55427
    (612) 542-9545
    Encoder to computer interfaces.

29. Teledyne Gurley
    P.O. Box 88-A
    Troy, NY 12181
    (800) 759-1844
    Incremental, absolute, rotary, and linear encoders.

30. Trans-Tek, Inc.
    Route 83
    Ellington, CT 06029
    (203) 872-8351
    Inexpensive high-accuracy variable capacitor; linear displacement and velocity transducers.

31. U.S. Digital Corp.
    8409 N.W. 15th Court
    Vancouver, WA 98665
    (800) 736-0194
    Non-contacting incremental and absolute optical shaft encoders and encoder interfaces.

Appendix D

Manufacturers of PC-Clone Products

Company names, addresses, and telephone numbers often change. To obtain more current information and other sources not listed here, consult the Thomas Register, available at many public libraries or from Thomas Publishing Company, One Penn Plaza, New York, NY 10001, (212) 695-0500.

This list contains manufacturers of boards and other products for popular PC buses and interfaces. It does not list manufacturers of desktop, deskside, laptop, or notebook PCs or servers.

1. Bit 3 Computer Corporation
   8120 Penn Avenue South
   Minneapolis, MN 55431-1393
   (612) 881-6955
   Memory mapped bus-to-bus adaptors for PCs, VMEbus, workstations, and several popular buses.

2. Byte Runner Technologies
   406 Monitor Lane
   Knoxville, TN 37922
   (800) 274-7897
   I/O boards for high speed serial or parallel communications.

3. ComputerBoards, Inc.
   125 High Street
   Mansfield, MA 02048
   (508) 261-1123
   PC I/O and data acquisition boards for ISA and PCMCIA interfaces.

4. CyberResearch, Inc.
   25 Business Park Drive
   Branford, CT 06405
   (203) 483-8815
   Data acquisition, A/D, D/A, serial and parallel I/O, digital oscilloscope, stepper motor controller, and GPIB boards and software.

5. Data Translation, Inc.
   100 Locke Drive
   Marlboro, MA 01752-1192
   (800) 525-8528
   Data acquisition, A/D, D/A, serial and parallel I/O, frame grabber, and video processing and display boards for PC and VMEbus.

6. Industrial Computer Source
   P.O. Box 910557
   San Diego, CA 92191-0557
   (800) 523-2320
   Rack mount PC (386 through Pentium) systems, single board PC computers, disks, I/O boards, A/D and D/A boards, and network boards.

7. Intelligent Instrumentation
   6550 South Bay Colony Drive, MS 130
   Tucson, AZ 85706
   (800) 685-9911
   Complete line of data acquisition, A/D, D/A, digital I/O, and digital signal processing (DSP) boards for ISA and EISA buses and software for DOS and Windows.

8. Jameco
   1355 Shoreway Road
   Belmont, CA 94002-4100
   (800) 831-4242
   Electronic components, PC ISA prototyping boards, PC motherboards, computer supplies.

9. Keithley Metrabyte
   440 Myles Standish Blvd.
   Taunton, MA 02780
   (800) 348-0033
   PC data acquisition, analog and digital I/O, PC instrument, and PC-to-GPIB boards and software.

10. National Instruments
    6504 Bridge Point Parkway
    Austin, TX 78730-5039
    (800) 433-3488
    Complete line of data acquisition, analog and digital I/O, VMEbus, VXIbus, and GPIB interface boards for PC, PS/2, Macintosh, and Sun SBus.

11. Oregon Micro Systems Inc.
 1800 N.W. 169th Place
 Suite A400
 Beaverton, OR 97006
 (503) 644-4999
 ISA and EISA PC motor controllers, DOS and Windows software.
12. Parker Compumotor Division
 5500 Business Park Drive
 Rohnert Park, CA 94928
 (800) 358-9068
 PC/XT/AT ISA bus single-axis step and direction indexer.
13. PCMCIA Headquarters
 1030G East Duane Ave.
 Sunnyvale, CA 94086
 (408) 720-0107
 Personal Computer Memory Card International Association to promote the use of its interface, found mostly on notebook computers; maintains a list of PCMCIA manufacturers.
14. Texas Micro Express
 5959 Corporate Drive
 Houston, TX 77036
 (800) 950-9199
 nuLogic servo motors and stepper motors, PC ISA controller cards, software.
15. Wind River Systems, Inc.
 1010 Atlantic Avenue
 Alameda, CA 94501
 (800) 545-WIND
 (415) 748-4100
 VxWorks UNIX-like real-time operating system for a variety of processors including Intel 80x86, Motorola 680x0, AMD 29000 series, SPARC, and MIPS R3000 and R4000 series.
16. Z-World Engineering
 1724 Picasso Avenue
 Davis, CA 95616
 (916) 757-3737
 Inexpensive single board computers programmable using Z-World C compilers running under DOS or Windows on a PC. One model is expandable to 400 digital I/O lines.

Appendix E

Manufacturers of Items Related to Telescope Control

Company names, addresses, and telephone numbers often change. To obtain more current information and other sources not listed here, consult the Thomas Register, available at many public libraries or from Thomas Publishing Company, One Penn Plaza, New York, NY 10001, (212) 695-0500.

Environmental Sensors

1. EG&G Environmental Equipment Division
   217 Middlesex Turnpike
   Burlington, MA 01803
   (617) 270-9100
   Humidity probes and meters.
2. Entran Devices, Inc.
   10 Washington Avenue
   Fairfield, NJ 07004
   (800) 635-0650
   High accuracy digital barometric pressure meter.
3. General Eastern Instruments Corporation
   20 Commerce Way
   Woburn, MA 01801
   (617) 938-7070
   Temperature and humidity sensors, including chilled mirror dewpoint sensors.

4. Hy-Cal Engineering
   9650 Telstar Avenue
   El Monte, CA 91731-3093
   (818) 444-4000
   Thin film humidity sensors and temperature sensors.

5. Hygrometrix Inc.
   7740 MacArthur Blvd.
   Oakland, CA 94605
   (510) 639-7800
   Humidity sensors

6. MesoTech International
   4670 Chancery Way
   Sacramento, CA 95864
   (916) 483-0600
   Rugged and reliable complete weather stations, cloud sensors.

7. Omega Technology, Inc.
   One Omega Drive
   Box 4047
   Stamford, CT 06907
   (800) 826-6342
   Temperature probes and systems.

8. Panametrics, Inc.
   Process Instrument Division
   221 Crescent Street
   Waltham, MA 02254
   (800) 848-5996
   Humidity sensors.

9. Pressure Systems, Inc.
   15 Research Drive
   Hampton, VA 23666
   (804) 865-1243
   Barometric pressure standard.

10. Protimeter, Inc.
    87 Modular Avenue
    Commack, NY 11725
    (800) 881-1089
    Very accurate chilled mirror dewpoint sensor.

11. Rotronic Instrument Corp.
    160 East Main Street
    Huntington, NY 11743
    (800) 899-6411
    Humidity and temperature sensors.

12. Sensor Instruments Co., Inc.
    41 Terrill Park Drive
    Concord, NH 03301
    (800) 633-1033
    Humidity, barometric pressure, and temperature sensors.
13. SenSym
    1244 Reamwood Avenue
    Sunnyvale, CA 94089
    (408) 744-1500
    Pressure transducers.
14. Setra Systems, Inc.
    45 Nagog Park
    Acton, MA 01720
    (508) 263-1400
    High accuracy barometric pressure transducers.
15. Vaisala Inc.
    100 Commerce Way
    Woburn, MA 01801
    (617) 933-4500
    Relative humidity, temperature, and barometric pressure sensors; meteorological instruments.

Global Positioning System Time and Location Receivers

1. Ashtech
   1170 Kifer Road
   Sunnyvale, CA 94086
   (408) 524-1400
2. Bancomm Division of Datum, Inc.
   6541 Via Del Oro
   San Jose, CA 95119
   (408) 578-4161
3. Garmin International, Inc.
   9875 Widmer Road
   Lenexa, KS 66215
   (800) 800-1020
4. Magellan Systems Corp.
   960 Overland Ct.
   San Dimas, CA 91773
   (714) 394-5000

5. Magnavox Advanced Products and Systems Company
   2829 Maricopa Street
   Torrance, CA 90503
   (213) 618-7319

6. Motorola, Inc.
   4000 Commercial Avenue
   Northbrook, IL 60062
   (800) 421-2477

7. Navstar Electronics, Inc.
   1500 N. Washington Blvd.
   Sarasota, FL 34236
   (800) 486-6338

8. Odetics - Precision Time Division
   1515 South Manchester Avenue
   Anaheim, CA 92802
   (800) 374-4783

9. Trimble Navigation
   645 North Mary Ave.
   P.O. Box 3642
   Sunnyvale, CA 94088-3642
   (800) TRIMBLE

10. True Time Instrument Company
    429 Olive Street
    Santa Rosa, CA 95401
    (707) 528-1230

In addition to these sources, consult marine supply and outdoor and camping equipment retailers. Hand-held units sell for less than $400.

VMEbus Sources

1. Bancomm Division of Datum, Inc.
   6541 Via Del Oro
   San Jose, CA 95119
   (408) 578-4161
   VMEbus GPS receiver

2. BICC-VERO Electronics Inc.
   1000 Sherman Avenue
   Hamden, CT 06514
   (800) BICCVME
   Very complete line of VMEbus racks, card cages, prototyping boards, power supplies, and hardware.
3. Bit3 Computer Corporation
   8120 Penn Avenue South
   Minneapolis, MN 55431-1393
   (612) 881-6955
   VMEbus adaptors for PCs (PCI, ISA, and EISA buses), Macintosh, DEC, SGI, and IBM computers.
4. Data Translation, Inc.
   100 Locke Drive
   Marlboro, MA 01752-1192
   (800) 525-8528
   Data acquisition, A/D, D/A, serial and parallel I/O, frame grabber, and video processing and display boards.
5. Electronic Solutions
   6790 Flanders Drive
   San Diego, CA 92121
   (619) 452-9333
   VMEbus racks, card cages, prototyping boards, extender boards, power supplies, and hardware.
6. Motorola Inc.
   Computer Group
   2900 South Diablo Way
   Dept. MD DW212 Tempe, AZ 85282
   (800) 234-4863
   VMEbus racks, card cages, processor boards, extender boards, power supplies, and hardware.
7. National Instruments
   6504 Bridge Point Parkway
   Austin, TX 78730-5039
   (512) 794-0100
   VMEbus adaptors for ISA and EISA PC buses, Microchannel (IBM PS/2), Macintosh NuBus, Sun Sparcstation, IBM RS/6000, DECstation 5000, and GPIB.
8. Radisys Corporation
   15025 SW Koll Parkway
   Beaverton, OR 97006
   (800) 950-0044
   VMEbus Intel 80386 and 80486 PC-compatible processors, keyboards, other peripherals, enclosures, power supplies, and software.

9. Schroff, Inc.
   170 Commerce Drive
   Warwick, RI 02886
   (800) 451-8755
   VMEbus racks, cabinets, cases, cards, and hardware.

10. VFEA International Trade Association
    10229 North Scottsdale Road
    Suite B
    Scottsdale, AZ 85253
    (602) 951-8866
    VMEbus Handbook, VMEbus Specification, and VMEbus Products Directory.

11. Wind River Systems, Inc.
    1010 Atlantic Avenue
    Alameda, CA 94501
    (800) 545-9463
    (415) 748-4100
    Real-time operating system for VMEbus processors.

Other Items

1. AB Engineering, Inc.
   5822 Kruse Drive
   Fort Wayne, IN 46818
   (219) 489-2845
   PC-based telescope control systems, and telescope mounts.

2. Astronomical Consultants & Equipment, Inc.
   P.O. Box 91946
   Tucson, AZ 85752-1946
   (520) 579-0698 voice or FAX
   Complete, turnkey PC-based telescope control systems.

3. COMSOFT
   1552 W. Chapala
   Tucson, AZ 85704
   PC-based telescope control systems.

4. Digi-Key Corporation
   701 Brooks Ave. South
   P.O. Box 677
   Thief River Falls, MN 56701-0677
   (800) 344-4539 (VISA, Master Card)
   Electronic components, including ICs, sockets, connectors, switches, LEDs, resistors, capacitors, transformers, surge suppressors, test equipment, and more.

5. Globe Electronic Hardware, Inc.
   P.O. Box 770727
   Woodside, NY 11377
   (800) 221-1505
   Electronic hardware, including fasteners, stand-offs, rack panel screws, washers, springs, and handles.
6. MultiFAX
   Rt. 1, Box 27
   Pulpwood Yard Road
   Peachland, NC 28133
   Weather satellite demodulator and image processing software.
7. Quadrant Systems
   P.O. Box 370
   12547 Regent Way
   Oregon House, CA 95962
   (916) 692-2563
   Portable and fixed-site commercial telescope control systems.
8. Soft-Tec Systems
   9333 Crowley Rd., Suite A
   Fort Worth, TX 76134-5904
   (817) 293-8446
   Complete telescope control systems.
9. Systems West, Inc.
   P.O. Box 222019
   Carmel, CA 93922
   (408) 625-6911
   Integrated turnkey weather satellite reception and processing systems.
10. Vanguard Electronic Labs.
    196–23 Jamaica Ave.
    Hollis, NY 11423
    Weather satellite antennas and receivers.

Appendix F

APT Control Algorithms

F.1 Introduction

For those who wish to develop their own APT control software, this appendix provides a number of useful algorithms or routines. Our special thanks to Louis Boyd for providing these. He first presented them at the 1984 IAPPP Big Bear Symposium.

F.2 Computer Mathematics

Some of the routines that follow require the use of integer division to force an integer output. In the algorithms in this article an integer constant is shown with no decimal point (e.g., 9) while a floating point number, even though it represents an integer value, will be shown with a decimal point (e.g., 9.). In addition, subtraction, and multiplication there is no difference in the result, but the division of two integers yields an integer that carries no fractional part. For example, $9./2. = 4.5$, but $9/2 = 4$.

If either number is floating point, the result will be floating point so that $9/2. = 4.5$. This should be kept in mind when converting these algorithms to a particular computer language.

Radians are used to measure angles to enhance the speed of computing trigonometric functions, of which there are many. Decimal hours are used instead of radians in some algorithms involving time calculations. It should be remembered that sidereal hours and solar hours are not the same.

F.3 Modified Julian Date

Modified Julian Date (MJD) is used in these calculations rather than Julian date because of the limited precision of BASIC09's real numbers.

The BASIC09 compiler has nine significant digits of precision, and the use of the full Julian date would cause unnecessary loss of resolution. The Julian date can easily be obtained by adding 2400000.5 to the MJD. The following algorithm is based on the Almanac for Computers. It is valid for dates between 1901 and 2099 as the year 2000 is a leap year.

Parameters	**Type**	**Values**
YEAR	integer	1909 – 2099 Universal Date
MONTH	integer	1 – 12 January = 1
DAY	integer	1 – 31 Day of month
HOUR	integer	0 – 23 Universal time
MINUTE	integer	0 – 59
SECOND	integer	0 – 59
MJUL	real	Modified Julian Date with fraction

```
MJUL = -678987.+367.*YEAR+INT(INT(275*MONTH/9)+DAY-INT(7*(YEAR+
       INT((MONTH+9)/12))/4))+(HOUR*3600.+MINUTE*60+SECOND)/86400.
```

F.4 Local Mean Sidereal Time

Local Sidereal Time is equivalent to the right ascension of a stellar object along the local meridian. Local Mean Sidereal Time (LMST) differs from Local Apparent Sidereal Time (LAST) in that it is not corrected to reflect the actual rotation rate of the Earth. The difference is at most about 13″. For the APT, LMST is sufficiently accurate and the calculation of nutation is complex. This algorithm is from the Almanac for Computers. Note that only the integer part of the modified Julian date is actually used in the calculation. While the LMST could be calculated from the decimal Julian date directly, accuracy would suffer because of the roundoff error in the computer. `MOD` stands for modulo, and `MOD(A,B)` is equivalent to `(A/B - INT(A/B))*B`.

Parameters	**Type**	**Values**
HOUR	integer	0 – 23 Universal Time
MINUTE	integer	0 – 59
SECOND	integer	0 – 59
MJUL	real	Modified decimal Julian Date
LMST	real	Local Mean Sidereal Time in decimal sidereal hours

Constants	**Type**	**Notes**
LONG	real	Local longitude in decimal sidereal hours

```
LMST = MOD(6.67170278+.065709823*(INT(MJUL)-33282.)+
       1.00273791*HOUR+MINUTE/60.+SECOND/3600.)-LONG,24)
```

F.5 Positon of the Sun

This is a relatively low precision calculation of the right ascension and declination of the Sun. It accounts for the tilt of the Earth's axis and uses an approximation to account for the ellipticity of the Earth's orbit. It will generally be correct to about one degree. Far better accuracy may be obtained from the trigonometric series calculations in the Almanac for Computers, but at the expense of greatly increased computation time. This calculation is good enough to calculate where the Sun is below the horizon to determine whether it is dark enough to operate the photometer. It could also be used for avoiding the Sun when doing infrared photometry in the daytime, and would be a good routine for Sun trackers on solar power devices.

Parameters	**Type**	**Values**
MJUL	real	MJD
RTA	real	RA in radians
DEC	real	Dec in radians
T	real	Julian centuries from 0 Jan 1900
MAOS	real	Mean anomaly of the Sun in radians
TLOS	real	True longitude of the Sun in radians

```
HALFPI=1.57079633
TWOPI =6.28318531
T=(MJUL-15019.5)*2.73785079E-5
MAOS=6.25658358+628.301946*T-2.61799388E-6*T*T
TLOS=MOD(MAOS+.03344051*SIN(MAOS)+.000349066*SIN(MAOS+MAOS)+
     4.93169,TWOPI)
RTA=ATN(.91746*TAN(TLOS))
(* QUADRANT CORRECTION *)
IF TLOS>HALFPI THEN RTA=RTA+PI
   IF TLOS>HALFPI+PI THEN RTA=RTA+PI
ENDIF
DEC=ASN(.39782*SIN(TLOS))
```

From the RA and Dec, the zenith angle may be calculated (see Zenith Angle) and when it exceeds a certain angle (100 degrees used in the Phoenix APT) it is dark enough to operate the telescope.

F.6 Position of the Moon

The calculation for the position of the Moon is simpler than that for the Sun, but has even less accuracy. It operates reliably for keeping the telescope 10° or more from the Moon when taking readings. This algorithm was extracted from *Celestial Basic* (Burgess, 1982) and compacted.

Parameters	Type	Values
MJUL	real	MJD + fraction
RTA	real	RA in radians
DEC	real	Dec in radians
BD	real	Base date
LM	real	Longitude of the Moon in radians

```
P2=6.28318531
BD=MJUL-36934.5
LM=5.43090264+.229971506*BD
RTA=MOD(LM+.109756775*SIN(LM-1.9443666E-3*BD-4.46356263),P2)
DEC=.403946*SIN(RTA)+.0398*SIN(RTA-3.11888592+9.24221652E-4*BD)
```

Avoidance of the Moon is accomplished as follows, if

```
SQ(RTASTAR - RTAMOON)+SQ(DECSTAR - DECMOON)<.03
```

then the star is within 10° of the Moon. Note that .03 is approximately 10° converted to radians squared. In calculating the RA difference of the Moon and the star, the shortest distance must be taken.

F.7 Zenith Angle

The zenith angle calculation is used in determining airmass for each observation and for determining whether twilight has occurred. This is a simple spherical trigonometry problem, and is limited in accuracy by the use of mean instead of apparent sidereal time. It ignores refraction, since the APT never operates near the horizon.

Parameters	Type	Values
RTA	real	RA of the object in radians
DEC	real	Dec of the object in radians
LMST	real	Local Mean Sidereal Time, 0–24 hours
COSZ	real	Cosine of the zenith angle

Constants	Type	Notes
SLAT	real	Sine of the local latitude
CLAT	real	Cosine of the local latitude

```
COSZ=SLAT*SIN(DEC)+CLAT*COS(DEC)*COS(.2618*LMST-RTA)
```

The cosine is chosen as the output for this routine as it saves a step in calculation, depending on whether conversion to airmass (which requires the secant of Z) or determining the zenith angle (which requires the arccosine of COSZ) is to be performed. Constants were chosen for the sine and cosine of the latitude to speed calculation by eliminating two trigonometric operations. These could be calculated if the program were to be made versatile for a portable instrument.

F.8 Heliocentric Correction

Several sources give algorithms for the calculation of heliocentric correction, but this one has been optimized for speed and should yield accuracy to a few seconds.

Parameters	Type	Notes
RTA	real	RA of the star in radians
DEC	real	Dec of the star in radians
MJUL	real	MJD with fraction
CORRECTION	real	Correction in decimal days
COSDC	real	Cosine of the star Dec
T	real	Base date
LOTS	real	Longitude of the Sun in radians
RADIUS	real	Radius vector of the Sun in mean Earth radii

```
COSDC=COS(DEC)
T=(MJUL-15019.5)*2.73785079E-5
LOTS=4.88162794+628.331951*T+5.279621E-6*T*T
RADIUS=.999720458/(1.+.016719*COS(LOTS+4.956105))
CORRECTION=-.0057755*(RADIUS*(COS(LOTS)*COS(RTA)*COSDC+
           SIN(LOTS)*(.39794248*SIN(DC)+.91741037*SIN(RTA)*COSDC)))
```

The correction should be added to the calculated Julian date to obtain the heliocentric corrected Julian date. Beware of limited computer precision when working with fractional Julian dates.

F.9 Precession to Current Coordinates

Precession is a large enough value (several minutes of arc) that it must be taken into account when hunting for navigation stars. It is easiest to work with star coordinates at an epoch for which there are good catalogs. As the Phoenix APT works stars of magnitude 8.0 or brighter, epoch J2000.0 coordinates were chosen because of the availability of the Sky Catalog 2000.0 which contains all stars which will be observed. This algorithm precesses from J2000.0 coordinates to current coordinates.

Parameters	Type	Notes
MJUL	real	MJD with fraction
RTA	real	Input and output RA in radians
DEC	real	Input and output Dec in radians
DAYS	real	Days from epoch 2000
RDIF	real	Change in RA
DDIF	real	Change in Dec

```
DAYS=51544.-MJUL
RDIF=DAYS*(6.112E-7+2.668E-7*SIN(RTA)*TAN(DEC))
DDIF=DAYS*(2.655E-7*COS(RTA)
```

```
RTA=RTA-RDIF
DEC=DEC-DDIF
```

These calculations do not take into account the change in direction of precession for extended periods of time, but they will suffice for APT use without correction well into the next century. Nutation, stellar aberration, and other apparent place corrections are small enough that they can be ignored.

F.10 Determination of Observability of a Star

The APT must make a decision as to whether a star is observable within a predetermined area above the telescope. A cone with a 45° zenith angle was selected to be the observable window. There may be other limitations, such as the limit in Dec placed by the selection of stars, and any other boundaries could be set. This algorithm calculates the rising and setting time for an object within a cone of a fixed zenith angle.

Parameters	**Type**	**Notes**
RTA	real	RA of the star in radians
DEC	real	Dec of the star in radians
RISE	real	Rising sidereal time
SET	real	Setting sidereal time
ANGLE	real	Absolute value of hour angle of rising or setting object

Constants	**Type**	**Notes**
SINLAT	real	Sine of the local latitude
COSLAT	real	Cosine of the local latitude
COSZMAX	real	Cosine of zenith angle of obs. cone
RAD_HRS	real	$12/\pi$ converts radians to hours

```
ANGLE=ABS(ACS((COSZMAX-SINLAT*SIN(DEC))/(COSLAT*COS(DEC))
RISE=MOD((RTA-ANGLE)*RAD_HRS+24,24)
SET=MOD((RTA+ANGLE)*RAD_HRS,24)
```

This algorithm has a number of uses in that it can calculate the time of sunrise, sunset, moonrise, and moonset if COSZMAX is set to zero. Note that the output times are in local apparent sidereal time which differs slightly (insignificantly) from local mean sidereal time.

These are all of the algorithms related to positional astronomy that are required to run an APT.

Appendix G

Automatic Photoelectric Telescope Software

This appendix contains all the software needed to operate an automatic photoelectric telescope of the type described in Chapter 15. All the procedures are written in Microware Basic09 except RAMP, which is in editor-assembler for the 6809 processor. Most of this software was originally written by Louis J. Boyd of the Fairborn Observatory. Portions of the software peculiar to the hardware were written by one of us (RG) and Lloyd W. Slonaker, Jr. The authors express their thanks to Louis Boyd for permission to use his programs in this book.

The intent in including this software is to provide complete working examples of control software at three levels of complexity. The simplest, the procedures QJOY2, DIGITAL, MOVE, and RAMP, provide a complete set of software to move the telescope with a control paddle. Only motions in three speeds in R.A. and Dec. are supported. See Chapter 16 for detailed descriptions.

Intermediate level procedures allow automatic initialization of the telescope and automatic slewing to stored or entered coordinates. These instructions should be of interest to the casual observer of double stars, deep sky objects, etc. To control the telescope to this extent is a fairly complex task, and a dozen procedures are involved as shown in the figure below. Although all of the various procedures (except JOY4) have been previously described, a brief description of how each works is given here to augment Figure G-1.

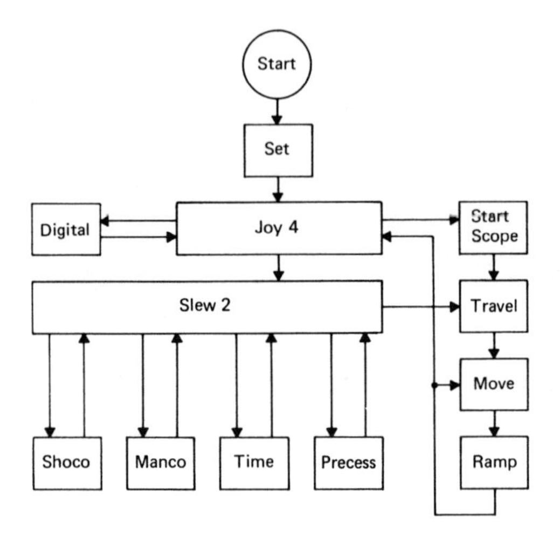

Fig. G-1

In the system for which this particular software was written the processor board has a built in clock calendar. While quite accurate, on occasion the time needs to be reset using SET (one second of error in time translates to a 15 arc second position error). JOY4 is the main control program for the slew to coordinates control program. To initialize the position of the telescope, STARTSCOPE moves the telescope to the southeast limit switches, an accurately known and calibrated position. SLEW2 is then used to provide choices of preprogramed objects or for manual entry via MANCO and SHOCO of new objects (Epoch 2000.0 coordinates). Coordinates are automatically brought to the current epoch with PRECESS, and TRAVEL, MOVE, and RAMP are used to actually determine the motion to be made and then make it. JOY4 also provides an option for direct manual control via DIGITAL and the control paddle.

The slew to coordinates ensemble has been found useful for two things: First, it can be used to align and check out the system. Second, one needs only to enter the coordinates of an object and the system slews to it and a precise scientific visual evaluation can be made by sorting the objects into such taxonomic classifications as:

1. Wow!,
2. Outstanding!,
3. You've seen one fuzzball you've seen them all, and
4. That faint smudge is supposed to be M xx?

Finally, there is the automatic system that finds and photometrically measures objects on its own. Here MAIN has overall control. This particular version of MAIN is written for a single photometric band ("V" of the Johnson UBV system) and was used by the Fairborn Observatory during the Fall of 1984 (concurrent with the writing of the first edition of this book). This section of the overall program works nicely, but its creation was a learning process. Since it is unlikely that anybody will duplicate our system exactly these listings make an excellent quide but should not be considered our (or your) final product.

The listings which follow are actual printouts of running, debugged systems. Experienced computer programmers will appreciate the fact that we have not attempted to "enhance" the quality of the listings by re-typing. We have also had enough experience with computers and software to feel it our duty to recognize that there are subtle differences from one situation to another that only become apparent when you make the new system try to run. Therefore, study these listings so that you know not only **what** is happening but **how** it is being done.

For ease in location, the listings are alphabetically ordered by procedure name and a summary list is:

1. BUILDFILE	9. LUNAR	17. RAMP	25. STOPSCOPE
2. COEFFICENTS	10. MAIN	18. SET	26. SUNANGLE
3. DATREAD	11. MANCO	19. SHOCO	27. THRESH
4. DIGITAL	12. MEAS	20. SHOWTIM	28. TILDARK
5. HELIO	13. MOVE	21. SLEW2	29. TIME
6. HUNT	14. PRECESS	22. SOLAR	30. TRANSFORM
7. JOY4	15. PTCLK	23. STARFILE	31. TRAVEL
8. LOCK	16. QJOY2	24. STARTSCOPE	32. ZENITH

Many of the above procedures can be used as listed in a simple manual, slew to coordinates, or automatic control system. In fact, about two-thirds of the procedures were used "as is" on a 6809 processor using Basic09 but with vastly different support hardware than the original system for which they were written.

1. BUILDFILE

```
(* THIS PROGRAM ALLOW MANUAL ENTRY OF NEW STARS INTO
(* THE FILE 'STARFILE' LOCATED ON THE 'CHD' DRIVE
(* THE STARS REMAIN IN THE ORDER ENTERED AS THERE IS NO
(* REQUIREMENT IN THE MAIN FROGRAM FOR A PARTICULAR ORDER
BASE 0
TYPE STARIN=INAM:STRING[10]; IRTA,IDEC,IMAG:REAL
TYPE GROUPIN=NAMEIN:STRING[10]; DIAPHRAGM:INTEGER; ISTAR(4)
 :STARIN
DIM FILE,OUTPATH: INTEGER
```

```
FILE=3
DIM STARNAM(5):STRING[5]
STARNAM(0)="CHECK"
STARNAM(1)=" SKY"
STARNAM(2)=" COMP"
STARNAM(3)=" VARI"
DIM STARG:GROUPIN
DIM NUMS:STRING[10]
NUMS="0123456789"
OPEN #FILE,"STARFILE"
LOOP
     PRINT "CURRENT GROUPS ON FILE:"
     MAXGROUP=0
     SEEK #FILE,0
     WHILE NOT(EOF(#FILE)) DO
          GET #FILE,STARG
          IF MOD(MAXGROUP,4)=0 THEN PRINT
          ENDIF
          PRINT USING "I6>,': ',S10<",MAXGROUP,STARG.NAMEIN;
          MAXGROUP=MAXGROUP+1
     ENDWHILE
     PRINT
     PRINT
     GROUP=MAXGROUP+1
     INPUT " QUIT ",A$
EXITIF A$="Y" THEN
ENDEXIT
     INPUT "PRINTOUT OF CURRENT FILES ",X$
     IF X$="Y" THEN
          OUTPATH=1
          SEEK #FILE,0
          CLOSE #1
          OPEN #OUTPATH,"/P1"
          GROUP=0
          WHILE NOT(EOF(#FILE)) DO
               GET #FILE,STARG
               GOSUB 10
               PRINT
               GROUP=GROUP+1
          ENDWHILE
          CLOSE #1
          OPEN #OUTPATH,"/TERM"
     ENDIF
     INPUT "ADD TO LIST (Y/N) ",X$
     IF X$<>"Y" THEN
          REPEAT
               INPUT "NUMBER OF GROUP TO MODIFY ",GROUP
          UNTIL GROUP>=0 AND GROUP<=MAXGROUP
          SEEK #FILE,GROUP*SIZE(STARG)
          GET #FILE,STARG
          SEEK #FILE,GROUP*SIZE(STARG)
          GOSUB 10 \(* DISPLAY CURRENT RECORD
     ENDIF
```

```
REPEAT
     PRINT "ENTER COMMON NAME OF STAR GROUP"
     PRINT "  "; NUMS
     IF X$<>"Y" THEN
          PRINT "  "; STARG.NAMEIN
     ENDIF
     INPUT K$
     IF K$<>"" THEN
          STARG.NAMEIN=K$
     ENDIF
     PRINT "DIAPHRAGM (0,1,2) ";
     IF X$>"Y" THEN
          PRINT "("; STARG.DIAPHRAGM; ")";
     ENDIF
     INPUT K$
     IF K$<>"" THEN
          STARG.DIAPHRAGM=VAL(K$)
     ENDIF
     FOR I=0 TO 3
          PRINT STARNAM(I); " :"
          IF I<>1 THEN PRINT "  STAR NAME [HDNNNNNNNN] OR [SAONNNNNNN]"
               PRINT "  "; NUMS
               IF X$<>"Y" THEN
                    PRINT "  "; STARG.ISTAR(I).INAM
               ENDIF
               INPUT K$
               IF K$<>"" THEN
                    STARG.ISTAR(I).INAM=K$
               ENDIF
          ELSE
               STARG.ISTAR(I).INAM=""
          ENDIF
          IF I=1 THEN PRINT "(ENTER ZERO'S FOR DEFAULT COMP-VAR CENTERING

          ENDIF
          IF X$<>"Y" THEN
               RTA=STARG.ISTAR(I).IRTA
               DEC=STARG.ISTAR(I).IDEC
          ENDIF
          RUN MANCO(RTA,DEC)
          STARG.ISTAR(I).IRTA=RTA
          STARG.ISTAR(I).IDEC=DEC
          IF I<>1 THEN
               PRINT "V MAGNITUDE = ";
               IF X$<>"Y" THEN
                              PRINT "("; STARG.ISTAR(I).IMAG; ")";
                         ENDIF
                         INPUT K$
                         IF K$<>"" THEN
                              STARG.ISTAR(I).IMAG=VAL(K$)
                         ENDIF
                    ELSE
                         STARG.ISTAR(I).IMAG=0
```

```
                ENDIF
            NEXT I
            IF STARG.ISTAR(1).IRTA=0 THEN
                STARG.ISTAR(1).IRTA=(STARG.ISTAR(2).IRTA+STARG.ISTAR(
                 3).IRTA)/2
            ENDIF
            IF STARG.ISTAR(1).IDEC=0 THEN
                STARG.ISTAR(1).IDEC=(STARG.ISTAR(2).IDEC+STARG.ISTAR(
                 3).IDEC)/2
            ENDIF
            GOSUB 10
            INPUT "IS THIS OK Y/N ",A$
        UNTIL A$="Y"
        IF X$<>"Y" THEN
            SEEK #FILE,GROUP*SIZE(STARG)
        ENDIF
        PUT #FILE,STARG
    ENDLOOP
    CLOSE #FILE
    END
10      \(* SUBROUTINE TO DISPLAY CURRENT GROUP
    PRINT USING "'CURRENT CONTENTS OF GROUP ',I14<"; GROUP
    PRINT "NAME: "; STARG.NAMEIN;
    PRINT "        DIAPHRAGM = "; STARG.DIAPHRAGM
    PRINT "      NAME        RIGHT ASCEN.  DECLIN.    V-MAG."
    FOR I=0 TO 3
        RTA=12/PI*STARG.ISTAR(I).IRTA
        DEC=180/PI*STARG.ISTAR(I).IDEC
        R1=INT(RTA)
        R2=INT(MOD(RTA*60,60))
        R3=MOD(RTA*3600,60)
        D1=INT(ABS(DEC))
        D2=INT(MOD(ABS(DEC)*60,60))
        D3=INT(MOD(ABS(DEC)*3600,60))
        IF DEC<0 THEN
            DSIGN$="-"
        ELSE
            DSIGN$="+"
        ENDIF
        MAG=STARG.ISTAR(I).IMAG
        PRINT USING "S6<",STARNAM(I);
        PRINT USING "S10<,' ,",STARG.ISTAR(I).INAM;
        PRINT USING "I3>,I3>,R5.1>,'    ',S1,I3>,I3>,I3>,R7.2>",R1
         ,R2,R3,DSIGN$,D1,D2,D3,MAG
    NEXT I
    RETURN
```

2. COEFFICENTS

```
0.36  0.0    '0.05
0.47 -0.036 '0.05
0.77 -0.036  0.01
KPRIME(VIS) KDOUBL(VIS) EPSILON(VIS)
```

```
KPRIME(BLU) KDOUBL(BLU) EPSILON(BLU)
KPRIME(ULT) KDOUBL(ULT) EPSILON(ULT)
```

3. DATREAD

```
(* DATREAD (NO PARAMETERS)
(* PROGRAM TO DISPLAY CONTENTS OF RAW DATA FILES FROM THE AUTOMATED
(* PHOTOMETER.
BASE 0
(* ISTAR(): 0=CHECK , 1=SKY , 2=COMPARISON , 3=VARIABLE
(* COLOR(): 0=ULTRAVIOLET , 1=BLUE , 2=VISUAL
TYPE STARIN=INAM:STRING[10]; IHOUR,COLOR(3),ISECZ:REAL
DATA 0,1,2,3,2,3,2,3,2,1,0
TYPE GROUPIN=NAMEIN:STRING[10]; CORRECT:REAL; ISTAR(11):STARIN
DIM GIN:GROUPIN
DIM FILE,MAXFILE,INFILE:INTEGER \(* FILE NUMBERS
DIM T: REAL
DIM T3,T4,T5:INTEGER
DIM NAME:STRING[10]
DIM INNAME:STRING[16]
DIM STARTYPE:INTEGER
INFILE=4
LOOP
     INPUT "FILE PATH (/D_/MJD_____) OR Q = ",P$
EXITIF P$="Q" OR P$="q" THEN
ENDEXIT
     OPEN #INFILE,P$
     FILE=0
     MAXFILE=0
     WHILE NOT(EOF(#INFILE)) DO
          MAXFILE=MAXFILE+1
          GET #INFILE,GIN
     ENDWHILE
     PRINT MAXFILE; " FILES FOUND"
     LOOP
          INPUT " DISPLAY FORWARD OR BACKWARD (F,B,Q)",A$
          IF FILE<MAXFILE AND A$="F" OR A$="f" THEN FILE=FILE+1
          ENDIF
          IF FILE>1 AND A$="B" OR A$="b" THEN FILE=FILE-1
          ENDIF
     EXITIF A$="Q'' OR A$="q" THEN
     ENDEXIT
          SEEK #INFILE,(FILE-1)*SIZE(GIN)
          GET #INFILE,GIN
          PRINT " FILE # "; FILE
          PRINT "GROUP = "; GIN.NAMEIN
          PRINT "HELIOCENTRIC CORRECITON = ";
          PRINT USING "R6.4>",GIN.CORRECT
          PRINT "TYPE      NAME         ULTRA   BLUE VISUAL   SECZ  HH
                                                           MM  SS"
          FOR ILINE=0 TO 10
               READ STARTYPE
               IF STARTYPE=0 THEN
```

```
                    PRINT "CHECK     ";
               ENDIF
               IF STARTYPE=1 THEN
                    PRINT "SKY       ";
               ENDIF
               IF STARTYPE=2 THEN
                    PRINT "COMP      ";
               ENDIF
               IF STARTYPE=3 THEN
                    PRINT "VARIABLE  ";
               ENDIF
               PRINT USING "S10<",GIN.ISTAR(ILINE).INAM;
               FOR FILTER=2 TO 0 STEP -1
               PRINT USING "R8.2>",GIN.ISTAR(ILINE).COLOR(FILTER);
               NEXT FILTER
               PRINT USING "R8.2>",GIN.ISTAR(ILINE).ISECZ;
               T=GIN.ISTAR(ILINE).IHOUR
               T3=INT(T)
               T=60*(T-T3)
               T4=INT(T)
               T=60*(T-T4)
               T5=INT(T)
               PRINT USING "3(I4>)'',T3,T4,T5
          NEXT ILINE
          RESTORE
     ENDLOOP
ENDLOOP
```

4. DIGITAL

```
(* digital (paddle:integer)
(* bit 0 west, bit 1 east
(* bit 2 north, bit 3 South
(* bit 4 s0, bit 5 s1
(* bit 6 end-flag--shared with cloud sense
(* bit 7 not used--rain sence  SSS
(* returns integer between 1 and 127
PARAM paddle:INTEGER
templ=PEEK($E011)
temp2=255-templ
IF temp2>127 THEN
     paddle=temp2-128
ELSE
     paddle=temp2
ENDIF
END
```

5. HELIO

```
(* HELIO(RTA,DEC,MJUL,CORRECTION:REAL)
(* RTA= RIGHT ASCENSION IN RADIANS
```

```
(* DEC= DECLINATION IN RADIANS
(* MJUL= MODIFIED JULIAN DATE IN DECIMAL DAYS
(* CORRECTION = HELIOCENTRIC CORRECTION IN DECIMAL DAYS
PARAM RTA,DEC,MJUL,CORRECTION:REAL
DIM T,LOTS,RADIUS,COSDC:REAL
RAD
COSDC:=COS(DEC)
T:=(MJUL-15019.5)*2.73785079E-05
LOTS:=4.88162794+628.331951*T+5.279621E-06*SQ(T)
RADIUS:=.999720458/(1.+.0I6719*COS(LOTS+4.956105))
CORRECTION:=-.0057755*(RADIUS*(COS(LOTS)*COS(RTA)*COSDC+SIN
     (LOTS)*(.39794248*SIN(DC)+.91741037*SIN(RTA)*COSDC)))
END
```

6. HUNT

```
(* HUNT(RADIUS,LOOPS,DURATION:INTEGER;THRESHOLD:REAL)
(* MODIFIED FOR USE WITH RAMP PROGRAM
(* THIS PROCEDURE WILL MOVE THE TELESCOPE IN A SQUARE SPIRAL CHECKING
(* for a star after move radius of diaphragm in steps
(* IF A STAR IS FOUND WHICH EXCEEDS THRESHOLD THEN THE LOOP IS ENDED.
(* IF NO STAR IS FOUND, THRESHOLD WILL REMAIN UNCHANGED.
BASE 0
PARAM RADIUS,LOOPS,duration:INTEGER
PARAM THRESHOLD:REAL
DIM I,Q,SIDE:INTEGER
TYPE ELEMENTS=LOW,MID,TOP,DSR,DSS,SNEW:BYTE
DIM X,Y:ELEMENTS
DIM COUNT:REAL
(* ENTER VALUES FOR RAMP AND SLEW HERE
X.DSR=200
X.DSS=20
GOSUB 200 \(* CHECK IF STAR ALREADY CENTERED
(* SET LENGTH OF MOVE (<32767 MAX)
X.LOW=MOD(RADIUS,256)
X.MID=RADIUS/256
X.TOP=0
(* BEGIN SQUARE SPIRAL SEARCH
(* LOOP PATTERN WEST,SOUTH,EAST,NORTH
SIDE=0
REPEAT
     SIDE=SIDE+1
     X.SNEW=1
     GOSUB 100
     X.SNEW=8
     GOSUB 100
     SIDE=SIDE+1
     X.SNEW=2
          GOSUB 100
          X.SNEW=4
          GOSUB 100
     UNTIL SIDE=LOOPS+LOOPS
```

```
     (* RETURN IN SOUTHEAST DIRECTION TO CENTER IF THRESHOLD NOT EXCEEDED
     X.LOW=MOD(RADIUS/2,256)
     X.MID=RADIUS/512
     X.SNEW=9
     GOSUB 100
     END  \(* END BECAUSE OF MAXIMUM # OF LOOPS
100  (* STEP THROUGH ONE SIDE
     FOR I=1 TO SIDE
          Y=X \(* TRANSFER REQUIRED BECAUSE RAMP MODIFIES PARAMETERS
          RUN RAMP(Y)
          GOSUB 200
     NEXT I
     RETURN
200  (* TEST FOR STAR IN FIELD
     RUN MEAS(duration,COUNT)
     PRINT COUNT
     IF COUNT>THRESHOLD THEN  \(* IF STAR IS FOUND !
          THRESHOLD=(THRESHOLD+COUNT)/2
          END  \(* EXIT THE PROGRAM WITH THRESHOLD SET TO STAR READING
     ENDIF
     RETURN
     END
```

7. JOY4

```
(* joy4(oldrta,olddec:real)
(* entry into joy4 allows the telescope to be moved using the
(* paddle. a cumulative position in right ascension and declination
(* will be returned as real parameters..
(* small revision   by lws  on 9/7/84
DIM oldrta,olddec:REAL
DIM TEMPA,TEMPD:REAL
DIM movex,movey,mult:REAL
DIM paddle,flag:INTEGER
deg_rad=PI/180.
rad_deg=180./PI
hrs_rad=PI/12.
rad_hrs=12./PI
olddec=.0
oldrta=.0
INPUT "would you like to run the telescope to start position (y/n) ?
                                                                   "
 ,a$
IF a$="y" OR a$="Y" THEN
     RUN startscope(oldrta,olddec)

          TEMPA=oldrta
          TEMPD=olddec
     ENDIF
     REPEAT
          RUN shoco(oldrta,olddec)
          INPUT "set coordinates (y/n)",a$
```

```
IF a$="Y" OR a$="y" THEN
     RUN manco(oldrta,olddec)
     TEMPA=oldrta
     TEMPD=olddec
ENDIF
LOOP
     POKE $E011,112
1    RUN digital(paddle)
     IF TEMPA<>oldrta OR TEMPD<>olddec THEN
          RUN shoco(oldrta,olddec)
          PRINT CHR$(11);
     ENDIF
     (* initialize to defaults
     movex=.0
     movey=.0
     mult=1
     (* check bit 6 for end flag
     IF paddle=64 THEN
          RUN SLEW2(oldrta,olddec)
          paddle=paddle-64
     ENDIF
     (* check bit 5 for slew speed
     IF paddle>=32 THEN
          mult=1000
          paddle=paddle-32
     ENDIF
     (* check bit 4 for set speed
     IF paddle>=16 THEN
          mult=50
          paddle=paddle-16
     ENDIF
     (* check bit 3 for south
     IF paddle>=8 THEN
          movey=-1.
          paddle=paddle-8
     ENDIF
     (* check bit 2 for north
     IF paddle>=4 THEN
          movey=1.
          paddle=paddle-4
     ENDIF
     (* check bit 1 for east
     IF paddle>=2 THEN
          movex=-1.
          paddle=paddle-2
     ENDIF
     (* check bit 0 for west
     IF paddle>=1 THEN
          movex=1.
          paddle=paddle-1
     ENDIF
     newrta=oldrta-movex*mult*hrs_rad/3600.
     newdec=olddec+movey*mult*deg_rad/240.
```

```
            TEMPA=oldrta
            TEMPD=olddec
            RUN travel(oldrta,olddec,newrta,newdec)
        ENDLOOP
        INPUT "Quit  (y/n)",B*
    UNTIL B$="Y" OR B$="y"
    END
```

8. LOCK

```
    (* LOCK(RADIUS:INTEGER;THRESHOLD:REAL)
    (* MODIFIED FOR USE WITH RAMP ( )
    (* READS FOUR POSITIONS IN A DIAMOND PATTERN AND MAKES LOGICAL MOVE B
    (* ON THE RESULT OF THE READINGS.
    (* ENTER WITH BRIGHTNESS DECISION THRESHOLD WHICH WILL REMAIN THE SAME
    (* IF NO STAR IS FOUND AND WILL BE SET TO THE READING OF THE STAR
    (* BRIGHTNESS IF IT IS.
    PARAM RADIUS:INTEGER
    PARAM THRESHOLD:REAL
    DIM SIDE,I,SUM,QRAD:INTEGER
    DIM THRESH,COUNT:REAL
    DIM TEST: BOOLEAN
    TYPE ELEMENTS=LOW,MID,TOP,DSR,DSS,SNEW:BYTE
    DIM X:ELEMENTS
    (* SET RAMP AND SLEW DELAY
    X.DSR=200
    X.DSS=20
    X.TOP=0 \(* NO MOVE ALLOWED > 32767 STEPS
    QRAD=RADIUS/4
    (* MAKE DIAMOND TEST : NE>R/2 , S>R , W>R , N>R , SE>R/2
    THRESH:=THRESHOLD
    SIDE:=QRAD*2 \X.SNEW:=6
100
    SUM:=0
    GOSUB 260
    IF TEST THEN SUM:=SUM+1
    ENDIF
    SIDE:=QRAD*4 \X.SNEW:=8
    GOSUB 260
    IF TEST THEN SUM:=SUM+2
    ENDIF
    SIDE:=QRAD*4 \X.SNEW:=1
    GOSUB 260
    IF TEST THEN SUM:=SUM+4
    ENDIF
    SIDE:=QRAD*4 \X.SNEW:=4
    GOSUB 260
    IF TEST THEN SUM:=SUM+8
    ENDIF
    ON SUM+1 GOTO 250,110,120,130,140,240,150,160,170,180,240,190
     ,200,210,220,250
    END
110 SIDE:=QRAD*4 \X.SNEW:=6
```

```
     GOTO 230
120  SIDE:=QRAD*4 \X.SNEW:=10
     GOTO 230
130  SIDE:=QRAD*2 \X.SNEW:=2
     GOTO 230
140  SIDE:=QRAD*4 \X.SNEW:=9
     GOTO 230
150  SIDE:=QRAD*2 \X.SNEW:=8
     GOTO 230
160  SIDE:=QRAD \X.SNEW:=10
     GOTO 230
170  SIDE:=QRAD*4 \X.SNEW:=5
     GOTO 230
180  SIDE:=QRAD*2 \X.SNEW:=4
     GOTO 230
190  SIDE:=QRAD \X.SNEW:=6
     GOTO 230
200  SIDE:=QRAD*2 \X.SNEW:=1
     GOTO 230
210  SIDE:=QRAD \X.SNEW:=5
     GOTO 230
220  SIDE:=QRAD \X.SNEW:=9
     GOTO 230
230
     GOSUB 260
240
     SIDE:=QRAD*4 \X.SNEW:=2
     GOTO 100
250  SIDE:=QRAD*2 \X.SNEW:=10
     GOSUB 260
     IF COUNT>THRESHOLD THEN
          THRESHOLD:=COUNT
     ENDIF
     END
260
     X.LOW=MOD(SIDE,256)
     X.MID=INT(SIDE/256)
     RUN RAMP(X)
     FOR I=1 TO 2000 \(* DAMPEN VIBRATION
     NEXT I
     RUN MEAS(2,COUNT)
     IF COUNT>THRESH+THRESH THEN
          THRESH=COUNT*.7
     ENDIF
     IF COUNT>THRESH THEN
          TEST:=TRUE
     ELSE
          TEST:=FALSE
     ENDIF
     RETURN
     END
```

9. LUNAR

```
(* LUNAR (MJUL,RTA,DEC:REAL)
(* ROUGH APPROXIMATION OF R.A. AND DEC. OF THE MOON
(* MJUL= MODIFIED JULIAN DATE (INPUT)
(* RTA= RIGHT ASCENSION OF THE MOON IN RADIANS (OUTPUT)
(* DEC= DECLINATION OF THE MOON IN RADIANS (OUTPUT)
(* DERIVED FROM "CELESTIAL BASIC" (SYBEX)
PARAM MJUL,RTA,DEC:REAL
DIM P2,BD,LM:REAL
P2=6.28318531 \(* TWO PI
BD=MJUL-36934.5 \(* BASE DAY
LM=5.43090264+.229971506*BD \(* LONGITUDE OF THE MOON
RTA=MOD(LM+.109756775*SIN(LM-1.9443666E-03*BD-4.46356263),P2)
DEC=.403946*SIN(RTA)+.0898*SIN(RTA-3.11888592+9.24221652E-04
 *BD)
END
```

10. MAIN

```
(* MAIN - Main telescope control subroutine.
(* FOR FOE         10/27/84
(* Reads the list of stars to be measured from a file named STARLIST.
(* Writes the readings to a file named MJD( ).
(*
(* Stars are always taken in groups of three and are read
(* in the following order : check,sky,comparison,variable,
(* comparison,variable,comparison,variable,
(* comparison,sky,check.
(* The check star should always be an easily locatable star as
(* it is acquired first.
(*
BASE 0
TYPE RISESET=RISE,SET:REAL; STATUS:INTEGER \(* RISE AND SET IN DECIMAL
                                                           L HOURS
TYPE STARIN=INAM:STRING[10]; IRTA,IDEC,IMAG:REAL
(* ISTAR(): 0=CHECK , 1=SKY , 2=COMPARISON , 3=VARIABLE
TYPE GROUPIN=NAMEIN:STRING[10]; PRIORITY:INTEGER; ISTAR(4):
 STARIN
TYPE STAROUT=ONAM:STRING[10]; OHOUR,READING(3),OSECZ:REAL
TYPE GROUPOUT=NAMEOUT:STRING[10]; CORRECT:REAL; OSTAR(11):STAROUT
DATA 0,1,2,3,2,3,2,3,2,1,0 \(* SEQUENCE FOR OUTPUT DATA
DIM TABLE(100):RISESET \(* INGROUP#, (RISETIME,SETTIME)
DIM GIN:GROUPIN
DIM GOUT:GROUPOUT
DIM GTEMP:GROUPOUT
DIM INFILE,OUTFILE:INTEGER \(* FILE NUMBERS
DIM MJUL,LMST,CORRECTION,PHOTONS,THRESHOLD:REAL
DIM RTA,DEC,OLDRTA,OLDDEC:REAL
DIM RTASUN,DECSUN,RTAMOON,DECMOON:REAL
DIM TWILIGHT:REAL
DIM ANGLE,MAXANGLE,COSZMAX:REAL
```

```
DIM TIM(6),TIMER:INTEGER
DIM DEPTH,GROUP,MAXGROUP,LASTGROUP:INTEGER
DIM skipgroup,paddle:INTEGER
DIM RADIUS(3),LOOPS,DURATION:INTEGER \(* FOR PROCEDURE HUNT AND LOCK
DIM OLINE,FILTER:INTEGER \(* FOR'NEXT COUNTERS
DIM VIS,BLU,ULT:INTEGER \(* FILTER NUMBERS
DIM NEWF,OLDF,NEWA,OLDA:INTEGER \(* FOR PROCEDURE APFILT
DIM ABORT:INTEGER
DIM VALCOUNT,VALTYPE,VALFILT,VALSTAR,VALLINE:INTEGER
DIM MEAN,MEANSQ,VALTEMP:REAL
DIM NAME:STRING[10]
DIM OUTNAME:STRING[16]
DIM TYPE$(4),FILT$(3):STRING[6]
DIM CLS:STRING[2]
DIM NIGHT,GROUPABORT:BOOLEAN
(* FIXED CONSTANTS
DEG_RAD=PI/180
RAD_DEG=180/PI
HRS_RAD=PI/12
RAD_HRS=12/PI
MIN_HRS=1/60.
SEC_HRS=1/3600.
VIS=0 \BLU=1 \ULT=2
TYPE$(0)="CHECK "
TYPE$(1)="SKY   "
TYPE$(2)="COMP  "
TYPE$(3)="VARI "
FILT$(0)="VISUAL"
FILT$(1)="BLUE  "
FILT$(2)="ULTRA "
(* ADJUSTABLE CONSTANTS
(* RADIUS OF DIAPHRAGMS IN HALF STEPS
RADIUS(0)=25
RADIUS(1)=25
RADIUS(2)=25
LATITUDE=39.80075*DEG_RAD
SINLAT=SIN(LATITUDE)
COSLAT=COS(LATITUDE)
TWILIGHT=100
TIMER=100 \(* 10 SECOND INTEGRATION TIME
(* ADJUST PHTONMULT FOR PHOTMETER IN HUNT AND LOCK BAND
PHOTONMULT=2500
(* ADJUST EXTINCTION FOR BAND USED IN HUNT AND LOCK
EXTINCTION=.24
CLS=CHR$($1B)+"K" \(* CLEAR SCREEN
INFILE=3 \(* INPUT FILE TO READ IN GROUP INFORMATION
OUTFILE=4 \(* OUTPUT FILE TO STORE MEASUREMENTS
RUN TIME(TIM,MJUL,LMST)
PRINT "START OF JULIAN DAY "; INT(MJUL)+2400000.5
PRINT "OPENING INPUT FILE AND BUILDING LOOKUP TABLE"
OPEN #INFILE,"/DO/STARFILE":READ
(* LOAD TABLE WITH SIDEREAL TIME WHICH EACH GROUP WILL RISE
(* .    AND SET IN THE OBSERVABLE CONE ABOVE THE OBSERVATORY.
```

```
COSZMAX=COS(55*DEG_RAD) \(* SET MAXIMUM ZENITH ANGLE TO 45 DEGREES
MAXANGLE=15*DEG_RAD \(* SET MAXIMUM EAST OR WEST HR ANGLE (MOUNT LIMI

      GROUP=0
      OPEN #OUTFILE,"/P1":WRITE
      WHILE NOT(EOF(#INFILE)) DO
           GET #INFILE,GIN
           RTA=GIN.ISTAR(1).IRTA
           DEC=GIN.ISTAR(1).IDEC
           ANGLE=ABS(ACS((COSZMAX-SINLAT*SIN(DEC))/(COSLAT*COS(DEC))
            ))
           IF ANGLE>MAXANGLE THEN  \(* LIMITED BY MOUNT ON PHX 10" TELESCOPE
             ANGLE=MAXANGLE
           ENDIF
           TABLE(GROUP).RISE=MOD((RTA-ANGLE)*RAD_HRS+24,24
           TABLE(GROUP).SET=MOD((RTA+ANGLE)*RAD_HRS,24)
           TABLE(GROUP).STATUS=0
           PRINT USING "S10<,R8.3>,R8.3>",GIN.NAMEIN,TABLE(GROUP).RISE
            ,TABLE(GROUP).SET
           PRINT #OUTFILE,"GROUP # "; GROUP
           PR1NT #OUTFILE,"RISE "; TABLE(GROUP).RISE,"SET "; TABLE(GROUP
            ).SET
            IF NOT(EOF(#INFILE)) THEN
                GROUP=GROUP+1
           ENDIF
      ENDWHILE
      CLOSE #OUTFILE
      MAXGROUP=GROUP
      (* OPEN OUTPUT FILE
      ON ERROR GOTO 20
      OUTNAME="/D1/MJD"+LEFT$(STR$(MJUL),5)
      PRINT "OPENING OUTPUT FILE "; OUTNAME
      OPEN #OUTFILE,OUTNAME \(* TEST TO SEE IF FILE EXISTS (ERROR IF FALSE)
10
      CLOSE #OUTFILE
      GOTO 30
20
      CREATE #OUTFILE,OUTNAME \(* MAKE NEW FILE IF NONE EXISTS
      GOTO 10
30
      ON ERROR  \(* RESET FOR NORMAL ERROR MESSAGES
      RUN STARTSCOPE(OLDRTA,OLDDEC)
      LASTGROUP=0
      DEPTH=0
      ABORT=0
      LOOP  \(* FOR EACH GROUP MEASURED
          (* CHECK LOOKUP TABLE FOR NEXT STAR TO RUN
          REPEAT  \(* UNTIL OBSERVABLE GROUP FOUND
               RUN TIME(TIM,MJUL,LMST)
               FIRSTSET=12.
               FOR TEMPGROUP=0 TO MAXGROUP
                    IF TABLE(TEMPGROUP).STATUS<=DEPTH THEN
                         TEMPRISE=LMST-TABLE(TEMPGROUP).RISE
```

```
                    IF TEMPRISE<0 THEN
                         TEMPRISE=TEMPRISE+24
                    ENDIF
                    TEMPSET=TABLE(TEMPGROUP).SET-LMST
                    IF TEMPSET<0 THEN
                         TEMPSET=TEMPSET+24
                    ENDIF
                    IF TEMPRISE<12 AND TEMPSET<12 AND TEMPSET<FIRSTSET THEN

                         FIRSTSET=TEMPSET
                         GROUP=TEMPGROUP
                    ENDIF
               ENDIF
          NEXT TEMPGROUP
          IF FIRSTSET=12 THEN
          DEPTH=DEPTH+1 \(* IF NO UNMEASURED STARS START REPEATING
ENDIF
RUN SUNANGLE(ANGLE)
(* IF ANGLE>TWILIGHT THEN
NIGHT=TRUE

(* ELSE
(* NIGHT=FALSE
(* ENDIF

WHILE LASTGROUP<>GROUP DO
     SEEK #INFILE,GROUP*SIZE(GIN)
     GET #INFILE,GIN
     LASTGROUP=GROUP
     ENDWHILE
     RTA=GIN.ISTAR(1).IRTA \(* USE SKY POSITION FOR TEST
     DEC=GIN.ISTAR(1).IDEC
     RUN LUNAR(MJUL,RTAMOON,DECMOON)
     RTAMOON=ABS(RTA-RTAMOON)
     IF RTAMOON>PI THEN
     RTAMOON=RTAMOON-(PI+PI)
     ENDIF
     DECMOON=ABS(DEC-DECMOON)
             IF SQ(RTAMOON)+SQ(DECMOON)<.03 THEN \(* LIMIT TO WITHIN 10 DEGRE
                                                          ES OF MOON
               PRINT "ATTEMPT TO MOVE NEAR MOON, DELETING GROUP "; GIN.NAMEIN
               TABLE(GROUP).STATUS=TABLE(GROUP).STATUS+1
               FIRSTSET=12.
          ENDIF
     UNTIL ABS(FIRSTSET-12.)>.00000001 OR NOT(NIGHT)
EXITIF NOT(NIGHT) THEN
     PRINT "NOT NIGHT - SOLAR ANGLE = "; ANGLE
ENDEXIT
     (* STAR CALCULATES TO BE OBSERVABLE SO MOVE TO FIRST STAR
     PRINT "MOVING TO GROUP "; TRIM$(GIN.NAMEIN); ", NUMBER = "
      ; GROUP
     PRINT " AT ";
     RUN SHOCO(RTA,DEC)
```

```
     RUN TRAVEL(OLDRTA,OLDDEC,RTA,DEC)
     (* FLIP STATEMENT REMOVED HERE AS NOT NEEDED DFM SYSTEM
     DURATION=1
     LOOPS=20 \(* 20 LOOPS FOR FIRST STAR OF A GROUP
     (* NEWF=0 STATEMENT SSS REMOVED HERE
     (* NEWA=GIN.DIAPHRAM*90 STATEMENT REMOVED HERE
     (* RUN APFILT(NEWF,OLDF,NEWA,OLDA) REMOVED HERE
     RUN THRESH(COUNT)
     PRINT "SKY BACKGROUND = ";  \ PRINT USING "R8.2<",COUNT
EXITIF COUNT>400 THEN  \(* INDICATES BRIGHT SKY OR CLOUDS
     PRINT "THRESHOLD COUNT > 400"
     ENDEXIT
     (* MEASUREMENT SEQUENCE
     RESTORE
     PRINT "TYPE    THRESH   NAME         ULTRA    BLUE  VISUAL      SECZ
                                                          HH  MM  SS"
     FOR OLINE=0 TO 10
          READ STAR
          NAME=GIN.ISTAR(STAR).INAM
          RTA=GIN.ISTAR(STAR).IRTA
          DEC=GIN.ISTAR(STAR).IDEC
          RUN PRECESS(MJUL,RTA,DEC)
          RUN TRAVEL(OLDRTA,OLDDEC,RTA,DEC)
          RUN digital(paddle)
          IF paddle>63 THEN
               RUN options(skipgroup)
          ENDIF
EXITIF skipgroup=1 THEN
END EXIT
     PRINT TYPE$(STAR);
     IF STAR<>1 THEN
          (* IF NOT SKY MEASUREMENT THEN SET THRESHOLD
          (* BASED ON THRESHOLD COUNT AND STAR MAGNITUDE
          RUN ZENITH(RTA,DEC,LMST,COSZ)
          THRESH1=COUNT+PHOTONMULT*2.511886^-(GIN.ISTAR(STAR).IMAG
           +EXTINCTION/COSZ)
          PRINT USING "R8.2>",THRESH1;
          PRINT "  ";
          LOOP

               THRESH2=THRESH1
               RUN HUNT(RADIUS,LOOPS,DURATION,THRESH2)
               DURATION=10
               LOOPS=8 \(* EIGHT LOOFS PER STAR ONCE IN A GROUP
          EXITIF THRESH2=THRESH1 THEN
               PRINT "UNABLE TO FIND STAR "; NAME
               PRINT "ABORTING GROUP "; GIN.NAMEIN
               GROUPABORT=TRUE
               TABLE(GROUP).STATUS=TABLE(GROUP).STATUS+1
          ENDEXIT
               PRINT "HUNT COUNT = ";  \ PRINT USING "R8.2<",THRESH2
               THRESH2=THRESH1
               RUN LOCK(RADIUS,THRESH2)
```

```
        EXITIF THRESH2<>THRESH1 THEN
            PRINT "LOCK COUNT = ";  \ PRINT USING "R8.2<",THRESH2
            GROUPABORT=FALSE
            ABORT=0
        ENDEXIT
        ENDLOOP
    ELSE
        PRINT "            ";
    ENDIF
EXITIF GROUPABORT THEN
    ABORT=ABORT+1
    RUN STARTSCOPE(OLDRTA,OLDDEC)
    (* OLDF,OLDA REMOVED FROM STARTSCOPE CALL ABOVE
ENDEXIT
    (* MAKE MEASUREMENTS IN THREE COLORS
    PRINT USING "S10<",NAME;
    FOR FILTER=ULT TO VIS STEP -1
        RUN MEAS(TIMER,PHOTONS)
        GOUT.OSTAR(OLINE).READING(FILTER)=PHOTONS
        PRINT USING "R8.2>",PHOTONS;
    NEXT FILTER
    GOUT.OSTAR(OLINE).ONAM=GIN.ISTAR(STAR).INAM
    RUN TIME(TIM,MJUL,LMST)
    GOUT.OSTAR(OLINE).OHOUR=TIM(3)+TIM(4)*MIN_HRS+TIM(5)*SEC_HRS
    RUN ZENITH(RTA,DEC,LMST,COSZ)
    GOUT.OSTAR(OLINE).OSECZ=1/COSZ
    PRINT USING "R8.2>",1/COSZ;
    PRINT USING "3(I4>)",TIM(3),TIM(4),TIM(5)
    IF OLINE=5 THEN
        GOUT.NAMEOUT=GIN.NAMEIN
        RUN HELIO(RTA,DEC,MJUL,CORRECTION)
        GOUT.CORRECT=CORRECTION
    ENDIF
NEXT OLINE
(* RUN FLIP(1) REMOVED HERE
IF NOT(GROUPABORT) THEN

PRINT "GROUPS ON RECORD "; OUTNAME
OPEN #OUTFILE,OUTNAME
SEEK #OUTFILE,0
I=0
WHILE NOT(EOF(#OUTFILE)) DO
            GET #OUTFILE,GTEMP
            I=I+1
            PRINT USING "I3>,' ',S10<",I,GTEMP.NAMEOUT;
            IF MOD(I,4)=0 OR EOF(#OUTFILE) THEN
                PRINT
            ENDIF
        ENDWHILE
        PUT #OUTFILE,GOUT
        CLOSE #OUTFILE
        PRINT "GROUP "; GOUT.NAMEOUT; " ADDED TO FILE"
        DEPTH=0
```

```
        (* VALIDATE DATA
        MAXDEVI=0
        PRINT "STATISTICS ( RMS/MEAN )"
        PRINT "TYPE     VISUAL  BLUE   ULTRA"
        FOR VALSTAR=0 TO 3
             PRINT TYPE$(VALSTAR);
             FOR VALFILT=VIS TO ULT
                  MEAN=0
                  MEANSQ=0
                  RESTORE
                  VALCOUNT=0
                  FOR VALLINE=0 TO 10
                       READ VALTYPE
                       IF VALSTAR=VALTYPE THEN
                            VALCOUNT=VALCOUNT+1
                            VALTEMP=GOUT.OSTAR(VALLINE).READING(VALFILT)
                            MEAN=MEAN+VALTEMP
                            MEANSQ=MEANSQ+VALTEMP*VALTEMP
                       ENDIF
                  NEXT VALLINE
                  DEVI=SQRT(MEANSQ/VALCOUNT)/(MEAN/VALCOUNT)
                  PRINT USING "R8.5>",DEVI;
                  IF DEVI>MAXDEVI AND VALSTAR<>1 THEN
                       MAXDEVI=DEVI
                  ENDIF
             NEXT VALFILT
             PRINT
        NEXT VALSTAR
        TABLE(GROUP).STATUS=TABLE(GROUP).STATUS+1
        (* TABLE INC. FROM MIDDLE OF IF - ELSE *)
        IF MAXDEVI<1.001 THEN  \(* REMOVE FROM TABLE IF DATA OK
             PR1NT " DATA VALID "
        ELSE
             PRINT "DATA NOT VALID, RMS/MEAN = ";
             PRINT USING "R8.5<",MAXDEVI
        ENDIF
   ENDIF
EXITIF ABORT>=4 THEN
   PRINT " FAILED FOUR TIMES TO FIND GROUP, NIGHTS RUN ABORTED"
ENDEXIT
ENDLOOP  \(* END MEASUREMENT LOOP
RUN STOPSCOPE
END
```

11. MANCO

```
   (* MANCO(RTA,DEC:REAL)
   (* GETS RIGHT ASCENSION AND DECLINATION IN RADIANS FROM
   (* FREE FORM KEYBOARD INPUT.  ALLOWS ENTRY IN FORMAT
   (* XX.XXX OR XX MM.MMM OR XX MM SS.SSS
   PARAM RTA,DEC:REAL
   DIM IN$,XX$(3):STRING[20]
   DIM XX:REAL
```

```
     DIM CHAR:STRING[1]
     DIM I,N,SIGN:INTEGER
     DIM DELIM:BOOLEAN
     HRS_RAD=PI/12.
     DEG_RAD=PI/180.
     PRINT "RIGHT ASCENSION IN HOURS = ";
     VALUE=RTA/HRS_RAD
     GOSUB 10
     RTA=HRS_RAD*VALUE
     PRINT ''  DECLINATION IN DEGREES = ";
     VALUE=DEC/DEG_RAD
     GOSUB 10
     DEC=DEG_RAD*VALUE
     END
10
     INPUT IN$
     IF IN$<>"" THEN
          FOR I=1 TO 3
               XX=.0
               XX$(I)="0"
          NEXT I
          N=1
          SIGN=1
          DELIM=TRUE
          FOR I=1 TO LEN(IN$)
               CHAR=MID$(IN$,I,1)
               IF CHAR<>" " AND CHAR<>"," AND CHAR<>"-" THEN
                    XX$(N)=XX$(N)+CHAR
                    DELIM=FALSE
               ELSE
                    IF CHAR="-" THEN
                         SIGN=-1
                    ELSE
                         IF NOT(DELIM) THEN
                              DELIM=TRUE
                              N=N+1
                         ENDIF
                    ENDIF
               ENDIF
          NEXT I
          VALUE=SIGN*(VAL(XX$(1))+VAL(XX$(2))/60.+VAL(XX$(3))/3600.
           )
     ENDIF
     RETURN
```

12. MEAS

```
(* MEAS(TIME:INTEGER,COUNT:REAL)
(* TIME IS THE INTEGRATION TIME IN TENTHS OF SECONDS,
(* AND COUNT IS ACTUAL COUNT NORMALIZED TO .1 SECONDS
PARAM TIME:INTEGER
     PARAM COUNT: REAL
     DIM TEMP:REAL
```

```
     DIM PORT,I,J:INTEGER
     PORT:=$E012
     COUNT:=.0
     FOR I:=1 TO TIME
          GOSUB 100
          COUNT:=COUNT+TEMP
     NEXT I
     COUNT:=COUNT/TIME
     END
100
     POKE PORT+1,0
     FOR J:=1 TO 510 \(* TRIM FOR 1/10 SECOND PER READING
     NEXT J
     TEMP:=.0
     FOR J:=1 TO 5
     TEMP:=TEMP*10+LAND(PEEK(PORT+1),$0F)
     NEXT J
     RETURN
```

13. MOVE

```
(* MOVE(MOVEX,MOVEY:REAL)
(* MODIFIED FOR USE WITH RAMP ( )
(* DRIVES TELESCOPE BY NUMBER OF STEPS GIVEN BY MOVEX AND MOVEY
(* POSITIVE X IS WEST, POSITIVE Y IS NORTH
PARAM MOVEX,MOVEY:REAL
DIM X,Y,VECTOR:REAL
DIM I:INTEGER
TYPE ELEMENTS=LOW,MID,TOP,DSR,DSS,SNEW:BYTE
DIM K:ELEMENTS
(* SET SLEW AND RAMP SPEEDS
K.DSR=100
K.DSS=12
X:=INT(MOVEX+.5*SGN(MOVEX))
Y:=INT(MOVEY+.5*SGN(MOVEY))
REPEAT
     IF ABS(X)>.0 AND ABS(Y)>.0 THEN
          VECTOR:=(ABS(X)+ABS(Y)-ABS(ABS(X)-ABS(Y)))/2.
     ELSE
          IF ABS(X)>ABS(Y) THEN
               VECTOR:=ABS(X)
          ELSE
               VECTOR:=ABS(Y)
          ENDIF
     ENDIF
     K.SNEW:=0
     IF X>.0 THEN K.SNEW:=LOR(K.SNEW,1)
     ENDIF
     IF X<.0 THEN K.SNEW:=LOR(K.SNEW,2)
     ENDIF
     IF Y>.0 THEN K.SNEW:=LOR(K.SNEW,4)
     ENDIF
     IF Y<.0 THEN K.SNEW:=LOR(K.SNEW,8)
```

```
     ENDIF
     K.LOW=INT(MOD(VECTOR,256))
     K.MID=INT(MOD(VECTOR/256,256))
     K.TOP=INT(VECTOR/65536.)
     RUN RAMP(K)
     X:=X-SGN(X)*VECTOR
     Y:=Y-SGN(Y)*VECTOR
UNTIL ABS(X)=.0 AND ABS(Y)=.0
END
```

14. PRECESS

```
(* PRECESS(MJUL,RTA,DEC:REAL)
(* PRECESSES FROM EPOCH 2000.0 TO PRESENT EPOCH
(* MJUL= MODIFIED JULIAN DATE IN DECIMAL DAYS (INPUT)
(* RTA= RIGHT ASCENSION IN RADIANS (INPUT/OUTPUT)
(* DEC= DECLINATION IN RADIANS (INPUT/OUTPUT)
PARAM MJUL,RTA,DEC:REAL
DIM RDIF,DDIF,DAYS:REAL
DAYS:=51544.-MJUL
RDIF:=DAYS*(6.112E-07+2.668E-07*SIN(RTA)*TAN(DEC))
DDIF:=DAYS*2.655E-07*COS(RTA)
RTA=RTA-RDIF
DEC=DEC-DDIF
END
```

15. PTCLK

```
(* PTCLK(TIM(6):INTEGER) - Reads the Peripheral Technology
(* MC146818 clock chip.  Returns an array TIM which contains
(* 0=year,1=month,2=day,3=hour,4=minute,5=second
(* Executes in about .01 second.
BASE 0
PARAM TIM(6):INTEGER
DIM DAT,ADR:INTEGER
DAT=$E01C
ADR=$E01D
REPEAT
     POKE ADR,$00
     TIM(5)=PEEK(DAT)
     POKE ADR,$02
     TIM(4)=PEEK(DAT)
     POKE ADR,*04
     TIM(3)=PEEK(DAT)
     POKE ADR,$07
     TIM(2)=PEEK(DAT)
     POKE ADR,$08
     TIM(1)=PEEK(DAT)
     POKE ADR,$09
     TIM(0)=PEEK(DAT)
     POKE ADR,$00
UNTIL PEEK(DAT)=TIM(5)
IF TIM(0)>80 THEN
```

```
     TIM(0)=TIM(0)+1900
ELSE
     TIM(0)=TIM(0)+2000
ENDIF
END
```

16. QJOY2

```
(* procedure qjoy2 (replaces qjoy)
(* qjoy2 (no parameters)
(* main program for manual control
(* runs digital, move2
(* does not display coordinates
(* west is plus movex

(* north is plus movey
DIM movex,movey,mult:REAL
DIM paddle,flag:INTEGER
POKE $E011,112
RUN digital(paddle)
(* initialize to defaults
movex=.0
movey=.0
mult=1
(* check bit 6 for end flag
IF paddle>=64 THEN
     PRINT "paddle>=64--end"
     END
ENDIF
(* check bit 5 for slew speed
IF paddle>=32 THEN
     mult=10000
     paddle=paddle-32
ENDIF
(* check bit 4 for set speed
IF paddle>=16 THEN
     mult=50
     paddle=paddle-16
ENDIF
(* check bit 3 for south
IF paddle>=8 THEN
     movey=-1.
     paddle=paddle-8
ENDIF
(* check bit 2 for north
IF peddle>=4 THEN
     movey=1.
     paddle=paddle-4
ENDIF
(* check bit 1 for east
IF paddle>=2 THEN
     movex=-1.
     paddle=paddle-2
```

```
ENDIF
(* check bit 0 for west
IF paddle>=1 THEN
     movex=1.
     paddle=paddle-1
ENDIF
(* calculate steps
movex=moves*mult
movey=movey*mult
RUN move2(movex,movey)
FOR i=1 TO 100
NEXT i
GOTO 1
END
```

17. RAMP

```
     * RAMP(ELEMENT(6):BYTE) -Called from Basic09 by STEPPER module.
     * There must be 6 bytes in the passed parameter which are:
     * ELEMENT(0):BYTE -LOW BYTE OF 24 BIT STEP COUNT
     * ELEMENT(1):BYTE -MID BYTE OF 24 BIT STEP COUNT
     * ELEMENT(2):BYTE - TOP BYTE OF 24 BIT STEP COUNT
     * ELEMENT(3):BYTE - DELAY FOR START OF RAMP
     * ELEMENT(4):BYTE - DELAY FOR SLEWING SPEED
     * ELEMENT(5):BYTE - DEFINES LEADS TO BE PULSED

     * Up to 4 motors may be operated simultaneously.
     * Bits 0 and 1 are for the first motor, 2 and 3 for second motor et
     * Directions to run should have the corresponding bit set to
     * one.  Pulsing both directions on one motor will produce
     * unpredictable results.
     *
     * Limit switches correspond to the output bits.  A limit switch
     * is normally high and must go low on reaching the limit.
     * Program will exit if all active limit switches have been
     * reached.  On a limit switch the limited motor stops immediately.
     * The residual count in ELEMENT 0,1,and 2 will indicate the
     * the remaining count if limit switches are encountered.
     * This can be used to detect a successful move.
     * The output port must consist of up/down counter inputs which
     * will respond to '0' level input.  Writing a '1' to the
     * output port should have no effect.
     * The drivers and limit switches share the write and read functions
     * of the same address, which in this case is set to $E010 which is
     * the low address of SWTPC port #4.
     *
                         NAM   RAMP
                         USE   /D0/DEFS/OS9DEFS
     *
     * OS-9 System Definition File Included
     *
                          opt   1
E010              PORT    EQU   $E010       ADDRESS OF MOTOR DRIVER AND LI
```

```
0021               TYPE    SET    SBRTN+OBJCT
0081               REVS    SET    REENT+1
0000 87CD0078              MOD    REND,RNAM,TYPE,REVS,RENT,0
000D 52414DD0     RNAM     FCS    /RAMP/
0011 EE64         RENT     LDU    4,S          GET ADDRESS OF ARRAY OF BYTES
0013 C600                  LDB    #0           SET RAMP DELAY TO MAXIMUM
     * BEGINNING OF MAIN LOOP
0015 E143         MLOOP    CMPB   3,U          CHECK IF AT SLEWING SPEED
0017 2701                  BEQ    TOPTEST
0019 5C                          INCB          DECREASE RAMP DELAY
001A 6D42         TOPTEST  TST    2,U
001C 2610                  BNE    DELAY1
001E 6D41                  TST    1,U
0020 2612                  BNE    DELAY2
0022 6DC4                  TST    0,U          LESS THAN 256 STEPS LEFT
0024 274D                  BEQ    EXIT         IF NO MORE STEPS LEFT
0026 E1C4                  CMPB   0,U          IF RAMPCOUNT > REMAINING STEPS
0028 2310                  BLS    RDEL         THEN DON'T BRANCH
002A E6C4                  LDB    0,U          AND SET RAMPCOUNT EQUAL TO REMAINI
002C 200C                  BRA    RDEL
     * EQUALIZATION DELAYS
002E 21FE         DELAY1   BRN    *             10 CLOCK CYCLES
0030 21FE                  BRN    *
0032 12                    NOP
0033 12                    NOP
0034 8D3C         DELAY2   BSR    RETURN        18 CLOCKCYCLES
0036 21FE                  BRN    *
0038 21FE                  BRN    *
     * VARIABLE DELAY FOR RAMP
003A 1F98         RDEL     TFR    B,A
003C A143         RLOOP    CMPA   3,U
003E 2703                  BEQ    SDEL
0040 4C                    INCA
0041 20F9                  BRA    RLOOP
     * VARIABLE DELAY FOR SLEW
0043 8600         SDEL     LDA    #0           COUNT UP FOR SLEW DELAY
0045 A144         SLOOP    CMPA   4,U
0047 2703                  BEQ    PULSE
0049 4C                    INCA
004A 20F9                  BRA    SLOOP
     * GENERATE STEP PULSES
004C A645         PULSE    LDA    5,U
004E 43                    COMA                 CONVERT TO NEGATIVE TRUE
004F BAE010                ORA    PORT          NO PULSE IF LIMIT SWITCH
0052 81FF                  CMPA   #$FF          NO PULSE TO OUTPUT ?
0054 271D                  BEQ    EXIT          QUIT IF ALL MOVES BLOCKED
0056 B7E010                STA    PORT          PULL SELECTED MOTOR COUNTERS L
0059 86FF                  LDA    #$FF
005B B7E010                STA    PORT          RESET MOTOR COUNTERS HIGH
     * DECREMENT STEP COUNTER
005E A6C4                  LDA    0,U
0060 8001                  SUBA   #1
0062 A7C4                  STA    0,U
```

```
0064 A641              LDA   1,U
0066 8200              SBCA  #0
0068 A741              STA   1,U
006A A642              LDA   2,U
006C 8200              SBCA  #0
006E A742              STA   2,U
0070 20A3              BRA   MLOOP
     * END OF MAINLOOP
0072 39       RETURN   RTS               FOR DELAYS
0073 5F       EXIT     CLRB              NO ERROR CODE
0074 39                RTS               RETURN TO CALLING PROGRAM
0075 179D02            EMOD
0078          REND     EQU   *
                       END
```

18. SET

```
(* SET (NO PARAMETERS) - Sets the Peripheral Technology PT-69
(* clock chip.
BASE 0
DIM TIM(6):INTEGER
DATA "   YEAR"."  MONTH","    DAY","   HOUR"," MINUTE"
DATA $09,$08,$07,$04,$02,$00
DAT=$E01C
ADR=$E01D
RUN PTCLK(TIM)
PRINT
FOR I=0 TO 4
     READ A$
     PRINT A$; " ":
     PRINT USING "I5>",TIM(I);
     INPUT " ",A$
     IF A$<>"" THEN
          TIM(I)=VAL(A$)
     ENDIF
NEXT I
TIM(5)=$00
TIM(0)=MOD(TIM(0),100)
(* SET DIVIDER RATIOS
POKE ADR,$0A
POKE DAT,$00
(* SET FORMAT REGISTER
POKE ADR,$0B
POKE DAT,$85
(* SET TIME-DATE REGISTERS
FOR I=0 TO 5
     READ REG
     POKE ADR,REG
     POKE DAT,TIM(I)
NEXT I
(* RELEASE SET BIT, 24HR FORMAT, NO DLST
POKE ADR,$0B
POKE DAT, $05
```

```
END
```

19. SHOCO

```
     (* SHOCO(RTA,DEC:REAL)
     (* DISPLAYS COORDINATES GIVEN RIGHT ASCENSION AND
     (* DECLINATION IN RADIANS.
     PARAM RTA,DEC:REAL
     DIM TEMP,RAD_SEC,RAD_ASEC:REAL
     DIM XX,MM,SS:INTEGER
     RAD_SEC=43200./PI
     RAD_ASEC=648000./PI
     TEMP=RTA*RAD_SEC+.5
     WHILE TEMP<0 DO
          TEMP=TEMP+86400.
     ENDWHILE
     GOSUB 10
     PRINT USING "'R.A. = ',3(I3>)",XX,MM,SS;
     TEMP=DEC*RAD_ASEC+.5
     IF TEMP>=0 THEN
          PRINT "  DEC. =  ";
     ELSE
          PRINT "  DEC. = -";
     ENDIF
     GOSUB 10
     PRINT USING "3(I3>)",XX,MM,SS
     END
10
     TEMP=ABS(TEMP)
     XX=INT(TEMP/3600.)
     TEMP=TEMP-XX*3600.
     MM=INT(TEMP/60.)
     SS=INT(TEMP-MM*60.)
     RETURN
     END
```

20. SHOWTIME

```
(* SHOWTIME - DISPLAYS TIME IN YY/MM/DD HH:MM:SS FORMAT
BASE 0
DIM TIM(6):INTEGER
RUN PTCLK(TIM)
PRINT TIM(0); "/"; TIM(1); "/"; TIM(2); "  ";
PRINT TIM(3); ":"; TIM(4); ":"; TIM(5);
END
```

21. SLEW2

```
(* slew to coordinates    v 0.1
(* BY P. SCOTT HAWTHORN & LLOYD SLONAKER
DIM oldrta.olddec:REAL
DIM TIM(6):INTEGER
```

```
DIM MJUL,LMST:REAL
DIM I,J:INTEGER
DIM COUNT:REAL
DIM CHAR:STRING[1]
PARAM rta,dec:REAL
J=1
RUN shoco(rta,dec)

olddec=dec
oldrta=rta
PRINT ""
PRINT "MENU FOR SLEW TO STAR"
PRINT ""
PRINT "1.    b peg"
PRINT "2.    ALTAIR"
PRINT "3.    B CYG"
PRINT "4.    A CYG"
PRINT "5.    SOME OTHER STAR"
PRINT "6.    TAKE READING ON STAR"
PRINT ""
PRINT "ENTER YOUR CHOICE"
GET #0,CHAR
IF CHAR="1" THEN
     rta=23.06289
     dec=55.23667
ENDIF
IF CHAR="2" THEN
     rta=5.19576519
     dec=.154781616
ENDIF
IF CHAR="3" THEN
     rta=5.1083751
     dec=.488086173
ENDIF
IF CHAR="4" THEN
     rta=5.41676024
     dec=.790289933
ENDIF
IF CHAR="5" THEN
     PRINT " ENTER NEW COORDINATES (2000.0 EPOCH) "
     RUN manco(rta.dec)
ENDIF
IF CHAR="6" THEN
     INPUT "NAME OF THE STAR ? ",A$
     PRINT "OUTPUT GOES TO PRINTER AND SCREEN"
     OPEN #J,"/P1"
     PRINT #J,""
     PRINT #J,"READING FOR STAR--- "; A$
     FOR I=1 TO 3
          RUN MEAS(10,COUNT)
          PRINT "LOOP # "; I; " COUNT = "; COUNT
          PRINT #J,"LOOP # "; I; " COUNT = "; COUNT
      NEXT I
```

```
      END
ENDIF
RUN TIME(TIM,MJUL,LMST)
RUN PRECESS(MJUL,rta,dec)
RUN travel(oldrta,olddec,rta,dec)
RUN shoco(rta,dec)
END
```

22. SOLAR

```
(* SOLAR  (MJUL,RTA,DEC:REAL)
(* MJUL= MODIFIED JULIAN DATE (INPUT)
(* RTA= RIGHT ASCENSION OF THE SUN IN RADIANS (OUTPUT)
(* DEC= DECLINATION OF THE SUN IN RADIANS (OUTPUT)
(* CALCULATIONS BASED ON US. NAVAL OBS. ALMANAC FOR COMPUTERS
PARAM MJUL,RTA,DEC:REAL
RAD
DIM HALFPI,TWOPI,T,LOTS:REAL
HALFPI=1.57079633
TWOPI=6.28318531
(* T=JULIAN CENTURIES FROM 0 JANUARY 1900 12H ET.
T=(MJUL-15019.5)*2.73785079E-05
(* MAOS=MEAN ANOMALY OF SUN, TLOS=TRUE LONGITUDE OF SUN
MAOS=6.25658358+628.301946*T-2.61799388E-06*T*T
TLOS=MOD(MAOS+.03344051*SIN(MAOS)+.000349066*SIN(MAOS+MAOS)
 +4.93169,TWOPI)
RTA=ATN(.91746*TAN(TLOS))
IF TLOS>HALFPI THEN RTA=RTA+PI
     IF TLOS>HALFPI+PI THEN RTA=RTA+PI
     ENDIF
ENDIF
DEC=ASN(.39782*SIN(TLOS))
END
```

23. STARFILE

```
CURRENT CONTENTS OF GROUP  0
NAME: LAM AND       DIAPHRAGM = 1
      NAME        RIGHT ASCEN.  DECLIN.     V-MAG.
CHECK HD 222439   23 40 24.4    + 44 20  2   4.14
  SKY             23 41 47.8    + 46 26 21    .00
 COMP HD 223047   23 46  1.9    + 46 25 13   4.95
 VARI HD 222107   23 37 33.7    + 46 27 29   3.83

CURRENT CONTENTS OF GROUP  1
NAME: 39 CET       DIAPHRAGM = 1
      NAME        RIGHT ASCEN.  DECLIN.     V-MAG.
CHECK HD 7476      1 14 49.1    -  0 58 26   5.70
  SKY              1 14  9.8    -  2 22 32    .00
 COMP HD 7147      1 11 43.4    -  2 15  4   5.94
 VARI HD 7672      1 16 36.2    -  2 30  1   5.41
CURRENT CONTENTS OF GROUP  2
NAME: SIGMA GEM        DIAPHRAGM = 1
```

```
     NAME       RIGHT ASCEN.  DECLIN.    V-MAG.
CHECK HD 60522    7 35 55.3   + 26 53 45   4.06
  SKY             7 39 13.7   + 29 55 20    .00
 COMP HD 60318    7 35  8.7   + 30 57 40   5.33
 VARI HD 62044    7 43 18.7   + 28 53  1   4.28
```

24. STARTSCOPE

```
(* STARTSCOPE (OLDRTA,OLDDEC:REAL)
(* THIS PROCEDURE IS USED TO INITIALIZE THE TELESCOPE.
(* IT MAKES SURE THAT THE TELESCOPE IS EXACTLY AT THE HOME
(* POSITION, THEN STARTS THE SIDEREAL CLOCK WITH THE
(* R.A. AND DEC. OF THE HOME POSITION SET FOR THE STARTING
(* LOCAL MEAN SIDEREAL TIME. IT ALSO INSURES THAT THE
(* DIAPHRAM FILTERS ARE INITIALIZED.
PARAM OLDRTA,OLDDEC:REAL
(* DELETED FROM ABOVE STATEMENT ";OLDF,OLDA:INTEGER"
DIM TIM(6):INTEGER: LMST,MJUL:REAL
(* DIM NEWF,NEWA:INTEGER DELETED HERE
(* DECLINATION OF HOME POSITION IN RADIANS
HOMEDEC=-.83761236
(* HOUR ANGLE OF HOME POSITION IN RADIANS
HOMERTA=5.4228765
HRS_RAD=PI/12
(* RESET DIAPHRAGM AND FILTER
(* THIS ENTIRE SECTION REMOVED--SEE ORIGINAL FOR OTHER SYSTEMS
(*
(* INITIALIZE $E011 PORT WITH SIDEREAL DRIVE OFF
POKE $E011,80
(*
(* INSURE TELESCOPE IS EXACTLY AT HOME POSITION
RUN MOVE(500.,500.)
RUN MOVE(-1000000.,-1000000.)
RUN MOVE(20.,20.)
RUN MOVE(-50.,-50.)
(* GET R.A. AND DEC. OF HOME POSITION AT THIS INSTANT
RUN TIME(TIM,MJUL,LMST)
OLDDEC=HOMEDEC
OLDRTA=LMST*HRS_RAD-HOMERTA
IF OLDRTA<0 THEN OLDRTA=OLDRTA+PI+PI
ENDIF
RUN SHOCO(OLDRTA,OLDDEC)
(*
(* CHANGES CONTROL PORT $E011 TO START SIDEREAL CLOCK
POKE $E011,112
END
```

25. STOPSCOPE

```
(* STOPSCOPE - NO PARAMETERS )
(* TURN OFF SIDEREAL RATE
POKE $E011,80
RUN MOVE(-1000000.,-1000000.) \(* MOVE TO SOUTHEAST LIMIT SWITCHES
```

```
END
```

26. SUNANGLE

```
(* SUNANGLE(ANGLE:REAL) - CALCULATES THE ZENITH ANGLE OF THE SUN
(* IN DEGREES
RAD_DEG=57.2957795
PARAM ANGLE:REAL
DIM TIM(6):INTEGER
DIM MJUL,LMST,RTA,DEC,COSZ:REAL
RUN TIME(TIM,MJUL,LMST)
RUN SOLAR(MJUL,RTA,DEC)
RUN ZENITH(RTA,DEC,LMST,COSZ)
ANGLE=RAD_DEG*ACS(COSZ)
END
```

27. THRESH

```
(* THRESH(COUNT:REAL)
(* MODIFIED FOR RAMP
(* MEASURES THE SKY IN FOUR LOCATIONS SURROUNDING THE CURRENT
(* POSITION, RETURNS A COUNT OF LOWEST READING FOUND
PARAM COUNT:REAL
DIM I,SIDE,LSIDE,MSIDE:INTEGER
DIM SPOT:REAL
TYPE ELEMENTS=LOW,MID,TOP,DSR,DSS,SNEW:BYTE
DIM X:ELEMENTS
SIDE=500 \(* LENGTH OF SIDE OF SEARCH PATTERN
LSIDE=MOD(SIDE,256)
MSIDE=INT(SIDE/256)
DATA 4,10,9,5,2
X.TOP=0
X.DSR=125 \(* SET RAMP DELAY
X.DSS=15 \(* set Slew Delay
COUNT=1.0E+30
FOR I=1 TO 5
     X.LOW=LSIDE
     X.MID=MSIDE
     READ X.SNEW
     RUN RAMP(X)
     RUN MEAS(10,SPOT)
     IF SPOT<COUNT THEN
          COUNT=SPOT
     ENDIF
NEXT I
END
```

28. TILDARK

```
(* TILDARK (NO PARAMETERS)
(* WAIT UNTIL DARK AND THEN RUN MAIN PROGRAM
REPEAT
```

```
     RUN SUNANGLE(ANGLE)
     ANGLE=INT(ANGLE*100)/100
     WHILE ANGLE<>OLDANGLE DO
          PRINT USING "'SUN ANGLE = ',R6.2>",ANGLE;
          PRINT CHR$(13);
          OLDANGLE=ANGLE
     ENDWHILE
UNTIL ANGLE>100
RUN MAIN
END
```

29. TIME

```
(* TIME(MJUL,LMST:REAL)
(* OUTPUT:
(* MJUL= MODIFIED JULIAN DATE (J.D.- 2400000.5) IN DECIMAL DAYS
(* LMST= LOCAL MEAN SIDEREAL TIME IN HOURS
BASE 0
PARAM TIM(6):INTEGER; MJUL,LMST:REAL
DIM MJUL0,FRACT,LONG:REAL
LONG:=5.598174 \(* LOCAL LONGITUDE IN HOURS
RUN PTCLK(TIM)
MJUL0:=-678987.+367.*TIM(0)+INT(INT(275*TIM(1)/9)+TIM(2)-INT
  (7*(TIM(0)+INT((TIM(1)+9)/12))/4))
FRACT=(TIM(3)*3600.+TIM(4)*60.+TIM(5))/86400.
MJUL:=MJUL0+FRACT
LMST:=MOD(6.67170278+.065709823*(MJUL0-33282.)+1.00273791*(
  FRACT*24)-LONG,24)
END
```

30. TRANSFORM

```
(* READS DATA FROM A DISK AND INFORMATION ON A PARTICULAR
(* STAR GROUP.  THE DATA IS REDUCED AND STORED ON ANOTHER DISK.
(* THE FILE NAME OF THE OUTPUT WILL BE THE HD NUMBER OF THE VARIABLE
(* CALLS A FILE NAME COEFFICIENTS TO GET CURRENT VALUES FOR
(* TRANSFORMATION AND EXTINCTION.
(* THIS PROGRAM DOES NOT ADJUST THE COEFFICIENTS.
BASE 0
TYPE STARIN=INAM:STRING[10]: IHOUR,READING(3),ISECZ:REAL
TYPE GROUPIN=NAMEIN:STRING[10]: CORRECT:REAL; ISTAR(11):STARIN
DIM GIN:GROUPIN
(* STAROUT(0,1,2,3)= VARI,VARI,VARI,CHECK
TYPE STAROUT=OHH,OMM,OSS:BYTE; DIFFMAG(3):REAL
TYPE GROUPOUT=HELIOJD:STRING[12]; PHASE,OMEAN(3),OSIGMA(3):
  REAL; OSTAR(4):STAROUT
DIM GOUT:GROUPOUT
DIM DATFILE,TEMPFILE,INFILE,OUTFILE:INTEGER \(* FILE NUMBERS
DIM STAR,DIR_COUNT:INTEGER
DIM MJDAY,SECZ,X1,X2,X3,AIRMASS:REAL
DIM T:REAL
DIM NAME(4):STRING[10]
```

```
DIM DI$,DO$:STRING[3]
DIM X$:STRING[64]
DIM DATNAME,OUTNAME:STRING[12]
DIM QUARTER:STRING[4]
DIM FILENAME(30):STRING[8]
DIM VIS,BLU,ULT:INTEGER
DIM CHECK,SKY,COMP,VARI:INTEGER
(* INFILE 0 1 2 3 4 5 6 7 8 9 10
(* STAR   0   1 2 3 4 5 6 7   8
(* TYPE   0 1 2 3 2 3 2 3 2 1 0
(* CHECK-COMP = (0+8)/2 - (1+3+5+7)/4
(* VARI-COMP(1) = 2 - (1+3)/2
(* VARI-COMP(2) = 4 - (3+5)/2
(* VARI-COMP(3) = 6 - (5-7)/2
DIM INSTMAG(11,3):REAL
DIM STDMAG(11,3):REAL
DIM KPRIME(3):REAL
DIM KDOUBL(3):REAL
DIM EPSILON(3):REAL
DIM AVGSKY(3):REAL
DIM VARI_COMP(3,3):REAL
DIM CHECK_COMP(3):REAL
DIM MEAN(3)
DIM SIGMA(3)
DIM FILT$(3):STRING[6]
DIM TYPE$(4):STRING[6]
DIM GROUP_OK:BOOLEAN
DATFILE=3 \(* HEADER DATA PATH
INFILE=4 \(* DATA INPUT PATH
OUTFILE=5 \(* OUTPUT DATA PATH
TEMPFILE=6 \(* MISC FILE PATH
CHCK=0 \SKY=1 \COMP=2 \VARI=3
VIS=0 \BLU=1 \ULT=2
FILT$(0)="VISUAL"
FILT$(1)="BLUE  "
FILT$(2)="ULTRA "
TYPE$(0)="CHECK "
TYPE$(1)="SKY   "
TYPE$(2)="COMP  "
TYPE$(3)="VARI  "
DATA 0,1,2,3,2,3,2,3,2,1,0
(* PROMPT FOR INPUT AND OUTPUT DRIVE
INPUT "Source drive number (0,1,2) ",DI$
INPUT "Destination drive   (0,1,2) ",DO$
DI$="/D"+DI$ \DO$="/D"+DO$
(*
(* GET NAME OF STAR TO BE RECORDED ON THIS PASS
(*
INPUT "Variable common name = ",N$
OUTNAME=""
FOR I=1 TO 10
     A$=MID$(N$,I,1)
     IF A$<>" " AND A$<>"" THEN
```

```
          OUTNAME=OUTNAME+A$
     ELSE
          OUTNAME=OUTNAME+"_"
     ENDIF
NEXT I
INPUT "NAME OF OUARTER i.e. 1Q84 ",QUARTER
(*
(* OPEN GENERAL DATA FILE
(*
ON ERROR GOTO 100
OPEN #DATFILE,DO$+"/d_"+OUTNAME:READ
FOR STAR=CHCK TO VARI
     READ #DATFILE,NAME(STAR)
     PRINT TYPE$(STAR),NAME(STAR)
NEXT STAR
PRINT "FILTER","K-PRIME","K-DOUBLE","EPSILON"
FOR FILTER=VIS TO ULT
     READ #DATFILE,KPRIME(FILTER),KDOUBL(FILTER),EPSILON(FILTER
      )
     PRINT FILT$(FILTER),KPRIME(FILTER),KDOUBL(FILTER),EPSILON
      (FILTER)
NEXT FILTER
READ #DATFILE,MJEPOCH,PERIOD
PRINT "EPOCH (MJD) "; MJEPOCH,"PERIOD (DAYS) "; PERIOD
READ #DATFILE,MJDAY
GOTO 110
100
     CREATE #DATFILE,DO$+"/d_"+OUTNAME:UPDATE
     PRINT "No data file exists on the output disk."
     INPUT "Enter reduction coeff. from /DO/CMDS/COEFFICIENTS (Y/N) "
      ,Q$
     IF Q$="Y" OR Q$="y" THEN
          OPEN #TEMPFILE,"/DO/CMDS/COEFFICIENTS":READ
          FOR FILTER=VIS TO ULT
               READ #TEMPFILE,KPRIME(FILTER),KDOUBL(FILTER),EPSILON(FILTER
                )
          NEXT FILTER
          CLOSE #TEMPFILE
     ELSE
          FOR FILTER=VIS TO ULT
               PRINT FILT$(FILTER);
               INPUT " KPRIME,KDOUBL,EPSILON ",KPRIME(FILTER),KDOUBL(FILTER
                ),EPSILON(FILTER)
          NEXT FILTER
     ENDIF
     INPUT "Epoch in MJD ",MJEPOCH
     INPUT "Period in days ",PERIOD
     FOR STAR=CHCK TO VARI
          NAME(STAR)=""
          NEXT STAR
          MJDAY=.0
110
     ON ERROR
```

```
     INPUT "Data OK (Y/N) ",Q$
     IF Q$="N" OR Q$="n" THEN
          CLOSE #DATFILE
          SHELL "DEL "+DO$+"/d_"+OUTNAME
          GOTO 100
     ENDIF
     CLOSE #DATFILE
     ON ERROR GOTO 120
     OPEN #OUTFILE,DO$+"/f_"+OUTNAME+OUARTER: WRITE
     GOTO 130
120
     CREATE #OUTFILE,DO$+"/f_"+OUTNAME+QUARTER:WRITE
130
     ON ERROR
     (*
     (* PROMPT FOR EACH DATA DISK
     (*
     LOOP
          PRINT "Last date reduced = MJD"; MJDAY
          PRINT "Insert data disk into drive "; DI$
          INPUT "Type 'Y' when ready ",Q$
     EXITIF Q$<>''Y" AND Q$<>"y" THEN
     ENDEXIT
          (*
          (* BUILD TABLE OF INPUT DATA FILE NAMES
          (*
          SHELL "DIR E "+DI$+" >"+DO$+"/INDIR"
          OPEN #TEMPFILE,DO$+"/INDIR":READ
          DIR_COUNT=0
          WHILE NOT(EOF(#TEMPFILE)) DO
               READ #TEMPFILE,X$
               IF MID$(X$,49,3)="MJD" THEN
                    DIR_COUNT=DIR_COUNT+1
                    FILENAME(DIR_COUNT)=MID$(X$,49,8)

                    ENDIF
               ENDWHILE
               CLOSE #TEMPFILE
               SHELL "DEL "+DO$+"/INDIR"
               FOR FCOUNT=1 TO DIR_COUNT
                    MJDAY=VAL(MID$(FILENAME(FCOUNT),4,5))
                    OPEN #INFILE,DI$+"/"+FILENAME(FCOUNT):READ
                    PRINT "FILE "; FILENAME(FCOUNT)
                    WHILE NOT(EOF(#INFILE)) DO
                         GET #INFILE,GIN
     IF GIN.NAMEIN=N$ THEN
          IF NAME(0)="" THEN
               FOR STAR=CHCK TO VARI
                    NAME(STAR)=GIN.ISTAR(STAR).INAM
               NEXT STAR
          ENDIF
          (*
          (* REDUCTION TO INSTRUMENTAL MAGNITUDES
```

```
(*
(* DETERMINE AVERAGE SKY BRIGHTNESS
FOR FILTER=VIS TO ULT
     AVGSKY(FILTER)=(GIN.ISTAR(1).READING(FILTER)+GIN.ISTAR
       (9).READING(FILTER))/2
     NEXT FILTER
     FOR STAR=0 TO 10
          IF STAR<>1 AND STAR<>9 THEN
               FOR FILTER=VIS TO ULT
                    IF GIN.ISTAR(STAR).READING(FILTER)>AVGSKY(FILTER
                     ) THEN
                         INSTMAG(STAR,FILTER)=-2.5*LOG10(GIN.ISTAR(STAR
                          ).READING(FILTER)-AVGSKY(FILTER))
                    ELSE
                         INSTMAG(STAR,FILTER)=1000
                    ENDIF
               NEXT FILTER
               (* BEMPORADS CORRECTION FOR AIRMASS
               X0=GIN.ISTAR(STAR).ISECZ
               X1=X0-1
               X2=X1*X1
               X3=X2*X1
               AIRMASS=X0-.0018167*X1-.002875*X2-.0008083*X3
                (* STANDARD MAGNITUDE CALCULATION
               BLU_VIS=(INSTMAG(STAR,BLU)-INSTMAG(STAR,VIS))*(
                1-(KDOUBL(BLU)-KDOUBL(VIS))*AIRMASS)-(KPRIME
                (BLU)-KPRIME(VIS))*AIRMASS
               ULT_BLU=(INSTMAG(STAR,ULT)-INSTMAG(STAR,BLU))*(
                1-(KDOUBL(ULT)-KDOUBL(BLU))*AIRMASS)-(KPRIME
                (ULT)-KPRIME(BLU))*AIRMASS
               STDMAG(STAR,VIS)=INSTMAG(STAR,VIS)-KPRIME(VIS)*
                AIRMASS+EPSILON(VIS)*BLU_VIS
               STDMAG(STAR,BLU)=INSTMAG(STAR,BLU)-KPRIME(BLU)*
                AIRMASS-KDOUBL(BLU)*AIRMASS*BLU_VIS+EPSILON
                (BLU)*BLU_VIS
               STDMAG(STAR,ULT)=INSTMAG(STAR,ULT)-KPRIME(ULT)*
                AIRMASS-KDOUBL(ULT)*AIRMASS*ULT_BLU+EPSILON
                (ULT)*ULT_BLU
          ENDIF
     NEXT STAR
     FOR FILTER=VIS TO ULT
          MEAN(FILTER)=0
     NEXT FILTER

FOR STAR=0 TO 2
     FOR FILTER=VIS TO ULT
       VARI_COMP(STAR,FILTER)=STDMAG(STAR*2+3,FILTER)-
         (STDMAG(STAR*2+2,FILTER)+STDMAG(STAR*2+4,
         FILTER))/2
         MEAN(FILTER)=VARI_COMP(STAR,FILTER)/3+MEAN(FILTER
          )
     NEXT FILTER
NEXT STAR
```

```
FOR FILTER=VIS TO ULT
     CHECK_COMP(FILTER)=(STDMAG(0,FILTER)+STDMAG(10,FILTER
      ))/2-(STDMAG(2,FILTER)+STDMAG(4,FILTER)+STDMAG
      (6,FILTER)+STDMAG(8,FILTER))/4
NEXT FILTER
(*
(* CALCULATE MEAN ERROR
(*
GROUP_OK=TRUE
FOR FILTER=VIS TO ULT
     SUMSQ=0
     FOR STAR=0 TO 2
          X_XBAR=VARI_COMP(STAR,FILTER)-MEAN(FILTER)
          SUMSQ=SUMSQ+X_XBAR*X_XBAR
     NEXT STAR
     SIGMA(FILTER)=SQRT(SUMQ/6)
     IF SIGMA(FILTER)>.02 THEN
          GROUP_OK=FALSE
     ENDIF
NEXT FILTER
IF GROUP_OK THEN (* DONT PRINT BAD DATA
   (*
   (* DEVELOP HELIOCENTRIC CORRECTED JULIAN DATE (TO BIG FOR R
                                                           EAL #)
   (*
   JDFRACT=.5+GIN.ISTAR(5).IHOUR/24+GIN.CORRECT
   JDINT=2400000.+MJDAY+INT(JDFRACT)
   JDFRACT=JDFRACT-INT(JDFRACT)+1 \(* FRACTIONAL PART OF JD+
    GOUT.HELIOJD=MID$(STR$(JDINT),1,8)+MID$(STR$(JDFRACT
    ),3,4)
   PRINT "      "; GOUT.HELIOJD
   (* DETERMINE PHASE
   IF PERIOD>0 THEN
        GOUT.PHASE=MOD((VAL(MID$(GOUT.HELIOJD,3,10))-.5-MJEPOCH)/PERIOD,1)
        ELSE
             GOUT.PHASE=.0
        ENDIF
        (* SET UP OUTPUT FOR MEAN AND SIGMA
        FOR FILTER=VIS TO ULT
             GOUT.OMEAN(FILTER)=MEAN(FILTER)
             GOUT.OSIGMA(FILTER)=SIGMA(FILTER)
        NEXT FILTER
        (* FOR VARI-COMP
        FOR STAR=0 TO 2
             FOR FILTER=VIS TO ULT
                  GOUT.OSTAR(STAR).DIFFMAG(FILTER)=VARI_COMP(STAR,FILTER)
                  NEXT FILTER
                  TTIME=GIN.ISTAR(STAR*2+3).IHOUR
                  GOUT.OSTAR(STAR).OHH=INT(TTIME)

                            TTIME=(TTIME-GOUT.OSTAR(STAR).OHH)*60
                            GOUT.OSTAR(STAR).OMM=INT(TTIME)
                            TTIME=(TTIME-GOUT.OSTAR(STAR).OMM)*60
```

```
                    GOUT.OSTAR(STAR).OSS=INT(TTIME)
               NEXT STAR
               (* FOR CHECK-COMP
               FOR FILTER=VIS TO ULT
                    GOUT.OSTAR(3).DIFFMAG(FILTER)=CHECK_COMP(FILTER
                     )
               NEXT FILTER
               TTIME=(GIN.ISTAR(0).IHOUR+GIN.ISTAR(10).IHOUR)/2
               GOUT.OSTAR(3).OHH=INT(TTIME)
               TTIME=(TTIME-GOUT.OSTAR(3).OHH)*60
               GOUT.OSTAR(3).OMM=INT(TTIME)
               TTIME=(TTIME-GOUT.OSTAR(3).OMM)*60
               GOUT.OSTAR(3).OSS=INT(TTIME)
               PUT #OUTFILE,GOUT
          ENDIF
     ENDIF
  ENDWHILE
  CLOSE #INFILE
NEXT FCOUNT
ENDLOOP
CLOSE #OUTFILE
DELETE DO$+"/d_"+OUTNAME
CREATE #DATFILE,DO$+"/d_"+OUTNAME:WRITE
FOR STAR=CHCK TO VARI
     WRITE #DATFILE,NAME(STAR)
NEXT STAR
FOR FILTER=VIS TO ULT
     WRITE #DATFILE,KPRIME(FILTER),KDOUBL(FILTER),EPSILON(FILTER
      )
NEXT FILTER
WRITE #DATFILE,MJEPOCH,PERIOD
WRITE #DATFILE,MJDAY
CLOSE #DATFILE
END
```

31. TRAVEL

```
(* TRAVEL(OLDRTA,OLDDEC,NEWRTA,NEWDEC:REAL)
(* CONVERT R.A. AND DEC. TO HALFSTEPS FOR STEPPER
(* DRIVER AND MAKE MOVE.  UPDATE CURRENT POSITION.
PARAM OLDRTA,OLDDEC,NEWRTA,NEWDEC:REAL
DIM MOVEX,MOVEY:REAL
MOVEX=NEWRTA-OLDRTA
IF MOVEX>PI THEN MOVEX=MOVEX-(PI+PI)
ENDIF
IF MOVEX<=-(PI) THEN MOVEX=MOVEX+(PI+PI)
ENDIF
MOVEY=NEWDEC-OLDDEC
IF MOVEY>PI THEN MOVEY=MOVEY-(PI+PI)
ENDIF
IF MOVEY<=-(PI) THEN MOVEY=MOVEY+(PI+PI)
ENDIF
(* REM EXACT NUMBER OF STEPS/RADIAN MAY NOT BE SAME
```

```
(* FOR X AND Y AXIS.  NOTE THAT INCREASING R.A. GIVES
(* NEGATIVE STEPS.
(* CONVERSION OF RADIANS TO HALF-STEPS (DEPENDS ON DRIVE RATIO)
MOVEX=MOVEX*-218139.
MOVEY=MOVEY*218139.
RUN MOVE(MOVEX,MOVEY)
OLDRTA=NEWRTA
OLDDEC=NEWDEC
END
```

32. ZENITH

```
(* ZENITH(RTA,DEC,LMST,COSZ,:REAL)
(* RTA= RIGHT ASCENSION OF OBJECT IN RADIANS (INPUT)
(* DEC= DECLINATION OF OBJECT IN RADIANS (INPUT)
(* LMST= LOCAL MEAN SIDEREAL TIME IN HOURS (INPUT)
(* COSZ= COSINE OF THE ZENITH ANGLE (OUTPUT)
PARAM RTA,DEC,LMST,COSZ:REAL
LATITUDE=39.80075*PI/180
SLAT:=SIN(LATITUDE)
CLAT:=COS(LATITUDE)
COSZ:=SLAT*SIN(DEC)+CLAT*COS(DEC)*COS(.2618*LMST-RTA)
END
```

Appendix H

Phoenix IV Control Software

```
1 KEY OFF:Z$ = "O":GOSUB 17000
2 LOCATE ,,0
3 HUE=2:COLOR HUE
4 SHADE=4:TINT=1
5 '
6 '                                   PIV = 0 ...... THE BIG SCOPE!
7 '                                   PIV = 1 ....  TRACKING MOUNT!
8 '
9 '
10  '                          ********************
20  '                          **    SKYFILE     **
30  '                          ********************
35 '
50 '                              9/15/84 by D.B.
55 '
60 '         Displays, scans, and selects objects from textfile of stars.
70 '         Selections stored in "BUFFER" file for use by P-IV.
80 '         RA$(I) =  MID$ (A$(I),13,8) : DEC$(I) =  MID$ (A$(I),26,9)
90 '
100  '    ************************************************************
101  '                                 SET-UP
102  '     ************************************************************
110  '             Dimension arrays, assign constants; BLOAD MAKSTR
120  '
125  CLS
130  DIM A$(275),BUF$(20),CON$(80)
140 RAS$=STRING$(79," ")
400  '    *************************************************************
410  '            Init Pager indices: PAGES, PG, LN, NDX
420  '
440 PAGES =  INT ((ENDFIL- 1) / 20) + 1
450 PG = 1:LN = 2
500  '    ************************************************************
501  '                               PAGER
502  '     ************************************************************
510  '         PRINT statline, up to 20 objects (LN inverse), comline
520  '
522  LOCATE 1,1
525  IF N$ = "" THEN 550
530 EP = 20: IF PG = PAGES THEN EP = ENDFIL - 20 * (PG - 1)
540  IF  LEN (A$(20 * (PG - 1) + EP)) < 5 THEN EP = EP - 1: GOTO 540
550  LOCATE 1,1:PRINT RAS$;
555 LINE (0,318)-(719,339),1,B
560  LOCATE 1,3:COLOR HUE,0,0,16:PRINT "SKYFILE:";:COLOR SHADE,0,0,0:
```

```
     PRINT " " N$;:LOCATE 1,17: COLOR HUE,0,0,16:PRINT  "BUFFER:";
     :COLOR SHADE,0,0,0
570  PRINT  ""NDX;:COLOR HUE,0,0,16:LOCATE 1,28:PRINT "TOTAL PAGES: ";
     :COLOR SHADE,0,0,0:PRINT "" PAGES;:COLOR HUE,0,0,16:LOCATE 1,44:
      PRINT "PAGE:";:COLOR SHADE,0,0,0:PRINT ""  PG;
571 LOCATE 1,55:COLOR 1,0,0,0 :PRINT LEFT$(DATE$,6)+RIGHT$(DATE$,2);:
      LOCATE 1,66:PRINT LEFT$(TIME$,5);:COLOR 4,0,0,64
572 IF PIV = 0 THEN LOCATE 1,74 : PRINT "*PIV*" : GOTO 576
573 LOCATE 1,74 : PRINT "XEROX"
576 COLOR 7:LINE(475,0)-(475,18):LINE (475,18)-(715,18):
    LINE (715,18)-(715,0)
577 COLOR 6:LINE (5,0)-(5,18):LINE (5,18)-(467,18):
    LINE (467,18)-(467,0)
578 COLOR HUE,0,0,0
580  FOR I = 20 * (PG - 1) + 1 TO 20 * (PG - 1) + 20
585  LOCATE ((I-1) MOD 20)+3,1:PRINT  RAS$;
590  IF I - 20 * (PG - 1) = LN THEN  COLOR SHADE
600  LOCATE CSRLIN,1:PRINT A$(I)
610  IF I - 20 * (PG - 1) = LN THEN  COLOR HUE
620  NEXT
640 LOCATE 24,1: PRINT "   CURSOR: [^] [v] [<] [>]   Buffer    Hunt
     Init   New_file   Scope   Quit";: COLOR SHADE: LOCATE 24,13:
     PRINT "^";: LOCATE 24,17: PRINT "v";: LOCATE 24,21: PRINT "<";
645 LOCATE 24,25: PRINT ">";: LOCATE 24,30: PRINT "B";: LOCATE 24,40:
     PRINT "H";:LOCATE 24,47: PRINT "I";: LOCATE 24,54: PRINT "N";:
     LOCATE 24,65: PRINT "S";: LOCATE 24,73: PRINT "Q";: COLOR HUE
700  '    ************************************************************
710  '                      GET A KEY
720  '
730  Z$=INKEY$:IF Z$="" THEN 730
740  IF LEN(Z$)=2 THEN Z$=RIGHT$(Z$,1) ELSE Z=ASC(Z$):GOTO 1103
745  Z=ASC(Z$)
750  '
800  '    ****************************
810  '            IF KEY = RIGHT
820  '
830  IF Z <  > 77 AND Z<> 81 THEN 910
840 PG = PG + 1
850  IF PG > PAGES THEN PG = 1
870  GOTO 500
880  '     ****************************
890  '           IF KEY = LEFT
900  '
910  IF Z <  >75 AND Z<> 73 THEN 980
920 PG = PG - 1: IF PG < 1 THEN PG = PAGES
940  GOTO 500
950  '      ****************************
960  '             IF KEY = DOWN
970  '
980  IF Z <  > 80 THEN 1060
990  LOCATE LN+2,1:PRINT A$(20 * (PG - 1) + LN)
1000 LN = LN + 1: IF LN > EP THEN LN = 2
1010  LOCATE LN+2,1:COLOR SHADE:PRINT A$(20 * (PG - 1) + LN):COLOR HUE
1020  GOTO 730
1030  '      ****************************
1040  '            IF KEY = UP
1050  '
1060  IF (Z <> 72) THEN 1103
1070  LOCATE LN+2,1:PRINT A$(20 * (PG - 1) + LN)
1080 LN = LN - 1: IF LN < 2 THEN LN = EP
1090  LOCATE LN+2,1:COLOR SHADE:PRINT A$(20 * (PG - 1) + LN):COLOR HUE
1100  GOTO 730
1101 '      ****************************
1102 '           IF KEY = "S"
1103 IF (Z <> 83 AND Z <> 115) THEN 1140
1104 IF (HNTFLAG = 0) THEN GOSUB 1900 : FOR ZZ = 1 TO 50 : NEXT ZZ :
```

```
     SOUND 440, 8 : GOTO 730
1105 LOCATE LN + 2, 1 : COLOR TINT : PRINT A$(20 * (PG - 1) + LN)
1106 SOUND 670, 3 : FOR ZZ = 1 TO 5:NEXT: SOUND 440,8
1107 RA$=MID$(A$(20*(PG-1)+LN),12,8): DEC$ = MID$( A$( 20 * (PG-1) + LN)
     ,25,9 )
1108 GOSUB 4000::REM'                    THE BIG BANANA, THE SCOPE MOVE
     CODE !!!!!!!
1109 LOCATE LN + 2, 1 : COLOR SHADE : PRINT A$(20 * (PG - 1) + LN) :
     COLOR HUE
1110 GOTO 730: '                  AND, FOR THE MINUTE, START OVER W/
     NEW KEY!!!
1119  '      ****************************
1120  '             IF KEY = "B"
1130  '
1140  IF (Z <> 66 AND Z <> 98) THEN 1154
1150  GOTO 2000
1151  ' ******************************
1152  '              IF KEY = "H"
1153  '
1154 IF (Z <> 72 AND Z <> 104) THEN 1170
1155 GOSUB 8000: '    Sector scan please, Scotty !!
1156 GOTO 730 : '     return to command line when done
1160 ' ********************************
1161 '              IF KEY = "I"
1162 '
1170 IF (Z <> 73 AND Z <> 105) THEN 1225
1175 RA$=MID$(A$(20*(PG-1)+LN),12,8):DEC$ = MID$(A$(20*(PG-1)+LN),25,9)
1180 GOSUB 10170 :  ' initialize, get bearings
1185 HNTFLAG = 1
1190 GOTO 730 :    '   and get a fresh character from keyboard
1200  '    *****************************
1210  '              IF KEY = "N"
1220  '
1225  IF (Z <> 78 AND Z <> 110) THEN 1530
1230  '         open text file
1240 GOSUB 3000
1250 GOTO 440
1500  '     *****************************
1510  '             IF KEY="Q"
1520  '
1530  IF (Z <> 81 AND Z <> 113) THEN GOSUB 1900 : GOTO 730 : '' illegal key
1540  COLOR 2 : CLS : KEY ON : COLOR 4
1550  LOCATE 12,30 : PRINT "PIV has ended. You're still in BASIC!" : END
1560  '                                                  (END PAGER)
1570  '
1580  '
1900  ''  *************************************************************
1910  ''                 E R R O R   display subroutine
1920  ''  *************************************************************
1930  FOR ZZ = 0 TO 32 :  0, (ZZ MOD 16) : ''  a visual "ouch!"
1940  SOUND (357 - 6 * ZZ), .4! : ''  ...and loud complaints
1950  NEXT ZZ :  : RETURN : ''   enough already
1960  ''
2000  ''  *************************************************************
2010  '                    BUFFER EDITOR
2020  '   *************************************************************
2025 '
2030  '                PRINT HEADER, COMLINE, SELECTION, BUFFER CONTENTS
2040  CLS
2050  LOCATE 1,17:COLOR SHADE,0,0,16:PRINT  "**************  BUFFER EDITOR
      **************";:COLOR HUE,0,0,0
2060 LOCATE 24,1 : PRINT" (I)nsert  (D)elete  (E)mpty  (K)yboard  (M)ove
     (R/W)rite  (S)kyfile  (X)chng";
2061 COLOR SHADE: LOCATE 24,3 : PRINT"I";: LOCATE ,13 : PRINT "D";: LOCATE
     ,23 : PRINT "E";: LOCATE ,32 : PRINT "K";: LOCATE ,43 : PRINT "M";:
      LOCATE ,51 : PRINT "R/W";: LOCATE ,62 : PRINT "S";: LOCATE ,73 :
```

```
      PRINT "X";: COLOR HUE
2065 LINE (0,320)-(719,336),1,B
2070  LOCATE 21,1 : PRINT " CURRENT SELECTION: "
2080  LOCATE 22,6 : COLOR SHADE : PRINT A$(20 * (PG - 1) + LN); : COLOR HUE
2090  LOCATE 3,1
2100  FOR I = 1 TO 16
2107 I$ = STR$(I)
2110  LOCATE CSRLIN,1 : PRINT RAS$; : LOCATE CSRLIN,1
2112 IF (I < 10) THEN PRINT " ";
2115 COLOR 4: PRINT I$". ";:COLOR 2 : PRINT LEFT$(BUF$(I),74)
2120  NEXT
2130  REM ********************************************************
2140  '                                         GET KEY
2150  '
2160  '
2170  Z$ = INPUT$(1)
2180  Z =  ASC(Z$)
2190  '    ****************************
2200  '                         IF (I)NSERT
2210  '
2220  IF (Z <> 73 AND Z <> 105) THEN 2340
2230  IF (LEN(BUF$(16)) > 0 OR A$(20 * (PG - 1) + LN) = "") THEN  PRINT
      CHR$(7);: GOTO 2160
2240  LOCATE 23,1 : PRINT RAS$ : LOCATE 23,10 : PRINT " INSERT WHERE
      (1-16) ";
2245  INPUT Z$ : LOCATE 23,1 : PRINT RAS$ : LOCATE 23,1
2250  IF (Z$ = "" OR VAL(Z$) < 1  OR  VAL(Z$) > 16) THEN PRINT CHR$(7);
      : GOTO 2160
2260  IF (VAL(Z$) <> 1  AND  LEN(BUF$(VAL(Z$) - 1)) = 0) THEN Z$ =
      STR$(VAL(Z$) - 1) : GOTO 2260
2270  FOR I = 16 TO VAL(Z$) STEP -1
2280 BUF$(I) = BUF$(I - 1) : IF (I = VAL(Z$)) THEN BUF$(I) = A$(20 *
     (PG - 1) + LN)
2290  NEXT
2300 NDX = NDX + 1 : GOTO 2070
2310  '    ******************************
2320  '                        IF (D)ELETE
2330  '
2340  IF (Z <> 68 AND Z <> 100) THEN 2420
2345  IF (BUF$(1) = "") THEN  PRINT CHR$(7); : GOTO 2160
2350  LOCATE 23,10 : PRINT " DELETE WHERE  (1-15) ";
2355  INPUT Z$ : LOCATE 23,1 : PRINT RAS$ : LOCATE 23,1
2360  IF (Z$ = "" OR  VAL(Z$) < 1  OR VAL(Z$) > 16) THEN  PRINT CHR$(7);
      : GOTO 2160
2370  IF (BUF$(VAL(Z$)) = "") THEN  PRINT CHR$(7); : GOTO 2160
2380  FOR I =  VAL(Z$) TO 16
2390  BUF$(I) = BUF$(I + 1)
2400  NEXT
2410 NDX = NDX - 1 : GOTO 2070
2420  '    ********************************
2430  '                       IF (E)MPTY
2435 '
2440  IF (Z <> 69 AND Z <> 101) THEN 2510
2445  IF (BUF$(1) = "") THEN  PRINT CHR$(7); : GOTO 2160
2450  LOCATE 23,1 : PRINT RAS$ : LOCATE 23,10 : PRINT " EMPTY THE
      BUFFER (Y/N)";
2455  INPUT Z$:LOCATE 23,1:PRINT RAS$:LOCATE 23,1
2460  IF (Z$ = "" OR (Z$ <> "y" AND Z$ <> "Y")) THEN 2160
2470  FOR I = 1 TO 16 : BUF$(I) = "" : NEXT : NDX = 0 : GOTO 2070
2480  '     ********************************
2490  '                               IF (X)CHNG
2500  '
2510  IF (Z <> 88 AND Z <> 120) THEN 2574
2515  IF (BUF$(2) = "") THEN  PRINT CHR$(7); : GOTO 2160
2520  LOCATE 23,1:PRINT RAS$:LOCATE 23,10:PRINT " SWAP WHICH TWO
       -- FIRST: ";
```

```
2525  INPUT Z$
2530  IF (Z$ = "" OR  VAL(Z$) < 1 OR  VAL(Z$) > 16 OR BUF$(VAL(Z$)) = "")
      THEN LOCATE 23,1:PRINT RAS$:LOCATE 23,1:BEEP: GOTO 2160
2532  LOCATE 23,1:PRINT RAS$:LOCATE 23,27:PRINT " SECOND: ";
2535 INPUT Z1$
2540  IF (Z1$ = "" OR BUF$(VAL(Z1$)) = "" OR VAL(Z1$) < 1 OR  VAL(Z1$)
      > 16) THEN  LOCATE 23,1:PRINT RAS$:BEEP: GOTO 2160
2550 LOCATE 23,1:PRINT RAS$
2560 BUF$(0) = BUF$( VAL (Z$)):BUF$( VAL (Z$)) = BUF$( VAL (Z1$)):
     BUF$( VAL (Z1$)) = BUF$(0)
2570  GOTO 2070
2571 ' *******************************
2572 '                    IF (M)OVE
2573 ' *******************************
2574 IF (Z <> 77 AND Z <> 109) THEN 2608
2575 LOCATE 23,1 : PRINT RAS$ : LOCATE 23,10 : INPUT "Move to which
     entry? ( 1 - 15 )",Z$
2576 LOCATE 23,1 : PRINT RAS$ : LOCATE 23,1
2578 IF (Z$ = "" OR VAL(Z$) < 1 OR  VAL(Z$) > 16) THEN  PRINT CHR$(7);
     : GOTO 2160
2579 IF (BUF$(VAL(Z$)) = "") THEN PRINT CHR$(7); : GOTO 2160
2580 IF (VAL(Z$) > 9) THEN QTY = 1 ELSE QTY = 0
2590 RA$ = MID$(BUF$(VAL(Z$)), 12, 8) : DEC$ = MID$(BUF$(VAL(Z$)), 25, 9)
2591 LOCATE 2+VAL(Z$),6: COLOR TINT : PRINT LEFT$(BUF$( VAL(Z$)),74);
     :COLOR HUE
2592 OLDZ$ = Z$
2595 GOSUB 4000 ::REM' the big banana
2596 Z$ = OLDZ$
2598 LOCATE 2+VAL(Z$),6: PRINT LEFT$(BUF$( VAL(Z$)),74);
2600 GOTO 2170
2602  '    *******************************
2604  '                  IF (R)EAD
2606  '
2608  IF (Z <> 82 AND Z <> 114) THEN 2740
2610  LOCATE 22,1:PRINT RAS$:LOCATE 22,10:COLOR TINT:PRINT "READING
      BUFFER";:COLOR HUE,0
2612  OPEN "BUFFER" FOR INPUT AS #1
2614 I = 0
2616 WHILE+ NOT EOF(1)
2618  LINE INPUT #1,A$
2620 I = I + 1:BUF$(I) = A$
2622 WEND
2624 NDX=I:CLOSE #1
2626  IF (NDX > 0 AND LEN(BUF$(NDX)) < 5) THEN NDX = NDX - 1 :
      GOTO 2626
2628  LOCATE 22,1:PRINT RAS$:LOCATE 23,1: GOTO 2070
2650 '  ********************************
2652 '                          IF KEYBOARD
2654 '
2660 '
2680 '
2690 '
2695 '
2698 '
2699 '
2710  '    *******************************
2720  '                     IF (W)RITE
2730  '
2740  IF (Z <> 87 AND Z <> 119) THEN 2880
2750  LOCATE 23,1:PRINT RAS$:LOCATE 23,10:COLOR TINT:PRINT
      "WRITING BUFFER ":COLOR HUE,0
2760  KILL "BUFFER"
2770  OPEN "BUFFER" FOR OUTPUT AS #1
2810  FOR I = 1 TO NDX: PRINT #1,BUF$(I): NEXT
2820  CLOSE #1
2830  LOCATE 23,1:PRINT RAS$
```

```
2840  GOTO 2070
2850  '    **********************************
2860  '                     IF (S)KYFILE
2870  '
2880  IF (Z <> 83 AND Z <> 115) THEN  PRINT CHR$(7); : GOTO 2160
2890  CLS: GOTO 500
2900  '                                           (RETURN TO PAGER)
3000 '           CONSTELLATION CHOICE SYSTEM
3001 CLS
3002 LOCATE 2,10:COLOR SHADE:PRINT "**** SKYFILES 2000: AN
     INTERACTIVE ASTRONOMICAL DATABASE! *****";:LINE (0,318)-(719,338),
     1,B:LINE (20,257)-(699,313),6,B:LINE (20,280)-(699,292),0,BF:LINE
     (0,8 )-(719,30),1,B:COLOR HUE
3003 LOCATE 24,10:PRINT "Use";:COLOR SHADE:PRINT" cursor keys ";:COLOR
     HUE:PRINT "to choose file, <";:COLOR SHADE:PRINT "CR";:COLOR
     HUE:PRINT"> to enter, <";:COLOR SHADE:PRINT "ESC";:COLOR
     HUE:PRINT"> to abort.";
3010 CON$="ANDCAMCMACVNHERLYRPSASEX..
     1..9ANTCANCMICYGHYDMISPUPSGR..
     2. 10AQLCAPCOLDELLACMONPYXTAU..
     3.11AQUCASCOMDRALEOOPHSAGTRI..
     4.12ARICAUCOREQULEPORISCLUMA..
     5halAURCENCRAERILIBPEGSCOUMI..
     6msrBOOCEPCRBFORLMIPERSCUVIR..
     7polCAECETCRTGEMLYNPISSERVUL..8pln"
3020 FOR I=1 TO 80
3030 CON$(I)=MID$(CON$,3*(I-1)+1,3)
3040 NEXT
3045 '
3047 '
3050 FOR I=1 TO 10
3060 FOR J=1 TO  8
3070 LOCATE 2*I+2,10*(J-1)+4
3080 PRINT CON$(10 *(J-1)+I);
3090 NEXT:NEXT
3094 '
3095 '
3100 LOCATE 4,4:COLOR SHADE:PRINT CON$(1);:COLOR HUE
3110 ROW=1:COL=1
3190 '
3191 '
3192 '
3200 '   GET KEY
3210 OLDROW=ROW:OLDCOL=COL
3220 Z$=INKEY$:IF Z$="" THEN 3220
3230 IF LEN(Z$)=2 THEN Z$=RIGHT$(Z$,1) ELSE GOTO 3275
3240 IF ASC(Z$)=72 THEN ROW=ROW-1:IF ROW=0 THEN ROW=10:GOTO 3300 ELSE
      GOTO 3300
3250 IF ASC(Z$)=80 THEN ROW=ROW+1:IF ROW=11 THEN ROW=1:GOTO 3300 ELSE
     GOTO 3300
3260 IF ASC(Z$)=75 THEN COL=COL-1:IF COL=0 THEN COL=8:GOTO 3300 ELSE
     GOTO 3300
3270 IF ASC(Z$)=77 THEN COL=COL+1:IF COL=9 THEN COL=1:GOTO 3300 ELSE
     GOTO 3300
3272 IF ASC(Z$)=71 THEN COL=1:ROW=1:GOTO 3300
3275 IF ASC(Z$)=27 THEN GOTO 3550
3280 IF ASC(Z$)=13 THEN 3400
3290 BEEP:GOTO 3200
3300 '                             PRINT "CURSOR" SUBROUTINE
3303 '
3310 LOCATE 2*OLDROW+2,10*(OLDCOL-1)+4
3320 PRINT CON$(10*(OLDCOL-1)+OLDROW);
3330 LOCATE 2*ROW+2,10*(COL-1)+4
3340 COLOR SHADE:PRINT CON$(10*(COL-1)+ROW);:COLOR HUE
3350 GOTO 3200
3400 '  OPEN AND READ THE SELECTED FILE
```

```
3410 N$ = CON$(10*(COL-1)+ROW)
3412 IF LEFT$(N$,1) = "." THEN 3200
3420 '
3422 LOCATE 2*ROW+2,10*(COL-1)+4
3424 COLOR TINT:PRINT CON$(10*(COL-1)+ROW);:COLOR HUE:SOUND 670,3:
     FOR ZZ = 1 TO 5:NEXT:SOUND 440,8
3425  OPEN N$ FOR INPUT AS #1
3430  '
3440  '          Read text file into A$ array via LINE INPUT
3450  '
3460  FOR I = 1 TO ENDFIL:A$(I) = "": NEXT
3470 I = 0
3480 WHILE+ NOT EOF(1)
3490  LINE INPUT #1,A$
3500 I = I + 1
3510 A$(I) = A$
3520  WEND
3530  ENDFIL = I:CLOSE #1
3540  IF  LEN (A$(ENDFIL)) < 5 THEN ENDFIL = ENDFIL - 1: GOTO 3540
3550 CLS:RETURN
3600 '
3700 '
4000 ' =================================================================
4010 '       TALK TO SCOPE, MOVE IN COORDINATE GRID, HARDWARE CODE
4012 '
4014 '
4015 ''  OLDDEC, OLDRA and DK have been previously obtained from "INIT" ...
4022 ''                      - or -
4023 ''  OLDDEC, OLDRA and DK have been preserved from last move
     "new" position
4027 ''
4030 GOSUB 10000 : ''  parse string into D, M, S & DL
4032 NEWDEC = (INT(DD * 1200) + INT(DM * (1200/60)) + INT(DS *
     (1200/3600)))
4033 IF (OLDDEC = NEWDEC) THEN 4520 : ''      ** FIX FOR = **
4035 '
4040 MNUM = 1 : ''          we'll address DECLINATION first
4050 '
4055 IF ((DK+DL) = 2) THEN  4114 : ''                type 1 move
4060 IF ((DK+DL) = 1) THEN  4120 : ''                type 2 move
4090 IF ((DK+DL) = 0) THEN  4200 : ''                type 3 move
4095 ''
4098 ' ............................................................
4100 '
4114 IF (OLDDEC > NEWDEC) THEN DIR =1 :MCNT = OLDDEC-NEWDEC:GOTO 4275:
     :REM' south
4115 '
4116 DIR = 0 : MCNT = NEWDEC - OLDDEC : ''      1
     north
4117  GOTO 4275
4118 ' ............................................................
4119 '
4120 IF (DK = 1) THEN  DIR = 1 : MCNT = OLDDEC + NEWDEC : GOTO 4275 : '
     south
4130 '
4140 DIR = 0 : MCNT = OLDDEC + NEWDEC : ''      2
     north
4150  GOTO 4275
4155 ' ............................................................
4170 '
4200 IF (OLDDEC < NEWDEC) THEN DIR = 1: MCNT = NEWDEC-OLDDEC : GOTO 4275
     ::REM'south
4210 '
4250 DIR = 0 : MCNT = OLDDEC - NEWDEC : ''      3
     north
4260 '
```

```
4265 ' ..............................................................
4268 '
4270 '                      MOTOR LOAD AREA
4274 '
4275 CMD = &H301 : ARG = &H300 : STATUS = &H301
4276 '
4277 IF (PIV = 1) THEN   MCNT = INT((MCNT + 3) / 6) : GOTO 4300 : '
     BRANCH !
4278 MCNT = MCNT - 1
4280 OUT CMD, (&H1A OR &H0) : '        RESET CHIP 0 TO *PIV* DEFAULTS
4281 OUT ARG, &HFF : '                 START SPEED
4282 OUT ARG, &H20 : '                 HI    SPEED
4283 OUT ARG, &H90 : '                 RAMP  LSB
4284 OUT ARG, &H3 : '                  RAMP  MSB
4290 GOSUB 5730
4294 GOTO 4505 : ' branching
4295 '
4296 '
4300 MCNT = MCNT - 1
4315 OUT CMD, (&H1A OR &H0) : '       RESET CHIP 0 TO *TRACKMOUNT* DEFAULTS
4320 OUT ARG, &HFF : '                START SPEED
4325 OUT ARG, &H14 : '                HI    SPEED
4330 OUT ARG, &H90 : '                RAMP  LSB
4335 OUT ARG, &H6 : '                 RAMP  MSB
4340 GOSUB 5730
4490 '
4496 ' ====================================================================
4498 '
4500 '                           RIGHT ASCENSION CASE
4502 '
4505 MNUM = 0
4512 '
4520 NEWRA = (INT(RH * 20000) + INT(RM * (20000/60)) + INT(RS * (20000/3600)))
4522 IF OLDRA = NEWRA THEN 4785 : '                        * FIX FOR = **
4530 '
4550 IF (OLDRA < NEWRA) THEN  4592
4560 IF ((OLDRA - NEWRA) < 240000!) THEN  4585
4570 DIR = 0 : MCNT = (480000! - OLDRA) + NEWRA : ''       equation 4
4580 GOTO 4640
4585 DIR = 1 : MCNT = OLDRA - NEWRA : ''                   equation 2
4590 GOTO 4640
4592 IF ((NEWRA - OLDRA) > 240000!) THEN  4605
4595 DIR = 0 : MCNT = NEWRA - OLDRA : ''                   equation 1
4600 GOTO 4640
4605 DIR = 1 : MCNT = (480000! - NEWRA) + OLDRA : ''       equation 3
4607 '
4608 ' ..................................................................
4610 '
4612 '                     MOTOR LOAD AREA
4614 '
4640 CMD  = &H303 : ARG = &H302 : STATUS = &H303
4642 '
4645 IF (PIV = 1) THEN  MCNT = INT(((MCNT*3)+10)/20) : GOTO 4700 : ' BRNCH!
4650 MCNT = MCNT - 1
4652 OUT CMD, (&H1A OR &H0) : '         RESET CHIP TO *PIV* DEFAULTS
4653 OUT ARG, &HFF : '                  START SPEED
4654 OUT ARG, &H20 : '                  HI SPEED
4655 OUT ARG, &H90 : '                  RAMP LSB
4656 OUT ARG, &H3 : '                   MSB
4660 GOSUB 5730
4665 GOTO 4785 : ' branching
4670 '
4675 '
4700 MCNT = MCNT - 1
4705 OUT CMD, (&H1A OR &H0) : '         RESET CHIP 1 TO *TRACKMOUNT*
     DEFAULTS
```

```
4710 OUT ARG, &HFF : '                      START SPEED
4715 OUT ARG, &H14 : '                      HI    SPEED
4720 OUT ARG, &H90 : '                      RAMP  LSB
4725 OUT ARG, &H6 : '                             MSB
4730 GOSUB 5730
4735 GOTO 4785
4750 '
4760 ' ..................................................................
4780 '                        "MOTOR WAIT" HERE !!!!
4782 '
4785 DECBZ = INP(&H301) AND &H4: IF DECBZ THEN 4785
4790 RABZ = INP(&H303) AND &H4: IF RABZ THEN 4790
4800 OLDDEC = NEWDEC : OLDRA = NEWRA : DK = DL : MCNT = 0 : '  update location
4810 '
4820 SOUND 1320,16: '                          MAKES NOISE on completion
4990 RETURN '                                  BACK TO skyfile or buffer
5000 '
5100 '
5200 '
5300 '
5680 ' """"""""""""""""""""""""""""""""""""""""""""""""""""""""""""""""""""""
5690 ' """""                                                            """""
5700 ' """""       ....YOU HAVE REACHED......THE MOVER ZONE....         """""
5710 ' """""                                                            """""
5720 ' """"""""""""""""""""""""""""""""""""""""""""""""""""""""""""""""""""""
5721 '
5730 ' counts fixed; do direction and low, mid, and high bytes outta here!!!
5735 '
5740 IF (DIR <> 0) THEN CDIR% = &H8 ELSE CDIR% = &H0 : ''  direction bit
5760 '
5770 COLOR 6 : LINE (5,0)-(5,18):LINE (5,18)-(470,18):LINE (470,18)-(470,0)
5775 '
5780 OUT CMD, (&H53 OR CDIR%) : ''  send cmd to request "RAMP U/D MOVE" . . .
5790 GOSUB 12000
5795 IF MCNT = 0 THEN MCNT = 1 : '              fixes a wrap-a-round prob!
5796 LOWBYE = FNLOWBY(MCNT)
5797 MIDBYE = FNLOWBY(INT(MCNT/256))
5798 HYEBYE = FNLOWBY(INT(MCNT/65536!))
5800 OUT ARG, LOWBYE : '                                   LSB
5810 GOSUB 12000
5820 OUT ARG, MIDBYE : '                                   MSB
5830 GOSUB 12000
5840 OUT ARG, HYEBYE : '                                   HSB
5900 RETURN
7000 '
8000 ' ====================================================================
8005 '
8010 '                SECTOR SCAN W/ LOW RANGE MOTORS      (not finished!)
8015 '
8020 LOCATE 24,1 : PRINT CLEER$:LOCATE 24,1:PRINT " Scan what size area
     (deg.) ";:INPUT SCNSZ;:LOCATE 24,1:PRINT CLEER$:PRINT " How many scans
     to be made ";:INPUT SCNNOS;:LOCATE 24,1:PRINT CLEER$
8025 LOCATE 24,1 : PRINT "SCAN NO. ";SCNCNT;" OF ";SCNNOS ";"IS DONE!";
8050 '
8055 '                                    offset to corner of scan pattern
8060 '
8070 '                                    pattern is x deg. BY x deg. big!
8072 '                                    to start, i sscans odd, or even?
9000 '
10000 ' ===================================================================
10010 '               PARSE SELECTED STRING TO MOTOR PULSES
10020 '
10050 RH=VAL(LEFT$(RA$,2)):RM=VAL(MID$(RA$,4,2)):RS=VAL(RIGHT$(RA$,2))
10060 DL = 1 : IF LEFT$(DEC$, 1) = "-" THEN DL = 0 : ''  got this straight
10070 DD=VAL(MID$(DEC$,2,2)):DM=VAL(MID$(DEC$,5,2)):DS=VAL(RIGHT$(DEC$,2))
10080 RETURN : ' end of  PARSE SEL STR etc .
```

```
10090 '
10140 ' ================================================================
10150 '                     INITIALIZE MOTOR POSITION(S)
10160 '
10170 IF A$(20 * (PG - 1) + LN) = "" THEN 10210 : ' take off, toad
10180 GOSUB 10000 : ' convert selected string to motor pulses
10185 DK = DL : ''  initialize sign of declination as computed in subr. 10000
10190 OLDRA = (INT(RH * 20000) + INT(RM * (20000/60)) + INT(RS * (20000/3600)))
10200 OLDDEC = INT(DD*1200) + (INT(DM*1200/60)) + (INT(RS*1200/3600))
10210 LOCATE LN + 2,1: COLOR TINT : PRINT A$(20 * (PG-1) + LN) : '      Flash!
10220 SOUND 670, 3 : FOR ZZ = 1 TO 100 : NEXT ZZ
10230 LOCATE LN+2, 1 : COLOR SHADE: PRINT A$(20 * (PG - 1) + LN) : ' FLASH!
10240 SOUND 440, 9 : COLOR HUE : ''  hue and only hue can print for-next files
10250 '
10290 RETURN : ' end of  INITIALIZE MOTOR POSITION(S) .
11000 '
12000 ' ===============================================================
12012 '              INPUT status, return when buffer-not-full
12014 '
12020 IBF = INP(STATUS) AND &H2 : IF (IBF) GOTO 12020 : '     busy loop!
12030 RETURN
13000 '
14000 '
15000 '
16000 '
17000 ' ================================================================
17010 '         CLEAR SCREEN, GET SCOPE TYPE, SOME DEFINITIONS, ETC
17011 '
17012 CLEER$ = "                                                              "
17020 '
17022 '
17024 HNTFLAG = 0: LOCATE ,,0::REM' cant't hunt cause not inited yet -
      cursor off!
17026 DEF FNLOWBY(V) = INT((V/256-INT(V/256))*256)
17028 CLS :  0,10:COLOR 7:LINE (0,0)-(719,349),1,B:COLOR 1: LINE (5,5)-(714,344)
      ,1,B:COLOR 6 :LINE (30,30)-(690,100),1,B:LOCATE 5,17:PRINT "************
      PIV Control Program ************";
17030 FOR ZZ = 1 TO 25 : NEXT
17032 SOUND 990, 12 : SOUND 1100, 12 : SOUND 880, 12
17034 SOUND 440, 12 : SOUND 660, 18
17036 FOR ZZ = 1 TO 25 : NEXT: COLOR 4
17038 LOCATE 10, 15:PRINT "1)  PLEASE - Reset Sil-Walker board w/ pushbutton
17040 LOCATE 12, 15:PRINT "2)  Subroutine (H)unt is not on line yet.........
17042 LOCATE 14,15 :PRINT "3)  (B)uffer is now interactive ................
17044 LOCATE 16,15 : PRINT "4)  Verify electrical connections at this time..
17068 FOR ZZ = 1 TO 2000 : NEXT : COLOR 7 :
17070 LOCATE 22,22 : COLOR 6 : PRINT "(Sets RATES, SPEED, RAMPS, etc.)"
17072 LOCATE 20, 14 : COLOR 7 : INPUT "WHAT SCOPE IS TO BE USED (PIV = 0,
      XEROX = 1) ";Z$
17074 IF ((LEN(Z$) < 1) OR (VAL(Z$) < 0) OR (VAL(Z$) > 1)) THEN 17072
17076 IF (VAL(Z$) = 0) THEN PIV = 0 ELSE PIV = 1
17078 FOR ZZ = 1 TO 200 : NEXT ZZ
17080  : RETURN
```

Appendix I

Calibrating Encoders Using Kalman Filtering

I.1 Weighted Least Squares

This appendix describes a method of obtaining small gains in pointing accuracy for large expenditures of effort. These gains are made by observing a large number of stars all over the sky, then using modern mathematical techniques to estimate the values of several error constants all at once. Only those with advanced mathematical skills and the need for very high pointing accuracy should continue reading. The rest of us are better advised to use special observations to measure each error source independently of the others, as discussed in Chapter 6.

To obtain estimates of the mechanical error constants identified in Chapter 13, a large number of stars can be used in conjunction with a digital filter designed to operate when there is an overabundance of data. There should be at least 5–10 times as many data points as there are parameters to be estimated. To provide a means of converting raw encoder readings to useful coordinates, stars of known position are aligned on the optical axis of the telescope, and the corresponding raw encoder readings are recorded. When high accuracy pointing is a requirement, the goal is to extract the most information possible about the error constants, that is, to obtain optimal estimates of the values of the constants.

One method of obtaining an optimal estimate of the error constants is to use a weighted least squares filter. Gelb (1974, p. 23) states the well-known theory of weighted least squares in the form used below. The notation is derived from Dunham (1983), and uses the convention that small letters denote differences between vectors, while names in capital letters denote the main vectors themselves. Another excellent source for digital filtering techniques is Wertz (1978).

To perform a weighted least squares estimate, the following information is needed:

1. The initial condition of the system. The system is described by the state vector X, whose m elements are the parameters being estimated and which uniquely determine the system. For the WMO alt-az mount, the azimuth parameters that are to be estimated are the following: $A1$ (zero offset), a and b (the azimuth tilt angles), p (the angle between the axes $-90°$), and C_{ew} (the east-west collimation error). The zenith distance parameters are the following: $Z1$ (zero offset), a, b, p, C_{ns} (the north-south collimation error), and $F(90°)$ (the flexure constant). The filter requires initial rough guesses of these values, such as those derived from special calibration observations.

2. A set of n measurements Y_i of one or more variables related to the system parameters by a set of observation equations. The measurements are the raw encoder readings from the azimuth and zenith distance axes. The observation equations are used to produce a matrix of observations of the state parameters $G(X_i)$. This generalization permits the estimation of parameters which are not observed directly.

3. An estimate of the errors in the measurements, given in the matrix R. The inverse of this matrix, R^{-1}, is the matrix of weights used to weight the effect of each measurement on the estimate that is made of the system parameters. It is assumed that the measurements are subject to Gaussian noise statistics.

Assuming there exists a set of "true" values of the state vector $[X]$, the filter is designed to estimate a correction vector x given by

$$x = [X] - X$$

which, when added to the original guess X, gives a "best estimate" of $[X]$.

To use this method, first a set of differences $Y_i = Y_i - G(X_i)$ is defined, which contains the observed values (O) minus the computed values (C), using the initial estimate X. In a linear system, there exists a matrix H of the observation partial derivatives

$$H_i = \frac{dG}{dX_i}$$

(where d/dX_i is the partial derivative with respect to X_i) such that $y_i = H_i x_i$. The H_i are evaluated using the initial estimate of the state of the system X. Next, the covariance matrix

$$P = (H^T R^{-1} H)^{-1}$$

and the estimate of the correction vector

$$x = PH^T R^{-1} y$$

are computed. The best estimate of the state of the system is then

$$X' = X + x.$$

This filter works on measurements taken together as a batch. The filter produces the same numerical estimates, regardless of how the measurements are ordered in the measurements vector Y (which is usually the order in which the measurements were taken). If the initial errors are unknown, the filter can be run once with $R = 1$, the unit matrix, then run again with a new initial state using $X(\text{run } 2) = X'(\text{run } 1)$ and using the O-Cs as the elements of R.

To apply this approach to the problem of estimating the error constants, the following corrections to azimuth and zenith distance are defined, using the equations given above:

$A1$ = azimuth encoder zero offset constant

$A2$ = azimuth tilt correction to azimuth

$$= \arccos\left\{\frac{\sin Z}{\sin z}(\cos A \cos b - \sin A \sin b)\right\} - A$$

where $\sin z = \sin[\arccos(\sin Z \cos Aa \sin b + \sin Z \sin Aa \cos b + \cos Z)]$ and (A, Z) are the coordinates in the system that is tilted through angle a and rotated through angle b

$A3$ = non-perpendicular axes error $= p \cot Z$

$A4$ = collimation error $= C_{\text{ew}} \csc Z$

$Z1$ = zenith distance encoder zero offset constant

$Z2$ = azimuth tilt correction to zenith distance
$= \arccos(\sin Z \cos Aa \sin b + \sin Z \sin Aa \cos b + \cos Z) - Z$

$Z3$ = non-perpendicular axes error $= p\Delta A$

$Z4$ = collimation error $= C_{\text{ns}}$

$Z5$ = tube flexure $= F(90^\circ) \sin Z$.

The servo lag error is a constant multiplied by the motor error command step, which is not directly a function of raw encoder readings. To keep the filter linear, it is ignored, and the servo time constant is obtained by direct measurement using a step function input.

The observation equations relating the observed quantities (raw encoder readings) to the computed quantities (the (A, Z) coordinates of the telescope) are as follows:

$$A = K_a E_a + A1 + A2 + A3 + A4$$

$$Z = K_z E_z + Z1 + Z2 + Z3 + Z4 + Z5$$

where A is the azimuth coordinate and Z is the zenith distance coordinate (obtained to high accuracy from a catalog, then reduced to topocentric place) of the star that is centered in the field of the telescope, E_a is the raw azimuth encoder reading, and E_z is the raw zenith distance encoder reading. The conversion factors K_a or K_z (in arc seconds per encoder count) are also constants, which are determined by autocollimating off the telescope axes and counting the encoder ticks while each axis is swung through precisely 360° back into autocollimation. The Y_i are the E_a and E_z.

The O-C equations are of the form

$$y_a = K_a E_a + A1 + A2 + A3 + A4 - A$$

$$y_z = K_z E_z + Z1 + Z2 + Z3 + Z4 + Z5 - Z.$$

The state parameters for the two axes overlap, in that a, b, and p are related to both axes, so that measurements from both encoders are coupled to each other through these state parameters. Therefore, the O-Cs are placed in a $2 \times n$ matrix y, and the partials matrix H contains the partial derivatives of the observation equations for both axes with respect to the eight state parameters $A1$, $Z1$, a, b, p, C_{ew}, C_{ns}, and $F_u(90°)$. Much of this H matrix consists of zeroes, which act as "place holders." The other matrices are as described above.

This method, although it provides optimal estimates for the correction constants, has three main drawbacks:

1. It is computationally inefficient, since matrix inversions need to be performed. As the number m of state parameters grows, the size of the covariance matrix P to be inverted grows by m^2, which increases the number of operations needed to perform the inversion by a factor of m^4.

2. The measurements must all be taken ahead of time, then processed as a batch. There is no way to perform the calculations while pointing

the telescope for the next error constant measurement. Thus there is no way to monitor the progression of the O-C matrix y as an increasing number of processed measurements diminishes the effects of the Gaussian measurement noise. Later, during observing operations, there is no way to take advantage of setting or guiding inputs made by the telescope operator to adjust the correction constants in real-time to improve pointing accuracy over the course of the first few hours of an observing run.

3. There is no computationally efficient way to extract more information from the observations, such as obtaining better estimates of the observer's longitude and latitude, since the resulting constraint equations are nonlinear.

I.2 Kalman Filter

The Kalman, or sequential, filter is mathematically identical to the weighted least squares filter, in that the same inputs produce a numerically identical optimal estimate (Gelb, 1974, p. 105). The Kalman filter is computationally more efficient, because it does not require the inversion of a matrix. Instead, the same basic information in the least squares filter has been re-arranged so that only a scalar is inverted.

The nature of the sequential filter algorithm requires initial best guesses of the state X, the error in the state x, and the covariance matrix P. Rather than treating all observations at once as a batch, the sequential filter processes each observation as it is received. To illustrate the procedure as an iterative process, the initial guesses will be treated as the outputs of the $i-1$-th step, and are labelled X_{i-1}, x_{i-1}, and P'_{i-1}.

When an observation Y_i is received by the sequential filter, the O-C

$$y_i = Y_i - G(X_{i-1})$$

and the partials matrix

$$H_i = \frac{dG_i}{dX_{i-1}}$$

are computed as before. Next, the Kalman gain

$$K_i = P_i H_i^T [H_i P_i H_i^T + R_i]^{-1}$$

is computed. R_i is the variance in the observation at the i-th step. Since the quantity to be inverted in the brackets is a scalar, no matrix inversions need be done. The covariance matrix is computed using K_i and the old covariance matrix P_i

$$P'_i = [I - K_i H_i] P_i$$

where I is the identity matrix, then the new state correction is computed

$$x'_i = x_{i-1} + K_i(y_i - H_i x_{i-1}).$$

The process is repeated using the most recent values of P' and x' for the new P and x, respectively, for each observation until all observations have been processed. Then the estimated state is found $X' = x' + X$ from the initial estimate of the state and the most recent state correction.

The Kalman filter has several advantages over the weighted least squares filter:

1. The results may be monitored while the measurements are being taken, since the filter is applied to each observation as it becomes available, not after all measurements have been completed.

2. It is computationally more efficient, since no matrices need to be inverted.

3. Using linearizing techniques (e.g., Gelb, 1974, Chapter 6), estimates of other parameters previously taken as known quantities can be made, including longitude, latitude, sidereal clock error, and the servo time constant. Although these parameters could be estimated using similar linearizing techniques in the weighted least squares filter, the resulting increase in the number of elements in the covariance matrix would have caused its inversion to take much longer.

The Kalman filter has two drawbacks. The first is that if the initial estimates of the state parameters and their errors are themselves too much in error, rather than converging on the true values, the filter will diverge to a very bad estimate of the state. This means that relatively good estimates of the state vector values must be made initially. This can be done most easily by making special calibrating observations to measure the constants first, then the Kalman filter can be used to refine these values.

The second problem is that if the estimate of the state deviation gets very good, the covariance matrix elements become very small and the resulting Kalman gain K becomes small, so that further measurements are ignored. This can be remedied by adding a process noise matrix Q to the covariance matrix P before computing the Kalman gain K. This represents an attempt to account for unmodeled errors in the system, and keeps the filter open enough to allow new measurements to affect the estimate of the state parameters. Note that if a process noise matrix is used, the numerical result is no longer identical to that given by the weighted least squares filter.

One may not always have a very good initial estimate of the state parameters, the measurement errors, or the state deviation, and one may have no idea what to use for the process noise values. Kalman filters require a good deal more tuning to implement than the weighted least squares filter, but often the extra effort is rewarded by the advantages of the Kalman filter approach.

I.3 Extended Kalman Filter

The extended Kalman filter is quite similar to the Kalman filter, with the difference being that the state vector is updated with every observation, rather than at the end, after all observations have been processed. This allows use of the estimated state quantities in real-time. For example, in a portable telescope system for which the observer's longitude and latitude may not be known to high accuracy, one could include them in the state vector and use the pointing residuals to obtain better estimates of the observer's location, which, in turn, give better pointing accuracy the next time a motor speed command is computed.

Again, initial best guesses of the state X and the covariance matrix P are needed, but since no state error matrix x is used in this filter, it is not needed. For the i-th step of the filter, the initial X and P are treated as X_{i-1} and P'_{i-1}.

As before, when an observation Y_i is received by the extended sequential filter, the O-C

$$y_i = Y_i - G(X_{i-1})$$

and the partials matrix

$$H_i = \frac{dG_i}{dX_{i-1}}$$

are computed. Also, as before, any process noise Q is added to the covariance matrix

$$P_i = P_{i-1} + Q$$

and the Kalman gain

$$K_i = P_i H_i^T [H_i P_i H_i^T + R_i]^{-1}$$

is computed. Again, R is the variance in the observation. The covariance matrix is updated

$$P'_i = [I - K_i H_i] P_i$$

and now the state vector itself is updated (rather than the state error vector)

$$X_i = X_{i-1} + K_i y_i.$$

The process is repeated for the next observation using X_i and P'_i.

For a dynamic system, such as an artificial satellite (e.g., Landsat 4), which contains an on-board computer to estimate its location by observing navigation satellites (e.g., Navstar GPS), the state vector contains Landsat's X, Y, and Z position vectors, and the observables are the arrival times and transmittal times of the GPS transmissions and the GPS satellite positions. The GPS positions and message transmit times are part of the GPS message received by the GPS user. In this system, the state itself is dynamic, and would have to be "propagated" to the next observation time by integrating the equations of motion between the last and the next observation, and updating the X and P arrays to the new integrated position before computing the Y_i in the first step. We have eliminated this complication by estimating the error coefficients of the telescope, which are static, rather than the telescope position itself (in A, Z), which is dynamic. This allows one to monitor the error coefficients during an evening to see how they behave at different telescope pointing angles.

The extended Kalman filter suffers from the same drawbacks as the Kalman filter, namely, the filter can diverge if a good initial value of the state vector is not known, and tuning the filter parameters, such as the process noise Q, is difficult. However, the extended Kalman filter offers the same advantage as the Kalman filter of providing an optimal estimate of the state parameters with a minimum of computation overhead.

One advantage of the extended filter is the accessability of the state parameters in real-time, as mentioned earlier. Whenever the telescope is commanded to point to a new location, the hand paddle inputs to do final centering or to re-center the star during tracking can be used as observation inputs into the extended filter. The result will be better estimates of the error coefficients, so that fewer and fewer hand paddle corrections will be needed as the evening progresses.

In developing a computerized project, the Kalman filter method of calibrating encoders should receive no attention until the system is functioning well. After that stage is reached, a Kalman filter can be installed to improve the estimates of the error coefficients.

Glossary

A azimuth

AAT Anglo-Australian Telescope

α right ascension

AC alternating current

A/D analog to digital

alt-az altitude-azimuth telescope mount

APT automatic photoelectric telescope

bus a digital signal pathway consisting of several parallel lines used to move data between different circuit boards which are connected to the pathway.

CCW counter-clockwise

chip integrated circuit

CMOS a technology for manufacturing integrated circuits using Complementary Metal Oxide and doped Silicon to achieve very low power drains and high noise immunity.

CPU Central Processing Unit (of a computer)

CW clockwise

D days

δ declination

D/A digital to analog

DC direct current

DEC registered trademark of the Digital Equipment Corporation

Dec declination

DMA direct memory access

D/S digital to synchro

E eccentric anomaly of the Earth in its orbit

e eccentricity of the Earth's orbit

ε mean obliquity of the ecliptic

η parallactic angle

ECL a standard family of very high speed logic devices using a non-saturating Emitter Coupled Logic technology.

EMI electromagnetic interference

EPROM Eraseable (usually with ultraviolet light) Programmable Read Only Memory

GPS Global Positioning System

h hour angle

I/O input/output

IR infra-red

KB kilobytes (1024 bytes)

Kb kilobits (1000 bits)

Kbaud kilobits per second along a single communications line

L_m geometric mean longitude of the Sun

LRC inductance-resistance-capacitance

L_t true longitude of the Sun

LVDT linear variable differential transformer

M months, or mean anomaly of the Sun

m minutes of time

$'$ minutes of arc

MB megabytes (2^{20} bytes)

Mb megabits (10^6 bits)

MOS a technology for manufacturing integrated circuits using Metal Oxide and doped Silicon to achieve very high component densities and relatively moderate power drains (used to make memories and CPUs).

mV millivolts (thousanths of a volt)

μV microvolts (millionths of a volt)

nV nanovolts (billionths of a volt)

PC personal computer

pc printed circuit

pixel picture element (a.k.a. pel or point)

PMT photomultiplier tube

pot potentiometer

PROM programmable read only memory

PWM pulse width modulation, in which information is conveyed by varying the width of each pulse in a constant stream of pulses.

RA right ascension

RAM random access memory (the main read/write memory used by the CPU)

RFI radio frequency interference

RMS root mean square

ROM read only memory

RSS root sum square

s, S seconds of time

$''$ seconds of arc

S/D synchro to digital

STD bus a standard bus developed by Pro-Log and Mostek

TTL a standard family of logic devices using a bipolar Transistor Transistor Logic technology

WMO Winer Mobile Observatory

Bibliography

Abdel-Gawad, M. K., *Analysis for Redesign of 150-inch Stellar Telescope Serrurier Truss Structure*, Kitt Peak National Observatory Engineering Department Technical Report No. 9. (Tucson:1969).

Abt, H.A., 1980, "The Cost-Effectiveness in Terms of Publications and Citations of Various Optical Telescopes at the Kitt Peak National Observatory," *PASP, 547*, 249.

A'Hearn, M.F., 1984a, private communication.

A'Hearn, M.F., 1984b, "When Not to Automate," talk delivered to the "Microcomputers in Astronomy II" Symposium, Fairborn, Ohio, July, 1984.

Almanac for Computers, published by the Nautical Almanac Office, United States Naval Observatory. (Washington:1979).

Arfken, G., *Mathematical Methods for Physicists*, Academic Press, Inc. (New York:1970).

Basili, V., 1980, private communication.

Beale, J.S., Gietzen, J. W., and Read, P. D., 1975, "Computer Control of the INT," in Huguenin and McCord (1975), 352.

Beaumont, G., and Wolfe, J., 1975, "Canada-France-Hawaii Telescope Computer and Control Systems," in Huguenin and McCord (1975), 275.

Bell, A.D., 1996, "Computer Controlled Drive System Design For Astronomical Telescopes," manuscript circulated privately.

Berry, R.L., 1983, "RE MEM BER THE YUMAN FAC TOR," in Genet (1983), 9.

Booch, G., *Software Engineering with Ada*, Benjamin/Cummings Publishing Company. (Menlo Park:1983).

Bothwell, G.W., 1975, "Development of the Computer System for the 3.9 Metre Anglo-Australian Telescope," in Huguenin and McCord (1975), 310.

Botlo, M., Jagielski, M., and Romero, A., 1993, "EPICS Performance Evaluation," copy available from `http://www.lanl.gov/epics/`.

Bourlon, P., and Vin, A., 1975, "Automation of a 40cm Telescope," in Huguenin and McCord (1975), 68.

Boyce, P., 1975, in Huguenin and McCord (1975), 39.
Boyd, L.J. and Genet, R.M. in Wolpert and Genet (1983).
Boyd, L.J., Genet, R.M., and Hall, D.S., 1984a, *IAPPP Communications, 15*, 20.
Boyd, L.J., Genet, R.M., and Hall, D.S., 1984b, in Wolpert and Genet (1984).
Boyd, L.J., Genet, R.M., and Hall, D.S., 1984c, in Genet and Genet (1984).
Boyd, L.J., Epand, D., Bresina, J., Drummond, M., Swanson, K., Crawford, D.L., Genet, D.R., Genet, R.M., Henry, G.W., McCook, G.P., Neely, W., Schmidtke, P., Smith, D.P., and Trueblood, M., 1993, *IAPPP Communications, 52*, 23.
Burgess, *Celestial Basic*, Sybex. (Berkeley:1982).
Burke, B.F., 1975, "Computer Control of Aperture Synthesis Interferometers," in Huguenin and McCord (1975), 378.
Colgate, S., 1975, in Huguenin and McCord (1975), 88.
Cox, R.E., and Sinnott, R. W., 1977, "Gleanings for ATM's," *Sky and Telescope, 54*, 4, 330.
Dijkstra, E.W., *A Discipline of Programming*, Prentice-Hall, Inc. (New Jersey:1976).
Dongarra, J.J., 1994, "Performance of Various Computers Using Standard Linear Equations Software," Computer Science Department, University of Tennessee, Knoxville, TN 37996-1301, dated April 24, 1994.
Dunham, D.W., 1984, private communication.
Dunham, J.B., 1983, private communication.
Dunham, T., 1978, "Is It Feasible to Provide a Simple But Effective Computer Drive for an Alt-Alt Telescope at Moderate Cost?," manuscript circulated privately.
Eisele, J.A. and Shannon, P. E. V., "Atmospheric Refraction Corrections for Optical Sightings of Astronomical Objects," NRL Memorandum Report 3058, Naval Research Laboratory. (Washington:1975).
Electronic Products, April, 1982.
Fridenberg, J.T., Westphal, J. A., and Kristian, J., 1975, "Microprocessors: A New Alternative for Automatic Telescope Control," in Huguenin and McCord (1975), 218.
Garfinkel, B., 1967, "Astronomical Refraction in a Polytropic Atmosphere," *AJ, 72*, 235.
Gelb, A., *Applied Optimal Estimation*, The M.I.T. Press. (Cambridge:1974).
Genet, R.M., 1979, *Telescope Making, 5*, 28.
Genet, R.M., 1980, *IAPPP Communications, 3*, 12.
Genet, R.M., Sauer, D.J., Kissell, K.E., and Roberts, G.C., 1982a, *IAPPP Communications, 9*, 43.

Genet, R.M., *Real-Time Control With the TRS-80*, H.W. Sams, Inc. (Indianapolis:1982b).

Genet, R.M., ed., *Microcomputers in Astronomy*, Fairborn Observatory. (Fairborn:1983).

Genet, R.M., Boyd, L.J., and Sauer, D.J., 1984, "Interfacing for Real-Time Control—An Astronomical Example," *Byte*, April, 1984.

Genet, R.M. and Genet, K.A., *Microcomputers in Astronomy II*, Fairborn Observatory. (Fairborn:1984).

Genet, R.M. and Hayes, D.S., *Robotic Observatories*, AutoScope Corporation. (Mesa:1989).

Ghedini, S., *Software for Photoelectric Photometry*, Willmann-Bell, Inc. (Richmond:1982).

Giovane, F., Wood, F.B., Oliver, J. P., and Chen, K.-Y., 1983, "Automatic South Pole Telescope," in Genet (1983), 86.

Goldberg, L., 1983, private communication.

Goldberg, *Automatic Controls: Principles of System Dynamics*, Allyn and Bacon, Inc. (Boston:1964).

Green, R.M., *Spherical Astronomy*, Cambridge University Press. (Cambridge:1985).

Gurnette, B.L. and Woolley, R.v.d.R., *Explanatory Supplement to The Astronomical Ephemeris and The American Ephemeris and Nautical Almanac*, Her Majesty's Stationery Office. (London:1961).

Hall, D.S., and Genet, R.M., *Photoelectric Photometry of Variable Stars*, Second Edition, Willmann-Bell, Inc. (Richmond, 1988).

Halliday, D. and Resnick, R., *Physics*, John Wiley and Sons, Inc. (New York:1966).

Harvey, J.W., 1994, private communication.

Harwood, J.V., 1975, "Institute for Astronomy 88-inch Telescope," in Huguenin and McCord (1975), 245.

Hearnshaw, J.B., 1995, "A Proposal for a Fiber-Fed Echelle Spectrograph for a Southern Hemisphere Robotic Telescope" in *Robotic Telescopes: Current Capabilities, Present Developments, and Future Prospects for Automated Astronomy*, eds. Henry, G.W. and Eaton, J.A., Astronomical Society of the Pacific. (Provo:1995).

Hill, S.J., 1975, in Huguenin and McCord (1975), 60.

Hirschfeld, A., and Sinnott, R.W., *Sky Catalog 2000.0*, Sky Publishing Corporation. (Cambridge, 1982).

Honeycutt, R.K., Kephart, J.E., and Hendon, A.A., 1977, *Sky and Telescope, 56*, 495.

Honeycutt, R.K. and Turner, G.W., 1992, "Architecture of the Software for the Indiana CCD Automated Telescope" in *Robotic Telescopes in the 1990s*, ed. Filippenko, A.V., Astronomical Society of the Pacific. (Provo:1992).

Huguenin, M.K. and McCord, T.B., 1975, Eds., *Telescope Automation: Proceedings of a Conference Held 29, 30 April, 1 May, 1975 at Massachusetts Institute of Technology, Cambridge, Massachusetts and Sponsored by the National Science Foundation.*

Ingalls, R., 1975, in Huguenin and McCord (1975), 24.

Jaworski, A., 1983, private communication.

Jelly, J.V., 1980, *Q. Jl. R. Astr. Soc., 21*, 14-31.

Kaplan, G.H., ed., *The IAU Resolution on Astronomical Constants, Time Scales, and the Fundamental Reference Frame*, United States Naval Observatory Circular No. 163, December 10, 1981.

Kibrick, R., 1984, in Genet and Genet (1984).

Klim, K. and Ziebell, M., 1985, "Coude Auxilliary Telescope: A low cost telescope at European Southern Observatory," *Optical Engineering, 24, 2,* 363.

Linnell, A.P., and Hill, S.J., 1975, "The MSU Computer Assisted Photometric System," in Huguenin and McCord (1975), 60.

Manly, P.L., *Unusual Telescopes*, Cambridge University Press. (Cambridge:1991).

Mansfield, A.W., 1984, private communication.

McCord, T.B., Paavola, S.H., and Snellen, G.H., 1975, "The MIT Automated Optical Telescope," in Huguenin and McCord (1975), 160.

McNall, J.F., Miedaner, T.L., and Code, A.D., 1968, *AJ, 73*, 756.

Meeks, M. L., 1975, in Huguenin and McCord (1975), 23.

Meeus, J., *Astronomical Formulae for Calculators, Second Edition*, Willmann-Bell, Inc. (Richmond:1982).

Meeus, J., *Astronomical Algorithms*, Willmann-Bell, Inc. (Richmond:1991).

Melsheimer, F., 1983, private communication.

Melsheimer, F., 1984, private communication.

Miller, R. W., *Servomechanisms: Devices and Fundamentals*, Reston Publishing Co., Inc. (Virginia:1977).

Moore, C.H., Rather, E.D., and Conklin, E.K., 1975, "Modular Software for On-Line Telescope Control," in Huguenin and McCord (1975), 1.

Moore, E., Merillat, P., Colgate, S., and Carlson, R., "Software for a Digitally Controlled Telescope," in Huguenin and McCord (1975), 95.

Murray, C.A., *Vectorial Astrometry*, Adam Hilger Ltd. (Bristol:1983).

Opal, C.B., 1980, "Rotary Dial Mechanism for Digitally Tuned Receivers," *Ham Radio*, July, 14.

Orlov, B.A., *Refraction Tables of Pulkovo Observatory.* Fourth edition (Pulkovo:1956).

Otis, M.G., 1983, *Sky and Telescope, 65*, 551.

Paavola, S.H., 1975, in Huguenin and McCord (1975), 160.

Paffrath, L., 1975, "Data Acquisition and Telescope Control Systems at Kitt Peak National Observatory," in Huguenin and McCord (1975), 392.

Petzold, Charles, *Programming Windows 3.1, Third Edition*, Microsoft Press. (Redmond:1992) ISBN 1-55615-395-3.

Racine, R., 1975, "Rampant Conservatism in Telescope Controls," in Huguenin and McCord (1975), 34.

Radick, R.R., Hartman, L., Mihalas, D., Worden, S. P., Africano, J. L., Klimke, A., and Tyson, E. T., 1982, *PASP, 94*, 934.

Rafert, J.B., and Cone, G., 1983, "Stepper Motor Controlled Telescope," in Genet (1983), 50.

Rather, E., 1975, in Huguenin and McCord (1975), 390.

Rodin, B., 1982, private communication.

Rosing, W., 1994, private communication.

Schnurr, R.G., and A'Hearn, M.F., 1983, "Apple-Based Occultation Photometer," in Genet (1983), 130.

Seidelmann, P.K., and Kaplan, G.H., 1982, *Sky and Telescope, 64*, 409.

Seidelmann, P.K., *Explanatory Supplement to the Astronomical Almanac*, University Science Books. (Mill Valley:1992).

Skillman, D.R., 1981, *Sky and Telescope, 61*, 71.

Smart, W.M., *Textbook on Spherical Astronomy*, (Sixth Edition, reprinted), Cambridge University Press. (Cambridge:1979).

Spaulding, C.P., *How to Use Shaft Encoders*, DATEX Corporation. (Monrovia:1965).

Stephenson, T.P., 1975, "The Multiple Mirror Telescope Mount Control System," in Huguenin and McCord (1975), 365.

Stoll, M., and Jenkner, H., 1983, "Vienna 150-cm Telescope Control," in Genet (1983), 33.

Taff, L.G., *Computational Spherical Astronomy*, Krieger Publishing Company. (Malabar:1991).

Tausworthe, R.C., *Standardized Development of Computer Software*, Prentice-Hall, Inc. (New Jersey:1977).

Taylor, J.H., 1975, "Automated Pulsar Observations at the Arecibo Observatory and the Five College Radio Astronomy Observatory," in Huguenin and McCord (1975), 28.

Titus, J., Titus, C., and Larson, J., *The STD Bus*, H.W. Sams, Inc. (Indianapolis:1982).

Tomer, A.J., and Bernstein, R., 1983, "Computer-Operated Portable Telescope," in Genet (1983), 74.

Tomer, A.J., 1984, "Towards Developing a Digital Telescope," manuscript circulated privately.

Trueblood, M., 1987, "A High-Speed Photometer Interface for the MicroVAX Q-Bus," p. 325 in Hayes, D.S., Genet, D.R., and Genet, R.M., eds., *New Generation Small Telescopes*, Fairborn Press (Mesa).

Trueblood, M., 1993, "In the Shadow of the Asteroid," *GPS World*, *4*, 11, 22.

Trueblood, M., 1995, "Avoiding Clouds: GPS Keeps Astronomers Under Clear Skies," *GPS World*, *6*, 6, 32.

Van der Lans, J., and Lorenson, S., 1975, "The ESO Telescope Control Systems," in Huguenin and McCord (1975), 253.

Wallace, P.T., 1975, "Programming the Control Computer of the Anglo-Australian 3.9 Metre Telescope," in Huguenin and McCord (1975), 284.

Wallace, Patrick, 1988, "Pointing and Tracking Algorithms for the Keck 10-Meter Telescope" in L.B. Robinson, Ed. *Instrumentation for Ground-Based Optical Astronomy*, Springer-Verlag. (New York).

Wasserman, L.H., 1986, private communication.

Wellnitz, D., 1983, private communication.

Wertz, J.R., ed., *Spacecraft Attitude Determination and Control*, D. Reidel Publishing Company. (Dordrecht:1978).

White, N., 1984, private communication.

Wolpert, R.C. and Genet, R.M., eds., *Advances in Photoelectric Photometry Volume I*, Fairborn Observatory. (Fairborn:1983).

Wolpert, R.C. and Genet, R.M., eds., *Advances in Photoelectric Photometry Volume II*, Fairborn Observatory. (Fairborn:1984).

Woolard, E.W. and Clemence, G.M., *Spherical Astronomy*, Academic Press. (New York:1966).

Worden, S. P., Schneeberger, T.J., Kuhn, J.R., and Africano, J.L., 1981, *Ap. J.*, *244*, 520.

Young, A.T., Genet, R.M., Boyd, L.J., Borucki, W.J., Lockwood, G.W., Henry, G.W., Hall, D.S., Smith, D.P., Baliunas, S.L., Donahue, R., and Epand, D.H., 1991, *PASP*, *103*, 221.

Index

U

V

W